Advances in
PARASITOLOGY

Opportunistic Protozoa in Humans

VOLUME 40

Editorial Board

C. Bryant Division of Biochemistry and Molecular Biology, The Australian National University, Canberra, ACT 0200, Australia

M. Coluzzi Director, Istituto di Parassitologia, Università Degli Studi di Roma "La Sapienza", P. le A. Moro 5, 00185 Roma, Italy

C. Combes Laboratoire de Biologie Animale, Université de Perpignan, Centre de Biologie et d'Ecologie Tropicale et Méditerranéenne, Avenue de Villeneuve, 66860 Perpignan Cedex, France

S.L. James Chief, Parasitology and Tropical Diseases Branch, Division of Microbiology and Infectious Diseases, National Institute for Allergy and Infectious Diseases, Bethesda, MD 20892-7630, USA

W.H.R. Lumsden 16A Merchiston Crescent, Edinburgh, EH10 5AX, UK

Lord Soulsby of Swaffham Prior Department of Clinical Veterinary Medicine, University of Cambridge, Madingley Road, Cambridge, CB3 0ES, UK

K. Tanabe Laboratory of Biology, Osaka Institute of Technology, Ohmiya, Asahi-Ku, Osaka 535, Japan

P. Wenk Institut für Tropenmedizin, Eberhard-Karls-Universität Tübingen, D7400 Tübingen 1, Wilhelmstrasse 31, Germany

Advances in
PARASITOLOGY
Opportunistic Protozoa in Humans

Series editors

J.R. BAKER, R. MULLER and D. ROLLINSON

Guest editor

S. TZIPORI

*Tufts University School of Veterinary Medicine,
North Grafton, MA, USA*

VOLUME 40

ACADEMIC PRESS
San Diego   London   Boston   New York
Sydney   Tokyo   Toronto

ACADEMIC PRESS
525 B Street, Suite 1900, San Diego,
California 92101–4495, USA
http://www.apnet.com

ACADEMIC PRESS LIMITED
24/28 Oval Road
LONDON NW1 7DX
http://www.hbuk.co.uk/ap/

Copyright © 1998, by
ACADEMIC PRESS LIMITED

This book is printed on acid-free paper

All Rights Reserved
No part of this publication may be reproduced or transmitted in any form
or by any means, electronic or mechanical, including photocopy,
recording, or any information storage and retrieval system,
without permission in writing from the publisher

A catalogue record for this book is available from the British Library

ISBN 0-12-031740-0

Typeset by J&L Composition Ltd, Filey, North Yorkshire
Printed in Great Britain by MPG Books Limited, Bodmin, Cornwall
98 99 00 01 02 03 MP 9 8 7 6 5 4 3 2 1

CONTRIBUTORS TO VOLUME 40

A.M. CEVALLOS, *Division of Geographic Medicine and Infectious Diseases, New England Medical Center, Tufts University School of Medicine, Boston, MA 02111, USA*

J.H. CRABB, *ImmuCell Corporation, Portland, ME 04103, USA*

E.S. DIDIER, *Department of Microbiology, Tulane Regional Primate Research Center, Covington, LA 70433, USA*

C.R. FRICKER, *Thames Water Utilities, Reading RG2 0JN, UK*

R.H. GILMAN, *Departamento de Patologia, Universidad Peruana Cayetano Heredia, Lima, Peru and Department of International Health, School of Hygiene, Johns Hopkins University, Baltimore MD21205, USA*

J.K. GRIFFITHS, *Division of Infectious Diseases, Tufts University School of Veterinary Medicine, North Grafton, MA 01536, USA and Department of Family Health and Community Medicine, Tufts University School of Medicine, Boston, MA 02111, USA*

D.P. KOTLER, *Gastrointestinal Division, Department of Medicine, St Lukes – Roosevelt Hospital Center, College of Physicians and Surgeons, New York, NY 10032, USA*

J.M. ORENSTEIN, *Department of Pathology, George Washington University School of Medicine, Washington, DC 20037, USA*

Y.R. ORTEGA, *Department of Veterinary Science and Microbiology, University of Arizona, Tucson, AZ 85721, USA*

J.A. SHADDUCK, *Department of Veterinary Pathobiology, Texas A&M University, College Station, TX 77843, USA*

K.F. SNOWDEN, *Department of Veterinary Parasitology, Texas A&M University, College Station, TX 77843, USA*

C.R. STERLING, *Department of Veterinary Science and Microbiology, University of Arizona, Tucson, AZ 85721, USA*

C.M. THEODOS, *Division of Infectious Diseases, Tufts University School of Veterinary Medicine, North Grafton, MA 01536, USA*

S. TZIPORI, *Division of Infectious Diseases, Tufts University School of Veterinary Medicine, North Grafton, MA 01536, USA*

C.R. VOSSBRINCK, *The Connecticut Agricultural Experiment Station, New Haven, CT 06504, USA*

H. WARD, *Division of Geographic Medicine and Infectious Diseases, New England Medical Center, Tufts University School of Medicine, Boston, MA 02111, USA*

L.M. WEISS, *Department of Pathology, Albert Einstein College of Medicine, Bronx, NY 10461, USA*

G. WIDMER, *Division of Infectious Diseases, Tufts University School of Veterinary Medicine, North Grafton, MA 01536, USA*

PREFACE

This volume represents a new departure for *Advances in Parasitology* – a special volume devoted to one topic. We normally try to include a mixture of topics to appeal to as wide a range of readers as possible, and this will remain our general policy. However, with the current surge of interest in opportunistic infections in persons with compromised immune systems – due largely to the advent of the human immunodeficiency viruses – it seemed important to devote a volume to protozoan parasites, which have become significant in this context. *Cryptosporidium parvum, Enterocytozoon bieneusi* and *Cyclospora cayetanensis* have emerged as important pathogens associated with human enteric disease. This volume provides a detailed account of the biology of these organisms, and hopefully will contribute to a better appreciation of cryptosporidiosis, microsporidiosis and cyclosporidiosis.

We are very fortunate in having as guest editor of this volume Dr Saul Tzipori, of Tufts University School of Veterinary Medicine in the USA, who has assembled from among his own colleagues and workers at other institutions a panel of distinguished authors, all of whom, like Dr Tzipori himself, have worked extensively on members of this group of parasites.

We believe that this volume will provide, as we intend all volumes in the series to provide, authoritative in-depth reviews of subjects that are currently in the vanguard of parasitological significance. We are very grateful to Dr Tzipori and the other authors for their hard work in compiling this excellent series of articles.

JOHN BAKER
RALPH MULLER
DAVID ROLLINSON

CONTENTS

CONTRIBUTORS TO VOLUME 40 v
PREFACE vii

PART 1 *CRYPTOSPORIDIUM PARVUM* AND RELATED GENERA

INTRODUCTION TO *CRYPTOSPORIDIUM PARVUM* 3

Natural History and Biology of *Cryptosporidium parvum*

S. Tzipori and J.K. Griffiths

1. Introduction 6
2. Natural History and Chronology 6
3. Life Cycle as Related to Control 8
4. Classification and Taxonomy 10
5. Risk to Humans from Infected Animals 17
6. The Unique Intracellular Niche of *Cryptosporidium* . . . 19
7. Concluding Remarks 28
   References 29

Human Cryptosporidiosis: Epidemiology, Transmission, Clinical Disease, Treatment, and Diagnosis

J.K. Griffiths

1. Introduction 38
2. Epidemiology 39
3. Transmission 50
4. Clinical Disease 55

5. Treatment 62
6. Diagnosis 70
   References 72

Innate and Cell-mediated Immune Responses to *Cryptosporidium parvum*

C.M. Theodos

1. Introduction 88
2. Innate Immunity 89
3. *Cryptosporidium parvum*-specific Cell-mediated Immunity . . 95
4. Immunotherapy 111
5. Concluding Remarks 113
   References 114

Antibody-based Immunotherapy of Cryptosporidiosis

J.H. Crabb

1. Introduction: Rationale for Therapy using Antibodies to Combat Cryptosporidiosis 122
2. Practical Considerations for Antibody Therapy 125
3. Early Studies of Immunotherapy for Cryptosporidiosis . . . 130
4. Laboratory Investigations of Anti-*Cryptosporidium* Antibodies 131
5. Studies in Humans using Antibodies 138
6. Future Prospects 141
   References 143

Cryptosporidium: Molecular Basis of Host–Parasite Interaction

H. Ward and A.M. Cevallos

1. Introduction 152
2. Oocysts and the Molecular Basis of Excystation and Oocyst Wall Formation 153
3. Invasive (Zoite) Stages and the Molecular Basis of Attachment, Invasion, and Parasitophorus Vacuole Formation 158

4. Intracellular Stages and the Molecular Basis of Stage
   Differentiation and Parasite Metabolism 173
5. Future Perspectives 177
6. Concluding Remarks 178
   References 178

Cryptosporidiosis: Laboratory Investigations and Chemotherapy

S. Tzipori

1. Introduction 188
2. Production of Oocysts for Biomedical Research 188
3. Propagation in Cell Culture 190
4. Animal Models 195
5. New Approaches to Drug Design 205
6. Laboratory Investigations on Drugs Presently Used in Humans 208
7. Concluding Remarks 212
   References 212

Genetic Heterogeneity and PCR Detection of *Cryptosporidium parvum*

G. Widmer

1. Relevance of Studying Polymorphism in *C. parvum* 224
2. Heterogeneity in *Cryptosporidium* 225
3. PCR Detection of *C. parvum* 233
   References 236

Water-borne Cryptosporidiosis: Detection Methods and Treatment Options

C.R. Fricker and J.H. Crabb

1. Water-borne Outbreaks: A Brief History 242
2. Occurrence of *Cryptosporidium* in Source Water 247
3. Detection of *Cryptosporidium* in Water 249
4. Regulatory Status and Environmental Laws 264
5. Water Treatment Options 267
   References 272

PART 2 ENTEROCYTOZOON BIENEUSI AND OTHER MICROSPORIDIA

INTRODUCTION TO MICROSPORIDIA 281

Biology of Microsporidian Species Infecting Mammals

E.S. Didier, K.F. Snowden and J.A. Shadduck

1. Introduction 284
2. Morphology 285
3. Life Cycle 290
4. Taxonomy 296
5. Mammalian Host–Parasite Relationships 303
6. Summary and Conclusions 309
   References 310

Clinical Syndromes Associated with Microsporidiosis

D.P. Kotler and J.M. Orenstein

1. Introduction 322
2. Microsporidial Species Causing Intestinal Disease in AIDS . 323
3. Epidemiology 323
4. Pathogenesis of Intestinal Injury 327
5. Immune Response to Microsporidia 330
6. Clinical Illness 330
7. Diagnosis 333
8. Treatment 341
9. Conclusion 342
   References 343

Microsporidiosis: Molecular and Diagnostic Aspects

L.M. Weiss and C.R. Vossbrinck

1. Introduction 352
2. Molecular Biology 353

3. Phylogeny 358
4. Diagnosis 369
5. Summary 385
   References 385

PART 3 CYCLOSPORA CAYETANENSIS AND RELATED SPECIES

Cyclospora cayetanensis

Y.R. Ortega, C.R. Sterling and R.H. Gilman

1. Introduction 400
2. Diagnosis and Purification 401
3. Molecular Biology 404
4. Life Cycle 405
5. Histopathology 406
6. Immunology 408
7. Clinical Signs of Infection 408
8. Treatment 408
9. Cyclosporiasis in AIDS Patients 409
10. Epidemiology 410
11. Pathogenesis 413
12. Cyclosporiasis in Peru 413
13. The Future 414
    References 414

Index 419

Color Plates are located between   338–9

PART 1

Cryptosporidium parvum and related genera

INTRODUCTION TO
CRYPTOSPORIDUM PARVUM

This special volume is devoted to three protozoa that are associated with human enteric disease. Less than two decades ago, they were either completely unknown (microsporidiosis and cyclosporidiosis) or not considered an important cause of disease (cryptosporidiosis). Among the three, *Cryptosporidium parvum* has been the most extensively investigated and is probably the most significant. Consequently, much more information is available for this pathogen than for the other two. Microsporidiosis of humans has emerged with AIDS, but may prove to be more common in the general population than currently realized. Cyclosporidiosis is the newest enteric disease and is the least known. The relative information available on the three protozoa is reflected in the number of chapters devoted to each in this volume.

The first eight chapters of this volume are concerned with *C. parvum*, a member of the enteric Apicomplexa. Although known since the turn of the century, its significance was recognized only in the last decade and a half. Each of the eight chapters addresses a specific aspect of *C. parvum* relevant to understanding the dynamics of the pathogen's interaction with the cell, and the clinical and immunological consequences in the host. The underlying theme that links the chapters is the development of strategies for therapy and prevention, and they offer a wealth of views and scientific interpretations regarding many pertinent aspects of *C. parvum* research.

The first chapter, by Tzipori and Griffiths, questions the current taxonomy of the parasite, and examines the intracellular location of *C. parvum* in the context of other closely related Apicomplexa, as it relates to trafficking of nutrients, and possibly drugs, between the parasite and the host cell. The chapter by Griffiths reviews the human disease in all its manifestations, highlighting the need for greater attention to be given to the role *C. parvum* plays in the development of chronic diarrhea and malnutrition of children in developing societies.

The chapter by Theodos is devoted entirely to the role of innate and cell-mediated immune responses of the host to *C. parvum*. Current knowledge, which is very fragmented, is reviewed and an attempt is made to delineate the relative role of each component in recovery, protection, and immunotherapy. Antibody-based therapy against chronic cryptosporidiosis has been rigorously pursued by several investigators over the last decade. Laboratory investigations and a few early stage clinical studies have at best been inconclusive and often conflicting. The chapter by Crabb analyzes this subject and critically examines the scientific rationale for the antibody-based approach.

Ward examines the host–parasite interface at the molecular level. This field is still in its infancy and the available knowledge at this stage is highly fragmented, despite a concerted effort by many investigators in recent years. The slow progress is largely attributed to the many technical and logistic problems associated with research on *C. parvum*. The chapter reviews the existing information on the molecular basis of various host–parasite interactions and the specific molecules involved. The chapter by Tzipori reviews the considerable progress made in the development of the methods, both *in vitro* and *in vivo*, necessary for laboratory investigations of the biology of the parasite and for testing the efficacy of potential immunotherapeutic and chemotherapeutic agents.

Widmer assesses the various methods that have been applied to the study of genotypic and phenotypic polymorphism in *C. parvum*. The chapter highlights the use of genetic fingerprinting to demonstrate for the first time the occurrence of unique *C. parvum* genotypes in humans that are distinct from those found in animals. The chapter by Flicker and Crabb reviews the most recent events that helped to establish cryptosporidiosis as a major waterborne disease. The chapter discusses the proposed regulations specifically designed to control this parasite, and highlights the new methods which are being developed to limit the threat to public health from contaminated water.

SAUL TZIPORI

Natural History and Biology of *Cryptosporidium parvum*

Saul Tzipori and Jeffrey K. Griffiths

Division of Infectious Diseases, Tufts University School of Veterinary Medicine, North Grafton, MA 01536, USA; Department of Family Health and Community Medicine, Tufts University School of Medicine, Boston, MA 02111, USA

1. Introduction 6
2. Natural History and Chronology 6
3. Life Cycle as Related to Control 8
4. Classification and Taxonomy 10
   4.1. The genus *Cryptosporidium* and speciation 10
   4.2. Preliminary molecular taxonomy 12
   4.3. Phenotypic and genotypic variation among *C. parvum* isolates 14
   4.4. The stability of genetic markers 16
5. Risk to Humans from Infected Animals 17
   5.1. Infections in domestic and wild ruminants 17
   5.2. Infections in swine 18
   5.3. Infections in domestic pets 18
   5.4. Infections in wild animals 19
6. The Unique Intracellular Niche of *Cryptosporidium* 19
   6.1. Intracellular locations of Apicomplexa 20
   6.2. Implications of the *Cryptosporidium* topology 26
7. Concluding Remarks 29
   References 29

The taxonomy of the genus Cryptosporidium *remains ambiguous, because the current criteria for speciation are insufficient to validate the 6–8 named species. Cross-transmission experiments have shown varying and conflicting results, and the limited genetic data available do not necessarily support currently proposed species designations. The reasons for this ambiguity lie with the ubiquitous nature of* Cryptosporidium, *probably infecting all vertebrates and variety of tissues therein, and the absence of reference strains with defined virulence attributes that can be linked to genetic markers for comparative analysis. The inability to*

ADVANCES IN PARASITOLOGY VOL 40
ISBN 0–12–031740–0

classify oocysts or confidently to identify their origin, implicate oocysts from all sources as hazardous to humans. Another major issue is the unusual degree of resistance that Cryptosporidium *has shown to antiprotozoan and antimicrobial agents. The intracellular but extracytoplasmic domain the parasite occupies is in itself a significant barrier to drug entry. In support of this we outline how the intracellular niche of this parasite differs from the related Apicomplexans,* Plasmodium *and* Toxoplasma, *and delineate why the feeder organelle membrane, rather than, or in addition to, the parasitophorous membrane, is the major portal of nutrient entry for* Cryptosporidium. *The broad conclusion is that anticryptosporidial agents will have to enter the parasite via the multiple apical membranes that camouflage the parasite, or via the host cell, possibly transported by vesicles to the feeder organelle membrane. This may have major implications for rational drug discovery and design.*

1. INTRODUCTION

This chapter provides a brief historical perspective on the natural history and chronology, taxonomy, and certain aspects of the biology of *Cryptosporidium parvum*, the major designated mammalian species. Other presumed species of *Cryptosporidium*, about which even less is known, will only be discussed in the context of classification and taxonomy. It is not intended to provide a comprehensive review of the literature regarding these topics. For this the reader is referred to recent published accounts (Dubey *et al.*, 1990; Current and Garcia, 1991; O'Donoghue, 1995). Instead, we will highlight some recent developments and address questions that are pertinent to the current knowledge and understanding of the parasite and its biology, which hopefully will lead readers to the specific topics addressed in subsequent chapters.

2. NATURAL HISTORY AND CHRONOLOGY

Taxonomically, *C. parvum* belongs to the phylum Apicomplexa (possessing apical complex), class Sporozoasida (reproduce by asexual and sexual cycles, with oocyst formation), subclass Coccidiasina (life cycle involving merogony, gametogony and sporogony), order Eucoccidiorida (schizogony occurs), suborder Eimeriorina (independent microgamy and macrogamy develop), family Cryptosporidiidae (four naked sporozoites within oocyst – sporocyst) (Levine, 1985). Like other enteric coccidia of vertebrates, *Cryptosporidium* has a monoxenous life cycle, which is primarily completed within the gastrointestinal tract of a single host. On the other hand, it has many unique characteristics that set it apart from other coccidia, of which the lack of host specificity, resistance to antimicrobial agents, ability for autoinfection, and the curious location it occupies within the host cell membrane, are the most obvious. Although first described in the gastric mucosa of the laboratory mouse in 1907 (Tyzzer, 1907), *Cryptosporidium* was first recognized as a potential cause of diarrhea in turkeys in 1955 (Slavin, 1955). *Cryptosporidium* was subsequently identified in other species of animals in which the infection was either thought to be opportunistic and harmless, or was shown to be associated with infrequent individual cases of enteric disease in calves (Panciera *et al.*, 1971; Meuten *et al.*, 1974), guinea pigs (Vetterling *et al.*, 1971a, b), and humans (Meisel *et al.*, 1976; Nime *et al.*, 1976).

Three broad entities of cryptosporidiosis became recognized between 1980 and 1993. The first was the recognition in 1980 that *C. parvum* was in fact a common, serious primary cause of outbreaks as well as sporadic cases of diarrhea in certain mammals. From 1983 onward, *C. parvum* emerged, with AIDS, as a life-threatening disease in this subpopulation. In 1993, *C. parvum* reached the public domain when it became widely recognized as the most serious, and difficult to control, cause of waterborne-related diarrhea.

The first glimpse of the seriousness of *Cryptosporidium* in mammals, namely calves, was provided in the late 1970s (Morin *et al.*, 1976; Pohlenz *et al.*, 1978). The later investigators successfully transmitted, for the first time, the infection between calves, but the resulting diarrhea was complicated by the presence of other enteric agents. Pohlenz *et al.* (1978) were also the first to show that cryptosporidiosis could be diagnosed by demonstrating the presence of oocysts in Giemsa-stained fecal smears. In 1980, *Cryptosporidium*, which was thought to be highly host specific, was successfully transmitted from calves, lambs and humans, to newborn mice, rats, guinea pigs, chicks, piglets, and lambs. Of the experimentally infected mammalian species, acute diarrhea was induced only in piglets, calves, and lambs, and a mild diarrhea was observed in infant rats (Tzipori *et al.*, 1980a, b).

Investigations of outbreaks of acute diarrhea in domestic herds in Scotland and in the USA during 1980 led to the recognition of *C. parvum* as a serious, highly contagious, primary cause of outbreaks of acute diarrhea of calves, lambs, red deer, and goat kids (Tzipori *et al.*, 1980c, 1981a, b, 1982a, b; Anderson, 1981, 1982; Angus *et al.*, 1982), and widespread infection in some 10 mammalian species tested serologically, including humans (Tzipori and Campbell, 1981).

The first reports linking *C. parvum* with AIDS appeared shortly after (Anon., 1982; Ma and Soave, 1982, 1983; Current *et al.*, 1983; Forgacs *et al.*, 1983). Almost simultaneously, the first two reports incriminating *C. parvum* as a common cause of acute diarrhea in a cluster of normal adults (Current *et al.*, 1983; Jokipii *et al.*, 1983) and in children (Tzipori *et al.*, 1983), were published. The chapter in this volume by Griffiths (1998) is devoted entirely to the spectrum of diseases and their manifestations in humans.

Contracting *C. parvum* by drinking contaminated water was recognized from earlier reports of human infections (Jokipii *et al.*, 1983; D'Antonio *et al.*, 1985; Fricker and Crabb, this volume). It was not until the major outbreak in Milwaukee in 1993 (MacKenzie *et al.*, 1994), however, that *C. parvum* was elevated to its current superstardom among water utilities and has since became a household name through the media frenzy that followed. The chapter in this volume by Fricker and Crabb (1998) is devoted entirely to the subject of waterborne cryptosporidiosis.

3. LIFE CYCLE AS RELATED TO CONTROL

The life cycle of *C. parvum* has been outlined in most reviews (Current and Garcia, 1991; O'Donoghue, 1995), and a simple schematic diagram is presented in chapter by Ward and Cevallos (1998) in this volume. Studies on the ultrastructure and morphogenesis of *C. parvum* have been few in recent years (Vetterling *et al.*, 1971a, b; Iseki, 1979; Bird and Smith, 1980; Göbel and Brändler, 1982; Marcial and Madara, 1986; Lumb *et al.*, 1988b). Certainly, few such studies have been published since the seminal and detailed observations made in 1986 on *C. parvum*, and on *Cryptosporidium baileyi* (Current and Reese, 1986; Current *et al.*, 1986).

Oocysts are ingested by the host, and the four naked sporozoites released in the gut infect epithelial cells and initiate asexual development. They become internalized and undergo two successive generations of merogeny, releasing eight and four merozoites, respectively. The four merozoites released from the second merogeny give rise to the sexual developmental stages, the micro- and macrogamonts. The release of microgametes, and their union with macrogametes gives rise to the zygote which, after two asexual divisions, forms the environmentally resistant oocyst containing four sporozoites, often while still inside the parasitophorous membrane.

The ability of the parasite to persist inside a single host is attributed to a repeated first-generation merogony, and the production of sporulated thin-wall oocysts, a characteristic quite distinct from other coccidia. The fact that most oocysts are sporulated upon release is probably the most important attribute, as thick wall oocysts, by far the predominant type (Current and Reese, 1986), readily excyst at body temperature (S. Tzipori, unpublished data), and are therefore likely to contribute considerably to autoinfection (Tyzzer, 1912). It is assumed that in the normal host the infection remains localized to the gastrointestinal tract. Extraintestinal phases, however, should not be ruled out, as oocysts injected into the blood stream of mice (Yang and Healey, 1994), or sporozoites into the peritoneal cavity (S. Tzipori, unpublished observations), lead to gut infection. The migration course of sporozoites from these sites into the gut is intriguing. An extragastrointestinal phase have been observed in *Eimeria tenella* and *Eimeria maxima* (Fernando *et al.*, 1987; Perry and Long, 1987). Aspects of the life cycle that are highly relevant to control, a major theme in this volume, are that it can be divided into (a) extracellular forms (sporozoites, two generations of merozoites, and microgametes), (b) the intracellular developmental forms (two generations of meronts and gamonts), and (c) the oocyst which is the environmentally resistant form, but nevertheless highly susceptible to excystation at body temperature.

The extracellular forms are highly vulnerable to the hostile gut environment. *In vitro* they have a short life span, depend on their own energy supply, and must reach and infect the host cell very quickly. They are readily inactivated by a variety of physical, chemical, and biological insults, including specific polyclonal antibodies, and extreme pH. Their survival in the gut depends on massive production, and on the fact that they do not have to travel far or for long. They do so often under the protective villus mucus gel, as they infect adjacent or even the same cell they were released from. The process of internalization is probably very rapid once the apical complex reaches the vicinity of the microvillus border, as the event is rarely witnessed microscopically in sections (Tzipori, 1988). For *Toxoplasma gondii*, invasion occurs within 10–30 seconds (Bommer, 1969; Mauel, 1996), and for *Plasmodium knowlesi* within 60 seconds (Bonnister *et al.*, 1975). Internalization of *C. parvum* sporozoites was completed in cell culture within 15 minutes after inoculation, but the time between attachment to the cell surface and internalization was not measured (Lumb *et al.*, 1988b). If the survival time of the extracellular forms in the gut lumen is, as suspected, very short, chemotherapeutic or immunotherapeutic agents directed against them are unlikely to be highly effective. Agents designed to inhibit parasite attachment and internalization—a process that is likely to be very rapid—have even less chance of being highly effective. For a drug or antibody to be effective it must be present in the gut lumen at the effective concentration at all times in all sites. While systemic drugs can achieve effective serum or intracellular concentrations for a period, this is near impossible in the gut, even with frequent dosing. This is because of the continuous flow of contents in the gut, which is much accelerated with diarrhea.

In contrast, the intracellular forms are well protected by the parasitophorous membrane originally derived from the host cell. The developmental process of the intracellular forms lasts 8–12 hours (Current and Reese, 1986). Therefore drugs that act against them have, theoretically, a better chance than do drugs that work against extracellular forms. Most effective drugs against malaria, *T. gondii*, and coccidia act against intracellular forms. Indeed, the most effective drugs against *C. parvum in vitro* also act against the intracellular forms. This includes nitazoxanide, paromomycin (see Tzipori, 1998), and thiostrepton (unpublished). Killing the intracellular forms of *C. parvum* has not been easy, however, as compared to other important Apicomplexan parasites, and may well be due to the unique intracellular niche that *Cryptosporidium* occupies.

4. CLASSIFICATION AND TAXONOMY

4.1. The Genus *Cryptosporidium* and Speciation

In 1980, *Cryptosporidium* isolates obtained from calves, lambs and a human adult with severe diarrhea, readily infected seven other species of animals (Tzipori *et al.*, 1980a, b). The transmission of the human isolate, which induced acute diarrhea in lambs indistinguishable from that caused by other animal isolates, strongly indicated the potential zoonotic nature of *Cryptosporidium*. Based on these early observations, the naming of *Cryptosporidium* species after their respective animal host (Levine, 1980) seemed questionable. Consequently, the question was whether the genus *Cryptosporidium* actually consists of a single species (Tzipori *et al.*, 1980a). While the answer to this question appears by consensus to be in the negative (see Current and Garcia, 1991; O'Donoghue, 1995), *C. parvum*, the named mammalian species (Tyzzer, 1912), nevertheless remains the single most important species perpetuating in mammals. The exact number of species is unclear, and is not likely to be resolved in the near future. Of the original 21 different *Cryptosporidium* species listed, the majority are now considered invalid. O'Donoghue (1995) lists six *Cryptosporidium* species, and quite correctly points out that further studies are required to confirm their validity. The six include two mammalian (*C. parvum* and *C. muris*), two avian (*C. melegridis* and *C. baileyi*), one reptilian (*C. serpentis*), and one fish (*C. nasorum*) species. Other workers list even more species (Fayer *et al.*, 1997).

The current differentiation of isolates into valid species is based on (a) host specificity, (b) oocyst morphology, and (c) site of infection. The concern is that the same variations in infectivity, oocyst morphology, and site of infection used to divide *Cryptosporidium* into species occur within species as well, namely *C. parvum*. For instance, human isolates of *C. parvum* that have been studied in our laboratory can be divided into those that infect mice, calves, and piglets, and those that infect piglets only. Yet, they all are clinical human isolates that induce diarrhea in piglets (S. Tzipori, unpublished). The failure to infect calves and mice with human isolates was also observed by others (M.J. Arrwood, personal communications).

Some of the evidence presented on host specificity in support of speciation are based on cross-transmission experiments performed with a limited number of *Cryptosporidium* isolates often tested in only one host, mostly in infant mice (Fayer *et al.*, 1995). The mouse is not representative of all mammals, nor presumably does a single *Cryptosporidium* field isolate represent all isolates that perpetuate in the class of vertebrates from which it was obtained. Some *C. parvum* and *C. baileyi* isolates are able to cross-infect avian and mammalian species, respectively (Tzipori *et al.*, 1980a; Lindsay *et al.*, 1987, Ditrich *et al.*, 1991), and most, if not all, *C. parvum* isolates readily infect chick embryos (Current and Long, 1983; Tzipori, 1988).

While the experiments described by Arcay *et al.* (1995) lack many key details, they successfully transmitted *C. parvum* to representative vertebrate animals (mammals, birds, reptiles, amphibians, and fish). In contrast, Graczyk *et al.* (1996b), in a better controlled and executed study, failed to transmit a different isolate of *C. parvum* to a somewhat similar set of representative vertebrates. The contrasting results could be because the studies were either not adequately controlled (Arcay *et al.*, 1995), too narrowly interpreted (Graczyk *et al.*, 1996b), or because the two *C. parvum* isolates were different. Perhaps the outcome would have been different had infant, not adult, representative vertebrate animals been used for the transmission experiments. These and other similar studies have major epidemiological and epizoolotiological implications relevant to public health, and their outcome should be taken at face value. The alternative can lead to a premature complacent view that humans are safe from exposure to *Cryptosporidium* isolates of lower vertebrates. From the biological standpoint, these studies also put in question the naming of new species based solely on the success or failure of transmission experiments in the absence of fully characterized reference *Cryptosporidium* strains. While some studies clearly show that *C. baileyi* (Lindsay *et al.*, 1986) and *C. serpentis* (Fayer *et al.*, 1995) do not infect a particular mammalian species, which may or may not be true for all other mammals, at issue here is whether results obtained with these specific isolates reflect all avian and all reptile isolates, respectively. The answer, at least with regard to *C. baileyi*, is clearly not (Ditrich *et al.*, 1991).

With regard to oocyst morphology, a recently carefully designed and executed study showed that *C. parvum* oocysts from calves were statistically significantly larger after recovery from experimentally infected baby rats (Beyer and Sidorenko, 1993). Finally, the ability of the parasite to infect a variety of sites, including the conjunctiva, upper and lower respiratory tracts, uterus, urinary bladder, and hepatobiliary and pancreatic systems, as well as every organ of the gastrointestinal tract, including the pharynx and esophagus, indicates that the site of infection is not a reliable taxonomic feature for speciation.

Only very few isolates of *C. muris*, *C. baileyi*, and *C. parvum* have so far been studied in any detail. Fewer have been subjected to rigorous comparative studies. A complicating factor is the inability at present to perform controlled comparative evaluation studies on isolates, and because of technical difficulties concerning growth and long-term preservation of the parasite, there are no clearly defined and fully characterized reference 'strains' of *Cryptosporidium*. Without fully characterized reference strains for comparative studies, naming of species based on studies performed on individual isolates without contrasting the spectrum of similarities and differences to reference strains is troublesome. Therefore the basis for speciation of *Cryptosporidium* remains ambiguous due to lack of clearly defined distinguishing phenotypic and genotypic parameters. In addition, because of the ability of *Cryptosporidium* to infect a variety of cells, tissues, organs, and species of vertebrate animals on one hand, and the confusing and often conflicting results of limited cross-transmission experiments on the other, it is probably premature to firmly divide *Cryptosporidium* into valid species without further studies.

Until some of the difficulties listed above can be overcome, *Cryptosporidium* should be viewed as a genus consisting of a very wide spectrum of isolates with differences, of which the host origin, the site of infection, and oocyst size are probably not the most critical features. Within this spectrum, we believe there are representative isolates, say of avian and mammalian origin, that are so closely related that the only differentiating feature between them is the origin of the host or the site of infection. It is more than likely that speciation in the future will not necessarily follow along classes of vertebrates, but rather be determined by tangible virulence and genetic attributes. Most observed differences that have been used to differentiate between *Cryptosporidium* species — species here being defined as a reproductively defined entity — can be explained through short- or long-term adaptation to a specific host or a site therein, with no clear evidence that these differences are permanent or temporary. Genotypic variation based on the few genes available in GenBank (see Section 4.2), would lend little support in favor of the currently named species, perhaps with the exception of *C. muris*, and may suggest that the observed differences are probably ecological in nature.

4.2. Preliminary Molecular Taxonomy

Pairwise comparisons were performed on the few small subunit (SSU) ribosomal RNA nucleotide sequences of *C. parvum*, *C. muris*, *C. baileyi* and *C. wrairi* genes available in GenBank. The analysis certainly lends little support to the notion of a separation into distinct, valid species. The rationale for this approach is that, if the current taxonomy of the genus is valid, then interspecies sequence similarity values should be greater than similarity values from intraspecies comparisons. SSU sequences available in GenBank

Table 1 Analysis of the few small subunit (SSU) ribosomal RNA nucleotide sequences of *C. parvum*, *C. muris*, *C. baileyi*, and *C. wrairi* genes available in the GenBank website for pairwise comparisons. SSU sequences available in the GenBank release generated the following similarity values using GCG's BESFIT program. The table reflects the intra- and interspecies percentage similarity among nine of the 18 SSU RNA sequences of the genus *Cryptosporidium*.

Species and access No.	*C. parvum* 1[a] X64340	*C. parvum* 2[a] X64341	*C. parvum* 3 L25642	*C. parvum* 4 L16996	*C. parvum* 5 L16997	*C. muris* 1 L19069	*C. muris* 2[a] X64342	*C. baileyi* 1 L19068	*C. wrairi* 1 U11440
C. parvum 2	99.620	N/A							
C. parvum 3	92.986	92.919	N/A						
C. parvum 4	93.298	93.124	99.734	N/A					
C. parvum 5	93.073	92.899	99.535	99.714	N/A				
C. muris 1	98.548	98.374	94.035	94.006	93.890	N/A			
C. muris 2	99.620	99.524	93.387	93.531	93.306	98.664	N/A		
C. baileyi 1	94.740	94.565	96.314	96.534	96.247	94.749	94.974	N/A	
C. wrairi 1	93.298	93.124	99.269	99.542	99.255	94.400	93.648	96.478	N/A

[a] Isolates *C. parvum* 1 and 2 are presumed to be identical with *C. muris* 2 (V. MacDonald, personal communication). Sequence homology among the three *Plasmodium* species, listed in the dendogram shown in Figure 1, were 88–89%, and 81–82% when compared with *Cryptosporidium* species. N/A = not applicable.

generated the similarity values summarized in Table 1, using the Genetics Computer Group's BESFIT program.

The analysis presented in Table 1, based on a small number of isolates, does not strongly support the current speciation of the genus, except perhaps for *C. muris*. The interspecies values are not appreciably different from intraspecies values. Cai *et al.* (1992) reported a 99% sequence homology between *C. parvum* and *C. muris*. It was revealed subsequently, however, that the two *C. parvum* isolates (accession Nos X64340 and X64341) listed in Table 1 are in fact *C. muris* 2 (accession No. X64342) (V. MacDonald, personal communication). The dendrogram in Figure 1 shows two minor groupings; the group that includes the two *C. muris* isolates (assuming that *C. parvum* 1 and 2 are the same as *C. muris* 2), and the rest. In contrast to the minuscule differences among the *Cryptosporidium* species, three valid species of *Plasmodium* included in the dendogram show a considerably greater degree of variation between them (R. Balakrishnan, G. Widmer, and S. Tzipori, unpublished). As with *P. falciparum* (Wilson *et al.*, 1996) and *T. gondii* (McFadden *et al.*, 1996), however, it is likely that extrachromosomal copies of genomic rRNA exist in *Cryptosporidium* as well. If multiple rRNA copies in *Cryptosporidium* are confirmed (Zamani *et al.*, 1996), the possible occurrence of different rRNA genes within each genome may be a factor to consider in future rRNA analyses.

Demonstration of significantly greater sequence homologies within species than among species, utilizing several highly conserved key genes, obtained from a large and diverse host origin and geography, could provide in the future a solid genetic basis for elucidating the taxonomy of the genus *Cryptosporidium*. An alternative approach could be based on the classical definition of species as a distinct reproductive entity. Once suitable molecular markers become available, oocysts obtained from a host infected with a mixed population (e.g. mice co-infected with *C. parvum* and *C. muris*, or chick embryos co-infected with *C. parvum* and *C. baileyi*) could be tested for the presence of hybrid oocysts. Absence of such hybrids would argue in favor of true species (G. Widmer and S. Tzipori, unpublished).

4.3. Phenotypic and Genotypic Variation among *C. parvum* Isolates

As mentioned earlier, considerable phenotypic and genotypic variations occur among and within *C. parvum* isolates. As the number of characterized isolates increases, and as new tools for differentiation become available, more such differences will be identified. Differences have been recognized based on infectivity to other animals (see O'Donoghue, 1995; Tzipori, 1983; S. Tzipori, unpublished), pathogenicity (Fayer and Ungar, 1986; Pozio *et al.*, 1992; Tzipori *et al.*, 1994), protein banding (Lumb *et al.*, 1988a; Mead *et al.*, 1990); antigenicity (McDonald *et al.*, 1991; Nichols *et al.*, 1991; Uhl *et al.*,

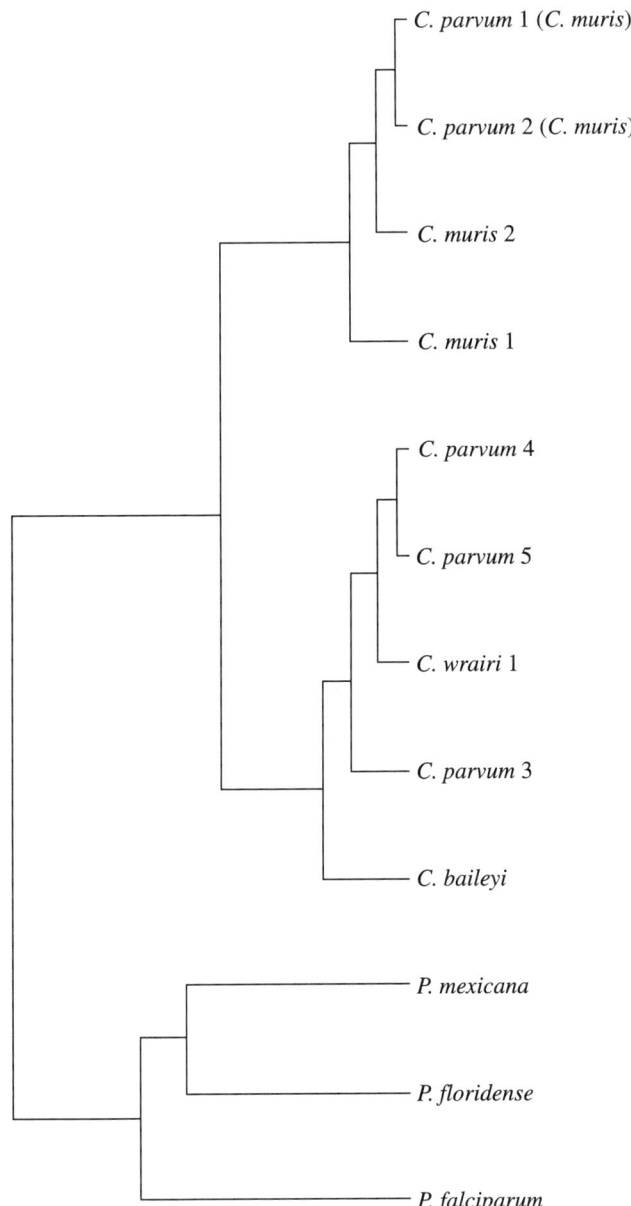

Figure 1 Dendrogram illustrating the interspecies versus intraspecies relationship among isolates of *Cryptosporidium*, as compared with three well-known valid species of *Plasmodium*.

1992), isoenzyme typing (O'Donoghue, 1992; Ogunkolade et al., 1993; Awad-El-Kariem et al., 1994), and genotyping (see the chapter by Widmer in this volume). Phenotypic (Pozio et al., 1992) and genotypic (see below) differences have been used to distinguish between human and animal (calf) isolates. There is clear evidence that among *C. parvum* isolates there is a strong preference for one host or a group of hosts. Some isolates have a clear preference for, and readily infect, adult rodents in which they perpetuate (Tyzzer, 1912; Webster and MacDonald, 1995). In contrast, most *C. parvum* isolated from humans and calves visibly infect only newborn rodents or immunodeficient adult rodents. The same probably applies to *Cryptosporidium* found in guinea pigs.

4.4. The Stability of Genetic Markers

Analysis of *C. parvum* genotypes using different genetic markers (Bonnin et al., 1996; Carraway et al., 1996) has demonstrated that certain genetic profiles from human and bovine isolates differ. This observation could imply the existence of *C. parvum* isolates that are transmitted exclusively from human to human or, alternatively, that the genetic profile of a specific isolate changes upon transmission to a different host. The natural heterogeneity of *C. parvum* populations infecting individual hosts (Carraway et al., 1994) would allow selection of different subpopulations, depending on the host species. Studies in human volunteers (DuPont et al., 1995) infected with oocysts of calf origin provided a unique opportunity to study the stability of genetic markers following transmission between different host species. DNA extracted from stool samples of six *C. parvum*-infected volunteers was genotyped in parallel with the calf inoculum using a ribosomal polymorphic marker (see the chapter by Widmer in this volume) and a newly identified restriction fragment length polymorphism (RFLP) located in a threonine-rich open-reading frame (G. Widmer, M. Carraway, and S. Tzipori, unpublished). In four out of six volunteers changes in the RFLP profile were detected, resulting in either a heterogeneous profile or a profile previously found in some of the human *C. parvum* isolates. Significantly, serial infections of neonatal pigs with calf-derived oocysts of isolate GCH1 (Tzipori et al., 1994) were successful in reproducing these genetic changes (G. Widmer, C. Chappell, L. Tchack, M. Carraway, and S. Tzipori, unpublished). Together, these observations suggest that genetic markers in *C. parvum* can change upon passage to a different host, possibly through a selective mechanism favoring different subpopulations.

The broader implications of these observations are difficult to predict at this stage. However, they may in the future provide a genetic basis to help explain the movement of *Cryptosporidium* among mammalian, or even vertebrate species, and shed new light on the speciation of the genus. Linking clearly recognized virulence attributes to stable genetic markers is a better way to define a species, which, for *Cryptosporidium*, one suspects will not necessarily be confined to a particular animal host, or even to a class of vertebrates. Until such time, however, it will be wise to regard *Cryptosporidium* isolates from all vertebrates as a potential risk to public health.

5. RISK TO HUMANS FROM INFECTED ANIMALS

Of the two, and some propose three (Chrisp *et al.*, 1992), mammalian *Cryptosporidium* species, only *C. parvum* appears to be associated with human disease. To date, *C. parvum* has been reported in some 79 mammalian species (O'Donoghue, 1995), and potentially all pose a health risk either by direct contact or indirectly through fecal contamination of food or water consumed by people. *C. muris*, the first mammalian species to be described (Tyzzer, 1907), was observed in the stomachs of a few animals, including mice, rats, cats, dogs, cattle, and camels. Oocysts of *C. muris* are larger than those of *C. parvum*, and the infection is asymptomatic (see O'Donoghue, 1995). *C. muris* has so far not been identified in humans.

5.1. Infections in Domestic and Wild Ruminants

The significance of *C. parvum* infection in domestic animals, newborn calves in particular, became evident in the early 1980s. The course of the infection and the disease it induces in a variety of small ruminants was reproduced experimentally, reported extensively, and reviewed many times (Angus, 1983; Tzipori, 1983, 1988; Current and Garcia, 1991; O'Donoghue, 1995).

Infections of domestic and wild ruminants provide the biggest source of environmental contamination. While *C. parvum* is probably present in every domestic cattle herd worldwide, the incidence of diarrhea varies considerably from none (Myers *et al.*, 1984), to more than 59% in a nationwide US survey involving 1103 farms (Garber *et al.*, 1994). The incidence of diarrhea in calves varies among farms and geographic locations, and between the ages of 1 and 6 weeks it can reach up to 75% of calves (Leek and Fayer, 1984), with individual calves excreting in excess of 10^{10} oocysts over the course of the infection. Apparently healthy calves can also become subclinically infected and contribute to oocyst excretion in feces (Tzipori, 1988). Asymptomatic infections and prolonged oocyst excretion by adult cattle (Nouri and Toroghi, 1991; Lorenzo-Lorenzo *et al.*, 1993; Kemp *et al.*, 1995; Scott *et al.*, 1995), have become recognized as another major and continuous source of environmental contamination, and clearly the source from which newborn calves contract the infection at a very young age.

C. parvum is also common in sheep and goat herds (Tzipori, 1988; O'Donoghue, 1995), but prevalence is not as well documented. Newborn lambs (Tzipori *et al.*, 1981a; Anderson, 1982; Angus *et al.*, 1982; Xiao *et al.*, 1994) and goat kids (Tzipori *et al.*, 1982b; Thamsborg, 1990) are susceptible to *C. parvum* and diarrhea, and undoubtedly contribute to oocyst dissemination, the extent of which is not clear.

5.2. Infections in Swine

Serological evidence suggests that the infection is widespread among swine herds (Tzipori and Campbell, 1981; Quilez *et al.*, 1996), although diarrhea is not a serious problem in young piglets. Nevertheless, the intensive nature of the industry and the methods currently practiced for effluent disposal must play some role in the generation and dissemination of oocysts from this source.

5.3. Infections in Domestic Pets

Infections in dogs, cats, and horses have been reported, and must be regarded as a potential source for human infection. However, *C. parvum* is not known to cause diarrhea in these animals (Tzipori, 1985), a vehicle normally responsible for massive production and environmental dissemination of oocysts. The prevalence of cryptosporidiosis in these species of animals is not extensively documented, but one serological survey indicates a high prevalence in all three (Tzipori and Campbell, 1981). Although most of the infections in horses are subclinical, in some herds they can be as high as 80% (Xiao and Herd, 1994). Experimental inoculation of newborn foals with *C. parvum* derived from calves induced only subclinical infections (Tzipori, 1985). Fatal cryptosporidiosis however has been reported in severely immunodeficient Arabian foals (Snyder *et al.*, 1978).

The presence of oocysts in surveyed dog feces has been detected in some populations (Johnston and Gasser, 1993), but not in others (Pojhola, 1984; Simpson *et al.*, 1988). In one serological study, cryptosporidiosis was reported to be widespread in cats (Tzipori and Campbell, 1981). However, the prevalence of *C. parvum* infections in cats is not well documented, and the presumed existence of another putative species, *C. felis*, further confuses the issue. Based on transmission experiments, it is thought that *C. felis* is only infective to other cats (Iseki, 1979; Asahi *et al.*, 1991). After oral challenge cats excrete *C. felis* for many weeks, though it is not clear whether this applies to *C. parvum* as well. Prolonged oocyst excretion by domestic pets has serious epidemiological implications.

5.4. Infections in Wild Animals

As mentioned earlier, some 79 animal species are known to be naturally susceptible to *Cryptosporidium*, the majority of which are in the wild. Until we have methods to identify key virulence factors associated with infectivity and pathogenicity of *Cryptosporidium* for humans, and until we are able confidently to speciate clinical isolates, *Cryptosporidium* from all sources, including birds and lower vertebrate animals, should be regarded as potentially hazardous to public health. Wild animals, whether clinically or subclinically infected with *Cryptosporidium*, contribute to the perpetuation of infections among mammals, as well as to the risk of environmental contamination in general. A report suggests that some animals may even disseminate oocysts by acting as mechanical vectors (Graczyk *et al.*, 1996a).

6. THE UNIQUE INTRACELLULAR NICHE OF *CRYPTOSPORIDIUM*

The degree of interaction between pathogens and the host cell varies from the benign (*Vibrio cholera*), to intimate without penetration (enteropathogenic *Escherichia coli*), and to actual invasion of the cell proper (enteroinvasive *Escherichia coli*, *Shigella*, *Salmonella*, viruses, *Toxoplasma*, *Plasmodium*). However, no organism other than *Cryptosporidium* so extensively alters the cell membrane to create a niche for itself between the cell membrane and the cell cytoplasm. Neither the mechanisms of this process, nor the implications in terms of accessibility to the parasite in this unusual location, are understood. We believe the location and the nature of this dual sequestration from the gut and from the cell cytoplasm hold the key to its enigmatic resistance to chemotherapy.

Cryptosporidium, like other coccidia, sequesters itself inside the host cell during development, protected from the host immune response and the hostile environment of the gut, and accessing the nutritional reservoirs of the host cell (Camus *et al.*, 1995). Again like other coccidia, it lies within a parasitophorous vacuole (PV) bounded by a parasitophorous vacuolar membrane (PVM), which in other coccidians is the portal through which nutrients from the host cytoplasm must flow into the parasite. Unlike any other organism, however, *Cryptosporidium* has in addition a unique structure known as the feeder organelle membrane (FOM), which directly separates the cell and parasite cytoplasms. It is assumed that the PVM in *Cryptosporidium* provides only a protective function, while the FOM is the entryway for nutrients derived from the host cell. It is conceivable, however, that the PVM is also selectively permeable to certain molecules from the gut lumen. This is based on the fact that the PVM, originally derived from the host cell membrane, may retain some of its absorptive and other functional activities. Considering the highly sequestered nature of the parasite, determining whether the PVM is permeable to certain molecules is not only a biological curiosity, but is also highly relevant to drug design and delivery. In this section the contrast between the extracytoplasmic form of *Cryptosporidium* and the intracytoplasmic forms of other, closely related location, coccidians will be highlighted with the aim of delineating structural and functional differences, and their likely consequences.

6.1. Intracellular Locations of Apicomplexa

6.1.1. Invasion and Intracellular Strategies

Apicomplexan parasites alter host cell membranes, via lipids and proteins found within parasite organelles, to assist in invasion and in the alteration of the subsequently formed PVM and the PV. All members of the phylum Apicomplexa have specialized tools for invading and altering host cells: rhoptries, which are tubulosaccular organelles that open at the apex of the parasite; micronemes, which are clustered at the anterior end of the parasite; and dense granules. Rhoptries arise from the endoplasmic reticulum and Golgi complex, and secrete membranous materials that alter the target host membrane. Micronemes and dense granules, which are not as well characterized, are also involved in the invasion process and/or changes in the PV and PVM. Other structures, including the polar rings, conoid, and microtubular protrusions are believed to provide a cytoskeletal framework during invasion (Sam-Yellowe, 1996). Apicomplexan parasites vary greatly in the specific intracellular niche that they occupy. Coccidians, unlike *Trypanosoma cruzi* (Rodriguez *et al.*, 1996), do not normally utilize host phagocytic processes.

6.1.2. Apicomplexan Parasites that Escape the PV

Theileria parva, after invading lymphocytes or erythrocytes, escapes its PVM by discharging its apical rhoptries and micronemes, and once intracytoplasmic it becomes enclosed in a microtubular meshwork (Shaw and Tilney, 1995). *Babesia* species discharge their rhoptries during erythrocyte invasion and PVM formation; subsequently, spherical bodies discharge through the lateral walls of the parasite and dissolve the *Babesia* PVM (Hines *et al.*, 1995). In both these examples the parasite is initially enclosed in a PVM and then escapes into the host cell cytoplasm, where nutrient uptake is direct.

6.1.3. Apicomplexans that Remain within Parasitophorous Vacuole

Intracellular coccidia and other related protozoa are, in general, surrounded by a PVM, which is a critical nutrient transport interface between the parasite and the host. *Plasmodium* and *Toxoplasma*, for instance, form and extensively modify their confining vacuoles. In marked contrast, *Cryptosporidium* is only partly surrounded by a PVM, and probably takes up most or all its nutrients and energy requirements via the FOM. Since the intracellular niche of *Cryptosporidium* is determined by the topology of invasion, it is worthwhile to outline these differences in delineating the particular site that *Cryptosporidium* occupies.

The invasion processes for *Toxoplasma*, *Plasmodium*, and *Cryptosporidium* are stereotypic for Apicomplexa (Aikawa *et al.*, 1977, Bannister *et al.*, 1975; Marcial and Madara, 1986; Lumb *et al.*, 1988b). Zoites (tachyzoites, sporozoites, or merozoites) locate a target cell; orient their anterior pole against the host; rhoptries and micronemes discharge their contents; and concurrently the host engulfs the parasite within a PVM. With *Plasmodium* and *Toxoplasma*, invasion via a sliding ring junction is accompanied by parasite entry into the host cell beyond the surface plane. The electron-dense ring junction moves posteriorly and circumferentially over the zoite during cell entry. *Cryptosporidium* does not penetrate beyond the apical surface, being engulfed by microvillar membranes that fuse over the parasite; indeed, no sliding ring junction is needed (Figure 2).

6.1.4. Formation of PV of Plasmodium, Toxoplasma, and Cryptosporidium

Plasmodium species extensively modify the intraerythrocytic PV and PVM (Elmendorf and Haldar, 1993; Haldar and Holder, 1993; Foley and Tilley, 1995). The malarial PVM is at least in part formed from parasite-derived lipids, which can exchange with host lipids during invasion (Miller *et al.*, 1979, Ward *et al.*, 1993; Dluzewski *et al.*, 1995). Freeze–fracture studies indicate a marked decrease in the number of intramembranous particles (IMPs) in the plasma face of the newly formed PVM, suggesting either exclusion (Aikawa *et al.*, 1981) or rapid removal (see Sibley, 1993) of host IMPs during PVM formation. After invasion, the parasite alters the PV by exporting numerous proteins into the vacuolar space (Sam-Yellowe, 1996). In addition, the malarial parasite exports proteins into the host erythrocyte (Elmendorf and Haldar, 1993; Haldar and Holder, 1993), via an extensive tubulovesicular membrane network that is continuous with the PVM. The *T. gondii* PVM appears to be formed from the host plasma membrane, and pinches off via a fission pore after invasion (Suss-Toby *et al.*, 1996).

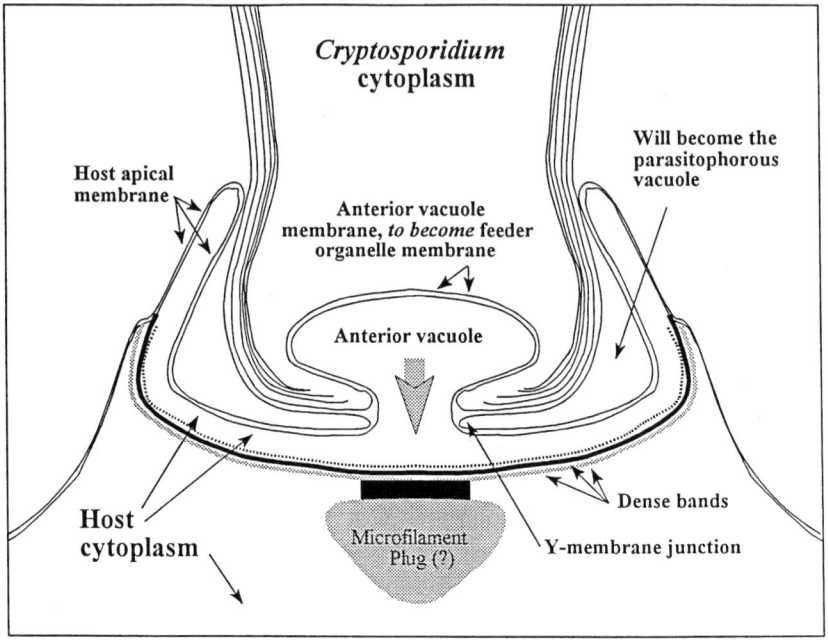

Figure 2 Schematic illustration of the membrane events at the time of invasion, as captured in photomicrograph A in Figure 4. The anterior vacuole membrane has fused with the host cell cytoplasm that lies on the apical surface of the dense bands. The two inner parasite pellicle membranes regress from the anterior pole of the parasite. The anterior vacuole membrane is destined to become the feeder organelle membrane. All membranes are shown as double lines, representing lipid bilayers. The cytoskeletal components of the plug-like structure beneath the dense bands are not known.

Rhoptry-associated lipids may contribute to the PVM (Joiner, 1991). Subsequently, the PVM is modified via dense granule discharge in a calcium-dependent process (Sibley and Boothroyd, 1991; Joiner et al., 1994). A complex intravacuolar membranous network forms inside the PV. Dense granule proteins become associated with the PVM and the vacuolar space (Sibley, 1993), as do rhoptry proteins (Beckers et al., 1994). The overarching message is that these parasites extensively remodel the PV and the PVM via organellar discharge and protein export.

Almost immediately after *Cryptosporidium* zoites contact the host cell, an electron-dense disk-like structure appears in the host cytoplasm (Marcial and Madara, 1986; Lumb et al., 1988b; Tzipori, 1988; Yoshikawa and Iseki, 1992). The structure can be resolved as a series of distinct layers by electron microscopy: three by transmission electron microscopy (Tzipori, 1988) and two by freeze–fracture electron microscopy (Yoshikawa and Iseki, 1992). The disk extends up to the level of the host cell surface membranes. A periodic structure to the dense bands has been seen (Marcial and Madara, 1986; Lumb et al., 1988b), suggesting subunit aggregation. Thus, before PVM or feeder organelle formation, the host cell cytoplasm is quickly segregated into two domains by the dense bands.

Unlike *Toxoplasma* and *Plasmodium*, *Cryptosporidium* is not submerged in the host cytoplasm, but is surrounded by host microvillar membranes that rise up and link over the top of the parasite (Bird and Smith, 1980; Tzipori, 1988). (see Figure 4). The PVM does not completely enclose *Cryptosporidium*, since at the attachment site the feeder organelle forms (see Section 6.1.5). It is believed that discharge of the rhoptries and micronemes induces this engulfment by the host. In freeze–fracture analysis, *Cryptosporidium* outer envelope membranes have IMP densities similar to that of adjacent host microvillus membranes (Marcial and Madara, 1986). In contrast, the inner envelope membrane, which forms the outer PVM becomes deficient in IMPs. Similarly, so does the inner PVM, which once formed the outermost parasite pellicular membrane (Yoshikawa and Iseki, 1992). Thus the two membranes bounding the PV become strikingly deficient in IMPs, suggesting (but not proving) that transport is not a dominant function of the PV or the PVMs. There is no published information to suggest that after invasion there is any discharge of a *Cryptosporidium* organelle. The relationship and the subsequent rearrangements between the cell host membranes and the parasite membranes are illustrated schematically in Figures 2 and 3 and in the corresponding micrographs in Figure 4.

Bonnin et al. (1991, 1993, 1995) have described microneme and dense granule antigens that become localizable on the PVM, supporting involvement of micronemes and dense granules in PV and PVM formation. These antigens were also present in detergent-extracted host cells, suggesting antigen attachment to the host cell cytoskeleton in the enveloping cytoplasmic rim that arches over the parasite. Of interest, none of these antigens were found on the FOM (Bonnin et al., 1995), reinforcing the concept that FOM formation occurs via a process that does not involve the usual invasion machinery. McDonald et al. (1991) have also described antisporozoite monoclonal antibodies that recognized antigens in the outer PVM and/or exterior envelope (villous) membrane.

6.1.5. Cryptosporidium: *Formation of the Feeder Organelle*

The FOM origin was elucidated by Lumb et al. (1988b) *in vitro*. After rhoptry and microneme discharge, a novel membrane-bound vacuole appears in the anterior third of the sporozoite. The outer sporozoite pellicle membrane fuses with the host cell membrane adjacent to the conoid process, and the new vacuolar membrane then fuses with the common host-sporozoite membrane, resulting in the release of vacuolar contents into the topological space between the FOM and the dense bands (Figures 2 and 4(A)). Simultaneously,

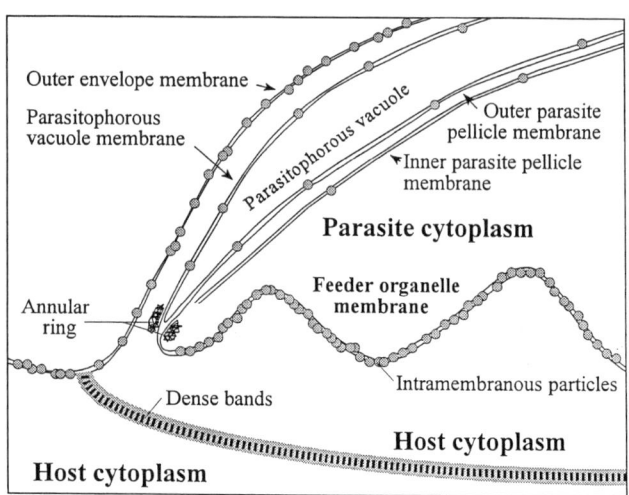

Figure 3 The outer envelope membrane and the parasitophorous vacule membrane are both derived from the host cell apical surface membrane. Note that the density of intramembranous particles (IMPs) changes at the inflection point marked by the junction of the dense bands and the surface membrane of the host cell. The density of IMPs on the parasitophorous vacuole membrane (derived from the host) and the outer parasite pellicle membrane (also called the inner parasitophorous vacuole membrane; derived from the parasite) is quite low; because of the unique Y-shaped membrane junction, these two membranes are continuous with one another. The feeder organelle membrane (FOM; derived from both the host and the parasite) has many IMPs. The annular ring is very electron dense and is located at the Y-shaped membrane junction of (a) the FOM, (b) the outer parasite pellicle membrane, and (c) the parasitophorous vacuole membrane, sometimes called the inner envelope membrane. The host cell cytoplasm extends up to the FOM, and in a thin arc over the parasite between the outer envelope membrane and the parasitophorous vacuole membrane. The inner parasite pellicle membrane disappears as a separate entity in the region of the FOM. Each of the membranes outlined above is drawn as two lines, representing lipid bilayers. (Not drawn to scale.)

the two inner unit membranes of the pellicle regress or dissolve, so that only the fused host/parasite-pellicle/vacuole membrane separates the parasite and host cytoplasms. A circular Y-shaped membrane junction is thus formed, with each branch of the Y forming a unit membrane. The FOM subsequently becomes redundant and folded, increasing its surface area tremendously (Marcial and Madara, 1986; Tzipori, 1988) (Figures 3 and 4(C)). At the Y-junction where the inner and outer PVMs join the FOM, an extremely electron-dense structure is found. This ring-like structure is called the annular ring. It appears to form or support the lateral boundaries of the feeder organelle, and the junction between the union of the host and parasite membranes. Its components are unknown. Science fiction could not produce cell biology as interesting as this!

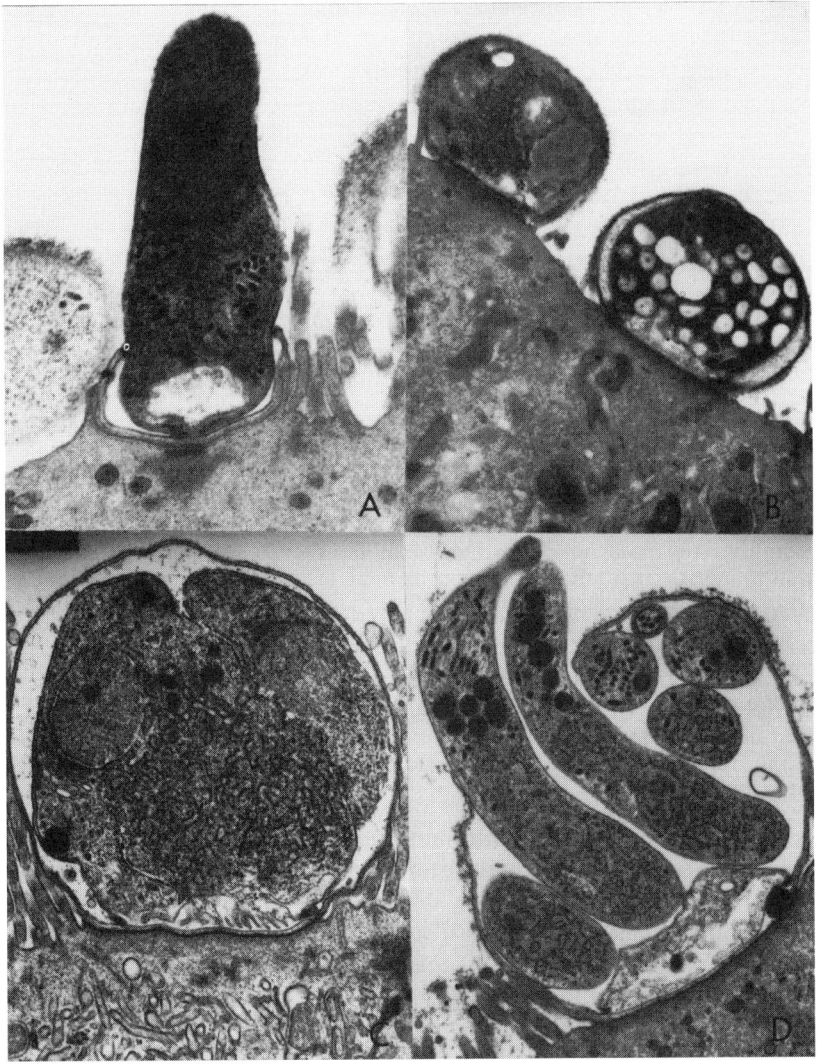

Figure 4 (A) An ultrathin section of mouse gut epithelium infected with C. parvum. The zoite, having formed a contact with the microvillus border of the cell is in the process of being internalized. For topological details see Figure 2. × 39 000. (B) A section as in (A), showing a fertilized macrogamate (right) just prior to sporulation and oocyst formation, with a thick wall. × 16 000. (C) An ultrathin section of a piglet gut epithelium infected with C. parvum. The trophozoite is in the process of dividing into merozoites. The various host–parasite membranes and their origin is illustrated in Figure 3. × 30 000. (D) A section as in (A) and (B), showing the maturation and rupture of the parasitophorous membrane for the release of merozoites into the lumen. × 30 000. Figure produced by Dr Barry Stein.

The FOM has an heterogeneous distribution of IMPs in freeze–fracture analysis (Marcial and Madara, 1986; Yoshikawa and Iseki, 1992). Intriguingly, vesicular budding and a Golgi-like stack of membranes have been associated with the FOM within the parasite cytosol (Yoshikawa and Iseki, 1992). Vetterling et al. (1971b) also found endocytic vesicles in the cytoplasm of intracellular *C. wrairi*.

6.2. Implications of the *Cryptosporidium* topology

6.2.1. *Trafficking Through the PV: Contrasts between* Plasmodium, Toxoplasma, *and* Cryptosporidium

The malarial and toxoplasmal PVMs are permeable to small molecules such as amino acids, purine nucleosides, and monosaccharides, and no specific PVM transporter is needed for these compounds to enter the PV (Desai et al., 1993; Schwab et al., 1994). Transporters on the *Toxoplasma* plasma membrane, or malarial cytosomes, can then complete the uptake process (Slomianny et al., 1985; Joiner et al., 1994). The toxoplasmal PVM is overcoated with vimentin, while anchored in an intermediate filament network (Halonen and Weidner, 1994). Host mitochondria closely appose the PVM, and host endoplasmic reticulum (ER) may incorporate into the PVM (de Melo et al., 1992). Thus *Toxoplasma* reorganizes the host respiratory machinery via close apposition of these host organelles, which provide it with energy and nutrients.

In contrast, the *Cryptosporidium* PVM has only the most tenuous of host cytoplasmic contact, via the thin remnant cytoplasmic rim arching over the parasite (Figure 3). As the parasite matures this remnant cytosol becomes less and less obvious (Marcial and Madara, 1986). There does not appear to be any membranous connection between the PVM and the cell surface (as in *Plasmodium*), nor are host cell organelles (e.g. mitochondria or ER as in *Toxoplasma*) attracted to it. The dearth of IMPs on the PVMs also implies a dearth of transmembrane trafficking. In sum, there is little structural evidence to suggest that the PVM is a gateway for nutrient uptake in *Cryptosporidium*, contrasting with the closely related coccidians *Toxoplasma* and *Plasmodium*.

6.2.2. *Structural Features of the Feeder Organelle Membrane that Suggest Transport*

First, the FOM is a direct path between the host and parasite cytoplasms. Transport of nutrients through the single *Cryptosporidium* FOM would serve the same function as transport through the PVM, PV, and tachyzoite outer membranes in *Toxoplasma*. Second, the FOM is highly redundant and folded with a greatly increased surface area, and transport is a function of surface area. Third, the FOM is rich in IMPs, whereas the PVM is deficient (Marcial and Madara, 1986; Yoshikawa and Iseki, 1992). In other biological systems transport can be related to the number of IMPs, since they represent transmembrane or intramembranous transporter molecules and recognition targets. Fourthly, the vesicular budding and Golgi-like stacks above the FOM directly suggest transport from this site (Vetterling *et al.*, 1971b; Yoshikawa and Iseki, 1992). Thus by every structural feature, the FOM is more likely than the PVM to be the site of nutrient transport. What is lacking at this time is confirmatory functional transport data to match these structural findings in *Cryptosporidium*.

6.2.3. *Implications for Drug Delivery*

We propose that the likely nutrient delivery route is: from the host cytosol across the dense bands; into the topological domain between the dense bands and the FOM; and then across the FOM via vesicular budding, assisted transport, or passive permeation. Without speculating on any potential barrier, anchoring, or sieving function that the dense bands perform, it should be noted that they clearly divide the host cell cytoplasm into two domains. Uptake across the FOM could be passive, via pores such as in the outer PVM of *Toxoplasma* (Schwab *et al.*, 1994). However, in unpublished microinjection experiments (J. K. Griffiths, R. Balakrishnan, K. Ebert, and S. Tzipori, unpublished), we have not consistently seen movement of fluorescently labeled compounds from the host cytoplasm into the parasite, indicating that, if such pores exist, their exclusion size may be quite small. In addition, a porous FOM would not explain the presence of vesicles seen in electron microscopy studies, unless there are two or more forms of nutrient uptake.

The failure of many antimicrobial agents that are active against other pathogenic Apicomplexa to inhibit the growth of *Cryptosporidium* suggests that perhaps the nature of parasite sequestration from the cell cytoplasm plays a major role. Intracellularly active drugs that inhibit the growth of related pathogenic protozoa are, somehow, less able to do so with *Cryptosporidium*.

6.2.4. *Studies of* Cryptosporidium *with Paromomycin and Geneticin*

Paromomycin, a nonabsorbed aminoglycoside, has been found to be somewhat effective in clinical cryptosporidiosis (see the chapters by Griffiths and Tzipori in this volume). We have studied the inhibitory action of paromomycin, and the related aminoglycoside geneticin, in a topologically restricted system (R. Balakrishnan, G. Widmer, S. Tzipori, and J. K. Griffiths, unpublished). Epithelial cells (Caco-2 or MDBK) grown on permeable filters form a monolayer joined by tight junctions, mimicking the intestinal epithelium.

Drugs (such as paromomycin) can be selectively added to the apical or the basolateral compartment of the cells (Griffiths *et al.*, 1994). We have consistently found that apical, but not basolateral, paromomycin is inhibitory to *Cryptosporidium*. Moreover, we have transformed Caco-2 cells with a gene that encodes the cytosolic enzyme aminoglycoside phosphotransferase (APH), which confers aminoglycoside resistance. Using these APH-encoding cell lines, apical paromomycin is still far more active than basolateral paromomycin. This implies that apical exposure is important to paromomycin delivery, and that delivery of paromomycin to the parasite does not require free trafficking through the host cytoplasm (since APH inactivates cytosolic paromomycin). Unlike paromomycin, the related aminoglycoside geneticin freely enters prokaryotic and eukaryotic cells and kills them, unless is inactivated by the enzyme APH. Using APH-transformed cell lines, geneticin was found also preferentially to inhibit *Cryptosporidium* from the apical domain. Timed drug exposure has been performed after invasion but before merogony, proving that paromomycin and geneticin are active against intracellular forms.

The logical conclusion is that these two inhibitory drugs either enter the parasite (a) through the PVM and PV, and/or (b) via the apical cell membrane, cell cytoplasm, and the FOM in a fashion that does not require cytoplasmic trafficking. Paromomycin has been found to accumulate in vesicles in some cell lines (Oshima *et al.*, 1989), and so paromomycin may be taken up directly from the cell membrane into vesicles that are directed to the FOM. The movement of radioactively labeled paromomycin through the membranes of the parasite and the host, and distribution in the cytoplasms of the parasite and the cell, including the PV, may help shed light on how drugs might reach the parasite.

7. CONCLUDING REMARKS

Since the first discovery of *Cryptosporidium*, the speciation of the genus has undergone several cycles. Ninety years later, based on preliminary genetic data analyzed in this chapter, *C. muris*, the first named species by Tyzzer in 1907, emerges at this time as the only distinct species. The actual number of species, and whether they will turn out to be divided largely along classes of vertebrate animals, as they are currently named, or ultimately be based on virulence attributes linked with specific genetic markers, remains to be seen. It is suggested that, until such time *Cryptosporidium* isolates from all animal sources, including other classes of vertebrates, be considered as potential risk to humans.

Cryptosporidium occupies a novel intracellular niche, and via fusion of its own membranes with those of the host creates an unique interface, the feeder organelle membrane of parasite origin. The structure and role of the disk-like dense bands that separate the host cell cytoplasm into two domains, is obscure, but it may serve anchoring or sieving functions. Unlike the closely related parasites *Toxoplasma* and *Plasmodium*, there is no evidence that the PV is the major site of nutrient exchange, and on the basis of structural information the feeder organelle appears to have supplanted the role of the PVM in *Cryptosporidium*. The astonishing lack of activity of a wide variety of antimicrobial agents effective against other related parasites may be due to obstacles in the delivery of these agents into this peculiar site.

REFERENCES

Aikawa, M., Komata, Y., Asai, T. and Midorikawa, O. (1977). Transmission and scanning electron microscopy of host cell entry by *Toxoplasma gondii*. *American Journal of Pathology* **47**, 223–233.
Aikawa, M., Miller, L.H., Johnson, J. and Rabbege, J. (1978). Erythrocyte entry by malarial parasites. A moving junction between erythrocyte and parasite. *Journal of Cell Biology* **77**, 72–82.
Aikawa, M., Miller, L.H., Rabbege, J.R. and Epstein, N. (1981). Freeze–fracture study on the erythrocyte membrane during malarial parasite invasion. *Journal of Cell Biology* **91**, 55–62.
Anderson, B.C. (1981). Pattern of shedding of cryptosporidial oocysts in Idaho calves. *Journal of the American Medical Association* **178**, 865–867.
Anderson, B.C. (1982). Cryptosporidiosis in Idaho lambs: natural and experimental infections. *Journal of the American Veterinary Medical Association* **181**, 151–153.
Angus, K.W. (1983). Cryptosporidiosis in man, domestic animals and birds: a review. *Journal of the Royal Society of Medicine* **76**, 62–70.
Angus, K.W., Appleyard, W.T., Menzies, J.P., Campbell, I. and Sherwood, D. (1982). An outbreak of diarrhoea associated with cryptosporidiosis in naturally reared lambs. *Veterinary Record* **110**, 129.
Anon. (1982). Human cryptosporidiosis – Alabama. *Morbidity and Mortality Weekly Report* **31**, 252–254.
Arcay, L., DeBorges, E.B. and Bruzual, E. (1995). Criptosporidiosis experimental en la escala de vertebrados. *Parasitol al Dia* **19**, 20–29.
Asahi, H., Koyama, T., Arai, H., Kunakoshi, Y., Yamaura, H., Sirasaka, R. and Okutoni, K. (1991). Biological nature of *Cryptosporidium* sp. isolated from a cat. *Parasitology Research* **77**, 237.
Awad-El-Kariem, F.M., Wahurst, D.C. and McDonald, V. (1994). Detection and species identification of *Cryptosporidium* oocyst using a system based on PCS and endonuclease restriction. *Parasitology* **109**, 19–22.
Bannister, L.H., Butcher, G.A., Dennis, E.D. and Mitchell, G.H. (1975). Structure and invasive behaviour of *Plasmodium knowlesi* merozoites *in vitro*. *Parasitology* **71**, 483–491.
Beckers, C.J.M., Dubremetz, J.-F., Mercereau-Puijalon, O. and Joiner, K.A. (1994). The *Toxoplasma gondii* rhoptry protein ROP2 is inserted into the parasitophorous vacuole membrane, surrounding the intracellular parasite, and is exposed to the host cell cytoplasm. *Journal of Cell Biology* **127**, 947–961.

Beyer, T.V. and Sidorenko, N.V. (1993). One more biological characteristic of coccidia in the genus *Cryptosporidium* (Sporozoa: Apicomplexa). *Parazitologiia* **27**, 309-319.

Bird, R.G. and Smith, M.D. (1980). Cryptosporidiosis in man: parasite life cycle and fine structural pathology. *Journal of Pathology* **132**, 217-233.

Bommer, W. (1969). The life cycle of virulent *Toxoplasma* in cell cultures. *Australian Journal of Experimental Biology and Medical Science* **47**, 505-512.

Bonnin, A., Durbremetz, J.-F. and Camerlynck, P. (1991). Characterization of microneme antigens of *Cryptosporidium parvum* (Protozoa, Apicomplexa). *Infection and Immunity* **59**, 1703-1708.

Bonnin, A., Drubremetz, J.-F. and Camerlynck, P. (1993). A new antigen of *Cryptosporidium parvum* micronemes possessing epitopes cross-reactive with macrogamete granules. *Parasitology Research* **79**, 8-14.

Bonnin, A., Gut, J., Durbremetz, J.F., Nelson, R.G., and Camberlynk, P. (1995). Monoclonal antibodies identify a subset of dense granules in *Cryptosporidium parvum* zoites and gamonts. *Journal of Eukaryotic Microbiology* **42**, 395-401.

Bonnin, A., Fourmaux, M.N., Dubremetz, J.F., Nelson, R.G., Gobet, P., Harley, G., Buisson, M., Puygauthier-Toubas, D., Gabriel-Pospisil, F., Naciri, M. and Camerlynck, P. (1996). Genotyping human and bovine isolates of *Cryptosporium parvum* by polymerase chain reaction-restriction fragment length polymorphism analysis of a repetitive DNA sequence. *FEMS Microbiological Letters* **137**, 207-211.

Cai, J., Collins, M.D., McDonald, V. and Thompson, D.E. (1992). PCR cloning and nucleotide sequence determination of the 18S rRNA genes and internal transcribed spacer 1 of the protozoan parasites *Cryptosporidium parvum* and *Cryptosporidium muris*. *Biochimica et Biophysica Acta* **1131**, 317-320.

Camus, D., Zalis, M.G., Vannier-Santos, M.A. and Banic, D.M. (1995) The art of parasite survival. *Brazilian Journal of Medical and Biological Research* **28**, 399-413.

Carraway, M., Widmer, G. and Tzipori, S. (1994). Genetic markers differentiate *Cryptosporidium parvum* isolates. *Journal of Eukaryotic Microbiology* **41**, 265.

Carraway, M., Tzipori, S. and Widmer, G. (1996). Identification of genetic heterogeneity in the *Cryptosporium parvum* ribosomal repeat. *Applied and Environmental Microbiology* **62**, 712-716.

Chrisp, C.E., Suckow, M.A., Fayer, R., Arrowood M.J., Healy M.C. and Sterling, C.R. (1992). Comparison of the host ranges and antigenicity of *Cryptosporidium wrairi* from guinea pigs. *Journal of Protozoology* **39**, 406-409.

Current, W.L. and Garcia, L.S. (1991). Cryptosporidiosis. *Clinical Microbiology Reviews* **4**, 325-358.

Current, W.L. and Haynes, T.B. (1984). Complete development of *Cryptosporidium* in cell cultures. *Science* **224**, 603-605.

Current, W.L. and Long, P.L. (1983). Development of human and calf *Cryptosporidium* in chicken embryos. *Journal of Infectious Diseases* **148**, 1108-1113.

Current, W.L. and Reese, N.C. (1986). A comparison of endogenous development of three isolates of *Cryptosporidium* in suckling mice. *Journal of Protozoology* **33**, 98-108.

Current, W.L., Reese, N.C., Ernst, J.V., Bailey, W.S., Heyman, M.B. and Weinstein, W.M. (1983). Human cryptosporidiosis in immunocompetent and immunodeficient persons. Studies of an outbreak and experimental transmission. *New England Journal of Medicine* **308**, 1252-1257.

Current, W.L., Upton, S.J. and Haynes, T.B. (1986). The life cycle of *Cryptosporidium*

baileyi n. sp. (Apicomplexa, Cryptosporidiae) infecting chickens. *Journal of Parasitology* **33**, 289–296.
D'Antonio, R.G., Winn, R.E., Taylor, J.P., Gustafson, T.L., Current, W.L., Rhodes, M.M., Gary, G.W. and Zajac, R.A. (1985). A waterborne outbreak of cryptosporidiosis in normal hosts. *Annals of Internal Medicine* **103**, 886–888.
de Melo E.J., de Carvalho, T.U. and de Souza, W. (1992). Penetration of *Toxoplasma gondii* into host cells induces changes in the distribution of the mitochondria and the endoplamsic reticulum. *Cell Structure and Function* **17**, 311–317.
Desai, S.A., Krogstad, D.J. and McCleskey, E.W. (1993). A nutrient-permeable channel on the intraerythrocytic malaria parasite. *Nature* **362**, 643–646.
Ditrich, O., Palkovic, L., Sterba, J., Prokopic J., Loudova, J. and Giboda, M. (1991). The first finding of *Cryptosporidium baileyi* in man. *Parasitology Research* **77**, 44–47.
Dluzewski, A.R., Zicha, D., Dunn, G.A. and Gratzer, W.B. (1995). Origins of the parasitophorous vacuole membrane of the malaria parasite: surface area of the parasitized red cell. *European Journal of Cell Biology* **68**, 446–449.
Dubey, J.P., Speer, C.A. and Fayer, R. (1990). *Cryptosporidiosis of Man and Animals*. Boston: CRC Press.
DuPont, H.L., Chappel, C.L., Sterling, C.R., Okhuysen, P.O., Rose, J.B. and Jakubowski, W. (1995). The infectivity of *Cryptosporidium parvum* in healthy volunteers. *New England Journal of Medicine* **332**, 855–859.
Elmendorf, H.G. and Haldar, K. (1993). Secretory transport in *Plasmodium*. *Parasitology Today* **9**, 98–102.
Fayer, R. and Ungar, B.L.P. (1986). *Cryptosporidium* spp. and cryptosporidiosis. *Microbiological Reviews* **50**, 458–483.
Fayer, R., Graczyk, T.K. and Cranfield, M.R. (1995). Multiple heterogenous isolates of *Cryptosporidium serpentis* from captive snakes are not transmissible to neonatal BALB/c mice (*Mus musculus*). *Journal of Parasitology* **81**, 482–484.
Fayer, R., Speer, C.A. and Dubey, J.P. (1997). The general biology of *Cryptosporidium*. In: Cryptosporidium *and Cryptosporidiosis* (R. Fayer, ed), pp. 1–41. Boca Raton, FL: CRC Press.
Fernando, M.A., Rose, M.E. and Millard, B.J. (1987). *Eimeria* spp. of domestic fowl: the migration of sporozoites intra- and extra-enterically. *Journal of Parasitology* **73**, 561–567.
Foley, M. and Tilley, L. (1995). Home improvements: malaria and the red blood cell. *Parasitology Today* **11**, 436–439.
Forgacs, P., Tarshis, A., Ma, P., Federman, M., Mele, L. Silverman, M.L. and Shea, J.A. (1983). Intestinal and bronchial cryptosporidiosis in an immunodeficient homosexual man. *Annals of Internal Medicine* **99**, 793–794.
Fricker, C.R. and Crabb, J.H. (1998). Water-borne cryptosporidiosis: detection methods and treatment options. *Advances in Parasitology* **40**, 241–278.
Garber, L.P., Salman, M., Hurd, H.S., Keefe, T. and Schlater, J.L. (1994). Potential risk factors for *Cryptosporium* infections in dairy calves. *Journal of the American Veterinary Medical Association* **205**, 86–89.
Göbel, E. and Brändler U. (1982). Ultrastructure of microgametogenesis, microgametes and gametogony of *Cryptosporidium* sp. In the small intestine of mice. *Protistologica* **18**, 331–344.
Graczyk, T.K., Cranfield, M.R., Fayer, R. and Anderson, M.S. (1996a). Viability and infectivity of *Cryptosporidium parvum* oocysts are retained upon intestinal passage through a refractory avian host. *Applied and Environmental Microbiology* **62**, 3234–3237.

Graczyk, T.K., Fayer, R. and Cranfield, M.R. (1996b). *Cryptosporidium parvum* is not transmissible to fish, amphibians, or reptiles. *Journal of Parasitology* **82**, 748–751.
Griffiths, J.K. (1998). Human cryptosporidiosis: epidemiology, transmission, treatment, and diagnosis. *Advances in Parasitology* **40**, 37–85.
Griffiths, J.K., Moore, R., Dooley, S., Keusch, G.T. and Tzipori, S. (1994). *Cryptosporidium parvum* infection of Caco-2 cell monolayers induces an apical monolayer defect, selectively increases transmonolayer permeability, and causes epithelial cell death. *Infection and Immunity* **62**, 4506–4514.
Haldar, K. and Holder, A.A. (1993). Export of parasite proteins to the erythrocyte in *Plasmodium falciparum*-infected cells. *Seminars in Cell Biology* **4**, 345–353.
Halonen, S.K. and Weidner, E. (1994). Overcoating of *Toxoplasma* parasitophorous vacuoles with host cell vimentin type intermediate filaments. *Journal of Eukaryotic Microbiology* **41**, 65–71.
Hines, S.A., Palmer, G.H., Brown, W.C., McElwain, T.F., Suarez, C.E., Vidotto, O. and Rice-Ficht, A.C. (1995). Genetic and antigenic characteriziation of *Babesia bovis* merozoite spherical body protein Bb-1. *Molecular and Biochemical Parasitology* **69**, 149–159.
Iseki, M. (1979). *Cryptosporidium felis* sp.n. (Protozoa: Eimeriorina) from the domestic cat. *Japanese Journal of Parasitology* **28**, 285–307.
Johnston, J. and Gasser, R.B. (1993). Copro-parasitological survey of dogs in southern Victoria. *Australian Veterinary Practice* **23**, 127.
Joiner, K.A. (1991). Rhoptry lipids and parasitophorous vacuole formation: a slippery issue. *Parasitology Today* **7**, 226–227.
Joiner, K.A., Beckers C.J.M., Bermudes, D., Ossorio, P.N., Schwab, J.C. and Dubremetz, J.F. (1994). Structure and function of the parasitophorous vacuole membrane surrounding *Toxoplasma gondii*. *Annals of the New York Academy of Sciences* **730**, 1–6.
Jokipii L., Pohjola S. and Jokipii, A.M.M. (1983). *Cryptosporidium*: a frequent finding in patients with gastrointestinal symptoms. *Lancet* **ii**, 358–361.
Kemp, J.S., Wright, S.E. and Bukhari, Z. (1995). On farm detection of *Cryptosporidium parvum* in cattle, calves and environmental samples. In: *Protozoan Parasites and Water* (W.B. Betts, D. Casemore, C. Fricker, H. Smith and J. Watkins, eds). London: Royal Society of Chemistry.
Leek, R.G. and Fayer, R. (1984). Prevalence of *Cryptosporidium* infections, and their relation to diarrhea in calves on 12 dairy farms in Maryland. *Proceedings of the Helminthological Society of Washington* **51**, 360.
Levine, N.D. (1980). Some corrections of coccidian (Apicomplexa: Protozoa) nomenclature. *Journal of Parasitology* **66**, 830–834.
Levine, N.D. (1985). Phylum II. Apicomplexa Levine, 1970. In: *An Illustrated Guide to the Protozoa* (J.J. Lee, S.H. Hutner and E.C. Bovee, eds), pp. 322–374. Lawrence, K.S.: Allen Press.
Lindsay, D.S., Blagburn B.L. and Sundermann, C.A. (1986). Host specificity of *Cryptosporidium* sp. isolated from chickens. *Journal of Parasitology* **72**, 565–568.
Lindsay, D.S., Blagburn B.L. and Ernest, J.A. (1987). Experimental *Cryptosporidium parvum* infections in chickens. *Journal of Parasitology* **73**, 242–244.
Lorenzo-Lorenzo, M.J., Ares-Mazas, E. and Villacorta Martinez de Maturana, I. (1993). Detection of oocysts and IgG antibodies to *Cryptosporidium parvum* in asymptomatic adult cattle. *Veterinary Parasitology* **47**, 9–15.
Lumb, R., Lanser, J.A. and O'Donoghue P.J. (1988a). Electrophoretic and immunoblot analysis of *Cryptosporidium* oocysts. *Immunology and Cell Biology* **66**, 369–376.

Lumb, R., Smith, K., O'Donoghue P.J. and Lanser, J.A. (1988b). Ultrastructure of the attachment of *Cryptosporidium* sporozoites to tissue culture cells. *Parasitology Research* **74**, 531–536.
Ma, P. and Soave, R. (1982). Cryptosporidiosis, a new enteric coccidiosis, emerged as an opportunistic infection in five homosexuals. In: *22nd Interscience Conference on Antimicrobial Agents and Chemotherapy*, p. 848. American Society for Microbiology.
Ma, P. and Soave, R. (1983). Three step stool examination for cryptosporidiosis in ten homosexual men with protracted watery diarrhea. *Journal of Infectious Diseases* **147**, 824–826.
MacKenzie, W.R., Hoxie, N.J., Proctor, M.E., Gradus, M.S., Blair, K.A., Peterson, D.E., Kazmierczak, J.J., Addiss, D.G., Fox, K.R., Rose, J.B. *et al.* (1994). A massive outbreak in Milwaukee of *Cryptosporidium* infection transmitted through the public water supply. *New England Journal of Medicine* **331**, 161–167.
Marcial, M.A. and Madara, J.L. (1986). *Cryptosporidium*: cellular localization, structural analysis of absorptive cell–parasite membrane–membrane interactions in guinea pigs, and suggestion of protozoan transport by M cells. *Gastroenterology* **90**, 583–584.
Mauel, J. (1996). Intracellular survival of protozoan parasites with special reference to *Leishmania* spp., *Toxoplasma gondii* and *Trypanosoma cruzii*. *Advances in Parasitology* **38**, 1–51.
McDonald, V., Deer, R.M.A., Nina, J.M.S., Wright, S., Chiodini, P.L. and McAdam, K.P.W.J. (1991). Characteristics and specificity of hybridoma antibodies against oocyst antigens of *Cryptosporidium parvum* from man. *Parasite Immunology* **13**, 251–259.
McFaddin, G.L., Reith, M.E., Munholland, J. and Lang-Unnasch, N. (1996). Plastids of human parasites. *Nature* **331**, 482.
Mead, J.R., Humphreys, R.C., Sammons, D.W. and Sterling, C.R. (1990). Identification of isolate-specific sporozoite proteins of *Cryptosporidium parvum* by two-dimensional gel electrophoresis. *Infection and Immunity* **58**, 2071–2075.
Meisel, J.L., Perera, D.R., Meligro, C. and Rubin, C.E. (1976). Overwhelming watery diarrhea associated with a *Cryptosporidium* in an immunosuppressed patient. *Gastroenterology* **70**, 1156–1160.
Meuten, D.J., Kruiningen, H.J. van and Lein, D.H. (1974). Cryptosporidiosis in a calf. *Journal of the American Veterinary Medical Association* **165**, 914–917.
Miller, L.H., Aikawa, M., Johnson, J.G. and Shiroishi, T. (1979). Interaction between cytochalasin B-treated malarial parasites and erythrocytes. Attachment and junction formation. *Journal of Experimental Medicine* **149**, 172–184.
Morin, M., Lariviere, S. and Lallier, R. (1976). Pathological and microbiological observations made on spontaneous cases of acute neonatal calf diarrhea. *Canadian Journal of Comparative Medicine* **40**, 228.
Myers, L.L., Firehammer, B.D., Border, M.M. and Shoop, D.S. (1984). Prevalence of enteric pathogens in the feces of healthy beef calves. *American Journal of Veterinary Research* **45**, 1544–1549.
Nichols, G.l., McLaughlin, J. and Samuel, D. (1991). A technique for typing *Cryptosporidium* isolates. *Journal of Protozoology* **38**, 237S–240S.
Nime, F.A., Burek, J.D., Page, D.L., Holscher, M.A. and Yardley, J.H. (1976). Acute enterocolitis in a human being infected with the protozoan *Cryptosporidium*. *Gastroenterology* **70**, 592–598.
Nouri, M. and Toroghi, R. (1991). Asymptomatic cryptosporidiosis in cattle and humans in Iran. *Veterinary Record* **128**, 358.
O'Donoghue, P.J. (1992). The cross-transmission potential of *Cryptosporidium* spp.

In: *Zoonoses (Australian Veterinarians in Public Health Symposium), Proceedings 194*, pp. 283–300. Sydney: Post Graduate Committee in Veterinary Science, University of Sydney.

O'Donoghue, P.J. (1995). *Cryptosporidium* and cryptosporidiosis in man and animals. *International Journal of Parasitology* **25**, 139–195.

Ogunkolade, B.W., Robinson, H.A., McDonald, V., Webster, K. and Evans, D.A. (1993). Isoenzyme variation with the genus *Cryptosporidium*. *Parasitology Research* **79**, 385–388.

Oshima, M., Hashiguchi, M., Nakasuji, M., Shindo, N. and Shibata, S. (1989). Biochemical mechanisms of aminoglycoside cell toxicity. II. Accumulation of phospholipids during myeloid body formation and histological studies on myeloid bodies using twelve aminoglycoside antibiotics. *Journal of Biochemistry* **106**, 794–797.

Panciera, R.J., Thomassen, R.W. and Garner, F.M. (1971). Cryptosporidial infection in a calf. *Veterinary Pathology* **8**, 479–484.

Perry, E.A. and Long, P.L. (1987). The extraintestinal stages of *Eimeria tenella* and *E. maxima* in the chicken. *Veterinary Parasitology* **25**, 9–17.

Pohjola, S. (1984). Survey of cryptosporidiosis in feces of normal healthy dogs. *Nordic Veterinary Medicine* **36**, 189–190.

Pohjola, S., Jokipii, L. and Jokipii, A. (1984). Dimethylsulphoxide Ziehl–Neelsen technique for detection of cryptosporidial oocysts. *Veterinary Record* **115**, 442–443.

Pohlenz, J., Moon, H.W., Cheville, N.F. and Bemrick, W.J. (1978). Cryptosporidiosis as a probable factor in neonatal diarrhea of calves. *Journal of American Veterinary Medical Association* **172**, 452.

Pozio, E., Gomez Morales, M.A., Barbieri, F.M. and La Rosa, G. (1992). *Cryptosporidium*: different behaviour in calves of isolates of human origin. *Transactions of the Royal Society of Tropical Medicine and Hygiene* **86**, 636–638.

Quilez, J., Ares-Mazas, E., Sanchez-Acedo, C., del Cacho, E., Clavel, A. and Causape, A.C. (1996). Comparison of oocyst shedding and the serum immune response to *Cryptosporidium parvum* in cattle and pigs. *Parasitology Resources* **82**, 529–234.

Rodriguez, A., Samoff, E., Rioult, M.G., Chung, A. and Andrews, N.W. (1996). Host cell invasion by trypanosomes requires lysosomes and microbutule/kinesin-mediated transport. *Journal of Cell Biology* **134**, 349–362.

Sam-Yellowe, T.Y. (1996). Rhaptory organelles of the Apicomplexa: their role in host cell invasion and intracellular survival. *Parasitology Today* **12**, 308–316.

Schwab, J.C., Beckers, C.J.M. and Joiner, K.A. (1994). The parasitophorous vacuole membrane surrounding intracellular *Toxoplasma gondii* functions as a molecular sieve. *Proceedings of the National Academy of Sciences of the USA* **B91**, 509–513.

Scott, C.A., Smith, H.V., Mtambo, M.M. and Gibbs, H.A. (1995). An epidemiological study of *Cryptosporidium parvum* in two herds of adult beef cattle. *Veterinary Parasitology* **57**, 277–288.

Shaw, M.K. and Tileny, L.G. (1995). The entry of *Theileria parva* merozoites into bovine erythrocytes occurs by a process similar to sporozoite invasion of lymphotcytes. *Journal of Cell Science* **111**, 455–461.

Sibley, L.D. (1993). Interactions between *Toxoplasma gondii* and its mammalian host cells. *Seminars in Cell Biology* **4**, 335–344.

Sibley, L.D. and Boothroyd, J.C. (1991). Calcium regulated secretion and modification of host endocytic compartments by *Toxoplasma*. *Journal of Cell Biology* **115**, 5a.

Simpson, J.W., Burnie, A.G., Miles, R.S., Scott, J.L. and Lindsay, D.I. (1988). Prevalence of *Giardia* and *Cryptosporidium* infection in dogs in Edinburgh. *Veterinary Record* **123**, 445.
Slomianny, C., Prensier, G. and Charet, P. (1985). Ingestion of erythrocytic stroma by *Plasmodium chabaudi* trophozoites: ultrastructural study by serial sectioning and 3-dimensional reconstruction. *Parasitology* **90**, 579–588.
Snyder, S.P., England, J.J. and McChesney, A.E. (1978). Cryptosporidiosis in immunodeficient Arabian foals. *Veterinary Pathology* **15**, 12–17.
Suss-Toby, E., Zimmerberg, J. and Ward, G.E. (1996). *Toxoplasma* invasion: the parsitophorous vacuole is formed from host cell plasma membrane and pinches off via a fission pore. *Proceedings of the National Academy of Sciences of the USA* **93**, 8413–8418.
Thamsborg, S.M. (1990). Cryptosporidiosis in kids of dairy goats. *Veterinary Record* **127**, 627.
Tyzzer, E.E. (1907). A sporozoon found in the peptic glands of the common mouse. *Proceedings of the Society for Experimental Biology and Medicine* **5**, 12–13.
Tyzzer, E.E. (1912). *Cryptosporidium parvum* (sp. nov.). A coccidium found in the small intestine of the common mouse. *Archiv für Protistenkunde* **26**, 394–412.
Tzipori, S. (1983). Cryptosporidiosis in animals and humans. *Microbiological Reviews* **47**, 84–96.
Tzipori, S. (1985). The relative importance of enteric pathogens affecting neonates of domestic animals. *Advances in Veterinary Science and Comparative Medicine* **29**, 103–206.
Tzipori, S. (1988). Cryptosporidiosis in perspective. *Advances in Parasitology* **27**, 63–129.
Tzipori, S. (1998). Cryptosporidiosis: laboratory investigations and chemotherapy. *Advances in Parasitology* **40**, 192–221.
Tzipori, S. and Campbell, I. (1981). Prevalence of *Cryptosporidium* antibodies in 10 animal species. *Journal of Clinical Microbiology* **14**, 455–456.
Tzipori, S., Angus, K.W., Campbell, I. and Gray, E.W. (1980a). *Cryptosporidium*: evidence for a single species genus. *Infection and Immunity* **30**, 884–886.
Tzipori, S., Angus, K.W., Gray, E.W. and Campbell, I. (1980b). Vomiting and diarrhea associated with cryptosporidial infection. *New England Journal of Medicine* **303**, 818.
Tzipori, S., Campbell, I., Sherwood, D., Snodgrass, D.R. and Whitelaw, A. (1980c). An outbreak of calf diarrhoea attributed to cryptosporidial infection. *Veterinary Record* **107**: 579–580.
Tzipori, S., Angus, K.W., Campbell, I. and Clerihew, L.W. (1981a). Diarrhea due to *Cryptosporidium* infection in artificially reared lambs. *Journal of Clinical Microbiology* **14**, 100–105.
Tzipori, S., Angus, K.W., Campbell, I. and Sherwood, D. (1981b). Diarrhea in young red deer associated with infection with *Cryptosporidium*. *Journal of Infectious Diseases* **144**, 170–175.
Tzipori, S., Angus, K.W., Campbell, I. and Gray, E.W. (1982a). Experimental infection of lambs with *Cryptosporidium* isolated from a human patient with diarrhoea. *Gut* **23**, 71–74.
Tzipori, S., Larsen, J., Smith, M. and Leufl, R. (1982b). Diarrhoea in goat kids attributed to *Cryptosporidium* infection. *Veterinary Record* **111**, 35–36.
Tzipori, S., Smith, M., Birch, C., Barnes, G. and Bishop, R. (1983). Cryptosporidiosis in hospital patients with gastroenteritis. *American Journal Of Tropical Medicine Hygiene* **32**, 931–934.
Tzipori, S., Rand, W., Griffiths, J.K., Widmer, G. and Crabb, J. (1994). Evaluation of

an animal model system for cryptosporidiosis: therapeutic efficacy of paromomycin and hyperimmune bovine colostrum-immunoglobulin. *Clinical and Diagnostic Laboratory Immunology* **1,** 450–463.

Uhl, E.W., O'Connor, R.M., Perryman, L.E. and Riggs, M.W. (1992). Neutralization-sensitive epitopes are conserved among geographically diverse isolates of *Cryptosporidium parvum*. *Infection and Immunity* **60,** 1703–1706.

Vetterling, J.M., Jervis, H.R., Merrill, T.G. and Sprinz, H. (1971a). *Cryptosporidium wrairi* from the guinea pig *Cavia porcellus* with an emendation of the genus. *Journal of Protozoology* **18,** 243–247.

Vetterling, J.M., Takeuchi, A. and Madden, P.A. (1971b). Ultrastructure of *Cryptosporidium wrairi* from the guinea pig. *Journal of Protozoology* **18,** 248–260.

Ward, H. and Cevallos, A.M. (1998). *Cryptosporidium*: molecular basis of host–parasite interaction. *Advances in Parasitology* **40,** 151–185.

Ward, G.E., Miller, L.H. and Dvorak, J.A. (1993). The origin of the paristophorous vacuole membrane lipids in malaria-infected erythrocytes. *Journal of Cell Science* **106,** 237–248.

Webster, J.P. and McDonald, D.W. (1995). Parasites of wild brown rats (*Rattus norvegicus*) on UK farms. *Parasitology* **111,** 247–255.

Wilson, R.J.M, Denny, P.W., Praiser, P.R., Rangachari, K., Roberts, K., Roy, A., Whyte, A., Strath, M., Moore, D.J., Moore P.W. and Williamson, D.H. (1996). Complete gene map of plastid-like DNA of the malaria parasite *Plasmodium falciparum*. *Journal of Molecular Biology* **261,** 155–172.

Xiao, L. and Herd, R.P. (1994). Review of equine *Cryptosporidium* infection. *Equine Veterinary Journal* **26,** 14–17.

Xiao, L., Herd, R.P. and Bowman, G.L. (1994a). Prevalence of *Cryptosporidium* and *Giardia* infections on two Ohio pig farms with different management systems. *Veterinary Parasitology* **52,** 331–336.

Xiao, L., Herd, R.P. and McClure, K.E. (1994b). Periparturient rise in the excretion of *Giardia* sp. cysts and *Cryptosporidium parvum* oocysts as a source of infection for lambs. *Journal of Parasitology* **80,** 55–59.

Yang, S. and Healey, M. (1994). Patent gut infections in immunosuppressed adult C57BL/6N mice following intraperitoneal injection of *Cryptosporidium parvum* oocysts. *Journal of Parasitology* **80,** 338–342.

Yoshikawa, H. and Iseki, M. (1992). Freeze–fracture study of the site of attachment of *Cryptosporidium muris* in gastric glands. *Journal of Protozoology* **39,** 539–544.

Zamani, F., Upton, S.J. and Le Blancq, S.M. (1996). Ribosomal RNA gene organization in *Cryptosporidium parvum*. In: *49th Annual Meeting of the American Society of Tropical Medicine and Hygiene*. Baltimore, MD abstract no. 455, p. 250.

Human Cryptosporidiosis: Epidemiology, Transmission, Clinical Disease, Treatment, and Diagnosis

Jeffrey K. Griffiths

Department of Family Medicine and Community Health, Tufts University School of Medicine, Boston MA 02111, USA and Division of Infectious Diseases, Tufts University School of Veterinary Medicine, North Grafton, MA 01536, USA

1. Introduction 38
2. Epidemiology. 39
   2.1. Global epidemiology. 39
   2.2. Cryptosporidiosis in people with human immunodeficiency virus infection 40
   2.3. Malnutrition and cryptosporidiosis 41
   2.4. Other groups at risk 49
3. Transmission 50
   3.1. Inoculum. 50
   3.2. Routes of transmission 50
4. Clinical Disease 55
   4.1. Background. 55
   4.2. Intestinal disease 55
   4.3. Respiratory disease. 59
   4.4. Hepatobiliary disease 61
5. Treatment 62
   5.1. Introducton 62
   5.2. Supportive care. 63
   5.3. Antiviral (anti-HIV) therapy 63
   5.4. Antimotility agents 65
   5.5. Immunotherapy (nonspecific immunomodulators) 66
   5.6. Specific anticryptosporidial agents 66
6. Diagnosis. 70
   6.1. The rationale for diagnosis 70
   6.2. Clinical diagnostic techniques. 71
   Note added in proof 71
   References 72

ADVANCES IN PARASITOLOGY VOL 40
ISBN 0–12–031740–0

Copyright © 1998 Academic Press Limited
All rights of reproduction in any form reserved

Cryptosporidiosis is now recognized as one of the most common human enteric infections. In this critical review, relatively unexplored details of transmission, the interaction with malnutrition and the development of chronic diarrhea, and the need for effective treatment are highlighted. Our inability to detect small numbers of foodborne oocysts limits our understanding of this transmission route, and the possibility of respiratory transmission is yet to be rigorously studied. The toll this disease imposes on children, especially the malnourished, has not been fully appreciated. Indeed, the dynamics of the progression from acute cryptosporidiosis to chronic diarrhea and death of malnourished children is still enigmatic. Our knowledge of the intestinal pathophysiology, while limited, is increasing. The lack of effective drug therapy is both remarkable and sobering. Overall, these unknown areas demonstrate how little we truly know about this parasite.

1. INTRODUCTION

Cryptosporidiosis is now established as one of the commonest human enteric infections (Current and Garcia, 1991). Essentially unrecognized before the acquired immune deficiency syndrome (AIDS) pandemic, it is a disease usually acquired by children, though adults are also at risk. It has gained notoriety because of its lethal consequences in the immunosuppressed person. In this chapter I will scrutinize specific aspects of the epidemiology, transmission, clinical aspects, physiology, and treatment of this infection, and attempt to highlight controversies and unanswered questions.

There have been some clear advances in our knowledge of human cryptosporidiosis, though in some areas the progress has been less prominent. Some epidemiological and transmission factors are better understood, and our knowledge of the pathophysiology of the disease is increasing. However, we do not yet have a sophisticated understanding of the role of food-borne cryptosporidiosis, in part due to the fact that we cannot detect small numbers of oocysts in foodstuffs. In addition, the role of cryptosporidiosis in malnutrition and chronic diarrhea has been only preliminarily explored. No truly remarkable breakthrough in drug therapy has been made, despite the efforts of many workers.

It appears that there are two species of *Cryptosporidium* that infect mammals (usually *C. parvum* and rarely *C. muris*), while other species that infect birds (e.g. *C. baileyi*), reptiles, or fish do not appear to infect humans (O'Donoghue *et al.*, 1987; Tzipori and Griffiths, this volume). *C. parvum* causes severe disease in humans and ruminants (O'Donoghue, 1995), on which this review will focus.

2. EPIDEMIOLOGY

Fundamentally, everyone is at some risk of acquiring cryptosporidiosis (Keusch et al., 1995). C. parvum is ubiquitous, infects most mammals, and is highly infectious. The close ecological association between humans and cattle, as well as peridomestic animals, fosters cross-contamination and sharing of parasites, and many human infections are zoonotic. Person-to-person spread is common, especially in crowded quarters. The required inoculum is small, and it is difficult to render water supplies free of the parasite. Since diagnosis requires specialized staining of fecal samples, cryptosporidiosis is underdiagnosed. All these facts conspire to make cryptosporidiosis a common infection.

2.1. Global Epidemiology

Recent studies have confirmed data acquired in the 1980s, e.g. that the infection is common, that childhood acquisition often occurs during or after weaning, and that episodic disease occurs throughout life. C. parvum ranks as one of the top three or four identified pathogens of human diarrheal disease (Current and Garcia, 1991; O'Donoghue, 1995). Since most reports have been based on laboratory surveys, or hospital data, reporting and sampling biases could be present. However, prospective, population-based studies have, in general, been consistent with the overall data available from laboratory surveys.

Prevalence rates of cryptosporidiosis in diarrheal illness range from a few per cent in cooler, more developed countries (0.1–2% overall, perhaps twice this number in children), to 0.5–10% in warmer, less developed countries. The peak incidence is in children aged 1–5 years (Casemore, 1990a). There is no apparent sex-related disposition to infection. Adults, especially those under 45 years of age, are also frequently infected. It has been theorized that, since this group often cares for young children, they are at risk of secondary transmission from ill children. The elderly do not appear to be at increased risk of infection, perhaps because of life-long exposure to the organism or age-related decreases in (fecal–oral) contact with oocysts.

While initial exposure is in early childhood in less developed countries, in more developed countries it may be linked to day care and schooling. In a recent fecal survey, Mercado and Garcia (1995) in Chile examined 4892 stool samples obtained from outpatients and patients at a nutritional recuperation center, many of whom did not have diarrhea; 0.5% of children under 2 years of age, 0.17% of children aged between 6 and 15 years, and none of the adults were infected with C. parvum, showing the age-related point-prevalence of the parasite. This also demonstrates that the point-prevalence of cryptosporidiosis in the general population (as in this study) is lower than that in people with diarrheal disease. In contrast, Kuhls et al. (1994) tested the sera of 803 children in Oklahoma, USA, using an enzyme-linked immune absorbent assay (ELISA). They found that seropositivity rates were higher in children who attended day-care centers than in children who did not, were equal in both sexes, were higher in groups with lower socioeconomic status, and were equal in rural and urban residents of Oklahoma. Thirteen per cent of children under 5 years old, 38% of those aged 5 to 13 years, and 58% of adolescents were seropositive for C. parvum. Thus antibody acquisition was substantially delayed until after the age of 5 years. In a comparative seroepidemiological study examining children and adults from Fortaleza in Brazil, rural Anhui in China, and Virginia in the USA, almost all children in Brazil had antibody to C. parvum by the age of 2 years. In contrast, in rural Anhui only half the children aged 5–7 years had antibody, and in Virginia only 16.9% of children and adults had antibody (Zu et al., 1994). The data point out the differences that the environment, water source, levels of hygiene, and income can make in the acquisition of, and re-exposure to, cryptosporidiosis, yet they must be interpreted cautiously, as systemic antibody reflects not only initial infection but also recurrent exposures.

Seasonal trends have been reported by a number of authors. The incidence of cryptosporidiosis is highest in warm or wet seasons; for example, Duong et al. (1995) found that the incidence in children under 2 years old was greatest in the rainy season in Gabon. In Aragon, Spain, Clavel et al. (1996) detected more cryptosporidiosis during the autumn–winter period than in the spring and summer, which they associated with attendance at day-care centers. Local transmission may be strongly affected by rainfall, farming practices such as the spreading of animal feces on fields, and human behavior, which may be more important than seasonal climatic factors (Casemore, 1990a). Since freezing or desiccation kill oocysts, transmission in very cold, or hot dry, weather is less likely than in warm, moist months. Fayer et al. (1996) have found that oocysts frozen and stored at $-10\,°C$ for a week are still infectious, but that heating oocysts to $72°C$ for 1 minute or longer renders them noninfectious. This implies that oocysts can survive in surface water that has frozen at mild temperatures, but that temperatures reached by common household water heaters may inactivate them.

2.2. Cryptosporidiosis in People with Human Immunodeficiency Virus Infection

Our knowledge of cryptosporidiosis has been strongly influenced by its association with advanced human immunodeficiency virus (HIV) infection.

A recent prospective long-term study from Europe suggested that 3–4% of those with HIV will have cryptosporidiosis when diagnosed with HIV infection, and that an equal number will develop it later in the course of their disease (Pedersen et al., 1996). No gender difference in the incidence of cryptosporidiosis was found in this large study. Data from the Centers for Disease Control and Prevention and from a study in Los Angeles (Sorvillo et al., 1994a) suggest that 3–4% of AIDS patients in the USA will become infected with cryptosporidiosis during their symptomatic period. In the developing world the equivalent numbers are much higher (up to 50% in hospital patients with AIDS) (reviewed by, among others, Angus, 1990; Casemore, 1990a; Chacin-Bonilla, 1995) (Table 1). Prospective cohort data from the developing world are needed to form a complete view of the importance of cryptosporidiosis and AIDS. The table summarizes some of the more recent papers regarding the incidence, or characteristics, of HIV-related cryptosporidiosis.

2.3. Malnutrition and Cryptosporidiosis

The toll that cryptosporidiosis takes on malnourished children is relatively unappreciated and unexplored. *C. parvum* is more common in malnourished than in well-nourished children with diarrhea. Garcia Velarde et al. (1991) in Mexico, Duong et al. (1991, 1995) in Gabon, and Sarabia-Arce et al. (1990) in Peru, have all reported higher incidence of cryptosporidiosis in malnourished children in either community-based epidemiological studies (Duong et al., 1991, 1995) or cohorts of children with diarrhea (Sarabia-Arce et al., 1990; Garcia Velarde, 1991). Cryptosporidiosis was the sole predictive factor for depressed nutritional status (weight/height) in a point-prevalence study of 205 Thai orphans (Janoff et al., 1990). In a prospective study in Gaza, Sallon et al. (1991) found cryptosporidiosis to be associated with depressed weight for age and dehydration. In Gabon, 31.8% of malnourished, but only 16.8% of normal, children had cryptosporidiosis (Duong et al., 1995).

Chronic diarrhea and associated malnutrition are major problems in emerging countries. Half the children observed by Jaggi et al. (1994) in India with acute, and 85% of those with chronic, diarrhea due to *Cryptosporidium* were malnourished. In Brazil, childhood chronic diarrhea and cryptosporidiosis are strongly associated (Lima et al., 1992; Fang et al., 1995). The effects of chronic cryptosporidiosis in this group can be severe. Vidal et al. (1991) found, in 23 malnourished children, that there was a mean weight loss of 25 g/day during an episode of cryptosporidiosis, together with vomiting and resistance to feeding.

Malnutrition affects cell-mediated immunity, and acute cryptosporidiosis can lead to malabsorption and anorexia, affecting nutritional status. Laxer et al. (1990) detected impaired cell-mediated immunity in all of 15 children with

Table 1 Representative studies published since 1990 on the incidence, prevalence, or specific characteristics of cryptosporidiosis in people with HIV.[a]

Location	Year	Population	Proportion infected	Comments	Reference
Brazil (São Paulo)	1991	271 HIV-infected people at a reference center for AIDS	14.3% of those with diarrhea, none of those without	542 stool samples from 271 HIV-positive individuals were studied — 100 with diarrhea, and 180 without (several counted once in each group with progression).	Rodrigues et al. (1991)
Brazil (São Paulo)	1993	131 people with AIDS, 81 presumed uninfected	19.1% of those with AIDS, none of uninfected persons	276 stool samples from 131 people with AIDS (2.1 per person) and from 81 presumed uninfected persons. Coccidia were found only in those with AIDS: *C. parvum* 25/131, *Isospora belli* 13/131	Sauda et al. (1993)
Cuba	1993	47 adults with HIV infection	38.3%	The group was in different stages of HIV progression, and about 45% of those with *C. parvum* were asymptomatic	Hadad Melendez et al. (1993)
Denmark (Copenhagen)	1991	Hospital patients exposed to nosocomial *C. parvum*	See text	60 HIV-seropositive and 73 HIV-seronegative patients in hospital developed cryptosporidiosis after an ice machine was contaminated by another patient with cryptosporidiosis. Of the 18 patients with AIDS, 5 recovered, 2 became carriers, 3 died of other causes, and 8 died of prolonged diarrhea	Ravn et al. (1991)
Ethiopia (Addis Ababa)	1994	Hospital patients (adults) with AIDS	40%	25 of 67 medical inpatients with AIDS who presented with chronic watery diarrhea and weight loss had cryptosporidiosis	Mengesha (1994)

Table 1 Continued

Location	Year	Population	Proportion infected	Comments	Reference
Europe (multi-national study)	1996	Cohort of 6548 AIDS patients	6.6% over time	An inception cohort of people with AIDS studied in 17 countries; 216 (3.3%) had cryptosporidiosis at the time of AIDS diagnosis, and 216 (3.3%) were subsequently positive for *C. parvum*; multivariate analysis showed that homosexual men were more likely to have cryptosporidiosis than intravenous drug users or women	Pedersen et al. (1996)
France (Lyons)	1993	81 people with AIDS	37.3% of those with diarrhea	Stool samples, duodenojejunal and/or colorectal biopsies from 81 people with AIDS and diarrhea or malabsorption. 13 of 81 had multiple pathogens. *Cryptosporidium* was the major pathogen	Cotte et al. (1993)
France (Nice)	1995	46 people with AIDS and chronic diarrhea	18.6%	43 people over 16 months were enrolled in a prospective study; *Enterocytozoon bieneusi* was found in 11 (24%), *C. parvum* in 8 (18.6%). Those with cryptosporidiosis were CDC stage C, had longer duration of diarrhea (19.6 vs 9.8 weeks, $P = 0.03$), greater weight loss (9.6 vs 2.1 kg, $P = 0.0003$) and lower Karnofsky scores (48% vs. 67%, $P = 0.01$)	Bernard et al. (1995)
Italy (Apulia)	1993	51 people with AIDS and diarrhea	33.3%	17 of 51 were positive for cryptosporidiosis by fecal samples, and it was the first clinical marker of AIDS in 7 of the 51	Brandonisio et al. (1993)

Table 1 Continued

Location	Year	Population	Proportion infected	Comments	Reference
Ivory Coast (Abidjan)	1993	217 adult AIDS patients with wasting	8.7%	In this group, 95% had lost weight, 89% had abdominal pain, 79% had chronic diarrhea, and 21% had nausea or vomiting. 19 of the 217 were infected with *C. parvum* as judged by stool sampling	Assoumou et al. (1993)
Malaysia	1994	HIV positive intravenous drug users	23%	Asymptomatic carriage was common in this somewhat diverse group	Kamel et al. (1994)
Mali (Bamako)	1990	Hospital patients (adults) with and without AIDS and with diarrheal disease	38% overall	In a hospital population with infectious diarrhea, 40% of whom were HIV-seropositive, 91% of the cases of cryptosporidiosis were in HIV-positive people, 40% of whom died during the first 2 weeks in hospital from chronic and profuse diarrhea. Cryptosporidiosis was the main cause of AIDS diarrhea in Mali	Pichard et al. (1990)
Mexico (Mexico City)	1994	Retrospective review of diarrhea and AIDS	30% of those with an identified etiology	Of 225 evaluable people with AIDS during 1983–1989, diarrhea was the most common symptom of AIDS. An etiology was found in 59%, and *C. parvum* (30%) was the most frequently found pathogen. Those with chronic diarrhea had higher mortality at 1 year (40%) than did those without (5–10%)	Sanchez-Mejorada and Ponce de Leon (1994)

Table 1 Continued

Location	Year	Population	Proportion infected	Comments	Reference
Spain (Barcelona)	1994	Prospective longitudinal study of 1456 people with AIDS	7.1% overall	253 of 1456 people had diarrhea for which a pathogen was found, or who had fever and a positive blood culture for an enteric pathogen. 26% of homosexual men had enteropathogens, compared with 12% of intravenous drug users	Moreno et al. (1994)
Spain (Madrid)	1995	Prospective study of 275 AIDS patients with diarrhea	15.6%	43 of 275 people with AIDS and chronic diarrhea were infected; homosexual men had higher incidence (33%) than did intravenous drug users (10.6%, $P < 0.001$). 13 of the 43 had extraintestinal infection – 8 biliary and 7 respiratory	Lopez-Velez et al. (1995)
Taiwan	1990	342 patients with gastrointestinal disease, with and without AIDS	19.8% stool acid-fast, and 29.8% seropositive overall	70 people with AIDS, and 272 without AIDS, were tested for *C. parvum*; 47.1% of AIDS patients and 12.9% of the others were positive by stool acid-fast staining, and 50% and 24.6% (respectively) by serum indirect fluorescent antibody (IFA) test. This study points out the differences that can be found between two different methods, one that measures current carriage (acid-fast stain) and the other exposure over a period of time (IFA).	Tsaihong and Ma (1990)
Thailand	1995	250 adults and children with HIV	8.8% overall	Retrospective study of patients diagnosed 1988–1993 in Nonthaburi, Thailand, presenting with chronic diarrhea. Prevalence rates were 7.9% of adults and 19% of children. 85% had chronic diarrhea and 100% had weight loss and malnutrition.	Moolasart et al. (1995)

Table 1 Continued

Location	Year	Population	Comments	Reference	
UK (London)	1992	HIV seropositive adults at a reference hospital	13.7% overall over 6 years	Retrospective report in which 5% of all those diagnosed with HIV and 21% of those with AIDS developed cryptosporidiosis. 28.7% had transient, 59.7% chronic, 7.8% fulminant, and 3.9% asymptomatic disease. Fulminant disease occurred only in those with CD4 cell counts $\leq 50/\text{mm}^3$, and their median survival was only 5 weeks, compared to 20 weeks for those with chronic, and 36 weeks for those with transient, infection. Only antiviral therapy affected survival.	Blanshard et al. (1992)
USA (Manhattan, NY)	1990	433 people with diarrhea (87 AIDS, 346 non-AIDS)	19% overall; about 4.6 times more common in people with AIDS.	45 of 87 (52%) stools from AIDS patients were positive, compared with only 39 of 346 (11.2%) of stools from subjects without AIDS. In this study, cryptosporidiosis was the leading cause of gastrointestinal illness in a New York City hospital.	Tsaihong et al. (1990)
USA (Los Angeles, CA)	1994	Retrospective report on 16 953 people with AIDS	3.8% overall	Data from an AIDS surveillance registry 1983–1992 were analyzed; prevalence of cryptosporidiosis was higher in those in whom the suspected route of HIV transmission was sexual (3.9%) than in those in other risk categories (2.6%, $P < 0.01$), or in those from Mexico (5.2%) compared to US born (3.8%, $P < 0.01$).	Sorvillo et al. (1994)

Table 1 Continued

Location	Year	Population	Proportion infected	Comments	Reference
USA (Milwaukee, WI)	1994–1996	City-wide exposure to water-borne *C. parvum*	See comments	In spring 1993, an estimated 403 000 people in Milwaukee developed cryptosporidiosis via contaminated water (MacKenzie *et al.*, 1994). Secondary epidemics associated with swimming-pool contamination occurred contemporaneously (MacKenzie *et al.*, 1995a), and the use of submicrometer filters appears to have been protective (Addiss *et al.*, 1996). In those with HIV, biliary carriage and low CD4 cell counts were associated with death within a year of the outbreak (Vakil *et al.*, 1996).	—
USA (Los Angeles, CA)	1995	Inner city HIV-seropositive adults	6%	In a survey of intestinal parasitosis in 100 people, 6% had *C. parvum*, 55% *Giardia duodenalis*, 10% *Isospora belli*, and 3% *Entamoeba histolytica*; these were all associated with anal–penile sex.	Esfandiari *et al.* (1995)
USA (Bronx, NY)	1995	Normal and immuno-compromised children	See text	6.4% (5 of 78) of normals and 22% (11 of 50) of immunodeficient, asymptomatic children were carriers.	Pettoello-Mantovani *et al.* (1995)
USA (Las Vegas, NV)	1996	City population	See text	An epidemic of cryptosporidiosis associated with drinking municipal water, despite 'state-of-the-art' water treatment. 61 of 78 laboratory-confirmed cases were in people with HIV, and 32 of the 61 (52.5%) diagnosed in the first quarter of 1994 were dead by June 30, 1994	Goldstein *et al.* (1996)

Table 1 Continued

Location	Year	Population	Proportion infected	Comments	Reference
Venezuela (Zulia State)	1992	29 children and adults with AIDS	41.3%	Three stools from 2 children and 27 adults, of whom 6 were women, were examined for parasites. 12 contained *C. parvum*, and most had diarrhea and weight loss.	Chacin-Bonilla *et al.* (1992)

[a] Other more narrowly focused studies are quoted and referenced in the text. This table is intended generally to illustrate a global perspective.

cryptosporidiosis who required admission to hospital, and who also had significantly depleted iron stores (Menorca et al., 1994). It is difficult to know if malnourished children who are constantly exposed to *Cryptosporidium* in the environment are simply at higher risk of chronic cryptosporidiosis because of their depressed immune systems, or if cryptosporidiosis is an independent risk factor for becoming malnourished. However, in a 3-year prospective community study conducted in Guinea Bissau (Molbak et al., 1993), cryptosporidiosis was very significantly linked to excess mortality in infants (2.9 times greater than in the control subjects), and this excess mortality persisted into the second year of life. The increased mortality could not be explained by malnutrition, socioeconomic factors, hygienic conditions, or breast feeding. This suggests that cryptosporidiosis is an independent risk factor for childhood death, though clearly disproportionately afflicting malnourished children.

No prospective trial of targeted nutritional support in children with cryptosporidiosis has been reported, nor do we know if premorbid nutritional status predicts the development of chronic diarrhea or whether *C. parvum* leads to chronic diarrhea more frequently than do other pathogens, such as *Shigella* or enterotoxigenic *Escherichia coli*.

There is a burgeoning realization that micronutrients are important in the resolution of infections as well as in malnutrition (Castillo-Duran et al., 1987). For example, zinc supplementation decreases the occurrence of infections in malnourished children (McClain, 1985). Zinc modulates cell-mediated immunity, and zinc supplementation decreases the incidence of opportunistic infections in AIDS (Mocchegiani et al., 1995). Given the malabsorption of nutrients and water-soluble vitamins in cryptosporidiosis, prospective studies are warranted on the influence of specific vitamins (A, B_{12}, E, etc.), micronutrients (iron, zinc, etc.) and caloric supplements in cryptosporidiosis. In my opinion, this is a major unaddressed area.

2.4. Other Groups at Risk

Other groups at risk for cryptosporidiosis are those living in communal circumstances such as orphanages and group homes (Janoff et al., 1990; Heald and Bartlett, 1994), secondary contacts of those with cryptosporidiosis (Adam et al., 1994; Current, 1994; Millard et al., 1994; Newman et al., 1994), travelers to endemic regions (Keusch et al., 1995), farm workers (Lengerich et al., 1993), and those with immunosuppression due to chemotherapy or (rarely) renal transplantation (Clifford et al., 1990) or diabetes (Trevino-Perez et al., 1995). Tanyuksel et al. (1995), studying diarrhea in people undergoing chemotherapy, found 17% (18 of 106) of those with diarrhea, but none of 60 without diarrhea, to have cryptosporidiosis. Foot et al. (1990) described six children undergoing chemotherapy for leukemia or lymphoma with cryptosporidiosis. Modification of the chemotherapeutic regimen in four allowed for parasite eradication and continued treatment of the malignancy. The other two died with persistent cryptosporidiosis. Selective immunoglobulin A (IgA) and saccharomyces opsonin deficiency has been associated with prolonged disease (Jacyna *et al.*, 1990), as have thalassemia major (Gledhill and Porter, 1990). Gomez Morales *et al.* (1996) in Rome, Italy reported chronic cryptosporidiosis in a child with interferon-γ deficiency (Theodos, 1998). Chronic *C. parvum* has also been linked to liver disease in a report from Nepal (Shrestha *et al.*, 1993). These groups are further discussed, as appropriate, below.

3. TRANSMISSION

3.1. Inoculum

Volunteer transmission studies have shown that immunocompetent adults can be infected by as few as 30 oocysts (DuPont *et al.*, 1995), though reinfection 1 year later requires about an order of magnitude more oocysts (C.L. Chappell, personal communication). In this work the median infective dose was 132 oocysts, 11 of 18 individuals had enteric symptoms, and 7 (39%) had clinical disease. Primates such as macaques can reliably be infected with as few as 10–30 oocysts (Miller *et al.*, 1990, 1991). In a recent epidemic of cryptosporidiosis in Las Vegas, USA, adults with HIV were infected with *Cryptosporidium* after drinking unboiled tap water despite 'state-of-the-art' water treatment including filtration, and an inability to detect any oocysts in treated water (Goldstein *et al.*, 1996). In summary, these experimental and epidemiological data strongly suggest that the minimum infectious dose of oocysts for humans is small and of the same order of magnitude as *Shigella* bacteria (e.g. 10–100). Reinfection is probably not difficult, given that the infectious dose after a recent primary infection may be only 100–1000 oocysts.

3.2. Routes of Transmission

Both water-borne and person to person transmission are important (Casemore, 1990a; Smith and Rose, 1990; Juranek, 1995), as are animal-to-animal, animal-to-human, and environmental transmission. Novel modes of transmission are being recognized, including via foodstuffs.

3.2.1. Water-borne Infection

Water-borne transmission plays a prominent role; most of it is probably relatively constant and therefore unnoticed. This is discussed extensively by Fricker and Crabb (1998), and here I only highlight specific illustrative points about transmission.

(a) *Fecal contamination (animal and human)*. Water contamination by cattle and other animals is an important factor in zoonotic transmission. Water draining off fields containing cattle feces can contaminate water supplies and lead to human outbreaks (Casemore, 1990a; Bridgman *et al.*, 1995). The massive 1993 Milwaukee (USA) outbreak of cryptosporidiosis (MacKenzie *et al.*, 1994) has not been proven to be zoonotic in origin, but the area is known for its dairy industry, and heavy rains preceded the outbreak. Moreover, contamination of the water supply by human sewerage or feces remains a constant problem. Joce *et al.* (1991) investigated an outbreak caused by plumbing defects in a swimming pool that allowed human sewage to enter the pool. Disease was associated with head immersion, and oocysts were detected in the water. In Fortaleza, Brazil, Newman *et al.* (1993) found that both stool samples from household animals and filtrates from water sources contained *Cryptosporidium* oocysts. Four of 18 water samples, including city water, contained oocysts, as did 6.3% of dry season and 14.3% of wet-season animal stools. Thus cattle, human, and other animal feces should all be viewed as potential sources of water contamination.

Other common epidemic point sources include fecally contaminated freshwater swimming pools, despite the routine use of chlorinating agents (Sorvillo *et al.*, 1992; MacKenzie *et al.*, 1995a). It may be prudent for immunocompromised individuals to avoid freshwater streams or pools where other humans have bathed, as direct fecal contamination may occur.

(b) *Water filtration for the prevention of cryptosporidiosis.* In the Las Vegas epidemic in the USA (Goldstein *et al.*, 1996), fatal cryptosporidiosis befell AIDS patients who drank municipal water, despite filtration, and water samples did not reveal oocysts. Similarly, no evidence could be found that filtration affected the age-standardized prevalence of cryptosporidiosis in AIDS patients in Los Angeles, USA (Sorvillo *et al.*, 1994b). These studies illustrated the point that filtration of municipal water sources (by current methods) does not rid them of this risk. Many epidemiological investigations have incriminated specific sources of water, yet oocysts have often been undetectable by conventional methods (e.g. Maguire *et al.*, 1995; Morgan *et al.*, 1995; Goldstein *et al.*, 1996). The failure to detect oocysts in water supplies should clearly not be taken as a guarantee of water safety. This issue is discussed at greater length by Fricker and Crabb (1998) [See also *Note added in proof* on pp. 71–72.]

Nonetheless, stringent filtration should prevent cryptosporidiosis. Addiss *et al.* (1996) reported that the use of filters with a pore size < 1 μm led to a substantial decrease in the risk of diarrheal disease during the Milwaukee, USA, outbreak. Only 2 of 11 users (18%) of such filters developed diarrhea, in contrast to 50–80% of those who drank unfiltered water or water passed through filter devices with pore sizes >1 μm ($P=0.02$). Severely immunocompromised people should consider using a household filter with pores below 1 μm in diameter; confirmatory studies are needed. This and other methods of prevention of transmission are discussed by Juranek (1995).

3.2.2. Person-to-person Spread

Day-care center epidemics are frequent and underdiagnosed (Cordell and Addiss, 1994). Prospective studies have shown that asymptomatic childhood carriage of *C. parvum* is common, and unsuspected child-to-child transmission may be important in endemic disease (Pettoello-Mantovani *et al.*, 1995). Household transmission has been examined in several recent studies (Current, 1994). Thirty-one households in Fortaleza, Brazil, with an index case of cryptosporidiosis in a child under 3 years of age were studied prospectively by Newman *et al.* (1994). Secondary cases occurred in 58% of the households, and the overall secondary transmission rate was 19%. In a food-borne epidemic in Maine, USA (Millard *et al.*, 1994), the secondary transmission rate was 15% (53 of 353 people), which is remarkably similar to the 19% rate from Brazil. In contrast, after the 1993 Milwaukee outbreak, the secondary transmission rate within a household was low (5%) when the index case was an adult (MacKenzie *et al.*, 1995b).

Outbreaks tend to occur in day-care centers, custodial institutions, and closed populations such as family units (Keusch *et al.*, 1995). Hospital employees and patients are at special risk of acquiring cryptosporidiosis, as AIDS patients in hospital may excrete large numbers of oocysts. Navarrete *et al.* (1991) described a pediatric hospital outbreak in which 82% of children exposed to an infant with cryptosporidiosis and AIDS developed cryptosporidiosis. In a sobering study from Denmark (Ravn *et al.*, 1991), 8 of 18 AIDS patients infected in hospital later died of cryptosporidiosis. Universal precautions are important, and immunocompromised patients should be protected from others with the disease.

3.2.3. Food-borne Disease

Point-source, localized epidemics of food-borne cryptosporidiosis are being recognized. After ingesting hand-pressed apple cider at an agricultural fair, 54% of 284 exposed people (compared with 2% of 292 unexposed persons) developed clinical cryptosporidiosis (Millard *et al.*, 1994). Cider, the cider press, and calf feces from the farm that supplied the apples contained oocysts. While this is the first outbreak so assuredly to associate apple cider with cryptosporidiosis, prior studies have incriminated uncooked sausage, offal, and raw milk as vehicles (Casemore, 1990b; Petersen, 1995). Molbak *et al.* (1994), working in Guinea-Bissau, found that the storage of cooked food was associated with childhood cryptosporidiosis.

An unquantified risk is that of low-level, unsuspected food-borne transmission. In 1996, the USA suffered a national epidemic of *Cyclospora cayetensis* diarrhea that may have been related to imported raspberries. *Cyclospora* is closely related to *Cryptosporidium,* and the infectious dose for *Cyclospora* cannot be much lower than that for *Cryptosporidium* (10–100 oocysts). Analogously, *Cryptosporidium* could contaminate foodtuffs at a central source (a farm or processing site) and then be distributed so widely that no one central authority would recognize a subsequent food-borne outbreak. It is not at present possible to detect small numbers of oocysts on foodstuffs. Using standardized methods, the recovery of experimentally added *Cryptosporidium* oocysts is only 1% from fruits and vegetables (Bier, 1991). Encouragingly, a polymerase chain reaction method has been published that can detect 1–20 oocysts in 20 ml of raw milk (Laberge *et al.*, 1996).

3.2.4. Sexual Transmission

Several recent large studies have illustrated the role of sexual transmission. In a prospective cohort study of 6548 people with AIDS (Pedersen *et al.*, 1996), homosexual men were more likely to develop cryptosporidiosis (4.1%) than were intravenous drug users (IVDUs) (1.3%, $P < 0.001$). The prevalence was again higher in gay men (33%) than in IVDUs (10.6%, $P < 0.001$) in a prospective study of AIDS and diarrhea in Spain (Lopez-Velez *et al.*, 1995). Again in Spain, Moreno *et al.* (1994) found 26% of homosexual men had enteropathogens (including *C. parvum*), compared to 12% of IVDUs. In a retrospective AIDS registry review of 16 953 people, the prevalence of cryptosporidiosis was higher in those whose suspected route of HIV transmission was sexual (3.9%) than in other risk groups (2.6%; $P < 0.01$) (Sorvillo *et al.*, 1994a). Esfandiari *et al.* (1995) found that intestinal parasitosis, including cryptosporidiosis, in adults with HIV infection was related to anal–penile sex. These studies very strongly suggest that cryptosporidiosis can be sexually acquired, but they do not conclusively prove it since transmission could, theoretically, be related to associated behaviors.

3.2.5. Zoonotic Transmission

Peridomestic animals, including pets, are reservoirs of infection. Over 40 mammalian species may harbor *C. parvum* (reviewed by Current and Garcia, 1991). Cryptosporidiosis can be acquired after visiting farms, and dairy farmers are known to have increased exposure to cryptosporidia (Lengerich *et al.*, 1993). It may be prudent for highly immunocompromised persons to avoid cattle. It is not known what proportion of human cryptosporidiosis is of zoonotic origin.

3.2.6. Air-borne Transmission

Respiratory symptoms are more common in cryptosporidiosis than in other diarrheal diseases. Numerous studies have suggested, but not proven, airborne transmission. Because the infectious inoculum is low, and the respiratory tract is easily infected, droplet transmission can easily be conjectured. This is reviewed in Section 4.3.

3.2.7. The Possible Protective Role of Breast Feeding in Cryptosporidiosis

There is little evidence that maternal colostral or breast milk antibody is protective against cryptosporidiosis. Sterling *et al.* (1991) have published a prospective cohort study which stratified mothers and newborn infants by maternal antibody to *C. parvum* sporozoites. In this community, breast feeding was strongly encouraged, and the study is best seen as a measure of the relative importance of breast-milk antibody within a cohort of breast-feeding mothers. No significant difference in the prevalence or duration of cryptosporidiosis among children grouped by maternal colostral antibody levels was found. This result is analogous to what has been found in pigtailed macaques and mice. Intentionally infected infants of disease-resistant macaque dams had similar disease whether the infant was fed on formula or breast milk (Miller *et al.*, 1990, 1991). Mouse pups suckling dams that had recovered from cryptosporidiosis, or that had recovered and been hyperimmunized, were not protected against intentional challenge (Moon *et al.*, 1988). Overall, it appears that colostral or breast-milk antibodies are not detectably protective. There are no data to support or refute the possibility that nonspecific breast-milk factors are protective.

Nonetheless, population-based studies (reviewed by Sterling *et al.*, 1991; Molbak *et al.*, 1994) suggest that breast feeding is protective against cryptosporidiosis. It can be argued that breast-fed infants are simply less likely to ingest oocysts. Once weaned, their exposure to oocysts in food and water increases, and so does their risk of infection (Sterling *et al.*, 1991). A reasonable view is that breast feeding separates infants from oocyst-contaminated foods and fluids and is thus protective, while breast milk itself is not protective after exposure.

3.2.8. Miscellaneous Routes

Insect-borne transmission has not been extensively studied. Cockroaches have been reported to carry *Cryptosporidium* oocysts (Zerpa and Huicho, 1994) and, in anectodal reports, so have flies (C. Sterling and R.H. Gilman, personal communication).

4. CLINICAL DISEASE

4.1. Background

Cryptosporidiosis of the normal host is primarily a time-limited diarrheal disease. Its clinical characteristics have been well described. Four forms of intestinal cryptosporidiosis are recognized: asymptomatic, acute self-limited or transient disease, chronic, and fulminant. Other major systemic complications include respiratory and hepatobiliary cryptosporidiosis. Treatment is described in Section 5, and clinical diagnosis in Section 6.

4.2. Intestinal Disease

4.2.1. Asymptomatic Carriage

Asymptomatic carriage may be an important part of endemic transmission, especially if the analogy to typhoidal carriage can be made (Griffiths and Gorbach, 1993). While the rate of asymptomatic cryptosporidiosis in normal populations is thought to be low, it may be much higher in those with HIV. In Malaysia, 23% of IVDUs with HIV in a drug rehabilitation center had asymptomatic cryptosporidiosis (Kamel *et al.*, 1994). In Cuba, about 45% of those with *C. parvum* and HIV were asymptomatic (Hadad Melendez *et al.*, 1993). There is little prospective information on silent carriage by HIV-seropositive individuals in developed countries. In asymptomatic children in the Bronx, New York, USA, 6.4% (5/78) of normal, and 22% (11/50) of immunocompromised, prospectively studied children had asymptomatic carriage (Pettoello-Mantovani *et al.*, 1995). Two individuals with leukemia have been reported to have been carriers (Gentile *et al.*, 1990). Additional studies are needed to place these works in context.

Duong *et al.* (1995), in Gabon, documented a 14.8% carriage rate in asymptomatic children ≤ 2 years of age (those with diarrhea had a 28% incidence rate). In Venezuela, Chacin-Bonilla *et al.* (1993) found that 15 of 21 positive individuals were asymptomatic in a point-prevalence survey. Asymptomatic carriage rates may be high in areas of low socioeconomic development and high exposure.

Asymptomatic biliary carriage of *C. parvum* theoretically represents a reservoir for sporadic transmission, but persuasive or confirmatory evidence for this reservoir is as yet lacking. For example, Roberts et al. (1989) studied 169 immunocompetent adults undergoing routine upper gastrointestinal endoscopy in New York, USA and found that 12.7% had *Cryptosporidium* in duodenal aspirates, though none had positive duodenal biopsies. This suggested that the biliary tract of these individuals was infected. Half of those with positive aspirates had oocysts in stool samples, though none had diarrhea. In contrast, we (T. Knox, J.K. Griffiths, and S. Tzipori, unpublished observations) have repeated this study at New England Medical Center in Boston, USA, and, in over 60 people, found no positive duodenal aspirate or biopsy. Further studies from a variety of geographic locales would assist in understanding the potential role of biliary carriage.

4.2.2. Acute Self-limited (Transient) Disease

Transient disease is common in the immunocompetent population. Numerous studies have shown an incubation period with a mean of approximately 6 days and a range of 2–30 days. The most frequent symptoms are watery diarrhea, anorexia, abdominal discomfort or pain, and fever or cough in a substantial minority (about 30%). It may be dismissed by medical workers as viral diarrhea. Mild malabsorption is common. Resolution begins a few days to a week after onset in most people. Though the data are scanty, the transmission study by DuPont et al. (1995) suggested that only some of those infected have symptomatic disease (see section 3.1).

Blanshard et al. (1992), reporting from a referral center in England, noted that transient disease often occurred in people with HIV; the illness was uniformly transient in persons with CD4 cell counts $\geq 250/mm^3$. In a retrospective review from the USA (Flanigan et al., 1992), HIV-seropositive people with CD4 cell counts $\geq 180/mm^3$ generally resolved the infection without sequelae. In the study by Blanshard (1992), a substantial but significant minority with CD4 cell counts $<50/mm^3$ had transient disease. Of 128 patients with 129 bouts of cryptosporidiosis, 42 (33%) had either transient (28%) or asymptomatic (4%) disease. Thus, placebo controls are needed in treatment studies, since cryptosporidiosis is not uniformly progressive or chronic in AIDS, and transient disease is common.

In well-conducted studies performed in West Africa, Guatemala, South America, and South East Asia, acute cryptosporidiosis is a predictor of childhood death, especially in those who are malnourished (reviewed by Molbak, 1993) (see Section 2.3). Summarizing these studies, children with cryptosporidiosis have a death rate approximately two- to four-fold higher than that seen in children with other diarrheal pathogens. The progression rate from acute to chronic cryptosporidiosis in children is unknown, and deserves prospective study (Lima et al., 1992; Fang et al., 1995) (see discussion in Section 2.3). The pathophysiology of acute disease is discussed in Section 4.2.5.

4.2.3. Chronic Disease

Chronic cryptosporial disease is commonest in patients with AIDS or malnutrition. In the series reported by Blanshard et al. (1992), it occurred in 56% of the 128 AIDS cases. Survival of persons with HIV infection and chronic cryptosporidiosis is significantly shorter than that of those with transient or asymptomatic disease (20 vs 36 weeks, $P < 0.05$; Blanshard et al., 1992). Probably as a result of wasting and dehydration, weight was also lower in the former group (64.7 vs 59.3 kg, $P < 0.01$). Most individuals were helped by antiviral therapy and/or opiates for their diarrhea. McGowan et al. (1993) reported that 27 of 38 subjects with cryptosporidiosis and AIDS had chronic diarrhea, with a mean survival time of only 11.5 weeks, compared to 66 weeks in those who remitted ($P < 0.001$). The data on chronic cryptosporidiosis and childhood malnutrition have been reviewed in Section 2.3.

If people with AIDS and chronic diarrhea are analyzed as a group, they have poorer survival than those without diarrhea. In the study by Sanchez-Mejorada and Ponce-de-Leon (1994), survival was only 60% at 1 year, compared with 90% for those with intermittent diarrhea and 95% for those with acute (limited) diarrhea. Cryptosporidiosis was the most common pathogen seen in their study. In a retrospective study of 250 Thai patients presenting with chronic diarrhea and HIV (Moolasart et al., 1995), 8% of adults and 19% of children had cryptosporidiosis; 85% had persistent disease and 100% had weight loss and malnutrition. Moss et al. (1995) reported severe chronic cryptosporidiosis at the same time as HIV seroconversion.

4.2.4. Fulminant Disease

Essentially exclusively seen in persons with AIDS or chemotherapy-induced immunosuppression, fuliminant cryptosporidiosis is a dramatic cholera-like illness with very high mortality. Profound hypovolemia and shock may require intensive care unit management of fluid and electrolyte balances. Diarrheal outputs of 1 l/h are not uncommon. A very short survival time (5 weeks according to Blanshard et al. (1992); 10.6 days in the report by Jordan (1996)) increased intercurrent infections and low weights at presentation are the rule.

4.2.5. Pathophysiology of Diarrheal Disease in Cryptosporidiosis

Most of our understanding of cryptosporidial pathophysiology is derived from acute infections in normal animals, not immunologically bereft people. This is evident in our inability to explain the astoundingly profuse diarrhea of fulminant disease (Sears and Guerrant, 1994; Clark and Sears, 1996). No cryptosporidial toxin has been isolated, though as yet incompletely characterized enterotoxic activities are present in cryptosporidial feces (Guarino et al.,

1994, 1995). The biological compound(s) responsible for this activity may be of host and not parasite origin.

Early in infection, the intestinal villus becomes blunted, while the intestinal crypts hyptertrophy. In animal studies, secretory (crypt) capacity is spared while absorptive function decreases (Argenzio et al., 1990; Moore et al., 1995). Within 36–48 hours, a neutrophilic infiltrate is usually seen (Moore et al., 1995). In the piglet, localized prostaglandins and tumor necrosis factor act to inhibit Na^+ absorption, inducing net chloride secretion (Argenzio et al., 1993; Kandil et al., 1994) via mechanisms that include the enteric nervous system (Argenzio et al., 1996). These studies suggest that NaCl transport is altered by stimulation of cholinergic interneurons that innervate cholinergic and vasoactive intestinal peptide (VIP) motor nerves, since these effects can be counteracted by the somatostatin analogs octreotide or clonidine. Luminal glutamine enhances Na^+ uptake far more than does glucose, suggesting that oral rehydration fluids containing glutamine (rather than glucose) might help to prevent dehydration from cryptosporidiosis (Argenzio et al., 1994), but human clinical evaluation has not been attempted, to my knowledge.

Overt disruption of the intestinal mucosa by cryptosporidiosis is rare. For example, in hyperinfected piglets only about 2% of villi were disrupted (Moore et al., 1995). Pneumatosis cystoides intestinalis, or air in the intestinal wall (signifying breakdown of the mucosa), has only rarely been reported (Naguib et al., 1997). This is curious, since *C. parvum* causes epithelial cell death *in vitro* (Griffiths et al., 1994), and other intracellular pathogens such as *Shigella* routinely broach the mucosa (Griffiths, 1997). Similarly, transepithelial electrical resistance, a measure of intestinal permeability to fluids and ions, is relatively unchanged *in vivo* during experimental infections of piglets and calves (Argenzio et al., 1990, 1993; Moore et al., 1995; J.K Griffiths and S. Tzipori, unpublished observations). In contrast, there is remarkable disruption *in vitro* of cell monolayers after infection (Adams et al., 1994; Griffiths et al., 1994). This suggests that the intact host (unlike a cell monolayer) is able to repair epithelial disruption (Clark and Sears, 1996).

Malabsorption of nutrients occurs, and there are decreased levels of intestinal brush-border digestive enzymes such as lactase and alkaline phosphatase (Kapembwa et al., 1990; Gardner et al., 1991; Handousa, 1991; Beier et al., 1995). Vitamin B_{12} and D-xylose absorption negatively correlate with the intensity of infection, with intestinal injury proportionate to the number of infecting organisms (Goodgame et al., 1995). Vitamin A malabsorption has been documented in calves (Holland et al., 1992). This has not been studied in humans (but should be, especially in malnutrition). The relationship between studies conducted in acutely infected normal, and chronically infected immunosuppressed, people is unclear.

Many reports suggest that inflammation occurs during infection, perhaps with production of tumor necrosis factor (TNF) (Genta et al., 1993; Kandil et al., 1994; Goodgame et al., 1995; Argenzio et al., 1996), and prospective studies have found fever present in a substantial minority of children with cryptosporidiosis (e.g. Duong et al., 1991). In contrast, Snijders et al. (1995) did not find mucosal cytokine-mediated inflammatory processes in cryptosporidial diarrhea, but their study omitted protease inhibitors from the intestinal tissue homogenates, perhaps falsely rendering the tissues free of cytokines or other mediators. There are other lines of evidence that secondary mediators are important. The somatostatin analog octreotide is often helpful in reducing diarrhea (see Section 5.4), and Argenzio et al. (1996) found that octreotide helped to normalize NaCl transport in the infected piglet ileum. In unpublished studies (J.K. Griffiths and S. Tzipori) we have found that site-specific infection of the calf ileum leads to decreased baseline short circuit currents distant from the infection, suggesting some linked or bystander mechanisms. Given the data suggesting that prostaglandins and TNF play roles in cryptosporidial diarrhea, we believe that focusing on this area might lead to therapeutic advances.

The extent of intestinal involvement is important, though not well studied. Clayton and colleagues (1994) have shown that severe disease correlates with proximal small intestine and colonic infection, whereas less severe disease correlates with the presence of parasites only in the colon or stool. Those with positive proximal small-bowel biopsy specimens had small-bowel crypt hyperplastic villous atrophy, lamina propria inflammatory infiltrates, poorer D-xylose absorption, greater weight loss, and shorter survival, and more often needed intravenous hydration or hyperalimentation. Patients with *Cryptosporidium* in small-bowel villi alone had milder disease than those with small-bowel crypt infection. A more sophisticated understanding of this facet awaits further studies.

4.3. Respiratory Disease

Respiratory infection is common but usually inapparent. In childhood cryptosporidiosis cough is frequent, being reported in one-fifth to one-third of normal children. For example, of 250 children in the Ivory Coast with cryptosporidial diarrhea, 77% had profuse diarrhea, 58% had fever, and 19% had pulmonary symptoms (Kone et al., 1992). Pulmonary symptoms are about three-fold more frequent in children admitted to hospital with cryptosporidial diarrhea that in children with other intestinal pathogens, a figure based on carefully conducted prospective case–control studies (Egger et al., 1990; Mausezahl et al., 1991; Sallon et al., 1991). It has been postulated that transient respiratory cryptosporidiosis is common in immunocompetent children (Egger et al., 1990). This is not surprising, given the tens or hundreds of billions of ocysts that may be excreted during infection, and the ability of the parasite to invade the respiratory tract epithelium. Air-borne transmission is an open question, and has been implicated in some prospective epidemiological studies (Molbak *et al.*, 1990).

In humans with AIDS, respiratory infection is marked by severe persistent cough and dyspnea (Brea Hernando *et al.*, 1993). In a series from Spain, 7 of 43 people with AIDS and cryptosporidiosis had respiratory infection (Lopez-Velez *et al.*, 1995). The major analogy is to avian cryptosporidiosis. In turkeys and chickens, lethal respiratory cryptosporidiosis caused by *C. baileyi* is common (Lindsay *et al.*, 1987). The large airways of the human, in contrast to the far narrower avian airways, are less prone to occlusion by secretions. One individual with AIDS and immunosuppression after kidney transplantation had infection with *C. baileyi*, with involvement of the trachea, larynx, lungs, and intestines at autopsy (Ditrich *et al.*, 1991). All other reports of respiratory disease in humans are believed to have been due to *C. parvum*.

Symptomatic respiratory cryptosporidiosis in people with AIDS (as in birds) is manifested by cough, copious tracheal secretions, and dyspnea. In a review from Spain, chronic cough was present in 91%, fever in 59%, and dyspnea in 64% of patients (Brea Hernando *et al.*, 1993). Respiratory tract infection preceded diarrhea in most, and was the root cause of death in the majority. Chest radiographs usually reveal diffuse interstitial infiltrates with bronchial accentuation, similar to *Pneumocystis carinii* pneumonia. Histological samples show parasites in the ciliated epithelium and a mononuclear cell infiltrate. Lethal respiratory cryptosporidiosis also occurs in other immunosuppressive conditions, such as malignant lymphoma (Travis *et al.*, 1990) or bone marrow transplantation (Kibbler *et al.*, 1987; Gentile *et al.*, 1991). In immunosuppressed rats, respiratory infection leads to respiratory distress, severe weight loss, and enlarged, elastic lungs (Meulbroek *et al.*, 1991). Increased mucus production and exfoliative epithelial necrosis result in the accumulation of extensive mucocellular exudates in the airways and patchy alveolitis. The scanty data extant suggest this is the case in humans as well; however, the pathobiology of human pulmonary cryptosporidiosis has been little studied.

In a report from Copenhagen, 8 of 86 diagnostic bronchoscopies in AIDS patients were positive for *Cryptosporidium*, as well as a medley of other (also unsuspected) pathogens (Jensen *et al.*, 1990). I suspect that, if bronchoscopy samples were routinely examined for *Cryptosporidium*, more cases would be reported in the literature. Only four of the seven people with AIDS and respiratory cryptosporidiosis in the study by Lopez-Velez *et al.* (1995) had abnormal chest radiographs. Respiratory cryptosporidiosis can be the presenting feature of AIDS (Mifsud *et al.*, 1994). There are case reports of the successful treatment of respiratory cryptosporidiosis in AIDS with azithromycin (Dupont *et al.*, 1996) and inhaled paromomycin (Mohri *et al.*, 1995).

In summary, respiratory cryptosporidiosis is probably common, is essentially uncharacterized as a factor in transmission, and can be fatal in the immuncompromised host.

4.4. Hepatobiliary Disease

Humans with persistent cryptosporidiosis may develop hepatobiliary or pancreatic duct infection. By way of background, acute infection in a variety of animal models usually shows disease limited to the intestinal tract, whereas chronic infection models also usually show involvement of the biliary system (Kuhls *et al.*, 1992; Brasseur *et al.*, 1994; Mead *et al.*, 1994; Rehg, 1994, 1996; Tzipori *et al.*, 1994, 1995b; Rehg, 1996) (see also Tzipori). Primates with simian immundeficiency virus infection also develop cryptosporidiosis of the biliary and pancreatic systems (Baskerville *et al.*, 1991; Kaup *et al.*, 1994). Biliary carriage represents a reservoir for intestinal relapse in persons with AIDS, even after treatment with luminal agents such as paromomycin, and carries its own risk.

Sclerosing cholangitis with papillary stenosis has been increasingly described in people with AIDS and biliary cryptosporidiosis (Kline *et al.*, 1993; Jablonowski, 1994). The process is marked by right upper quadrant pain, cholestasis, and other signs of cholangitis. Sphincterotomy ameliorates the disease. Benhamou and colleagues (1993) in Paris, France, reported that biliary disease in 26 consecutive patients with AIDS presented in two radiographic fashions. The first pattern was gradual and regular stenosis of the common bile duct without irregularity of the intrahepatic duct (present in 7 of the 26 cases), and the second was distal extrahepatic biliary duct stenosis with diffuse irregularity of the larger intrahepatic ducts (present in 19 of the 26). *Cryptosporidium* or cytomegalovirus infection was linked to the second pattern, and sphincterotomy ameliorated pain. Similar radiological findings were found in a review from New York, USA (Teixidor *et al.*, 1991). Microsporidial biliary system infection may be associated with unsuspected biliary cryptosporidiosis (Pol *et al.*, 1993).

The natural history of biliary disease in cohort groups with AIDS and cryptosporidiosis (as opposed to anectodal case reports) has only infrequently been addressed. Blanshard *et al.* (1992) noted that 18% of individuals in their series had documented, and 26% had probable, sclerosing cholangitis. McGowan and colleagues (1993) reported that 10 of 38 people with cryptosporidiosis developed sclerosing cholangitis, one had an episode of pancreatitis, and another had acute cholecystitis. In a series of 30 patients, 4 had proven and 13 had probable biliary disease, but this did not appear to affect survival (Gunthard *et al.*, 1996). From a somewhat different viewpoint, Forbes *et al.* (1993) reported that 13 of 20 people with sclerosing cholangitis in AIDS had cryptosporidiosis, and that survival was not statistically affected by cholangitis. Curiously, they found a striking reversal of the usual inverse correlation of age and survival in AIDS, which could not be explained by immunosuppression or opportunistic infection.

After the Milwaukee, USA, outbreak in 1993, biliary disease had ominous overtones (Vakil *et al.*, 1996). Of 82 HIV-seropositive individuals who developed cryptosporidiosis, only 4 of 24 (17%) with biliary symptoms were alive 1 year after the outbreak, compared with 30 of 58 (52%) without biliary symptoms ($P = 0.003$). However, those with biliary symptoms were more likely to have CD4 cell counts $\leq 50/mm^3$ (21 of 24 (88%) vs 36 of 57 (63%) in those without; $P = 0.03$). Seventy-two per cent of those with CD4 cell counts $\leq 50/mm^3$ had died in 1 year, compared with 25% of those with higher counts ($P < 0.001$). Thus biliary disease was associated with a more advanced immunocompromised state, and might not have been an independent marker for death. Further prospective studies are warranted to delineate the risk factors for, and implications of, biliary disease.

Case reports of biliary or pancreatic disease in people without AIDS, but with other immunocompromising conditions, exist (e.g. Mendez and Bosch, 1995). An Australian child with hypogammaglobulinemia who was treated with the original hyperimmune bovine colostrum preparation (Tzipori et al., 1986) later died of persistent biliary disease (S.Tzipori, personal communication). Thus this is a complication that can occur in anyone with chronic cryptosporidiosis, whatever the underlying immunosuppressive condition.

In summary, hepatobiliary disease is common in chronic cryptosporidiosis, and symptoms of cholangitis can be ameliorated with sphincterotomy or analgesics. Sclerosing cholangitis does not appear independently to lead to accelerated death in AIDS. Biliary carriage may be a significant reason for relapse in patients with AIDS whose intestinal tracts have been decolonized by agents such as paromomycin. Luminal agents such as paromomycin do not treat biliary disease, presumably because the drug cannot retrogressively enter the common bile duct.

5. TREATMENT

5.1. Introduction

No consistently effective treatment exists for cryptosporidiosis. An intact immune system is the major factor in resolving this disease, which is cold comfort to those with AIDS or an impaired immune system. Recent reviews have collectively delineated the dozens if not hundreds of unsuccessful antibiotic and antiprotozoan agents that have been clinically or experimentally assessed, and we will not repeat this information *ad infinitum*. The vast majority of tested agents have shown no benefit in humans. I strongly believe this almost unprecedented failure to find effective treatment is in fact useful information, that may prove of great importance in ultimately developing novel and effective agents (see Section 5.6.1.). An overview of treatment is provided in Table 2.

5.2. Supportive Care

Robust hydration and nutritional support is the mainstay of treatment, especially for people with fulminant, cholera-like disease. Most people with HIV and transient cryptosporidiosis do not require intravenous fluids or nutrition, but unhappily it is essential in some. Careful attention must be paid to electrolyte balance and fluid shifts, and aggressive nutritional support is warranted. Little other than generic advice has been published on this subject.

5.3. Antiviral (Anti-HIV) Therapy

The importance of adequate antiviral therapy in those with cryptosporidiosis and AIDS cannot be overstated (Angus, 1990; Soave, 1990; Blanshard et al., 1992; Petersen, 1992; Wittner et al., 1993; Chacin-Bonilla, 1995; Hoepelman, 1996). Given the paucity of direct therapies for this protozoan infection, every effort must be made to improve the host's immunological status via HIV suppression. The removal of other immunodepressant factors (e.g. IVDU or alcohol abuse) may also help some individuals. Optimization of antiviral therapy is crucial not only for the individual with HIV and cryptosporidiosis, but also for the conduct of clinical trials investigating new anticryptosporidial therapies. Shifting antiviral therapies can confound or bias these studies.

Blanshard et al. (1992) found that survival was significantly correlated with the initiation of zidovudine therapy when cryptosporidiosis was first diagnosed. The median survival time was 15.2 weeks in those who never received or discontinued zidovudine, 29.3 weeks in those who had taken it for at least 2 months before onset, and 98.7 weeks in those who commenced therapy at the time of diagnosis. The obvious interpretation is that, in those naive to zidovudine, treatment led to viral suppression and a rebound in cell-mediated immunity, lengthening survival. Given the profound recent advances in HIV treatment, these gains should be taken advantage of. For example, T.P. Flanigan, H. Grube, M. Milano and colleagues in Rhode Island, USA (personal communication) have treated a person with AIDS and unremitting cryptosporidiosis with combined antiviral drugs including a protease inhibitor, resulting in apparent cure of the cryptosporidiosis.

One practical difficulty is the difficulty in absorbing some oral agents, such as the protease inhibitor saquinavir, given the malabsorption characteristic of cryptosporidiosis. The 1997 guidelines for HIV treatment included the use of two or three antiviral agents. Should one or several antivirals be taken but not absorbed, this may result in monotherapy or suboptimal drug levels, with the attendant risks of HIV resistance. Thus my opinion is that easily absorbed agents are preferable to poorly absorbed drugs. Since zidovudine is available in an intravenous form, its initial use in the naive host with malabsorption and diarrhea should be considered.

Table 2 Treatment of cryptosporidiosis.

Host type	Treatment
Immunocompetent hosts	
Normal child or adult	No specific therapy needed in most cases, as self-limited disease is the rule. In areas such as the developing world where incipient malnutrition is rife, careful attention to nutrition may be warranted.
Malnourished	Unknown: increased mortality in this group, and the rate of progression from acute disease to chronic diarrheal disease and wasting is unknown. Recommend aggressive caloric supplementation; role of vitamin or micronutrient supplementation has not been studied. Malnourished individuals may develop acquired immunological deficiencies, such as T-cell deficiencies related to zinc, and this group may be considered at high risk of immunodeficiency.
Normal host yet disease persists	Look for unsuspected immunodeficiencies, such as HIV, congenital deficiencies (e.g. of IgA), congenital or acquired T-cell abnormalities, idiopathic states with low CD4 cell counts.
Metabolic problems that are conducive to persistent disease	These include renal failure, liver disease, etc. Optimization of the underlying problem, and perhaps a course of either paromomycin or azithromycin, may be warranted (see below).
All immunocompetent patients	In any person who is believed to be immunocompetent, yet in whom disease persists, therapy with a macrolide or paromomycin is warranted, as these drugs are safe and the chance of cure appears to be higher than in those with HIV-related disease. Even the fully immunocompetent child can die of dehydration, so dehydration must not be allowed to occur.
Immunocompromised hosts	
HIV	1. Aggressive attention to hydration, nutrition. Intravenous hydration and nutrition can be life-saving or life-preserving in those with fulminant disease. 2. Antiviral (anti-HIV) therapy with multiple drugs if possible, being wary of malabsorption with oral agents. Consider the use of intravenous zidovudine (AZT) in those naive to this drug so as to avoid malabsorption. Initiation of antiviral drugs has been shown to be the most important predictor of prolonged survival. The goal is to suppress viremia and boost CD4 cell counts.

Table 2 Continued.

Host type	Treatment
	3. Antimotility agents such as opiates and somatostatin analogues (e.g. 50–500 μg octreotide every 6–8 hours). 4. Consideration of immunotherapy with agents such as bovine hyperimmune colostrum, if available. 5. Use of specific agents such as paromomycin (500 mg four times daily or 1 g twice daily) and azithromycin or another macrolide has the highest chance of success.
Malignancy, undergoing chemotherapy; transplantation	1. Stop chemotherapy and/or immunosuppressive agents 2. Other steps as outlined above

5.4. Antimotility Agents

A decrease in diarrheal flux improves the quality of life for people with cryptosporidiosis and may (theoretically) improve nutrient exchange through longer intestinal contact time (Nousbaum et al., 1991). Opiates and other nonspecific drugs are the mainstay of antimotility agent treatment (Soave, 1990). Somatostatin analogs, principally the long-acting agent octreotide, are also useful (Fanning et al., 1991; Friedman, 1991; Kreinik et al., 1991; Nousbaum et al., 1991; Romeu et al., 1991; Girard et al., 1992; Liberti et al., 1992; Moroni et al., 1993; Oehler and Loos, 1993; Ritchie and Becker, 1994). Octreotide therapy does not have any effect on parasite shedding (Liberti et al., 1992), and cessation of octreotide in responders results in relapse of the diarrhea. Somatostatin analogs have become an agent of last resort for those with persistent cryptosporidiosis (Table 2). I recommend an initial trial of 50 μg every 6–8 hours, with rapid escalation up to a maximum of 500 μg every 6 hours (Romeu, 1991; Ritchie and Becker, 1994). It would be useful if prospectively acquired data on changes in intestinal function related to these agents could be acquired, since so little information exists.

5.5. Immunotherapy (Nonspecific Immunomodulators)

This topic is dealt with in greater detail Theodos (1998) and Crabb (1998). Rasmussen et al. (1992, 1995), Leitch and He (1994) and Rehg (1996b) have found that the immunomodulators dehydroepiandrosterone and diethyldithiocarbamate modestly reduce infection in immunosuppressed rodents. While these results, in my view, do not support the use of these agents in humans, they do reinforce the general point about improving the immune systems of people with AIDS and cryptosporidiosis, e.g. by suppressing HIV viremia with combination drug therapy.

5.6. Specific Anticryptosporidial Agents

5.6.1. Theoretical Considerations in the Failure to Find Effective Treatment

It is biologically surprising that drugs successful against *Toxoplasma, Plasmodium, Eimeria,* and *Cyclospora* are essentially ineffective for human cryptosporidiosis. Reasons for this may include the following, perhaps in conjunction.

(i) Parasite enzymes that are profoundly different from those of other closely related coccidians, so that these therapeutic targets are not inhibited by the same agents that inhibit *C. parvum*'s closest relatives. While not impossible, it seems unlikely that this range of ineffective drugs is explained by an equally great range of divergent enzymatic targets.

(ii) Acquired resistance of the parasite *C. parvum* is ubiquitous and infects both animals and humans. Because of this, widespread exposure to antibiotics and biological agents used in humans and animals has undoubtedly occurred. Thus *C. parvum* may have globally acquired resistance to a very broad range of agents. Only by testing organisms from different locations for antiprotozoan drug resistance, or having the ability to test the specific parasite biochemical pathways, can this possibility be systematically addressed.

(iii) Structural impediments to drug entry. The parasite lies in an unusual and perhaps priveleged site, beneath the host cell apical membranes and separated from the host-cell cytoplasm by the feeder organelle membrane, thus being intracellular but extracytoplasmic. The feeder organelle membrane may block the transfer of molecules — such as therapeutic drugs — into the parasite cytoplasm from the host cell. Nothing is known about the flow of nutrients or drugs from the host into the parasite, though on a structural basis there is evidence that the feeder organelle is a likely site of transmembrane transport (see the chapter by Tzipori and Griffiths in this volume).

5.6.2. Specific Agents with Potential Utility Against C. parvum

(a) *Macrolides.* Erythromycin, spiramycin, clarithromycin, and azithromycin display varying activity against *Cryptosporidium*. Oral erythromycin has not proven helpful, primarily because of gastrointestinal intolerance (Connolly et al., 1988); I am unaware of any prospective trial of its intravenous use. Spiramycin, in double-blind trials, has not proven a reliable agent (Wittenberg et al., 1989), and may be unacceptably toxic in people with AIDS (Weikel et al., 1991). Its activity in rodent models of cryptosporidiosis has been disappointing (Brasseur et al., 1991; Rehg 1991b). Nonetheless, anectodal reports of its utility in asymptomatic or mild disease persist (Pettoello-Mantovani et al., 1995).

In both tissue culture and experimental rodent infections, azithromycin is a potent inhibitor of *C. parvum* (see Rehg, 1994), though not curative. It decreases biliary disease in the immunosuppressed rat model (Rehg, 1994) and, when administered 1 hour before infection, patent infection is prevented (Rehg, 1991c). Its efficacy in humans has not been defined by published prospective double-blind studies, though there are encouraging anectodal reports (Vargas et al., 1993; Dupont et al., 1996; Hicks et al., 1996). The results of an open-label trial conducted by the AIDS Clinical Trials Group* have not yet been published, and I am informed it was not found to be helpful (C. Fichtenbaum, personal communication). The administration of oral agents to people with malabsorption and diarrhea may be far less secure than parenteral administration, and this study (like other studies of oral agents in diarrheal disease) may have been marred by this problem.

Clarithromycin has been evaluated in a rodent model (Cama et al., 1994) and was found to produce significant reductions in parasite burden, albeit not eradication. Jordan (1996) retrospectively reviewed the records of 471 AIDS patients in Los Angeles, USA, and found that 7 (1.5%) had developed clinical cryptosporidiosis before the introduction of clarithromycin. All 7 (8%) were in a subgroup of 86 people with CD4 cell counts $<25/mm^3$. During the evaluation of clarithromycin (500 mg twice daily) as prophylaxis against *Mycobacterium aviumintracellulare* complex disease, there was a decrease in the incidence of cryptosporidiosis in the subgroup with CD4 cell counts $<25/mm^3$ (0 of 38 receiving clarithromycin, 4 of 48 not receiving clarithromycin). In a 2-year follow-up of 217 patients with CD4 cell counts $<50/mm^3$ taking clarithromycin, none developed clinical cryptosporidiosis. This intriguing retrospective information suggests that pretreatment with a macrolide may prevent patent disease.

In summary, these data suggest that macrolides have a variable but positive effect on *C. parvum*. Azithromycin and clarithromycin are notable for achieving high intracellular levels, perhaps consistent with their activity

---

* M.W. Dunne and colleagues in Study 066-167 (1993). Open label azithromycin in treatment of cryptosporidiosis. Paper presented at the Ninth International Conference on AIDS, Berlin, 6–11 June 1993.

against *C. parvum*. High intracellular macrolide levels may be associated with entry into the parasite and the prevention of clinical cryptosporidiosis (Rehg, 1991c; Jordan, 1996); furthermore, if the host is able to absorb the drug it may be helpful in treatment. Failures of macrolides in treatment could potentially be related to malabsorption, warranting investigation of systemic (intravenous) administration.

(b) *Paromomycin*. Paromomycin, an aminoglycoside antibiotic with broad antiprotozoan and antibacterial activity, is currently the most consistently effective anti-*Cryptosporidium* agent. Paromomycin has shown variable but generally positive activity in model systems (e.g. Marshall and Flanigan, 1992; Rehg, 1994; Tzipori *et al.*, 1994), and in humans when given orally (e.g. Fichtenbaum *et al.*, 1993; Bisseul *et al.*, 1994b; Scaglia *et al.*, 1994; White *et al.*, 1994; Flanigan *et al.*, 1996; Gunthard *et al.*, 1996; Stefani *et al.*, 1996). It is usually not curative in people with AIDS, relapse is common, and it is ineffective against biliary or pancreatic infection. In the study by Stefani *et al.* (1996), for example, 19 of 25 people (76%) with AIDS and cryptosporidiosis improved when given 500 mg of paromomycin four times daily, but parasitological cure occurred in only 6, and recrudescence after cessation of treatment was common. Other aminoglycosides are, in general, ineffective in model systems (Rehg, 1994; O'Donoghue, 1995). Inhalation therapy with paromomycin has been reported to be successful in human respiratory tract cryptosporidiosis (Mohri *et al.*, 1995).

A conundrum about paromomycin is that it is not significantly absorbed, yet acts on the intracellular parasite. Workers in our laboratory (Balakrishnan *et al.*, unpublished) have recently shown that paromomycin affects intracellular but not extracellular parasites via a route that does not require host cytoplasmic trafficking. We believe that paromomycin may enter the intracellular parasite via the overlying apical host membranes. This mode of drug entry may explain why paromomycin is most effective in people with scanty diarrhea, as outlined below.

A common regimen of paromomycin for cryptosporidiosis in adults is 500 mg four times daily (1500–2000 mg/day, 25–35 mg/kg per day). Thus 'waves' of high concentration intestinal juices may alternate with subsequent waves of lower concentration. Exposure of infected cells to paromomycin is increasingly brief as the diarrheal flux worsens and transit time decreases. People with fulminant cryptosporidiosis may have diarrheal fluxes of 1 litre/h. If 2 g/day of paromomycin were evenly distributed in this, the average concentration would only be 2 g/24 litres or about 83 µg/ml. In contrast, in the laboratory we use 2000 µg/ml to achieve an 80% reduction in the number of parasites in tissue culture.

Using gnotobiotic piglets infected with two different isolates of *Cryptosporidium*, we have shown that paromomycin was less effective in animals with more diarrhea (Tzipori *et al.*, 1994). In this model, high doses of paromomycin (500 mg/kg per day) were well tolerated, 250 mg/kg per day were less effective, and 125 mg/kg per day were ineffective. Verdon and colleagues (1994) and Fayer and Ellis (1993) found 100 mg/kg per day to be effective as treatment or prophylaxis in animal models. These doses are 3–20 times the human doses. Bissuel and colleagues (1994a) found that long-term high-dose treatment in humans did not lead to significant systemic absorption. Thus there may be merit in attempting to use higher doses of paromomycin in humans (Tzipori *et al.*, 1995a; Verdon *et al.*, 1995). White and colleagues (1995) have noted that humans do not necessarily tolerate paromomycin well. Vertigo, diarrhea, nausea, vomiting, and abdominal pain occur with doses > 3 g/day, and pancreatitis has been reported (Tan *et al.*, 1995). The latter implies some degree of systemic absorption, contrary to the results of Bissuel *et al.* (1994a).

There are several possible directions for exploring the role of paromomycin therapy. One is examining higher total daily dosages in people with severe or fulminant disease, realizing that in this group the benefits may outweigh the risks. Another possible direction is the investigation of alternate dosing schedules that lead to either higher peak luminal levels, such as 2 g orally at one time, or studying continuous luminal exposure, perhaps via constant irrigation with a duodenal tube. A third direction is the chemical modification of paromomycin in the quest for more active agents. Lastly, paromomycin therapy in combination with other potentially active agents (such as intravenous macrolides) could be studied.

(c) *Ionophores* Polyether ionophores are anticoccidial compounds that permeabilize target cell membranes. Lasalocid has been shown to have inconsistent activity in rodent models (Kimata *et al.*, 1991; Rehg, 1993; Leitch and He, 1994). Maduramycin and alborixin are polyether ionophores that have been studied in a variety of model systems (Mead *et al.*, 1995), producing 96% and 71% decreases (respectively) in oocyst excretion in SCID mice. These agents may best be viewed as prototypes for further modification, given their side effects. Letrazuril displayed some modest activity in human trials, which were limited by drug-related rashes and relapses (Harris *et al.*, 1994; Loeb *et al.*, 1995). The manufacturer of letrazuril has reportedly stopped further clinical trials of it.

(d) *Hyperimmune bovine colostrum.* The use of hyperimmune bovine colostrum is dealt with in depth by Crabb (1998). There is now a wealth of clinical experience suggesting that oral immunotherapy for cryptosporidiosis can be effective and safe. We do not know if its variable success is because of antigenic variation amongst isolates, or for other reasons. As it is a luminal agent, profuse diarrhea may dilute the antibody and reduce its effectiveness *in vivo*.

(e) *Sulfonamides.* Sulfa drugs, useful in the treatment of other closely related coccidians, have not proved effective in the treatment of cryptosporidiosis in humans. Moreover, trimethoprim–sulfamethoxazole prophylaxis for *Pneumocystis carinii* pneumonia does not prevent cryptosporidiosis. Since sulfonamides show activity in a rat model (Rehg, 1991a), there may be a rationale for chemical modification of these agents in the search for more active drugs.

(f) *Miscellaneous agents.* Antimalarial drugs have been used without particular success in cryptosporidiosis. Mefloquine caused hypoglycemia requiring intravenous glucose therapy in a cachectic person with AIDS (Assan *et al.*, 1995). Qinghaosu (artemisinin) and derivatives are not protective in mice (Fayer and Ellis, 1994), and to my knowledge have not been assessed in humans. Atovoquone did not prevent death from cryptosporidiosis in SCID mice, and its manufacturer has abandoned human clinical trials (Rohlman *et al.*, 1993).

Nitazoxanide is an agent with wide antiparasitic actions, and in unpublished studies has been effective in a wide array of people with cryptosporidiosis. In people with AIDS, diarrhea decreased in approximately one-third (R. Soave, personal communication) in preliminary studies. The AIDS Clinical Trials Groups is currently (1997) conducting a randomized trial of nitazoxanide (unpublished information).

6. DIAGNOSIS

The most crucial element in the chain of events needed to diagnose cryptosporidiosis is for the clinician to think of cryptosporidiosis, and not to rely on routine fecal culture and sensitivity tests to provide all the answers.

6.1. The Rationale for Diagnosis

Arguments have been made that it is pointless to diagnose cryptosporidiosis in the normal host, since no treatment exists. This is a false argument in my view. One reason to make the diagnosis is economic. A parent who contracts cryptosporidiosis from an ill child and has to miss work, and thus suffers economic loss, might have been more scrupulous about avoiding fecal–oral contamination had she or he known the risk. The economic consequences of this disease have not been studied except for detailing the costs of making the water supply safe (see Fricker and Crabb, 1998). A second argument for making a diagnosis is the public health benefit. Cryptosporidiosis in the normal host may imply water-supply contamination, which is important to immunocompromised people at risk and to the public health authorities. Massachusetts is one of a number of states in the USA that has made cryptosporidiosis a reportable disease on this basis.

6.2. Clinical Diagnostic Techniques

There has been little progress in devising novel diagnostic techniques (Casemore, 1991). Commonly, fecal detection is via modified Ziehl–Neelsen (acid-fast) or auramine staining. Fluorescently labeled anti-*Cryptosporidium* monoclonal antibodies are available, and have been reported to be more sensitive than modified acid-fast stains (see e.g. Tee *et al.*, 1993; Alles *et al.*, 1995). In contrast, in my experience and that of others (Weitz and Astorga, 1993; Kehl *et al.*, 1995) the modified Ziehl–Neelsen stain is as sensitive and specific as the monoclonal antibody detection method. The monoclonal antibody test — which, while more expensive is easier than the acid fast stain, and requires less expertise by the viewer — is most helpful where the diagnosis of cryptosporidiosis is uncommon, or when the density of oocysts may be low, and when the expense and specialized equipment needed are not an economic barrier. Entrala *et al.* (1995) have reported that hydrogen peroxide treatment of fresh feces (10 minutes with a 5 volume final concentration) substantially improved detection with acid-fast stains and monoclonal antibodies, and if this result is confirmed it may be of general use.

Sucrose flotation methods can also be used to diagnose cryptosporidiosis (Current & Garcia, 1991). Histological diagnosis can be made by standard hematoxylin and eosin, or thin-section toluidine blue, staining. It is important to view sections using a high-power oil immersion objective, as the parasites are small ($c.$ 5µm) and easily escape notice. Electron microscopy is an elegant but costly way to make this diagnosis. Immunological methods of diagnosis, such as ELISA-based techniques using fecal samples, have been described but have not been used in settings other than the research laboratory.

Note added in proof

Intuitively, it makes sense that the filtration of particulates (including oocysts) out of water sources should decrease the transmission of cryptosporidiosis. However, the potential value of municipal water filtration in decreasing transmission has been questioned. Sorvillo *et al.* (1994b) (see p. 47) reported that the rates of cryptosporidiosis in people with AIDS did not differ between two areas served by a filtered, or a nonfiltered, water source in Los Angeles, USA either before or after filtration began at the second source. Their conclusion that filtration did not decrease the incidence of cryptosporidiosis depends on the assumption that cryptosporidiosis introduced by the unfiltered supply did not lead to cryptosporidiosis in the filtered area. However, this assumption seems suspect, given the possibility of sexually transmitted secondary cases, and the reality of mobility in the population. Indeed, simple reanalysis of the data provided by Sorvillo *et al.*

leads to the opposite conclusion, namely that filtration was associated with a decrease in the occurrence of cryptosporidiosis in people with AIDS in Los Angeles (Robert Morris, personal communication). In this reanalysis, no assumption is made that the areas are epidemiologically exclusive. If one does not treat the two areas as separate, but rather simply analyzes the rates of cryptosporidiosis before and after filtration became complete for the entire city, then the differences in pre-filtration and post-filtration are very highly significant (125 cases in 2479 people with AIDS (5.04%) pre-filtration, versus 285 cases in 8509 people with AIDS (3.25%) post-filtration, $P = 0.00009$). Sorvillo *et al.* (1994b) have argued, since the rates went down in both communities, that filtration had no impact. However, it may well be that filtration did reduce waterborne transmission of cryptosporidiosis, and that as the introduction of filtration reduced cryptosporidiosis in the previously unfiltered area, it also reduced person-to-person transmission from the unfiltered community to the community that had always had filtration. Thus, a reduction in waterborne cryptosporidiosis also resulted in a reduction in person-to-person spread.

Overall, I believe that the removal or inactivation of oocysts in public waters will decrease both the number of primary infections, and secondary person-to-person transmissions due to the initial introduction of waterborne *C. parvum* into a population.

REFERENCES

Adam, A.A., Hassan, H.S., Shears, P. and Elshibly, E. (1994). Cryptosporidium in Khartoum, Sudan. *East African Medical Journal* **71**, 745–6.

Adams, R.O.B., Guerrant, R.L., Zu, S., Fang, G. and Roche, J.K. (1994). *Cryptosporidium parvum* infection of intestinal epithelium: morphological and functional studies in an *in vitro* model. *Journal of Infectious Diseases* **169**, 170–177.

Addiss, D.G., Pond, R.S., Remshak, M., Juranek, D.D., Stokes, S. and Davis, J.P. (1996). Reduction of risk of watery diarrhea with point-of-use water filters during a massive outbreak of waterborne *Cryptosporidium* infection in Milwaukee, Wisconsin, 1993. *American Journal of Tropical Medicine and Hygiene* **54**, 549–553.

Alles, A.J., Waldron, M.A., Sierra, L.S. and Matia, A.R. (1995). Prospective comparison of direct immunofluorescence and convention staining methods for detection *Giardia* and *Cryptosporidium* spp. in human fecal specimens. *Journal of Clincial Microbiology* **33**, 1632–1634.

Angus, K.W. (1990). Cryptosporidiosis and AIDS. *Baillière's Clinical Gastroenterology* **4**, 425–441.

Argenzio, R.A., Liacos, J., Levy, M., Meuten, D., Lecce, J. and Powell, D.W. (1990). Villous atrophy, crypt hyperplasia, cellular infiltration, and impaired glucose–Na absorption in enteric cryptosporidiosis of pigs. *Gastroenterology* **98**, 1129–1140.

Argenzio, R.A., Leece, J. and Powell, D.W. (1993). Prostanoids inhibit intestinal NaCl

absorption in experimental porcine cryptosporidiosis. *Gastroenterology* **104**, 440–447.
Argenzio, R.A., Rhoads, J.M., Armstrong, M. and Gomez, G. (1994). Glutamine stimulates prostaglandin-sensitive Na^+–H^+ exchange in experimental porcine cryptosporidiosis. *Gastroenterology* **106**, 1418–1428.
Argenzio, R.A., Armstrong, M. and Rhoads, J.M. (1996). Role of the enteric nervous system in piglet cryptosporidiosis. *Journal of Pharmacology and Experimental Therapeutics* **279**, 1109–1115.
Assan, R., Perronne, C., Chotard, L., Larger, E. and Vilde, J.L. (1995). Mefloquine-associated hypoglycaemia in a cachectic AIDS patient. *Diabète et Métabolisme* **21**, 54–58.
Assoumou, A., Kone, M., Penali, L.K., Coulibaly, M. and N'Draman, A.A. (1993). Cryptosporidiose et VIH à Abidjan (Côte-d'Ivoire). *Bulletin de la Société de Pathologie Exotique* **86**, 85–86.
Baskerville, A., Ramsay, A.D., Millward-Sadler, G.H., Cook, R.W. and Cranage, M.P. (1991). Chronic pancreatitits and biliary fibrosis associated with cryptosporidiosis in simian AIDS. *Journal of Comparative Pathology* **105**, 415–421.
Beier, T.V., Sidorenko, N.V. and Svezhova, N.V. (1995). Kletochnye vzaimodeistviia pri vnutrikletochnom parazitirovanii kriptosporidii. I. Vliianie kriptosporidii *Cryptosporidium parvum* na fosfataznuiu aktivnost' v enterotsitakh tonkoi kishki eksperimental'no zarazhennykh novorozhdennykh krysiat. *Tsitologiia* **37**, 829–837.
Benhamou, Y., Caumes, E., Gerosa, Y., Cadranel, J.F., Dohin, E., Katlama, C., Amouyal, P., Canard, J.M., Azar, N., Hoang, C., Le Charpentiér, Y., Gentilini, M., Opalon, P. and Valla, D. (1993). AIDS-related cholangiopathy. Critical analysis of a prospective series of 26 patients. *Digestive Diseases and Sciences* **38**, 1113–1118.
Bernard, E., Carles, M., Pradier, C., Boissy, C., Roger, P.M., Hébuterne, X., Mondain, V., Michiels, J.F., Le Fichoux, Y. and Dellamonica, P. (1995). Diarrhée persistante chez les patients infectés par le VIH: place d'*Enterocytozoon bieneusi*. *La Presse Médicale* **24**, 671–674.
Bier, J.W. (1991). Isolation of parasites on fruits and vegetables. *Southeast Asian Journal of Tropical Medicine and Public Health* **22**, supplement, 144–145.
Bissuel, F., Cotte, L., De Montclos, M., Ragbodonirina, M. and Trepo, C. (1994a). Absence of systemic absorption of oral paromomycin during long-term, high-dose treatment for cryptosporidiosis in AIDS. *Journal of Infectious Diseases* **170**, 749–750.
Bisseul, F., Cotte, L., Rabodonirina, M., Rougier, P., Piens, M.A. and Trepo, C. (1994b). Paromomycin: an effective treatment for cryptosporidial diarrhea in patients with AIDS. *Clinical Infectious Diseases* **18**, 447–449.
Blanshard, C., Jackson, A.M., Shanson, D.C., Francis, N. and Gazzard, B.G. (1992). Cryptosporidiosis in HIV-seropositive patients. *Quarterly Journal of Medicine* **85**, 813–823.
Brandonisio, O., Maggi, P., Panaro, M.A., Bramante, L.A., Di Coste, A. and Angarano, G. (1993). Prevalence of cryptosporidiosis in HIV-infected patients with diarrhoeal illnesss. *European Journal of Epidemiology* **9**, 190–194.
Brasseur, P., Lemeteil, D. and Ballet, J.-J. (1991). Anti-cryptosporidial drug activity screened with an immunosuppressed rat model. *Journal of Protozoology* **38**, 230S–231S.
Brasseur, P., Favennec, L., Lemeteil, D., Roussel, F. and Ballet, J.J. (1994). An immunosuppressed rat model for evaluation of anti-*Cryptosporidium* activity of sinefungin. *Folia Parasitologica* **41**, 13–16.

Brea Hernando, A.J., Bandres Franco, E., Mosquera Lozano, J.D., Lantero Benedito, M. and Ezquerra Lezcano, M. (1993). Criptosporidiasis pulmonary SIDA. Presentación de un caso y revisión de la literatura. *Anales de Medicina Interna* **10**, 232–236.
Bridgman, S.A., Robertson, R.M., Syed, Q., Speed, N., Andrews, N. and Hunter, P.R. (1995). Outbreak of cryptosporidiosis associated with a disinfected groundwater supply. *Epidemiology and Infection* **115**, 555–566.
Cama, V.A., Marshall, M.M., Shubitz, L.F., Ortega, Y.R. and Sterling, C.R. (1994). Treatment of acute and chronic *Cryptosporidium parvum* infections in mice using clarithromycin and 14-OH clarithromycin. *Journal of Eukaryotic Microbiology* **41**, 25S.
Casemore, D.P. (1990a) Epidemiological aspects of human cryptosporidiosis. *Epidemiology and Infection* **104**, 1–28.
Casemore, D.P. (1990b). Food borne protozoal infection. *Lancet* **335**, 1427–1432.
Casemore, D.P. (1991). Laboratory methods for diagnosing cryptosporidiosis. (*ACP Broadsheet 128.*) *Journal of Clinical Pathology* **44**, 445–451.
Castillo-Duran, C., Heresi, G., Fisberg, M. and Vauy, R. (1987). Controlled trial of zinc supplementation during recovery from malnutrition: effects on growth and immune function. *American Journal of Clinical Nutrition* **45**, 602–608.
Chacin-Bonilla, L. (1995). Criptosporidiosis en humanos. Revisión. *Investigación Clínica* **36**, 207–250.
Chacin-Bonilla, L., Guanipa, N., Cano, G., Raleigh, X. and Quijada, L. (1992). Cryptosporidiosis among patients with acquired immunodeficiency syndrome in Zulia State, Venezuela. *American Journal of Tropical Medicine and Hygiene* **47**, 582–586.
Chacin-Bonilla, L., Mejia de Young, M., Cano, G., Guanipa, N., Estevez, J. and Bonilla, E. (1993). *Cryptosporidium* infections in a suburban community in Maracaibo, Venezuela. *American Journal of Tropical Medicine and Hygiene* **49**, 63–67.
Clark, D.P. and Sears, C.L. (1996). The pathogenesis of cryptosporidiosis. *Parasitology Today* **12**, 221–225.
Clavel, A., Olivares, J.L., Fleta, J., Castillo, J., Varea, M., Ramos, FJ., Arnal, A.C. and Quilez, J. (1996). Seasonality of cryptosporidiosis in children. *European Journal of Clinical Microbiology and Infectious Diseases* **15**, 77–79.
Clayton, F., Heller, T. and Kotler, D.P. (1994). Variation in the enteric distribution of cryptosporidia in acquired immunodeficiency syndrome. *American Journal of Clinical Pathology* **102**, 420–425.
Clifford, C.P., Crook, E.W., Conlon, C.P., Fraise, A.P., Day, D.G. and Peto, T.E. (1990). Impact of waterborne outbreak of cryptosporidiosis on AIDS and renal transplant patients. *Lancet* **335**, 1455–1456.
Connolly, G.M., Dryden, M.S., Shanson, D.C. and Gazzard, B.G. (1988). Cryptosporidial diarrhoea in AIDS and its treatment. *Gut* **29**, 593–597.
Cordell, R. and Addiss, D. (1994). Cryptosporidiosis in child care settings: a review of the literature and recommendations for prevention and control. *Pediatric Infectious Disease Journal* **13**, 310–317.
Cotte, L., Rabodonirina, M., Piens, M.A., Perreard, M., Mojon, M. and Trepo, C. (1993). Prevalence of intestinal protozoans in French patients infected with HIV. *Journal of Acquired Immune Deficiency Syndromes* **6**, 1024–1029.
Crabb, J.H. (1998). Antibody-based chemotherapy of cryptosporidiosis. *Advances in Parasitology* **40**, 121–149.
Current, W.L. (1994). *Cryptosporidium parvum*: household transmission. *Annals of Internal Medicine* **120**, 518–519.
Current, W.L. and Garcia, L.S. (1991) Cryptosporidiosis. *Clinical Microbiology Reviews* **4**, 325–358.

Ditrich, O., Palkovic, L., Sterba, J., Prokopic, J., Loudova, J. and Givoda, M. (1991). The first finding of *Cryptosporidium baileyi* in man. *Parasitology Research* **77**, 44–47.
Duong, T.H., Kombila, M., Dufillot, D., Richard-Lenoble, D., Owono Medang, M., Martz, M., Gendrel, D., Engohan, E. and Moreno, J.L. (1991). Place de la cryptosporidiose chez l'enfant au Gabon. Résultats de deux enquêtes prospectives. *Bulletin de la Société de Pathologie Exotique* **84**, 635–644.
Duong, T.H., Dufillot, D., Kojko, J., Nze-Eyo'o, R., Thuilliez, V., Richard-Lenoble, D. and Kombila, M. (1995). Cryptosporidiose digestive chez le jeune enfant en zone urbaine au Gabon. *Santé* **5**, 185–188.
Dupont, C., Bougnoux, M.E., Turner, L., Rouveix, E. and Dorra, M. (1996). Microbiological findings about pulmonary cryptosporidiosis in two AIDS patients. *Journal of Clinical Microbiology* **34**, 227–229.
Dupont, H.L., Chappell, C.L., Sterling, C.R., Okhuysen, P.C., Rose, J.B. and Jakubowski, W. (1995). The infectivity of *Cryptosporidium parvum* in healthy volunteers. *New England Journal of Medicine* **332**, 855–859.
Egger, M., Mausezahl, D., Odermatt, P., Marti, H.P. and Tanner, M. (1990). Symptoms and transmission of intestinal cryptosporidiosis. *Archives of Disease in Childhood* **65**, 445–447.
Entrala, E., Rueda-Rubio, M., Janssen, D. and Mascaro, C. (1995). Influence of hydrogen peroxide on acid-fast staining of *Cryptosporidium parvum* oocysts. *International Journal for Parasitology* **25**, 1473–1477.
Esfandiari, A., Jordan, W.C. and Brown, C.P. (1995). Prevalence of enteric parasitic infection among HIV-infected attendees of an inner city AIDS clinic. *Cellular and Molecular Biology* **41**, Supplement 1, S19–S23.
Fang, G.D., Lima, A.A., Martins, C.V., Nataro, J.P. and Guerrant, R.L. (1995). Etiology and epidemiology of persistent diarrhea in northeastern Brazil: a hospital-based, prospective, case–control study. *Journal of Pediatric Gastroenterology and Nutrition* **21**, 137–144.
Fanning, M., Monte, M., Sutherland, L.R., Lloyd, R., Broadhead, M., Murphy, G.F. and Harris, A.G. (1991). Pilot study of Sandostatin (octeotride) therapy of refractory HIV-associated diarrhea. *Digestive Diseases Science* **36**, 476–480.
Fayer, R. and Ellis, W. (1993). Paromomycin is effective as prophylaxis for cryptosporidiosis in dairy calves. *Journal of Parasitology* **79**, 771–774.
Fayer, R. and Ellis, W. (1994). Qinghaosu (artemisinin) and derivatives fail to protect neonatal BALB/c mice against *Cryptosporidium parvum* (Cp) infection. *Journal of Eukaryotic Microbiology* **41**, 41S.
Fayer, R., Trout, J. and Nerad, T. (1996). Effects of a wide range of temperatures of infectivity of *Cryptosporidium parvum* oocysts. *Journal of Eukaryotic Microbiology* **43**, 64S.
Fichtenbaum, C.J., Ritchie, D.J. and Powederly, E.G. (1993). Treatment of cryptosporidiosis in patients with AIDS with paromomycin. *Clinical Infectious Diseases* **16**, 298–300.
Flanigan, T.P., Whalen, C., Turner, J., Soave, R., Toerner, J., Havlir, D. and Kotler, D. (1992). *Cryptosporidium* infection and CD4 counts. *Annals of Internal Medicine* **116**, 840–842.
Flanigan, T.P., Ramratnam, B., Graeber, C., Hellinger, J., Smith, D., Wheeler, D., Hawley, P., Heath-Chiozzi, M., Ward, D.J., Brummitt, C. and Turner, J. (1996). Prospective trial of paromomycin for cryptosporidiosis in AIDS. *American Journal of Medicine* **100**, 370–372.
Foot, A.B., Oakhill, A. and Mott, M.G. (1990). Cryptosporidiosis and acute leukaemia. *Archives of Disease in Childhood* **65**, 236–237.

Forbes, A., Blanshard, C. and Gazzard, B. (1993). Natural history of AIDS related sclerosing cholangitis: a study of 20 cases. *Gut* **34**, 116–121.
Fricker, C.R. and Crabb, J.H. (1998). Water-borne cryptosporidiosis: detection methods and treatment options. *Advances in Parasitology* **40**, 241–278.
Friedman, L.S. (1991). Somatostatin therapy for AIDS diarrhea: muddy waters. *Gastroenterology* **101**, 1446–1448.
Garcia Velarde, E., Chavez Legaspi, M., Coello Ramirez, P., Gonzalez, J. and Aguilar Benavides, S. (1991). *Cryptosporidium* sp. in 300 children with and without diarrhea. *Archivos de Investigación Médica* **22**, 329–332.
Gardner, A.L., Roche, J.K., Weikel, C.S. and Guerrant, R.L. (1991). Intestinal cryptosporidiosis: pathophysiologic alterations and specific cellular and humoral immune respones in rnu/+ and rnu/rnu (athymic) mice. *American Journal of Tropical Medicine and Hygiene* **44**, 49–62.
Genta, R.M., Chappell, C.L., White, A.C., jr, Kimball, K.T. and Goodgame, R.W. (1993). Duodenal morphology and intensity of infection in AIDS-related intestinal cryptosporidiosis. *Gastroenterology* **105**, 1769–1775.
Gentile, G., Caprioli, A., Donelli, G., Venditti, M., Mandelli, F. and Martino, P. (1990). Asymptomatic carriage of *Cryptosporidium* in two patients with leukemia. *American Journal of Infection Control* **18**, 127–128.
Gentile, G., Venditti, M., Micozzi, A., Caprioli, A., Donelli, G., Tirindelli, C., Meloni, G., Arcese, W. and Martino, P. (1991). Cryptosporidiosis in patients with hematologic malignancies. *Reviews of Infectious Disease* **13**, 842–846.
Girard, P.M., Goldschmidt, E., Vittecoq, D., Massip, P., Gastiaburu, J., Meyohas, M.C., Coulaud, J.P. and Schally A.V. (1992). Vapreotide, a somatostatin analogue, in cryptosporidiosis and other AIDS-related diarrhoeal diseases. *AIDS* **6**, 715–718.
Gledhill, J.A. and Porter, J. (1990). Diarrhoea due to *Cryptosporidium* infection in thalassemia major. *Brisith Medical Journal* **301**, 212–213.
Goldstein, S.T., Juranek, D.D., Ravenholt, O., Hightower, A.W., Martin, D.G., Mesnik, J.L., Griffiths, S.D., Bryant, A.J., Reich, R.R. and Herwaldt, B.L. (1996). Cryptosporidiosis: an outbreak associated with drinking water despite state-of-the art water treatment. *Annals of Internal Medicine* **124**, 459–468.
Gomez Morales, M.A., Ausiello, C.M., Guarino, A., Urbani, F., Spagnuolo, M.I., Pignata, C. and Pozio, E. (1996). Severe, protracted intestinal cryptosporidiosis associated with inteferon gamma deficiency: pediatric case report. *Clinical Infectious Diseases* **22**, 848–850.
Goodgame, R.W., Kimball, K., Ou, C.N., White, A.C., jr, Genta, R.M., Lifschitz, C.H. and Chappell, C.L. (1995). Intestinal function and injury in acquired immonodeficiency syndrome-related cryptosporidiosis. *Gastroenterology* **108**, 1075–1082.
Griffiths, J.K. (1997). Commentary: how humbling to understand so little. The importance of a rare observation. *Infectious Diseases in Clincial Practice* **6**, 3–5.
Griffiths, J.K. and Gorbach, S.L. (1993). Other bacterial diarrhoeas. *Baillière's Clinical Gastroenterology* **7**, 263–305.
Griffiths, J.K., Moore, R., Dooley, S., Keusch, G.T. and Tzipori, S. (1994). *Cryptosporidium parvum* infection of Caco-2 cell monolayers induces an apical monolayer defect, selectively increases transmonolayer permeability, and causes epithelial cell death, *Infection and Immunity* **62**, 4506–4514.
Guarino, A., Canani, R.B., Pozio, E., Terracciano, L., Albano, F. and Mazzeo, M. (1994). Enterotoxic effect of stool supernatant of *Cryptosporidium*-infected calves on human jejunum. *Gasteroenterology* **106**, 28–34.
Guarino, A., Canani, R.B., Casola, A., Poizio, E., Russo, R., Bruzzese, E., Fontana, M. and Rubino, A.E. (1995). Human intestinal cryptosporidiosis: secretory diarrhea

and enterotoxic activity in Caco-2 cells. *Journal of Infectious Diseases* **171**, 976–983.
Gunthard, M., Meister, T., Luthy, R. and Weber, R. (1996). Intestinale Kryptosporidiose bei HIV-Infecktion. Krankheitsbild, Verlauf und Therapie. *Deutsche Medizinische Wochenschrift* **121**, 686–692.
Hadad Melendez, P., Fernandez Abascal, H., Millan Marcelo, J.C., Ramos Garcia, A. and Nuñez Fernandez, F. (1993). Infección por *Cryptosporidium* sp. en individuos cubanos infectados por el VIH. *Revista Cubana de Medicina Tropical* **45**, 55–58.
Handousa, A.E., el-Sahzly, A.M., el-Nashaar, N.M. and Hamouda, M.M. (1991). Malabsorption syndrome in patients with cryptosporidiosis. *Journal of the Egyptian Society of Parasitology* **21**, 791–796.
Harris, M., Deutsch, G., MacLean, J.D. and Tsoukas, C.M. (1994). A phase I study of letrazuril in AIDS-related cryptosporidiosis. *AIDS* **8**, 1109–1113.
Heald, A.E. and Bartlett, J.A. (1994). *Cryptosporidium* spread in a group residential home. *Annals of Internal Medicine* **121**, 467–468.
Hicks, P., Zweiner, R.J., Squires, J. and Savell, V. (1996). Azithromycin therapy for *Cryptosporidium parvum* infection in four children infected with human immunodeficiency virus. *Journal of Pediatrics* **129**, 297–300.
Hoepelman, I.M. (1996). Human cryptosporidiosis. *International Journal of STD and AIDS* **7**, supplement 1, 28–33.
Holland, R.E., Boyle, S.M., Herdt, T.H., Grimes, S.D. and Walker, R.D. (1992). Malabsorption of vitamin A in preruminating calves infected with *Cryptosporidium parvum*. *American Journal of Veterinary Research* **53**, 1947–1952.
Jablonowski, H., Szelenyi, H., Becker, K., Lubke, H.J., Borchard, F., Strohmeyer, G. and Hengels, K.J. (1994). Sklerosierende Cholangitis mit Papillenstenose bei einem HIV-infizierten Patienten mit Cryptosporidieninfektion. *Zeitschrift für Gastroenterologie* **32**, 441–443.
Jacyna, M.R., Parkin, J., Goldin, R. and Baron, J.H. (1990). Protracted enteric cryptosporidial infection in selective immunoglobulin A and *Saccharomyces* opsonin deficiencies. *Gut* **31**, 714–716.
Jaggi, N., Rajeshwari, S., Mittal, S.K., Mathur, M.D. and Baveja, U.K. (1994). Assessment of the immune and nutritional status of the host in childhood diarrhoea due to *Cryptosporidium*. *Journal of Communicable Disease* **26**, 181–185.
Janoff, E.N., Mead, P.S., Mead, J.R., Echeverria, P., Bodhidatta, L., Bhaibulaya, M., Sterling, C.R. and Taylor, D.N. (1990). Endemic *Cryptosporidium* and *Giardia lamblia* infections in a Thai orphanage. *American Journal of Tropical Medicine and Hygiene* **43**, 248–256.
Jensen, B.N.A., Gerstoft, J., Hojlyng, N., Backer, V., Paaske, M., Gomme, G. and Skinhoj, P. (1990). Pulmonary pathogens in HIV-infected patients. *Scandinavian Journal of Infectious Disease* **22**, 413–420.
Joce, R.E., Bruce, J., Kiely, D., Noah, N.D., Dempster, W.B., Stalker, R., Gumsley, P., Chapman, P.A., Norman, P., Watkins, J., Smith, H.V., Price, T.J. and Watts, D. (1991). An outbreak of cryptosporidiosis associated with a swimming pool. *Epidemiology and Infection* **107**, 497–508.
Jordan, E.C. (1996). Clarithromycin prophylaxis against *Cryptosporidium* enteritis in patients with AIDS. *Journal of the National Medical Association* **88**, 425–427.
Juranek, D.D. (1995). Cryptosporidiosis: sources of infection and guidelines for prevention. *Clinical Infectious Diseases* **21**, S57–S61.
Kamel, A.G., Maning, N. and Arulmainathan, S. (1994). Cryptosporidiosis among HIV positive intravenous drugs users in Malaysia. *Southeast Asian Journal of Tropical Medicine and Public Health* **25**, 650–653.

Kandil, H.M., Berschneider, H.M. and Argenzio, R.A. (1994). Tumour necrosis factor alpha changes porcine intestinal ion transport through a paracrine mechanism involving prostaglandins. *Gut* **35**, 934–940.

Kapembwa, M.S., Bridges, C., Joseph, A.E., Fleming, S.C., Batman, P. and Griffin, G.E. (1990). Ileal and jejunal absorptive function in patients with AIDS and enterococcidial infection. *Journal of Infection* **21**, 43–53.

Kaup, F.J., Kuhn, E.M., Makoschey, B. and Hunsmann, G. (1994). Cryptosporidiosis of liver and pancreas in rhesus monkeys with experimental SIV infection. *Journal of Medical Primatology* **23**, 304–308.

Kehl, K.S., Cicirello, H. and Havens, P.L. (1995). Comparison of four different methods for detection of *Cryptosoporidium* species. *Journal of Clinical Microbiology* **33**, 416–418.

Keusch, G.T., Hamer, D., Joe, A., Kelley, M., Griffiths, J.K. and Ward, H. (1995). *Cryptosporidia* – who is at risk? *Schwiezerische Medizinische Wochenschrift* **125**, 899–908.

Kibbler, C.C., Smith, A., Hamilton-Dutoit, S.J., Milburn, H., Pattinson, J.K. and Prentice, H.G. (1987). Pulmonary cryptosporidiosis occurring in a bone marrow transplant patient. *Scandinavian Journal of Infectious Diseases* **19**, 581–584.

Kimata, I., Uni, S. and Iseki, M. (1991). Chemotherapeutic effect of azithromycin and lasalocid on *Cryptosporidium* infection in mice. *Journal of Protozoology* **38**, 232S–233S.

Kline, T.J., de las Morenas, T., O'Brien, M., Smith, B.F. and Afdhal, N.H. (1993). Squamous metaplasia of extrahepatic biliary system in an AIDS patient with cryptosporidia and cholangitis. *Digestive Diseases and Sciences* **38**, 960-962.

Kone, M., Penali, L.K., Enoh, S., Gershy-Damet, G.M. and Anderson, M. (1992). La cryptosporidiose chez les enfants ivoiriens de Yopougon. *Bulletin de la Société de Pathologie Exotique* **85**, 167–169.

Kreinik, G., Burstein, O., Landor, M., Bernstein, L., Weiss, L.M. and Wittner, M. (1991). Successful management of intractable cryptosporidial diarrhea with intravenous octreotide, a somatostatin analogue. *AIDS* **5**, 765–767.

Kuhls, T.L., Greenfield, R.A., Mosier, D.A., Crawford, D.L. and Joyce, W.A. (1992). Cryptosporidiosis in adult and neonatal mice with severe combined immunodeficiency. *Journal of Comparative Pathology* **106**, 399–410.

Kuhls, T.L., Mosier, D.A., Crawford, D.L. and Griffis, J. (1994). Seroprevalence of cryptosporidial antibodies during infancy, childhood, and adolescence. *Clinical Infectious Diseases* **18**, 731–735.

Laberge, I., Ibrahim, A., Barta, J.R. and Griffiths, M.W. (1996). Detection of *Cryptosporidum parvum* in raw milk by PCR and oligonucleotide probe hybridization. *Applied and Environmental Microbiology* **62**, 3259–3264.

Laxer, M.A., Alcantara, A.K., Javato-Laxer, M., Menorca, D.M., Fernando, M.T. and Ranoa, C.P. (1990). Immune response to cryptosporidiosis in Philippine children. *American Journal of Tropical Medicine and Hygiene* **42**, 131–139.

Leitch, G.J. and He, Q. (1994). Putative anticryptosporidial agents tested with an immunodeficient mouse model. *Antimicrobial Agents and Chemotherapy* **38**, 865–867.

Lengerich, E.J., Addiss, D.G., Marx, J.J., Ungar, B.L. and Juranek, D.D. (1993). Increased exposure to cryptosporidia among dairy farmers in Wisconsin. *Journal of Infectious Diseases* **167**, 1252–1255.

Liberti, A., Bisogno, A. and Izzo, E. (1992). Octreotide treatment in secretory and cryptosporidial diarrhea in patients with acquired immunodeficiency syndrome (AIDS): clinical evaluation. *Journal of Chemotherapy* **4**, 303–305.

Lima, A.A., Fang, G., Schorling, J.B., de Albuquerque, L., McAuliffe, J.F., Mota, S.,

Leite, R. and Guerrant, R.L. (1992). Persistent diarrhea in northeast Brazil: etiologies and interactions with malnutrition. *Acta Paediatrica* **381**, 39S–44S.
Lindsay, D.S., Blagburn, B.L. and Hoerr, F.J. (1987). Experimentally induced infections in turkeys with *Cryptosporidium baileyi* isolated from chickens. *American Journal of Veterinary Research* **48**: 104–108.
Loeb, M., Walach, C., Phillips, J., Fong, I., Salit, I., Rachlis, A. and Walmsley, S. (1995). Treatment with letrazuril of refractory cryptosporidial diarrhea complicating AIDS. *Journal of Acquired Immune Deficiency Syndromes and Human Retrovirology* **10**, 48–53.
Lopez-Velez, R., Tarazona, R., Garcia Camacho, A., Gomez-Mampaso, E., Guerror, A., Moreira, V. and Villanueva, R. (1995). Intestinal and extraintestinal cryptosporidiosis in AIDS patients. *European Journal of Clinical Microbiology and Infectious Diseases* **14**, 677–681.
MacKenzie, W.R., Hoxie, N.J., Proctor, M.E., Gradus, M.S., Blair, K.A., Peterson, D.E., Kazmierczak, J.J., Addiss, D.G., Fox, K.R., Rose J.B. and Davis, J.P. (1994). A massive outbreak in Milwaukee of *Cryptosporidium* infection transmitted through the public water supply. *New England Journal of Medicine* **331**, 161–167.
MacKenzie, W.R., Kazmierczak, J.J. and Davis, J.P. (1995a). An outbreak of cryptosporidiosis associated with a resort swimming pool. *Epidemiology and Infection* **115**, 545–553.
MacKenzie, W.R., Schell, W.L., Blair, K.A., Addiss, D.G., Peterson, D.E., Hoxie, N.J., Kazmierczak, J.J. and Davis, J.P. (1995b). Massive outbreak of waterborne *Cryptosporidium* infection in Milwaukee, Wisconsin: recurrence of illness and risk of secondary transmission. *Clinical Infectious Diseases* **21**, 57–62.
Maguire, H.C., Holmes, E., Hollyer, J., Strangeways, J.E., Foster, P., Holliman, R.E. and Stanwell-Smith, R. (1995). An outbreak of cryptosporidiosis in south London: what value the *p* value? *Epidemiology and Infection* **115**, 279–287.
Marshall, R.J. and Flanigan, T.P. (1992). Paromomycin inhibits *Cryptosporidium* infection of a human enterocyte cell line. *Journal of Infectious Diseases* **165**, 772–774.
Mausezahl, D., Egger, M., Odermatt, P. and Tanner, M. (1991). Klinik und Epidemiologie der Kryptosporidiose bei immunkompetenten Kindern. *Schweizerische Rundschau für Medizin Praxis* **80**, 936–940.
McClain, C.S. (1985). Zinc metabolism in malabsorption syndromes. *Journal of the American College of Nutrition* **4**, 49–64.
McGowan, I., Hawkins, A.S. and Weller, I.V. (1993). The natural history of cryptosporidial diarrhoea in HIV-infected patients. *AIDS* **7**, 349–354.
Mead, J.R., Ilksoy, N., You, X., Belenkaya, Y., Arrowood, M.J., Fallon, M.T. and Schinazi, R.F. (1994). Infection dynamics and clinical features of cryptosporidiosis in SCID mice. *Infection and Immunity* **62**, 1691–1695.
Mead, J.R., You, X., Pharr, J.E., Belenkaya, Y., Arrowood, M.J., Fallon, M.T. and Schinazi, R.F. (1995). Evaluation of maduramicin and alborixin in a SCID mouse model of chronic cryptosporidiosis. *Antimicrobial Agents and Chemotherapy* **39**, 854–858.
Mendez, M. and Bosch, R. (1995). Fiebre, anemia y descompensación de una hepatopatia en un niño de nueve años con una inmunodeficiencia congenita. *Medicina Clinica* **105**, 549–556.
Mengesha, B. (1994). Cryptosporidiosis among medical patients with the acquired immunodeficiency syndrome in Tikur Anbessa Teaching Hospital, Ethiopia. *East African Medical Journal* **71**, 376–378.
Menorca, D.M., Laxer, M.A., Alcantara, A.K., Javato-Laxer, M., Fernando, M.T. and Gonzales, V. (1994). Statistical analysis of clinical, immunological and nutritional

factors in pediatric cryptosporidiosis in the Philippines. *Southeast Asian Journal of Tropical Medicine and Public Health* **25**, 300–304.
Mercado, R. and Garcia, M. (1995). Frecuencia anual de las infecciónes por *Cryptosporidium parvum* en pacientes niños y adultos ambulatorios, y adultos infectados por el virus de la inmunodeficiencia humana. *Revista Médica de Chile* **123**, 479–484.
Meulbroek, J.A., Novella, M.N. and Current, W.L. (1991). An immunosuppressed rat model of respiratory cryptosporidiosis. *Journal of Protozoology* **38**, 113S–115S.
Mifsud, A.J., Bell, D. and Shafi, M.S. (1994). Respiratory cryptosporidiosis as a presenting feature of AIDS. *Journal of Infection* **28**, 227–229.
Millard, P.S., Gensheimer, K.F., Addiss, D.G., Sosin, D.M., Beckett, G.A., Houck-Jankoski, A. and Hudson, A. (1994). An outbreak of cryptosporidiosis from fresh-pressed apple cider. *Journal of the American Medical Association* **272**, 1592–1596.
Miller, R.A., Bronsdon, M.A. and Morton, W.R. (1990). Experimental cryptosporidiosis in a primate model. *Journal of Infectious Diseases* **161**, 312–315.
Miller, R.A., Bronsdon, M.A. and Morton, W.R. (1991). Failure of breast-feeding to prevent *Cryptosporidium* infection in a primate model. *Journal of Infectious Diseases* **164**, 826–827.
Mocchegiani, E., Veccia, S., Ancarani, F., Scalise, G. and Fabris, N. (1995). Benefit of oral zinc supplementation as an adjunct to zidovudine (AZT) therapy against opportunistic infections in AIDS. *International Journal of Immunopharmacology* **17**, 719–727.
Mohri, H., Fujita, H., Asakura, Y., Katoh, K., Okamoto, R., Tanabe, J., Harano, H., Noguchi, T., Inayama, Y., Amano, T. and Okubo, T. (1995). Case report: inhalation therapy of paromomycin is effective for respiratory infection and hypoxia by [*sic*] *Cryptosporidium* with AIDS. *American Journal of the Medical Sciences* **309**, 60–62.
Molbak, K., Hojlyng, N., Ingholt, L., Da Silva, A.P., Jepsen, S. and Aaby, P. (1990). An epidemic outbreak of cryptosporidiosis: a prospective community study from Guinea Bissau. *Pediatric Infectious Disease Journal* **9**, 566–570.
Molbak, K., Hojlyng, N., Gottschau, A., Sa, J.C., Ingholt, L., Da Silva, A.P. and Aaby, P. (1993). Cryptosporidiosis in infancy and childhood mortality in Guinea Bissau, West Africa. *British Medical Journal* **307**, 417–420.
Molbak, K., Aaby, P., Hojlyng, N. and Da Silva, A.P. (1994). Risk factors for *Cryptosporidium* diarrhea in early childhood: a case–control study from Guinea-Bissau, West Africa. *American Journal of Epidemiology* **139**, 734–740.
Moolasart, P., Eampokalap, B., Ratanasrithong, M., Kanthasing, P., Tansupaswaskul, S. and Tanchanpong, C. (1995). Cryptosporidiosis in HIV infected patients in Thailand. *Southeast Asian Journal of Tropical Medicine and Public Health* **26**, 335–338.
Moon, H.W., Woodmansee, D.B., Harp, J.A., Abel, S. and Ungar, B.L.P. (1988). Lacteal immunity to enteric cryptosporidiosis in mice: immune dams do not protect their suckling pups. *Infection and Immunity* **56**, 649–653.
Moore, R., Tzipori, S., Griffiths, J.K., Johnson, K., de Montigny, L. and Lomakina, I. (1995). Temporal changes in permeability and structure of piglet ileum after site-specific infection by *Cryptosporidium parvum*. *Gastroenterology* **108**, 1030–1039.
Moreno, A., Gatell, J.M., Mensa, J., Valls, M.E., Vila, J., Claramonte, X., Miro, J.M., Mallolas, J., Zamora, L., Lozano, L., Trilla, A. and Soriano, E. (1994). Incidencia de enteropatogenos en pacientes con infección por el virus de la inmunodeficiencia humana. *Medicina Clínica* **102**, 205–208.
Morgan, D., Allaby, M., Crook, S., Casemore, D., Healing, T.D., Soltanpoor, N., Hill,

S. and Hooper, W. (1995). Waterborne cryptosporidiosis associated with a borehole supply. *Communicable Disease Report CDC Review* **5**, R93–R97.
Moroni, M., Esposito, R., Cernuschi, M., Franzetti, F., Carosi, G.P. and Fiori, G.P. (1993). Treatment of AIDS-related refractory diarrhoea with octreotide. *Digestion* **54**, 30S–32S.
Moss, P.J., Read, R.C., Kudesia, G. and McKendrick, M.W. (1995). Prolonged cryptosporidiosis during primary HIV infection. *Journal of Infection* **30**, 51–53.
Naguib, M.T., Kipgen, W.M., Slater, L.N., Kuhls, T.L. and Greenfield, R.A. (1997). AIDS-associated cryptosporidiosis presenting with pneumatosis intestinalis. *Infectious Diseases in Clinical Practice* **6**, 1–2.
Navarrete, S., Stetler, H.C., Avila, C., Garcia Aranda, J.A. and Santos-Preciado, J.I. (1991). An outbreak of *Cryptosporidium* diarrhea in a pediatric hospital. *Pediatric Infectious Disease Journal* **10**, 248–250.
Newman, R.D., Wuhib, T., Lima, A.A., Guerrant, R.L. and Sears, C.L. (1993). Environmental sources of *Cryptosporidium* in an urban slum in northeastern Brazil. *American Journal of Tropical Medicine and Hygiene* **49**, 270–275.
Newman, R.D., Zu, S.X., Wuhib, T., Lima, A.A., Guerrant, R.L. and Sears, C.L. (1994). Household epidemiology of *Cryptosporidium parvum* infection in an urban community in northeast Brazil. *Annals of Internal Medicine* **120**, 500–505.
Nousbaum, J.B., Robaszkiewicz, M., Cauvin, J.M., Garre, M. and Gouerou, H. (1991). Treatment of intestinal cryptosporidiosis with zidovudine and SMS 201–995, a somatostatin analog. *Gastroenterology* **101**, 874.
O'Donoghue, P.J. (1995). *Cryptosporidium* and cryptosporidiosis in man and animals. *International Journal for Parasitology* **25**, 139–195.
O'Donoghue, P.J., Tham, V.L., de Saram, W.G., Paull, K.L. and McDermott, S. (1987). *Cryptosporidium* infections in birds and mammals and attempted cross-transmission studies. *Veterinary Parasitology* **26**, 1–11.
Oehler, R. and Loos, U. (1993). Therapie der schweren AIDS-assoziierten Diarrho mit dem Somatostatinanalogon Octreotid. *Medizinische Klinik* **88**, 45–47.
Pedersen, C., Danner, S., Lazzarin, A., Glauser, M.P., Weber, R., Katlama, C., Barton, S.E. and Lundgren, J.D. (1996). Epidemiology of cryptosporidiosis among European AIDS patients. *Genitourinary Medicine* **72**, 128–131.
Petersen, C. (1992). Cryptosporidiosis in patients infected with the human immunodeficiency virus. *Clinical Infectious Diseases* **15**, 903–909.
Petersen, C. (1995). *Cryptosporidium* and the food supply. *Lancet* **345**, 1128–1129.
Pettoello-Mantovani, M., Di Martino, L., Dettori, G., Vajro, P., Scotti, S., Ditullio, M.T. and Guandalini, S. (1995). Asymptomatic carriage of intestinal *Cryptosporidium* in immunocompetent and immunodeficient children: a prospective study. *Pediatric Infectious Disease Journal* **14**, 1042–1047.
Pichard, E., Doumbo, O., Minta, D. and Traore, H.A. (1990). Place de la cryptosporidiose au cours des diarrhées chez les adultes hospitalisés à Bamako. *Bulletin de la Société de Pathologie Exotique* **83**, 473–478.
Pol, S., Romana, C.A., Richard, S., Amouyal, P., Desportes-Livage, I., Carnot, F., Pays, J.F. and Berthelot, P. (1993). Microsporidia infection in patients with the human immunodeficiency virus and unexplained cholangitis. *New England Journal of Medicine* **328**, 95–99.
Rasmussen, K.R., Arrowood, M.J. and Healey, M.C. (1992). Effectiveness of dehydroepiandrosterone in reduction of cryptosporidial activity in immunosuppressed rats. *Antimicrobial Agents and Chemotherapy* **36**, 220–222.
Rasmussen, K.R., Healey, M.C., Cheng, L. and Yang, S. (1995). Effects of dehydroepidandrosterone in immunosuppressed adult mice infected with *Cryptosporidium parvum*. *Journal of Parasitology* **81**, 429–433.

Ravn, P., Lundgren, J.D., Kjaeldgaard, P., Holten-Anderson, W., Hojlyng, N., Nielsen, J.O. and Gaub, J. (1991). Nosocomial outbreak of cryptosporidiosis in AIDS patients. *British Medical Journal* **302**, 277–280.
Rehg, J.E. (1991a). Anticryptosporidial activity is associated with specific sulfonamides in immunosuppressed rats. *Journal of Parasitology* **77**, 238–240.
Rehg, J.E. (1991b). Anti-cryptosporidial activity of macrolides in immunosuppressed rats. *Journal of Protozoology* **38**, 228S–230S.
Rehg, J.E. (1991c). Activity of azithromycin against cryptosporidia in immunosuppressed rats. *Journal of Infectious Diseases* **163**, 1293–1296.
Rehg, J.E. (1993). Anticryptosporidial activity of lasalocid and other ionophorous antibiotics in immunosuppressed rats. *Journal of Infectious Diseases* **168**, 1293–1296.
Rehg, J.E. (1994). A comparison of anticryptosporidial activity of paromomycin with that of other aminoglycosides and azithromycin in immunosuppressed rats. *Journal of Infectious Diseases* **170**, 934–938.
Rehg, J.E. (1996). Effect of diethyldithiocarbamate on *Cryptosporidium parvum* infection in immunosuppressed rats. *Journal of Parasitology* **82**, 158–162.
Ritchie, D.J. and Becker, E.S. (1994). Update on the management of intestinal cryptosporidiosis in AIDS. *Annals of Pharmacotherapy* **28**, 767–778.
Roberts, W.G., Green, P.H.R., Ma, J., Carr, M. and Ginsberg, A.M. (1989). Prevalence of cryptosporidiosis in patients undergoing endoscopy: evidence for an asymptomatic carrier state. *American Journal of Medicine* **87**, 537–539.
Rodrigues, J.L., Leser, P., Silva, T. do M., dos Santos, M.I., Dalboni, M.A., Acceturi, C.A. and Castelo Filho, A. (1991). Prevalência de criptosporidiose na sindrome diarreica do paciente HIV positivo. *AMB; Revista Da Associação Médica Brasileira* **37**, 79–84.
Rohlman, V.C., Kuhls, T.L., Mosier, D.A., Crawford, D.L., Hawkins, D.R., Abrams, V.L. and Greenfield, R.L. (1993). Therapy with atovaquone for *Cryptosporidium parvum* infection in neonatal severe combined immunodeficiency mice. *Journal of Infectious Diseases* **168**, 58–60.
Romeu, J., Miró, J.M., Sirera, G., Mallolas, J., Arnal, J., Valls, M.E., Tortosa, F., Clotet, B. and Foz, M. (1991). Efficacy of octreotide in the management of chronic diarrhoea in AIDS. *AIDS* **5**, 1495–1499.
Sallon, S., el Showwa, R., el Masri M., Khalil, M., Blundell, N. and Hart, C.A. (1991). Cryptosporidiosis in children in Gaza. *Annals of Tropical Paediatrics* **11**, 277–281.
Sanchez-Mejorada, G. and Ponce-de-Leon, S. (1994). Clinical patterns of diarrhea in AIDS: etiology and prognosis. *Revista de Investigación Clínica* **46**, 187–196.
Sarabia-Arce, S., Salazar-Lindo, E., Gilman, R.H., Naranjo, J. and Miranda, E. (1990). Case–control study of *Cryptosporidium parvum* infection in Peruvian children hospitalized for diarrhea: possible association with malnutrition and nosocomial infection. *Pediatric Infectious Disease Journal* **9**, 627–631.
Sauda, F.C., Zamarioli, L.A., Ebner Filho, W. and Mello, L. de B. (1993). Prevalence of *Cryptosporidium* sp. and *Isospora belli* among AIDS patients attending Santos Reference Center for AIDS, São Paulo, Brazil. *Journal of Parasitology* **79**, 454–456.
Scaglia, M., Atzori, C., Marchetti, G., Orso, M., Maserati, R., Ornai, A., Novati, S. and Olliaro, P. (1994). Effectiveness of aminosidine (paromomycin) sulfate in chronic *Cryptosporidium* diarrhea in AIDS patients: an open, uncontrolled, prospective, clinical trial. *Journal of Infectious Diseases* **170**, 1349–1350.
Sears, C.L. and Guerrant, R.L. (1994). Cryptosporidiosis: the complexity of intestinal pathophysiology. *Gastroenterology* **106**, 252–254.
Shrestha, S., Larasson, S., Serchand, J. and Shrestha, S. (1993). Bacterial and cryp-

tosporidial infection as the cause of chronic diarrhoea in patients with liver disease in Nepal. *Tropical Gastroenterology* **14**, 55–58.
Smith, H.V. and Rose, J.B. (1990). Waterborne cryptosporidiosis. *Parasitology Today* **6**, 8–12.
Snijders, F., van Deventer, S.J., Bartelsman, J.F., den Otter, P., Jansen, J., Mevissen, M.L., van Gool, T., Danner, S.A. and Reiss, P. (1995). Diarrhoea in HIV-infected patients: no evidence of cytokine-mediated inflammation in jejunal mucosa. *AIDS* **9**, 367–373.
Soave, R. (1990). Treatment strategies for cryptosporidiosis. *Annals of the New York Academy of Sciences* **616**, 442–451.
Sorvillo, F.J., Fujioka, K., Nahlen, B., Tormey, M.P., Kebabjian, R. and Mascola, L. (1992). Swimming-associated cryptosporidiosis. *American Journal of Public Health* **82**, 742–744.
Sorvillo, F.J., Lieb, L.E., Kerndt, P.R. and Ash, L.R. (1994a) Epidemiology of cryptosporidiosis among persons with acquired immunodeficiency syndrome in Los Angeles County. *American Journal of Tropical Medicine and Hygiene* **51**, 326–331.
Sorvillo, F., Lieb, L.E., Nahlen, B., Miller, J., Mascola, L. and Ash, L.R. (1994b). Municipal drinking water and cryptosporidiosis among persons with AIDS in Los Angeles County. *Epidemiology and Infection* **113**, 313–320.
Stefani, H.N., Levi, G.C., Amato, Neto V., Braz, L.M., Azevedo, H.D., Possa, T.A., Silva, N.F., de Mendonca, J.S. and Fernandes, A.O. (1996). Tratamento da criptosporidiase em pacientes com AIDS, por meio da paromomicina. *Revista da Sociedade Brasileira de Medicina Tropical* **29**, 355–357.
Sterling, C.R., Gilman, R.H., Sinclair, N.V., Cama, V., Castillo, R. and Diaz, F. (1991). The role of breast milk in protecting urban Peruvian children against cryptosporidiosis. *Journal of Protozoology* **38**, 23S–25S.
Tan, W.W., Chapnick, E.K., Abter, E.I., Haddad, S., Zimbalist, E.H. and Lutwick, L.I. (1995). Paromomycin-associated pancreatitis in HIV-related cryptosporidiosis. *Annals of Pharmacotherapy* **29**, 22–24.
Tanyuksel, M., Gun, H. and Doganci, L. (1995). Prevalence of *Cryptosporidium* sp. in patients with neoplasia and diarrhea. *Scandinavian Journal of Infectious Diseases* **27**, 69–70.
Tee, G.H., Moody, A.H., Cooke, A.H. and Chiodini, P.L. (1993). Comparison of techniques for detecting antigens of *Giardia lamblia* and *Cryptosporidium parvum* in faeces. *Journal of Clinical Pathology* **46**, 555–558.
Teixidor, H.S., Godwin, T.A. and Ramirez, E.A. (1991). Cryptosporidiosis of the biliary tract in AIDS. *Radiology* **180**, 51–56.
Theodos, C.M. (1998). Innate and cell-mediated responses to *Cryptosporidium parvum*. *Advances in Parasitology* **40**, 87–119.
Travis, W.D., Schmidt, K., MacLowry, J.D., Masur, H., Condron, K.S. and Fojo, A.T. (1990). Respiratory cryptosporidiosis in a patient with malignant lymphoma. Report of a case and review of the literature. *Archives of Pathology and Laboratory Medicine* **114**, 519–522.
Trevino-Perez, S., Luna-Castanos, G., Matilla-Matilla, A. and Nieto-Cisneros, L. (1995). Diarrhea cronica y *Cryptosporidium* en pacientes diabeticos con subpoblación linfocitaira normal. Informe de dos casos. *Gaceta Médica de México* **131**, 219–222.
Tsaihong, J.C. and Ma, P. (1990). Comparison of an indirect fluorescent antibody test and stool examination for the diagnosis of cryptosporidiosis. *European Journal of Clinical Microbiology and Infectious Diseases* **9**, 770–773.
Tsaihong, J.C., Guiguli, P., Liou, M.Y. and Ma, P. (1990). Cryptosporidiosis in AIDS

and non-AIDS patients. *Chinese Journal of Microbiology and Immunology* **23**, 155–161.
Tzipori, S. (1998). Cryptosporidiosis: laboratory investigations and chemotherapy. *Advances in Parasitology* **40**, 187–221.
Tzipori, S. and Griffiths, J.K. (1998). Natural history and biology of *Cryptosporidium parvum*. *Advances in Parasitology* **40**, 5–36.
Tzipori, S., Robertson, D. and Chapman, C. (1986). Remission of diarrhoea due to cryptosporidiosis in an immunodeficient child treated with hyperimmune bovine colostrum. *British Medical Journal* **293**, 1276–1277.
Tzipori, S., Rand, W., Griffiths, J.K., Widmer, G. and Crabb, J. (1994). Evaluation of an animal model system for cryptosporidiosis: therapeutic efficacy of paromomycin and hyperimmune bovine colostrum-immunoglobulin. *Clinical and Diagnostic Laboratory Immunology* **1**, 450–463.
Tzipori, S., Griffiths, J. and Theodos, C. (1995a). Paromomycin treatment against cryptosporidiosis in patients with AIDS. *Journal of Infectious Diseases* **171**, 1069–1070.
Tzipori, S., Rand, W. and Theodos, C. (1995b). Evaluation of a two-phase scid mouse model preconditioned with anti-interferon-γ monoclonal antibody for drug testing against *Cryptosporidium parvum*. *Journal of Infectious Diseases* **172**, 1160–1164.
Vakil, N.B., Schwartz, S.M., Buggy, B.P., Brummitt, C.F., Kherellah, M., Letzer, D.M., Gilson, I.H. and Jones, P.G. (1996). Biliary cryptosporidiosis in HIV-infected people after the waterborne outbreak of cryptosporidiosis in Milwaukee. *New England Journal of Medicine* **334**, 19–23.
Vargas, S.L., Shenep, J.L., Flynn, P.M., Pui, C.H., Santana, V.M. and Hughes, W.T. (1993). Azithromycin for treatment of severe *Cryptosporidium* diarrhea in two children with cancer. *Journal of Pediatrics* **123**, 154–156.
Verdon, R., Polianski, J., Gaudebout, C., Marche, C., Garry, L. and Pocidalo, J. (1994). Evaluation of curative anticryptosporidial activity of paromomycin in a dexamethasone-treated rat model. *Antimicrobial Agents and Chemotherapy* **38**, 1681–1682.
Verdon, R., Polianski, J., Gaudebout, C. and Pocidalo, J.-J. (1995). Paromomycin for cryptosporidiosis in AIDS. *Journal of Infectious Diseases* **171**, 1070.
Vidal, T., Gamboa, C., Henriquez, M.I. and Biolley, A. (1991). *Cryptosporidium*: brote epidemico en un centro de recuperación nutricional, Temuco. *Revista Medica de Chile* **119**, 1136–1139.
Weikel, C., Lazenby, A., Belitsos, P., McDewitt, M., Fleming, H.E., jr, and Barbacci, M. (1991). Intestinal injury associated with spiramycin therapy of *Cryptosporidium* infection in AIDS. *Journal of Protozoology* **38**, 147S.
Weitz, J.C. and Astorga, B. (1993). *Cryptosporidium parvum* en pacientes con diarrea cronica y SIDA: diagnostico mediante inmunofluorescencía indirecta con anticuerpos monoclonales. *Revista Medica de Chile* **121**, 923–926.
White, A.C., jr, Chappell, C.L., Hayat, C.S., Kimball, K.T., Flanigan, T.P. and Goodgame, R.W. (1994). Paromomycin for cryptosporidiosis in AIDS: a prospective, double-blind, trial. *Journal of Infectious Diseases* **170**, 419–424.
White, A.C., jr, Goodgame, R.W. and Chappell, C.L. (1995). Reply. *Journal of Infectious Diseases* **171**, 1071.
Wittenberg, D.F., Miller, N.M. and Van den Ende, J. (1989). Spiramycin is not effective in treating *Cryptosporidium* in infants: results of a double-blind randomized trial. *Journal of Infectious Diseases* **159**, 131–132.
Wittner, M., Tanowitz, H.B. and Weiss, L.M. (1993). Parasitic infections in AIDS

patients. Cryptosporidiosis, isosporiasis, microsporidiosis, cyclosporiasis. *Infectious Diseases Clinics of North America* **7**, 569–586.

Zerpa, R. and Huicho, L. (1994). Childhood cryptosporidial diarrhea associated with identification of *Cryptosporidium* sp. in the cockroach *Periplaneta americana*. *Pediatric Infectious Disease Journal* **13**, 546–548.

Zu, S.X., Li, J.F., Barrett, L.J., Fayer, R., Shu, S.Y., McAuliffe, J.F., Roche, J.K. and Guerrant, R.L. (1994). Seroepidemiologic study of *Cryptosporidium* infection in children from rural communities of Anhui, China and Fortaleza, Brazil. *American Journal of Tropical Medicine and Hygiene* **51**, 1–10.

Innate and Cell-mediated Immune Responses to *Cryptosporidium parvum*

Cynthia M. Theodos

Division of Infectious Diseases, Tufts University School of Veterinary Medicine, North Grafton, MA 01536, USA

1. Introduction 88
2. Innate Immunity. 89
   2.1. Resistance to infection 89
   2.2. Cytokines 89
   2.3. Cellular components of innate immunity 92
3. *Cryptosporidium parvum*-specific Cell-mediated Immunity 95
   3.1. General evidence for acquired immunity in the resolution of cryptosporidiosis .95
   3.2. The role of T lymphocytes in protective immunity 96
   3.3. The role of cytokines in protective immunity 101
   3.4. Nitric oxide 110
4. Immunotherapy 111
   4.1. Cytokines 111
   4.2. Other modulators of cellular immunity 112
5. Concluding Remarks 113
   References 114

Cryptosporidium parvum *has gained much attention as a major cause of diarrhea in the world. Knowledge of the host immune mechanisms responsible for the clearance of this parasite from the gastrointestinal tract may prove to be vital for successful therapeutic treatment of cryptosporidiosis, particularly in the immunodeficient host. This chapter focuses on the innate and cell-mediated immune mechanisms associated with resistance to and resolution of a* C. parvum *infection. Much of the work in these areas is still in its infancy. Despite this, general consensus supports a role for interferon-γ (IFNγ) in mediating the initial resistance to* C. parvum, *although the mechanism by which this cytokine imparts resistance is unclear. It is also generally agreed that CD4+ T lymphocytes are required for the resolution of both acute and chronic cryptosporidiosis. However, the effector mechanism is again unclear. Several studies suggest that IFNγ may also be involved in the resolution of cryptosporidiosis. However, the extent to which this cytokine is involved in the actual resolution of infection has been debated. Less extensive studies investigating the participation of other cells and cytokines in the innate and cell-mediated immune responses to C. parvum are also discussed.*

1. INTRODUCTION

The increased severity and duration of illness observed in immunocompromised individuals (Petersen, 1992) suggests that resistance to and resolution of a *Cryptosporidium parvum* infection requires the activation of the host immune system. However, our current understanding of the immune response to *C. parvum* remains limited. Advancements in this area have in large part been hampered by the lack of suitable murine models of infection. The existence of genetically inbred murine strains and the abundant availability of immunological reagents for the murine immune system have made these models invaluable for evaluating immune responses to infectious diseases. With the exception of a single report that described persistent infection of Peyer's patches in adult, immunocompetent C57BL/6 mice (Johansen and Sterling, 1994) it has been extremely difficult to establish an acute, detectable infection in immunocompetent mice 3 weeks of age or older (Sherwood *et al.*, 1982; Enriquez and Sterling, 1991). As a consequence, most of what is known about the immune response to *C. parvum* is the result of studies performed in susceptible neonatal mice (Heine *et al.*, 1984; Ernest *et al.*, 1986; Novak and Sterling, 1991), adult congenitally T-cell deficient or T- and B-cell deficient mice (Heine *et al.*, 1984; Ungar *et al.*, 1990, 1991; Mead *et al.*, 1991a,b, 1994; Kuhls *et al.*, 1992; McDonald *et al.*, 1992; Chen *et al.*, 1993a,b; McDonald and Bancroft, 1994; Perryman *et al.*, 1994; Aguirre *et al.*, 1994; Waters and Harp, 1996), immunosuppressed rodents (Rehg *et al.*, 1987, 1988; Brasseur *et al.*, 1988; Rossi *et al.*, 1990; Rasmussen and Healey, 1992a; Rasmussen *et al.*, 1992, 1993; Yang and Healey, 1993, 1994; Petry *et al.*, 1995; Cheng *et al.*, 1996) or mice in which various components of the immune system were depleted by *in vivo* antibody treatment. (Ungar *et al.*, 1990, 1991; McDonald *et al.*, Chen *et al.*, 1993a,b; 1992; McDonald and Bancroft, 1994; Tzipori *et al.*, 1995). One caveat in the use of immunocompromised models is that the precise characterization of parasite-specific acquired immunity is very difficult due to the lack of functional lymphocytes. Nevertheless, studies utilizing these models have provided us with important insights into the host immune mechanisms that may be involved following challenge with *C. parvum*. This chapter will focus on what is known concerning immune mechanisms associated with innate resistance to infection and cell-mediated immunity associated with the resolution of cryptosporidiosis.

2. INNATE IMMUNITY

2.1. Resistance to Infection

As previously stated, it has been difficult to establish a detectable, acute *C. parvum* infection in immunocompetent mice 3 weeks of age or older. While the mechanism(s) responsible for innate resistance to infection is not clearly understood, a study by Harp *et al.* (1988), suggested that resistance may be linked in part to the presence of acquired intestinal microflora. The authors formulated the hypothesis for this study based upon the observation that infant mice spontaneously cleared their *C. parvum* infections at 3 weeks of age (Heine *et al.*, 1984; Novak and Sterling, 1991), the time at which the gastrointestinal (GI) tract becomes colonized with resident microflora (Savage *et al.*, 1968; Davis *et al.*, 1973). Harp and colleagues found that, while *C. parvum* parasites were generally undetectable in 92% of infected conventional CD-1 mice and 80% of infected BALB/c mice, they were readily evident in the gastrointestinal tract of all infected, germ-free CD-1 and BALB/c mice. The authors concluded that resistance to *C. parvum* infection was due to activation of the immune system by existing intestinal microflora.

2.2. Cytokines

2.2.1. Interferon-γ

Determination of the innate immune mechanism(s) responsible for resistance to infection can best be addressed in immunodeficient strains of mice that lack acquired immune systems. Although ultimately susceptible, all adult immunodeficient SCID and nude mice displayed an initial level of resistance to *C. parvum* during the first 3–5 weeks of infection (Ungar *et al.*, 1990; Harp *et al.*, 1992; Kuhls, *et al.*, 1992; McDonald *et al.*, 1992; McDonald and Bancroft, 1994; Mead *et al.*, 1994; Tzipori *et al.*, 1995). As with immunocompetent mice, studies performed in germ-free and conventional C.B-17/Icr Tac-scid mice (Harp *et al.*, 1992) revealed that this resistance was associated with the presence of intestinal microflora. Since SCID mice lack an acquired immune system, the authors concluded that this resistance was due to the activation of an innate immune response, with interferon-γ (IFNγ) production by natural killer (NK) cells as the most likely explanation. This hypothesis was based upon results from a previous study by Ungar *et al.* (1991) in which treatment of *C. parvum*-infected, adult BALB/c nude mice with the IFNγ-neutralizing monoclonal antibody (mAb) XMG-6 significantly enhanced the level of oocyst shedding in treated mice when compared to untreated, *C. parvum*-infected adult BALB/c nude mice. The participation of IFNγ in mediating innate immunity to *C. parvum* was confirmed in several studies in which IFNγ-neutralizing mAbs were administered prior to and after infection (Chen *et al.*, 1993a; Kuhls *et al.*, 1994; McDonald and Bancroft, 1994; Urban *et al.*, 1996). These studies confirmed that an enhanced level of oocyst shedding (McDonald and Bancroft, 1994) and a greater extent of mucosal infection (Chen *et al.*, 1993a; Kuhls *et al.*, 1994 and McDonald and Bancroft, 1994) occurred in anti-IFNγ-treated adult SCID mice. Urban *et al.* (1996) extended the role of IFN-γ in mediating innate resistance to include *C. parvum* infection of neonatal SCID mice.

Enhanced susceptibility of adult SCID mice to infection with *C. parvum* was also achieved with a single injection of an IFNγ-neutralizing mAb. This was first reported by Chen *et al.* (1993a), who demonstrated that a single injection of the IFNγ-specific neutralizing mAb R46A2 either 2 hours before or 18 hours after infection enhanced the extent of mucosal infection in adult SCID mice. This observation was extended by Tzipori *et al.* (1995). A single injection of 1 mg of the IFNγ-neutralizing mAb XMG1.2 2 hours prior to infection shortened the prepatent period and significantly increased the level of oocyst shedding ($P<0.001$ by two-way ANOVA) when compared to SCID mice treated with an isotype-matched, anti-β-galactosidase control mAb (Figure 1). The extent of mucosal infection was also significantly enhanced ($P<0.0001$ by one-way ANOVA, data not shown).

In a study by Chen *et al.* (1993a), splenocytes isolated from *C. parvum*-infected or uninfected SCID mice secreted IFNγ following *ex vivo* restimulation with a soluble parasite antigen. The cellular source of this cytokine, and the mechanism for production by splenocytes from infected and uninfected SCID mice were not directly investigated. Nevertheless, the data discussed in this section collectively support a role for IFNγ in mediating innate resistance to *C. parvum*.

2.2.2. *Other Cytokines*

(a) *Interleukin-2* Interleukin-2 (IL-2) can induce NK-cell proliferation, cytotoxic activity and IFNγ secretion (Kawase *et al.*, 1983; Baume *et al.*, 1992). To investigate the role of this cytokine in the induction of IFNγ-mediated innate resistance, Ungar *et al.* (1991) treated adult BALB/c mice with the neutralizing anti-IL-2 mAb S4B6 or anti-IL-2 plus the anti-IL-2 receptor mAb PC-61 prior to and after oral challenge with *C. parvum*. Neither treatment protocol enhanced the susceptibility of BALB/c mice to infection. Moreover, treatment with anti-IL-2 mAb plus the anti-IFNγ mAb XMG 6 did not enhance the severity of infection over that observed in BALB/c mice treated with the anti-IFNγ mAb alone. These results suggested that IL-2 may not be involved in the induction of innate resistance to *C. parvum*.

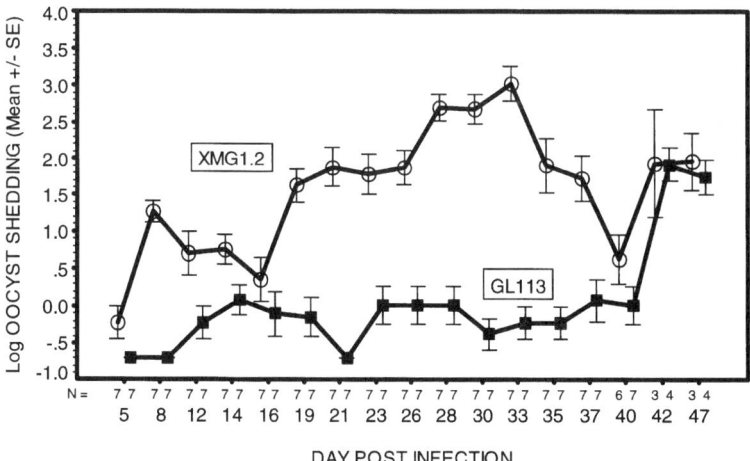

Figure 1 Oocyst shedding from *C. parvum*-infected SCID mice. Following randomization, 3-week-old male SCID mice received a single intraperitoneal injection of 1 mg anti-IFNγ mAb XMG 1.2 (○, n = 7) or 1 mg isotype-matched control mAb GL113 (■, n=7). Two hours later, all mice were inoculated orally with 1×10^7 oocysts of the GCH1 isolate. The number of oocysts in 30 high power fields of a Kinyoun carbol fuchsin stained fecal smear was determined for each mouse. Results are presented as the mean number of oocysts shed per group ± the standard error (SE).

(b) *Interleukin-12* Produced by macrophages, dendritic cells and B cells, interleukin-12 (IL-12) induces IFNγ secretion from NK cells and T cells (Wolf *et al.*, 1994). Interleukin-12 is an important mediator of IFNγ-dependent innate resistance to a number of infectious microorganisms, including the Apicomplexan protozoan parasite *Toxoplasma gondii* (Gazinelli *et al.*, 1993; Hunter *et al.*, 1994; Khan *et al.*, 1994). This cytokine can also modulate antigen-specific acquired immunity by mediating preferential induction of CD4+ T helper 1 (Th1) cells (Sypek *et al.*, 1993; Wolf *et al.*, 1994). Results from a study by Chehimi *et al.* (1994), suggested that the defect in immune status of individuals infected with the human immunodeficiency virus (HIV) may be linked to a decreased production of IL-12. Recently, Urban *et al.* (1996) investigated the role of IL-12 in modulating the susceptibility to *C. parvum* infection in immunocompetent and immunodeficient mice. The administration of a polyclonal anti-IL-12 antibody prior to, and after oral challenge with *C. parvum* significantly enhanced mucosal infection in neonatal BALB/c and C.B-17 SCID mice. These data suggest that endogenously produced IL-12 participates in the innate resistance to *C. parvum* infection observed in both immunocompetent and immunodeficient mice.

(c) *Tumor necrosis factor α* Tumor necrosis factor (TNFα) is a cytokine involved in protective immunity to many pathogenic microorganisms

(Vassalli, 1992). One mechanism by which TNFα participates in antimicrobial immune responses is by augmenting the secretion of IFNγ from NK cells. In 1991, Wherry *et al.* demonstrated that T-cell-independent immunity to *Listeria monocytogenes* relied upon TNFα-induced IFNγ secretion from NK cells. A study by Hunter *et al.* (1994) revealed that T-cell-independent resistance to *Toxoplasma gondii* observed in SCID mice was in part mediated by the TNFα-augmented production of IFNγ by NK cells. Moreover, in this system, TNFα synergized with IL-12 and significantly enhanced the *in vitro* secretion of IFNγ from SCID spleen cells. The role of TNFα in establishing innate immunity to *C. parvum* was investigated by evaluating the *in vivo* production of TNFα. Administration of the TNFα neutralizing mAb TN3-19.12 (McDonald and Bancroft, 1994) or a rabbit polyclonal anti-TNFα antibody preparation (Chen *et al.*, 1993a) prior to and during infection failed to enhance the susceptibility of SCID mice to *C. parvum*. Further indirect evidence to refute a role for TNFα in the immune response to *C. parvum* was provided in a recent study by Wyatt *et al.* (1997). In this study, *ex vivo* reverse transcriptase PCR analysis revealed a decrease in TNFα-specific mRNA in intraepithelial lymphocytes (IELs) isolated from *C. parvum*-infected calves when compared to uninfected control calves. Collectively, the data presented in this section suggest that TNFα was not an important mediator of innate resistance to *C. parvum*.

2.3. Cellular Components of Innate Immunity

Several studies have investigated the role of various cells of the immune system in establishing innate resistance to *C. parvum*. Most of the work to date has focused on the role of NK cells, γδ+ T cells, and mast cells.

2.3.1. General Evidence

The adoptive transfer of naive, immunocompetent BALB/c spleen cells by intraperitoneal injection completely protected adult SCID mice when challenged with *C. parvum* oocysts 24 hours later (McDonald *et al.*, 1992). These results suggested that lymphoid cells may play a role in the establishment of resistance to *C. parvum*.

In contrast to the results of McDonald *et al.* (1992), the adoptive transfer of immune spleen cells 7 days prior to oral challenge with *C. parvum* could not prevent the establishment of infection in SCID mice (Tatalick and Perryman, 1995). In a study by Harp and Whitmire (1991), spleen or mesenteric lymph node cells isolated from adult BALB/c mice that cleared a *C. parvum* infection were adoptively transferred into infant mice by intraperitoneal injection either (a) immediately upon removal or (b) following *in vitro* cultivation with concanavalin A for 72 hours. One day later, the infant mice were orally challenged with oocysts. Results indicated that the adoptive transfer of immune cells prior to challenge did not significantly reduce either the number of mice infected with *C. parvum* or the extent of infection.

2.3.2. *Natural Killer Cells*

NK cells are widely accepted as important components of innate resistance to a number of infectious microorganisms (Hunter, 1996). As discussed above, the absence of T cells in SCID and nude mice led to the hypothesis that NK cells were responsible for the production of the IFNγ that mediated innate resistance to *C. parvum*. To address this hypothesis, two separate approaches were taken. The first approach involved determination of susceptibility to *C. parvum* infection using mice homozygous for the beige mutation (Rasmussen and Healey, 1992a; Enriquez and Sterling, 1991; Mead *et al.*, 1991a). Results from these studies were varied. A study by Rasmussen and Healey (1992a) revealed that adult C3H/HeJ/beige mice were not susceptible to infection. However, an investigation conducted by Enriquez and Sterling (1991) detected a slight increased susceptibility in adult C57BL/6J-bg^J on days 7 and 14 post-infection as determined by oocyst shedding and histological analysis. Finally, NIH-III mice (*bg/bg, nu/nu, xid/xid*) examined by Mead *et al.* (1991a) were found to have more severe (lethal) cryptosporidial infections than did SCID mice. One caveat to this approach is that mice bearing the beige mutation are not truly 'deficient' in NK cells. While the NK cells in beige mice have decreased cytotoxic activity (Roder *et al.*, 1979), they were still able to secrete IFNγ following activation (Kawase *et al.*, 1983). It is possible, therefore, that IFNγ secretion from NK cells, even at a potentially reduced level, may be responsible for the varied results obtained using *C. parvum*-infected mice bearing the beige mutation.

A second approach taken to study the role of NK cells during an immune response to *C. parvum* involved the *in vivo* depletion of NK cells by administration of an antibody directed against asialo-GM1. Treatment of nude (Ungar *et al.*, 1991), SCID (Rohlman *et al.*, 1993; McDonald and Bancroft, 1994) and BALB/c (Rohlman *et al.*, 1993) mice with this antibody did not enhance their level of *C. parvum* infection. While splenic NK cells were readily depleted by this protocol (Rohlman *et al.*, 1993), depletion of NK cells in the intestinal mucosa was not assessed in any of these studies. One limitation to functional studies that rely on *in vivo* antibody depletion is that the interpretation of the data can be complicated by the possibility that some activity may remain due to incomplete depletion by the antibody. With respect to the depletion of NK cells by anti-asialo GM1 treatment, there is precedence in the literature indicating that this protocol may result in the incomplete depletion of NK cells, particularly in the intestinal mucosa. A study by Tagliabue *et al.* (1982) revealed that the activity of splenic NK cells was readily depleted by anti-asialo GM1 treatment, while NK-cell cytotoxicity of IELs from the intestinal mucosa remained largely unaffected. In a study by Chen *et al.* (1993b), *C. parvum*-infected SCID mice reconstituted with naive, immunocompetent spleen cells were able to resolve their infection despite treatment with a polyclonal anti-asialo-GM1 antibody. Analysis of the mesenteric lymph nodes of these mice revealed that a significant, albeit reduced, number of asialo-GM1+ cells were still present. In a study by Hunter *et al.* (1994), the incomplete depletion of NK cells following treatment with anti-asialo GM1 resulted in enough residual IFN-γ activity to mediate resistance to *T. gondii* infection in SCID mice.

Finally, a study by Rasmussen and Healey (1992b) demonstrated that splenic NK-cell activity was decreased in aged Syrian hamsters that displayed enhanced susceptibility to infection with *C. parvum*. It should be noted however, that decreased B- and T-lymphocyte responses were also observed in these animals (see below). In summary, the data presented in this section do not uniformly support or refute the participation of NK cells in the immune response to *C. parvum*. It would appear that further studies are necessary in order to define better the contribution of NK cells.

2.3.3. γδ *T Cells*

Our appreciation of γδ+ T cells as contributors of early innate resistance to microbial infections has grown as we have come to understand better this unique population of cells (Chien *et al.*, 1996). One way in which γδ+ T cells may contribute to innate immunity is through the production of cytokines. In this regard, activated γδ+ T cells were found to secrete a variety of different cytokines, including IFNγ (Haas *et al.*, 1993). In 1995, a study by Harp *et al.* examined the phenotype of cells present in various lymphoid tissues from *C. parvum*-infected calves. One striking difference between uninfected and infected animals was the marked increase in the number of cells expressing the null (N2) antigen. Since this antigen was found on a subset of bovine γδ+ cells, the authors concluded that these cells may be important in immunity to *C. parvum*. The contribution of γδ+ T cells in the immune response to *C. parvum* was directly investigated in a study by Waters and Harp (1996). In this study, γδ+ T-cell-deficient mice were examined for their susceptibility to infection with *C. parvum*. When infected as neonates, TCR-γδ-deficient mice displayed an enhanced level of infection, but ultimately cleared *C. parvum* from the gastrointestinal tract by 7 weeks of age (6 weeks after challenge). When infected as adults (8 weeks old), roughly 20% of these mice shed oocysts for 1 week after challenge. Based upon these results, the authors concluded that γδ T cells may be involved in the initial resistance to *C. parvum*.

2.3.4 Mast Cells

The ability of mast cells to secrete a variety of cytokines following activation has increased the interest in these cells as possible mediators of innate immunity to infectious microorganisms (Wershil et al., 1994; Galli and Wershil, 1995). The role of mast cells in the establishment of innate resistance to C. parvum was investigated by Harp and Moon (1991) using mast-cell-deficient W/W^v mice on a C57BL/6J background. In this study, no difference in susceptibility to infection was observed between W/W^v mice and their mast cell bearing littermates when infected at 1 week of age. In contrast, W/W^v mice were found to be slightly more susceptible than control mice when infected at 7 weeks of age. At 1 week of infection, a two-fold increase in the infectivity score was observed in the gastrointestinal tract of C. parvum-infected, adult W/W^v mast-cell-deficient mice when compared to control mice. Parasites were not evident in the gastrointestinal tract of either group by 2 weeks of infection. These results suggested a role for mast cells in the innate resistance of adult mice to C. parvum. However, direct proof of this was not provided since mast cells were not detected at the site of infection in control mice.

3. CRYPTOSPORIDIUM PARVUM-SPECIFIC CELL-MEDIATED IMMUNITY

3.1. General Evidence for Acquired Immunity in the Resolution of Cryptosporidiosis

The progressive course of infection observed in individuals with AIDS (Petersen, 1992), immunocompromised SCID mice (Mead et al., 1991a, 1994; Harp et al., 1992; Kuhls et al., 1992; McDonald et al., 1992; McDonald and Bancroft, 1994; Tzipori et al., 1995), foals with SCID (Bjorneby et al., 1991), nude mice (Sherwood et al., 1982; Heine et al., 1984, Ungar et al., 1990; Mead et al., 1991a), nude rats (Gardner et al., 1991), NIH-III mice (Mead et al., 1991a), immunosuppressed rodents (Rehg et al., 1987, 1988; Brasseur et al., 1988; Rossi et al., 1990; Rasmussen and Healey, 1992a; Rasmussen et al., 1992, 1993; Yang and Healey, 1993, 1994; Petry et al., 1995; Cheng et al., 1996), and mice infected with the murine retrovirus LP-BM5 (Darban et al., 1991; Alak et al., 1993) indicated that resolution of a C. parvum infection required the participation of the host acquired immune system. This was given indirect support in a study by Marcial and Madara (1986). In this study, transmission electron micrographs revealed the presence of Cryptosporidium parasites within the cytoplasm of mature and immature Peyer's patch M cells from spontaneously infected guinea pigs. Intact and partially digested parasites were also present with M-cell-associated macrophages, a cell that functions as an antigen presenting cell for T-lymphocyte activation. More recently, an *in vitro* study by Martinez *et al.* (1992) demonstrated that both *C. parvum* oocysts and sporozoites infected murine peritoneal exudate-derived macrophages. Collectively, these results indirectly suggest that *C. parvum* parasites can be taken up by the mucosal immune system and that they may be processed by antigen presenting cells for T-cell activation.

3.1.1. Adoptive Transfer of Naive Immunocompetent Cells

Participation of acquired immunity has been confirmed by numerous experiments in which clearance of *C. parvum* from the gastrointestinal tract was achieved following the adoptive transfer of naive lymphoid cells into genetically immunodeficient mouse strains. The adoptive transfer of naive, immunocompetent spleen cells by intravenous injection resulted in resolution of acute (Chen *et al.*, 1993b), and chronic (McDonald and Bancroft, 1994) cryptosporidiosis in adult SCID mice. Resolution of acute cryptosporidiosis was also achieved by the transfer of a combination of naive, immunocompetent bone marrow cells, thymocytes, and spleen cells into *C. parvum*-infected SCID mice (Mead *et al.*, 1991b). This observation was later extended by Perryman *et al.* (1994), who demonstrated that the adoptive transfer of either naive bone marrow cells, thymocytes, or splenocytes ameliorated acute cryptosporidiosis in *C. parvum*-infected, adult SCID mice. Collectively, these data support a role for lymphoid cells in the resolution of both acute and chronic *C. parvum* infections.

3.1.2. Adoptive Transfer of Immune Cells

In addition to naive immunocompetent cells, 'immune' lymphoid cells isolated from mice that resolved a cryptosporidial infection also conferred protection in immunodeficient mice. In a study by Ungar *et al.* (1990), adult BALB/c nude mice infected with *C. parvum* for 1 month cleared their infection following the intravenous adoptive transfer of spleen and mesenteric lymph node (MLN) cells from adult immunocompetent BALB/c mice that resolved a neonatal infection with *C. parvum*.

3.2. The Role of T Lymphocytes in Protective Immunity

3.2.1. General Evidence

As previously stated, the inability of immunocompromised SCID (Mead *et al.*, 1991a, 1994; Harp *et al.*, 1992; Kuhls *et al.*, 1992; McDonald *et al.*, 1992;

McDonald and Bancroft, 1994; Tzipori et al., 1995) and nude (Heine et al., 1984, Mead et al., 1991a) mice to eliminate C. parvum from the gastrointestinal tract indicated that B and/or T lymphocytes were required for parasite clearance. This theory was supported in a study by Rasmussen and Healey (1992b), wherein the increased susceptibility of aged Syrian hamsters to C. parvum infection was associated with a decreased T- and B-lymphocyte mitogenic proliferative responses. General support for the requirement of αβ TCR+ T cells in cryptosporidial-specific protective immunity was offered by Waters and Harp (1996). In this study, the authors analyzed C. parvum infections in mice with a disrupted TCR-constant-α gene, which rendered them deficient in αβ TCR+ T cells. When infected as neonates, TCR-αβ deficient mice developed chronic C. parvum infections. When infected as adults, 60–70% of αβ-TCR deficient mice displayed patent infections when the experiment was terminated 7 weeks after challenge. From this study, the authors concluded that αβ TCR$^+$ T cells were required for resolution of a C. parvum infection.

3.2.2. CD4+ T Cells

The identification of cryptosporidiosis as one of the opportunistic infections of individuals with AIDS suggested that immunity to this parasite required CD4+ T cells. Data to support this theory were provided by Blanshard et al. (1992). In this study, the authors described a spectrum of cryptosporidiosis in patients with acquired immunodeficiency syndrome (AIDS) ranging from asymptomatic to fulminant disease. Analysis of the CD4+ T cell count revealed that fulminant disease only occurred in individuals with a CD4 cell count of less than 50 cells/mm^3. Individuals with a CD4+ T cell count of 200 cells/mm^3 or more displayed only transient disease. These results were in accordance with those from several other studies. In 1992, a study by Flannigan et al. reported that individuals with a CD4+ T cell count of 180 cells/mm^3 displayed self-limited disease, while individuals with lower CD4+ T cells counts (mean of 57 cells/mm^3) had protracted cryptosporidiosis. A study by Vakil et al. (1996) analyzed the incidence of biliary disease and rate of mortality due to cryptosporidiosis in HIV-infected individuals following a water-borne outbreak in Milwaukee, Wisconsin. Results indicated that individuals with a CD4+ T cell count of < 50 cells/mm^3 had an increased incidence of biliary disease, an indicator of chronic cryptosporidiosis, and a decreased survival time when compared to those individuals with higher CD4+ T cell counts. While the studies presented in the preceding paragraph provided evidence linking CD4+ T cells counts with the severity of cryptosporidiosis, it should be noted that there are several reports which failed to demonstrate such a correlation (Genta et al., 1994; Molbak et al., 1994). Therefore, many researchers turned to experimental models of cryptosporidiosis to address this question.

Indirect evidence for the activation of CD4+ T cells during cryptosporidiosis was provided in 1991 by Gardner et al., who demonstrated a positive delayed-type hypersensitivity response to cryptosporidial antigens in rats that had recovered from a *C. parvum* infection. More direct evidence for the contribution of CD4+ T cells in the immune response to *C. parvum* was initially provided in a series of studies by Ungar et al. (1990, 1991). In the first study (Ungar et al., 1990), treatment with the anti-CD8 mAb 2.43 or normal rat immunoglobulin G (IgG) beginning on day 5 of age did not alter the ability of neonatally infected BALB/c mice to resolve their *C. parvum* infections when compared to unmanipulated control mice. In contrast, the administration of the anti-CD4 mAb GK 1.5 or the co-administration of GK 1.5 and 2.43 mAbs abrogated the ability of these mice to clear this parasite. In a subsequent study, Ungar et al. (1991) demonstrated that adult immunocompetent BALB/c mice developed chronic *C. parvum* infections following the administration of GK 1.5 or both GK 1.5 and 2.43 mAbs. In this study, the antibody injections were administered one day before *C. parvum* challenge and weekly thereafter.

The role of CD4+ T cells in an immune response to *C. parvum* was also investigated using SCID mice that received an adoptive transfer of lymphoid cells depleted of, or enriched for, CD4+ T cells by the *in vitro* or *in vivo* treatment with anti-CD4 or anti-CD8 mAbs. Using this approach, CD4+ T cells were found to mediate the resolution of both acute and chronic *C. parvum* infections. With respect to acute cryptosporidiosis, a study by Perryman et al. (1994) demonstrated that the adoptive transfer of naive, immunocompetent spleen cells enriched for CD4+ T cells into adult SCID mice 2 weeks after infection with *C. parvum* significantly lowered mucosal infection scores and the level of oocyst shedding when compared to infected, unreconstituted SCID mice. A similar conclusion was drawn by Chen et al. (1993b). As previously mentioned, these authors demonstrated that the reconstitution of SCID mice with naive, immunocompetent BALB/c spleen cells mediated resolution of an acute *C. parvum* infection. This resolution of infection was abrogated by the *in vivo* administration of the mAb GK 1.5 to deplete the CD4+ T cells following reconstitution.

Experimental models have also demonstrated that CD4+ T cells were involved in the resolution of chronic cryptosporidiosis. In a study by McDonald and Bancroft (1994), naive, immunocompetent BALB/c spleen cells were treated with either the anti-CD4 mAb YTS 191.1 or the anti-CD8 mAb YTS 169.4 prior to intravenous adoptive transfer into adult SCID mice that were infected with *C. parvum* for 8 weeks. This protocol resulted in the elimination of CD4+ or CD8+ T cells by *in vivo* opsonization. Depletion of CD4+ T cells resulted in a persistent and unaltered level of oocyst shedding when compared to infected, nonreconstituted SCID mice. In contrast, *C. parvum*-infected adult SCID mice reconstituted with unfractionated splenocytes, or splenocytes depleted of CD8+ T cells *in vivo*, cleared the parasite by day 21 after reconstitution. In a study by Tatalick and Perryman (1995), the adoptive transfer of CD4+ T cells, with or without B cells, from naive or *C. parvum*-immunized mice enhanced parasite clearance from persistently infected SCID mice.

The participation of CD4+ T cells in *C. parvum* protective immunity was also investigated using mice genetically deficient in major histrocompatibility complex (MHC) class II proteins (Aguirre *et al.*, 1994). These mice fail to develop CD4+ T cells as a consequence of their deficiency. In this study, MHC class II-deficient C57BL/6J neonatal and adult mice displayed enhanced *C. parvum* infections, as measured by the extent of mucosal infection at 4 (neonatal and adult) and 8 (neonatal) weeks after infection.

While most of the studies discussed above used splenocytes, two reports addressed the role of CD4+ IELs in the immune response to *Cryptosporidium*. In a study by Wyatt *et al.* (1997), calves acutely infected with *C. parvum* had an increased number of CD4+ IEL expressing CD25 (the IL-2Rα chain) when compared to uninfected control calves. This result suggested the presence of activated CD4+ lymphocytes within the IEL compartment of infected calves. A study by McDonald *et al.* (1996) demonstrated that the adoptive transfer of IEL from *C. muris*-immune BALB/c mice resulted in parasite clearance from infected SCID mice. This protective effect was abrogated by the depletion of CD4+ cells from the IEL preparation prior to adoptive transfer.

The results presented in this section collectively support a role for CD4+ T cells in the clearance of both acute and chronic cryptosporidiosis. At the present time, however, the primary effector function of these cells and, more specifically, the relative contribution of the Th1 and Th2 subsets of CD4+ T cells to *C. parvum* protective immunity remains to be determined.

3.2.3. *CD8+ T Cells*

The role of CD8+ T lymphocytes in the resolution of a *C. parvum* infection is largely unresolved. Studies designed to address this question utilized the same approaches outlined above for CD4+ T cells and were typically done in parallel with those studies. In general, most data supported the view that CD8+ T cells were not primary participants in the immune response to *C. parvum*. Administration of the anti-CD8 mAb 2.43 to neonatal BALB/c mice did not inhibit the resolution of a *C. parvum* infection (Ungar *et al.*, 1990). Moreover, treatment of adult BALB/c mice with the mAb 2.43 had no effect on their relative resistance to infection (Ungar *et al.*, 1991). In a study by Chen *et al.* (1993b), administration of the anti-CD8 mAb TIB-210 had no effect on the resolution of a *C. parvum* infection in SCID mice that received an intravenous adoptive transfer of naive, immunocompetent spleen cells on day 25 of infection. In contrast to the results presented above for MHC class II-deficient mice, neonatal and adult MHC class I-deficient C57BL/6J mice were no more susceptible to infection than age-matched control mice (Aguirre *et al.*, 1994).

While the results discussed in the preceding paragraph suggest that CD8+ T cells were not involved in cryptosporidial immunity, a few studies have supported a protective, albeit minor role for these cells. In a study by Perryman *et al.* (1994) naive, immunocompetent spleen cells enriched for CD8+ T cells *in vitro* were adoptively transferred into adult SCID mice infected with *C. parvum* for 2 weeks. This treatment significantly lowered the mucosal infection score and fecal oocyst shedding of these mice when compared to infected, unreconstituted SCID mice. However, this protective effect was less pronounced than that obtained from SCID mice that received spleen cells enriched for CD4+ T cells. One caveat was that the authors were unable to demonstrate substantial engraftment of CD8+ T cells in the spleens of reconstituted SCID mice. It was possible, as the authors stated, that the CD8+ T cells migrated to the mucosal immune system, and therefore escaped detection in the spleen. However, it was equally possible that reconstitution was low, and therefore the full contribution of CD8+ T cells to *C. parvum*-specific immunity could not be assessed. In a study by McDonald and Bancroft (1994), the *in vivo* depletion of CD8+ T from adoptively transferred, naive immunocompetent BALB/c splenocytes did not inhibit the clearance of *C. parvum* from the gastrointestinal tract of infected SCID mice. However, the rate of clearance was slower than that observed in SCID mice that received unfractionated spleen cells. This suggested a small, but nevertheless noticeable, contribution of CD8+ T cells to the clearance of *C. parvum*. Similar results were obtained with *C. muris* infected mice (McDonald *et al.*, 1996). In this study, the adoptive transfer of IELs depleted of CD8+ cells slowed the resolution of *C. muris* infection in SCID mice when compared to mice that received unfractionated IELs.

A few studies have examined CD8+ T cells in *C. parvum*-infected calves. An increased number of CD8+ IELs was observed in calves acutely infected with *C. parvum* (Wyatt *et al.*, 1997). This observation indirectly suggested the activation of these cells during cryptosporidiosis. Similar results were obtained by Harp *et al.* (1995), who demonstrated an increased percentage of CD8+ T lymphocytes in the spleens of *C. parvum*-infected calves. Whether these cells are actually contributing to *C. parvum* protective immunity is unclear at present. In summary, the varied results obtained from the studies presented in this section indicate that the role of CD8+ T cells in the immune response to *C. parvum* still needs to be defined.

3.3. The Role of Cytokines in Protective Immunity

3.3.1. *Interferon-γ*

One effector mechanism by which T cells participate in protective immunity is through the secretion of cytokines. With respect to *C. parvum*, most research efforts have focused on the role of IFNγ. As discussed above, IFNγ was shown to be important in the establishment of innate resistance to infection. The role of this cytokine in the resolution of infection is less clear. Two reports in the literature, one using *C. parvum* (Ungar *et al.*, 1991) and a second using *C. muris* (McDonald *et al.*, 1992), suggested that IFNγ had limited efficacy in the resolution of cryptosporidiosis. In the study by Ungar *et al.* (1991), weekly administration of the neutralizing IFNγ-specific mAb XMG 6, beginning the day before infection, enhanced the level of oocyst shedding from *C. parvum*-infected adult BALB/c mice, but the infection remained self-limited. Similar results were obtained by McDonald *et al.* (1992) using C57BL/6 and BALB/c mice infected with *C. muris*. Based upon these results, the authors of both studies concluded that clearance of a *Cryptosporidium* infection occurred via IFNγ-independent effector mechanisms.

To delineate the role of CD4+ T cells and IFNγ in *C. parvum* protective immunity, Ungar *et al.* (1991) initiated a series of experiments in which both anti-CD4 and anti-IFNγ mAbs were administered to *C. parvum*-infected adult immunocompetent BALB/c mice by various treatment schedules. The general outcome of these studies indicated that co-administration of anti-CD4 and anti-IFNγ mAbs resulted in a chronic *C. parvum* infection, with high numbers of oocysts shed in the feces. Removal of the anti-IFNγ mAb from the treatment schedule at any time point reduced the number of oocysts shed. Maintenance of chronic oocyst shedding required treatment with anti-CD4, confirming the importance of CD4+ T cells in the clearance of this parasite. However, mice treated with multiple injections of a neutralizing IFNγ-specific antibody still cleared their infections. From this result, the authors concluded that IFNγ limited the severity of infection but was not responsible for the ultimate resolution.

In contrast to the studies presented above, several reports have indicated a role for IFNγ in the resolution of a *C. parvum* infection. In a study by Chen *et al.* (1993b) the administration of the neutralizing IFNγ-specific mAb R46A2 to *C. parvum*-infected SCID mice reconstituted with immunocompetent spleen cells abrogated the resolution of infection imparted by the adoptively transferred cells. Recent results from our laboratory indicated that the extent of IFNγ neutralization was important for determining infection outcome (Theodos *et al.*, 1997). In this study, IFNγ activity was temporarily neutralized in 4- to 6-week old C57BL/6J mice by a single intraperitoneal injection of the murine IFNγ-specific mAb XMG 1.2 at 2 hours prior to oral challenge with *C. parvum* oocysts. Oocyst shedding was detectable in all XMG 1.2–treated mice by day 5 of infection (Figure 2). Two peaks of shedding were typically observed; the first peak occurred between days 7 and 10 of infection and the second peak around days 15 and 16 of infection. Oocysts were undetectable in the feces of all XMG 1.2–treated mice by day 25 of infection. Control mice treated with 1 mg normal rat IgG prior to challenge failed to develop patent *C. parvum* infections.

In contrast to temporary IFN-γ neutralization, the complete lack of this cytokine in mice with a targeted disruption of the IFNγ gene (gene knockout (GKO), 97% homologous to C57BL/6) resulted in uncontrolled *C. parvum* infections (Theodos *et al.*, 1997). Infected GKO mice began shedding oocysts by days 4–5 of infection, with extremely high levels of shedding evident by days 8–11 of infection (Figure 3). These mice generally succumbed within 2–3 weeks, with acute unresolved cryptosporidiosis.

Our oocyst shedding results were paralleled by the level of mucosal infection. Overwhelming infection and profound mucosal destruction of the gastrointestinal tract was observed in GKO mice infected with *C. parvum* for 2 weeks. Unlike most rodent models (Rehg *et al.*, 1988; Mead *et al.*, 1991a), the entire small intestine of GKO mice became infected with multiple parasite forms per cell. Hepatobiliary tract infection, which ultimately leads to death in chronically infected SCIDs (Mead *et al.*, 1991a, 1994) was not observed in infected GKO mice.

The small intestine was also the primary site of infection in XMG 1.2-treated C57BL/6J mice. Although extensive infection and changes in the mucosa were observed on days 8 and 17 of infection, the overall degree of both was less than that observed in infected GKO mice. Parasitization subsided considerably by day 25 of infection. The mucosa was free of parasites and appeared essentially normal by day 30 of infection.

As discussed above, the use of antibodies to neutralize the activity of a cytokine *in vivo* cannot insure that complete neutralization has occurred. In the studies by Ungar *et al.* (1991) and McDonald *et al.* (1992), it is possible that IFNγ activity in the gastrointestinal mucosa was incompletely neutralized, particularly in those mice in which CD4+ T cells were functional and may have been secreting IFNγ as a consequence of the infection. The residual IFNγ activity may have been sufficient to contribute to the clearance of this parasite from the gastrointestinal tract. As previously mentioned, residual IFNγ activity due to incomplete NK-cell depletion was sufficient to mediate resistance to *T. gondii* infection in SCID mice (Hunter *et al.*, 1994). Our results suggested that the extent of neutralization of IFNγ activity can determine the outcome of a *C. parvum* infection. Moreover, results obtained with *C. parvum*-infected GKO mice supported the involvement of IFNγ, either directly or indirectly, in determining both the severity and duration of a *C. parvum* infection. The inability of GKO mice to resolve cryptosporidial infections further suggested

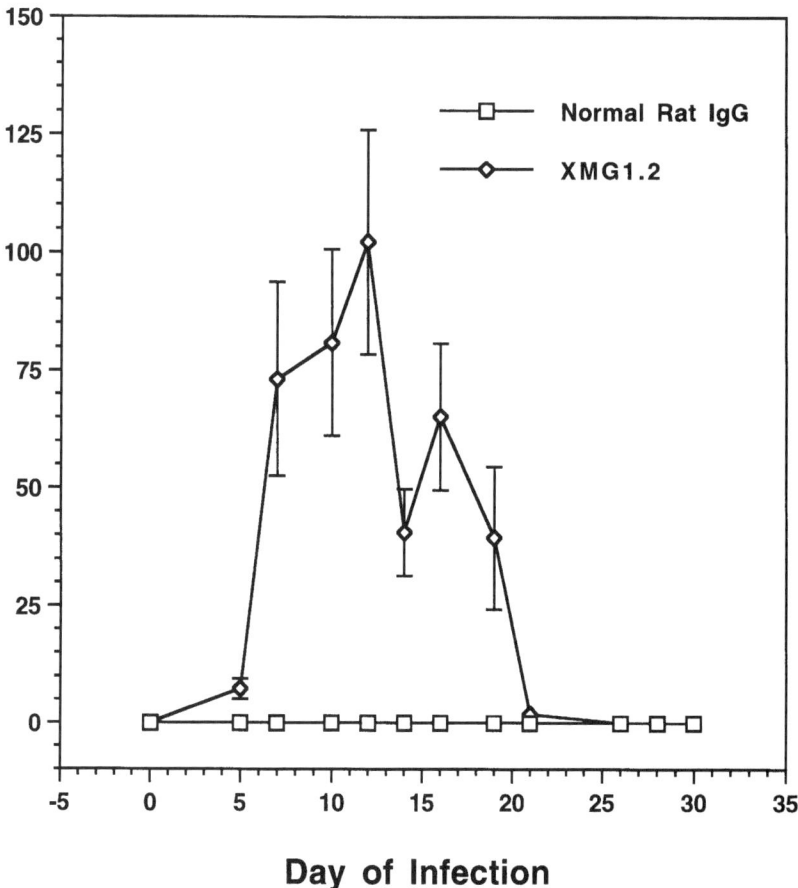

Figure 2 Oocyst shedding from *C. parvum*-infected C57BL/6J mice. Four- to 6-week-old female C57BL/6J mice received a single injection of 1 mg XMG 1.2 (*n* = 10) or normal rat IgG (*n* = 10) 2 hours prior to oral inoculation with 1×10^7 oocysts of the GCH1 isolate. Oocyst shedding was determined as described in Figure 1. Results are presented as the mean number of oocysts shed per group ± the standard error (SE).

that, in the absence of IFNγ, CD4+ T cells were not able to eliminate this parasite.

The association of IFNγ with the clearance of *C. parvum* received indirect support from several studies. In a study by Gomez Morales *et al.* (1996), the peripheral blood mononuclear cells (PBMCs) from a 2-year-old child recovered from cryptosporidiosis secreted IFNγ and IL-10 when restimulated *in vitro* with a crude antigen extract of *C. parvum* oocysts. In contrast, PBMCs obtained from a 1-month-old infant chronically infected with *C. parvum* secreted IL-10, but not IFNγ, following *in vitro* restimulation. Kapel *et al.*

(1996) examined the kinetics of IFNγ production in the ileum of neonatal NMR1 mice infected with *C. parvum*. IFNγ was detected in ileal homogenates by day 3 after challenge; peak levels occurred on day 6, and then declined on days 9 and 13 after challenge. This kinetics of IFNγ production directly correlated with the presence of *C. parvum* parasites in the homogenates. Parasites were first detected at very low levels on day 3 after challenge, peaked on day 6, and declined thereafter. These results indirectly suggested that IFNγ may be involved in the clearance of *C. parvum* from the gastrointestinal tract. However, the presence of other cytokines in the ileal homogenates was not assessed. Finally, a study by Urban *et al.* (1996) demonstrated an increase in mRNA specific for IFNγ in the mesenteric lymph nodes and ileum of neonatal BALB/c mice infected with *C. parvum* for 3 days.

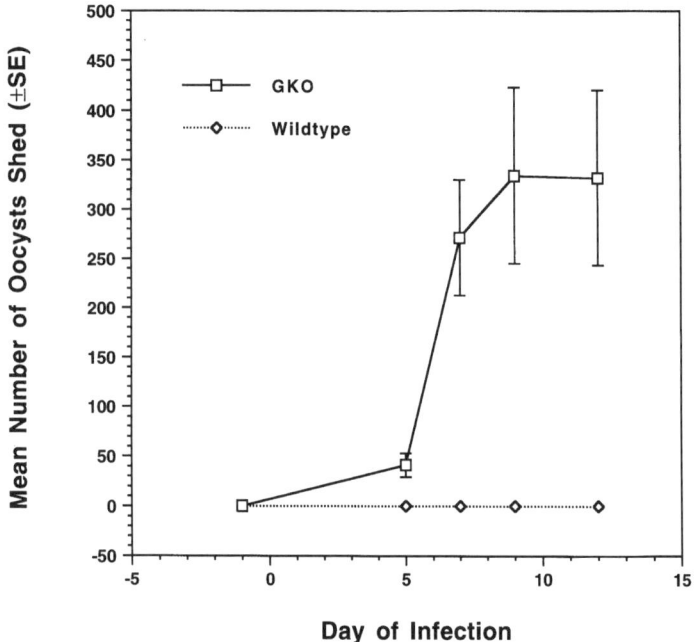

Figure 3 Oocysts shedding from *C. parvum*-infected GKO mice and their wild-type littermates. Following randomization, 4- to 7-week-old female GKO mice ($n = 4$) and their wild-type littermates ($n = 4$) were inoculated orally with 1×10^7 oocysts of the GCH1 isolate. Oocyst shedding was determined as described in Figure 1. Results are presented as the mean number of oocysts shed per group ± the standard error (SE).

3.3.2. Other Cytokines

Little evidence exists to suggest a role for cytokines other than IFNγ in a protective immune response to *C. parvum*. Most of what is known was derived from studies involving (a) neutralization of the *in vivo* activity of particular cytokines by the specific antibodies, (b) analysis of mRNA levels for specific cytokines in lymphoid tissues, or (c) analysis of cytokine secretion following *ex vivo* restimulation of cells primed *in vivo* for *C. parvum*.

(a) In vivo *neutralization* In a study by Enriquez and Sterling (1993), the anti-murine IL-5 mAb TRFK 5 or a combination of anti-IL-5 and anti-murine IL-4 (11B11) mAbs was administered to adult BALB/c mice prior to and 5 days after challenge with *C. parvum*. Results indicated an increased level of oocyst shedding and infection of the terminal ileum when compared to mice treated with a control mAb directed against *Heligmosomoides polygyrus*. In contrast, treatment with either the anti-murine IL-2 mAb G4G11 or anti-IL-4 only enhanced oocyst shedding. The authors concluded from this study that IL-2 and IL-4 may be involved in the initial control of infection, while IL-5, or IL-5 and IL-4, may be responsible for controlling the chronicity of infection.

(b) *Messenger RNA analysis* In a study by Urban *et al.* (1996), mRNA levels for IL-2, IL-4, and IL-10 in the mesenteric lymph nodes and ileum of *C. parvum*-infected neonatal BALB/c mice were comparable to those seen in uninfected mice. These data suggested that these particular cytokines were not involved in the neonatal immune response to *C. parvum*. This conclusion differed from that of Enriquez and Sterling (1993) above, who used adult BALB/c mice. These results may reflect an age-related difference in the immune response to *C. parvum*.

(c) Ex vivo *Analysis of T lymphocytes*

(i) *Murine studies*. An additional method by which T-cell responses can be analyzed involves the *in vitro* restimulation of T cells primed *in vivo* as a consequence of infection or immunization. With respect to *C. parvum*, there have been several reports that attempted to address this issue using mice orally challenged with *C. parvum* oocysts (Whitmire and Harp, 1990; Moss and Lammie, 1993; Harp *et al.*, 1994; Harp and Sacco, 1996; Cheng *et al.*, 1996; Theodos *et al.*, 1997), mice immunized with *C. parvum* antigens (Tatalick and Perryman, 1995), and individuals with a history of cryptosporidial infections (Gomez Morales *et al.*, 1995). Most of these studies analyzed splenic T-cell responses. The findings of Whitmire and Harp (1990) suggested that an *ex vivo* proliferative response to a *C. parvum* antigen extract was only obtained using spleen cells from BALB/c mice after five weekly oral challenges with *C. parvum* oocysts beginning at 1 week of age. Proliferative responses were not observed when spleen cells were obtained from mice that received a single oral challenge (1 week of age), a single intraperitoneal challenge (4 weeks of age), or a single oral challenge (1 week) followed by a single intraperitoneal challenge (4 weeks). Proliferative responses were also not obtained when MLN cells were used.

In a subsequent study, Harp *et al.* (1994) examined the proliferative response of purified populations of splenic T cells obtained from BALB/c mice following multiple oral challenges with *C. parvum* oocysts beginning at 1 week of age. Enrichment for CD4+ T cells resulted in a 4- to 14-fold increase in the proliferative response to soluble *C. parvum* antigen when compared to unenriched cells. Enhanced proliferative responses were also associated with increased age and exposure to *C. parvum* oocysts. In contrast, splenocytes enriched for CD8+ T cells failed to proliferate following *ex vivo* restimulation with *C. parvum* antigen. Splenocytes enriched for total T cells or CD4+ cells, but not CD8+ cells, secreted IFNγ in response to *ex vivo* stimulation with soluble *C. parvum* antigen. Curiously, the unenriched splenic cell population displayed the greatest amount of IFNγ secretion, despite a poor proliferative response. IL-2 secretion was only observed in cultures enriched for total T cells. IL-4 production was never observed.

Similar results were obtained in a latter study by Harp and Sacco (1996). Splenocytes were isolated at weekly intervals from BALB/c mice orally challenged with *C. parvum* oocysts at 1 week of age. When restimulated with a soluble parasite antigen extract *ex vivo*, these cells failed to demonstrate a proliferative response that differed from splenocytes of naive mice. Moreover, no difference in IFNγ secretion was observed between these two groups. One interesting observation from these studies, was that splenocytes from 3- to 4-week-old mice (both *C. parvum* challenged and naive) secreted more IFNγ than splenocytes isolated from 1- to 2-week-old mice. This result may explain, in part, the enhanced age-related resistance to *C. parvum* infection.

In a study by Tatalick and Perryman (1995), adult BALB/c mice were immunized by multiple subcutaneous injections with either excysted, solubilized *C. parvum* sporozoites and oocyst shells or a purified antigen from this preparation (SA-1). Splenocytes were harvested 7–10 days after the final immunization and restimulated *ex vivo* with excysted *C. parvum* oocysts and sporozoites. An increased stimulation index was observed with splenocytes isolated from immunized mice when compared to control mice. This suggested that *C. parvum*-reactive cells were present in the spleens of immunized mice.

In general, the proliferative responses and cytokine secretion following *ex vivo* restimulation with *C. parvum* antigen in many of these studies was low. This suggested that the *in vivo* priming of *C. parvum*-reactive T cells was potentially minimal. An additional caveat to the above studies comes from reports showing compartmentalization of immune responses, with different lymphoid organs presenting different responding T cell phenotypes (Caulada-Benedetti *et al.*, 1991; Kelly *et al.*, 1991; Chardes *et al.*, 1993; Boher *et al.*, 1994). In a study by Caulada-Benedetti *et al.* (1991), splenocytes displayed a predominant Th1 phenotype in response to vaccination with irradiated *Schistosoma mansoni* cercariae, while the response in the draining lymph nodes was of a Th2 phenotype. Similar results were obtained by Kelly *et al.* (1991) in an analysis of the immune response to *Trichinella spiralis*, with parasite-specific Th1-like cells detected in the spleen and Th2-like cells detected from MLN. A comparison of mucosal and systemic cellular immune responses following oral challenge with *T. gondii* also revealed a predominant Th2-like response in the MLN and a Th1-like response in the spleen (Chardes *et al.*, 1993). With respect to cryptosporidiosis, differences were observed in the populations of cells found in the ileal and jejunal Peyer's patches of BALB/c mice. When infected with *C. parvum* as neonates (Boher *et al.*, 1994), an increased number of CD8+ T cells, monocytes, macrophages, and dendritic cells were evident in the ileal Peyer's patches. In contrast, an increased number of CD4+ T cells, monocytes, macrophages, and dendritic cells was observed in the jejunal Peyer's patches. These results suggested the activation of both CD4+ and CD8+ T cells, with potential functional differences in the immune responses generated in ileal and jejunal Peyer's patches from *C. parvum*-infected neonatal mice. It remains to be determined if these differences are found in infected adult mice. In summary, these studies indicated that the site of analysis can be extremely important in determining the ultimate outcome of a study. Moreover, the lymphoid tissue draining the site of infection should preferentially be analyzed, particularly when the mucosal immune system is involved.

Two reports in the literature analyzed the proliferative response of MLN cells to *C. parvum*. In a study by Moss and Lammie (1993), inbred SWR/J mice were infected with *C. parvum* at 3 days of age. The proliferative response of MLN and spleen cells was determined every 3 days, beginning on day 10 of infection. Proliferative responses were obtained at isolated time points from MLN cells restimulated with soluble or insoluble oocyst antigen extracts. However, increased proliferative responses to these antigens were also observed when MLN cells isolated from uninfected mice were assayed. The reason for this was unclear. As in the study by Whitmire and Harp (1990), splenocytes did not proliferate at any time tested, corroborating the conclusion that multiple oral challenge may be necessary to prime splenocytes in immunocompetent mice.

Recently, we reported on a parasite-specific proliferative response obtained with MLN cells isolated on days 14 and 17 of infection from *C. parvum*-infected nonhealing GKO and healing XMG 1.2–treated C57BL/6J mice, respectively (Theodos *et al.*, 1997). These time points were selected since they corresponded to the terminal (day 14, GKO) and healing (day 17, XMG 1.2–treated C57BL/6J) phases of each model. These cells proliferated in response to both freshly excysted *C. parvum* sporozoites and a freeze–thaw sporozoite antigen preparation following *ex vivo* restimulation (Figure 4). This response was parasite-specific in that proliferation was not observed following stimulation with ovalbumin, a *C. parvum* unrelated antigen. The fact that *C. parvum*-infected GKO mice mounted a parasite-specific immune response but failed to control their infection indicated that the immune response generated in these mice was not protective. Preliminary evidence indicated that MLN cells isolated from *C. parvum*-infected non-healing GKO mice secreted IL-2 following *in vitro* restimulation with *C. parvum*, while cells isolated from healing C57BL/6 mice secreted IL-2 and IFNγ (author's unpublished observation). Production of IL-4 was not observed from either cell type. The phenotypic and functional properties of these cells, as well as further characterization of the cytokines they are secreting, are currently under investigation.

(ii) *Calf studies*. Proliferative responses to *C. parvum* antigens were also observed using PBMCs isolated from infected calves (Whitmire and Harp, 1993). In this study, PBMCs isolated from calves 16–23 days after infection proliferated when restimulated *ex vivo* with a soluble *C. parvum* antigen preparation obtained from disrupted oocysts. In contrast, proliferation was not observed with PBMCs isolated from uninfected calves. In a subsequent study, Harp *et al.* (1995) demonstrated that this proliferative response was restricted to PBMCs isolated from infected calves. Proliferation was not observed with MLN cells, and prescapular lymph node cells isolated from these calves were restimulated *ex vivo* with soluble *C. parvum* antigen. Curiously, splenocytes from either uninfected or infected calves displayed a significant proliferative response in the absence of antigen. This response decreased when *C. parvum* antigen was added.

(iii) *Human studies*. A study by Gomez Morales *et al.* (1995) evaluated the presence of *C. parvum*-reactive cells in the peripheral blood of immunocompetent individuals and HIV-infected patients. PBMCs isolated from immunocompetent individuals with a history of transient cryptosporidiosis proliferated to a *C. parvum* oocyst wall protein (COWP) and a crude extract of *C. parvum* oocysts (CCE) following *ex vivo* restimulation. These cells secreted IL-10 and IFNγ following restimulation with CCE. Estimates of the amount of each cytokine produced indicated a 10-fold increase in IFNγ levels when compared to IL-10. Moreover, PBMCs from individuals who had recovered from cryptosporidiosis secreted only IFNγ following restimulation *ex vivo*. These results also correlated the production of IFNγ with the resolution of cryptosporidiosis. In contrast, PBMCs isolated from HIV-infected individuals with active cryptosporidiosis did not proliferate following stimulation with CCE. An observation of note was that PBMCs from presumably healthy individuals also proliferated in response to CCE and secreted IL-10, but not IFNγ, following restimulation. The authors concluded that either *C. parvum* antigens were cross-reactive with antigens found on other common microorganisms, or that the incidence of *C. parvum* exposure in this population was high. These results emphasized the inherent complications with studies on the human population

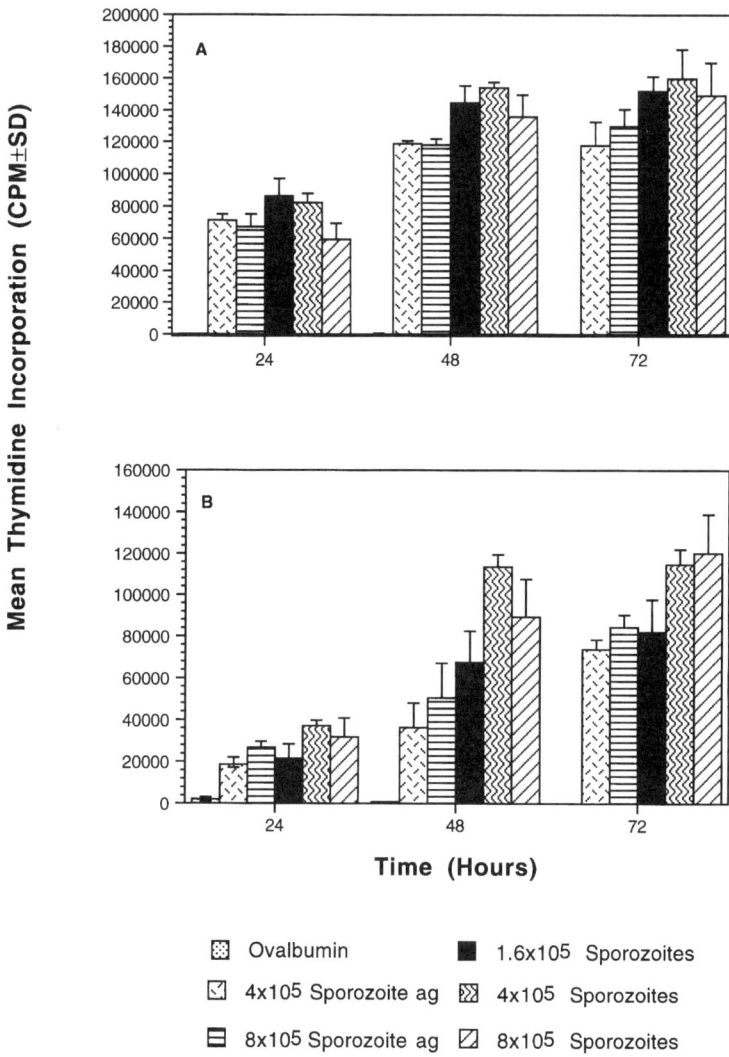

Figure 4 MLN cells were isolated from (A) infected GKO and (B) anti-IFNγ-treated C57BL/6J mice at 14 and 17 days of infection, respectively, and restimulated *ex vivo* with the indicated antigen. Proliferation was assessed by the addition of [^{3}H]thymidine at 24, 48, or 72 hours after restimulation. Results are presented as the mean thymidine incorporation (counts per minute (CPM)) of quadruplicate wells stimulated with antigen ± the standard deviation (SD). The mean background thymidine incorporation (CPM obtained with culture medium alone) was subtracted from all results.

when the precise medical history of the individuals enrolled may not be completely known.

The data presented in the preceding sections collectively support a role for CD4+ T cells and T-cell-associated cytokines in the resolution of a *C. parvum* infection. While most evidence suggests that IFNγ may be important in the clearance of *C. parvum* from the gastrointestinal tract, the precise mechanism by which this cytokine contributes to cryptosporidial protective immunity is unknown. Moreover, preliminary results indicated that Th2 associated cytokines such as IL-5 and IL-10 may also be associated with *C. parvum*-specific protective immunity. Comparable results were obtained by Tilley *et al.* (1995) using a *C. Muris* model of infection. In this study, splenocytes from *C. muris*-infected BALB/c mice secreted IFNγ, IL-2, and IL-4 following restimulation with parasite antigen *ex vivo*. In summary, these results collectively suggested that both Th1- and Th2-associated cytokines were involved in immunity to *Cryptosporidium*. However, more extensive analysis is required to identify fully the cytokines involved in protective immunity to this parasite.

3.4. Nitric Oxide

One mechanism by which IFNγ may mediate protective immunity against *C. parvum* would be via the generation of nitric oxide. Nitric oxide has potent antimicrobial activity against many extracellular and intracellular microorganisms (Nussler and Billiar, 1993). The role of nitric oxide in immunity to *C. parvum* was investigated by indirect approaches. In one study, adult BALB/c and SCID mice were treated with the nitric oxide synthase inhibitor aminoguanidine prior to and after oral challenge with *C. parvum* oocysts (Kuhls *et al.*, 1994). This treatment did not alter the course of cryptosporidiosis in either mouse strain when compared to control mice. From these results, the authors concluded that nitric oxide was not involved in immunity to *C. parvum*. In contrast, results from a study by Leitch and He (1994) suggested that nitric oxide may be involved in the clearance of *C. parvum*. In this study, supplementation of the diet with L-arginine decreased the shedding of oocysts from chronically infected adult nude mice. Administration of the nitric oxide synthase inhibitor *N*-nitro-L-arginine methyl ester abrogated the protective effect of the supplement and significantly enhanced oocyst shedding from chronically infected mice. While this result suggested that nitric oxide may be involved in *C. parvum* protective immunity, more extensive analysis is needed to confirm a role for this important antimicrobial mediator.

4. IMMUNOTHERAPY

The inability to identify chemotherapeutic agents that effectively eliminate *C. parvum* from the gastrointestinal tract (Peterson, 1992) suggests that immunotherapeutic or combined chemotherapeutic/immunotherapeutic approaches may be necessary for the successful treatment of cryptosporidiosis. To date, many of the immune modulators tested have not shown therapeutic efficacy against cryptosporidiosis.

4.1. Cytokines

4.1.1. Interferon-γ

Several immunotherapy trials aimed at enhancing cellular immunity have focused on IFNγ. The effectiveness of this cytokine has yielded varying results to date. In a study by Kuhls *et al.* (1994) treatment of neonatal SCID mice with recombinant murine IFNγ prior to infection with *C. parvum* did not enhance their initial resistance to infection. Moreover, the treatment of *C. parvum*-infected neonatal SCID mice at weekly intervals beginning prior to infection did not reduce their level of mucosal infection or improve their survival time when compared to control mice. These results suggested that exogenously administered IFNγ does not function either prophylactically or therapeutically against *C. parvum*. A study by Rehg (1996a) investigated the prophylactic and therapeutic efficacy of IFNγ in an immunosuppressed rat model of cryptosporidiosis. In contrast to the results of Kuhls *et al.* (1994), a significant reduction in the extent of infection of the small intestine was observed with both treatment schedules. However, only prophylactic treatment with IFNγ resulted in significantly fewer parasites in the colon, cecum, and common bile duct when compared to control rats. One obvious difference between these two studies is that the latter involved daily administration of IFNγ. This could have accounted for the different results, since the *in vivo* half-life of exogenously administered IFNγ is extremely rapid (Kurzrock *et al.*, 1985).

4.1.2. Interleukin-12

The administration of recombinant murine IL-12 to neonatal BALB/c mice prior to and after challenge with *C. parvum* resulted in a marked decrease in infection (Urban *et al.*, 1996). This protective affect of IL-12 correlated with the induction of IFNγ since 1) an increase in IFNγ-specific mRNA was observed in the ileum and mesenteric lymph nodes following the administration of IL-12 and 2) the co-administration of an anti-IFNγ mAb along with recombinant murine IL-12 blocked the inhibition of infection due to IL-12 administration. A significant outcome of this study was the observation that IL-12 must be administered prior to oral challenge with oocysts in order to completely inhibit a *C. parvum* infection. The administration of IL-12 to neonatal BALB/c mice after infection did not augment resolution, despite the enhanced level of IFNγ mRNA observed.

4.2. Additional Modulators of Cellular Immunity

4.2.1. Dehydroepiandrosterone

Administration of the immune modulator dehydroepiandrosterone (DHEA) to *C. parvum*-infected aged Syrian golden hamsters (Rasmussen and Healey, 1992c) and dexamethasone treated rats (Rasmussen et al., 1992) significantly reduced the intensity of infection. Further analysis of this immune modulator revealed that DHEA enhanced T- and B-lymphocyte mitogenic responses in the dexamethasone suppressed rat model of cryptosporidiosis (Rasmussen et al., 1993). Moreover, the intensity of *C. parvum* infection was lower in DHEA-treated rats when compared to the untreated, immunosuppressed control group. No enhancement of NK-cell activity was observed following DHEA treatment, suggesting that this cell may not be involved in the immunological control of cryptosporidiosis. In all these studies, DHEA was given either prior to infection or beginning on the day of infection and daily thereafter. It remains to be determined whether DHEA would function therapeutically against an already established *C. parvum* infection.

4.2.2. Diethyldithiocarbamate

A second immune modulator, diethyldithiocarbamate, was tested in the dexamethasone suppressed rat model of cryptosporidiosis (Rehg, 1996b). Like DHEA, diethyldithiocarbamate reduced the severity of ileal infection when given to these rats prophylactically. Moreover, therapeutic treatment with diethyldithiocarbamate also reduced the extent of ileal infection. Only prophylactic treatment inhibited infection of the biliary tract, a consequence of chronic cryptosporidiosis. It is unclear, however, whether diethyldithiocarbamate functioned through augmentation of the immune response of these rats.

4.2.3. Dialyzable Leukocyte Extract

In 1990, a study by McMeeking et al. examined the effectiveness of a dialyzable leukocyte extract (DLE) produced from the lymph node cells of *C. parvum* immune calves in treatment of cryptosporidiosis in patients with AIDS. The authors chose this modulator since studies from other laboratories indicated that DLEs augmented cellular immunity in a variety of clinical settings. In this study, 75% of those individuals who received the DLE from *C. parvum*-immune calves showed marked improvement in their cryptosporidiosis. Administration of a DLE from nonimmune calves had no effect. The authors hypothesized that this effect was due to increased cell-mediated immunity to *C. parvum*, although no direct evidence was provided. It is of interest to note that the administration of DLE had no effect on the CD4:CD8 cell ratio in these individuals, nor were their proliferative responses to recall antigens enhanced.

5. CONCLUDING REMARKS

In summary, the results presented in this chapter indicate that both innate and parasite specific-cell-mediated immune responses are involved in immunity to cryptosporidiosis. However, most of the specific components of these responses have not been identified completely. There is general consensus in the literature to support a role for IFNγ in the initial resistance to *C. parvum*. However, the mechanism by which IFN-γ imparts resistance is unclear. There is also general agreement that CD4+ T cells are important in the resolution of both acute and chronic cryptosporidiosis. Again, the effector mechanism by which these cells eliminate *C. parvum* from the gastrointestinal tract remains to be determined. There is also some agreement on IFNγ-mediated control or resolution of cryptosporidiosis. However, the extent to which this cytokine is involved in actual resolution of infection has been debated. While several studies have investigated the role of other cells and cytokines in the innate and acquired immune responses to *C. parvum*, these studies have not been extensive. Much more work needs to be done to clarify the components and the mechanisms responsible for immunity to *C. parvum*. While we have made significant strides in the past 15 years, we have only begun to determine the immune mechanisms responsible for resistance to and elimination of this most unusual parasite.

ADDENDUM

As mentioned in the text of this chapter, McDonald *et al.* (1996) observed that the adoptive transfer of IEL from *C. muris*-immune BALB/c mice resulted in parasite clearance from infected SCID mice. In an extension of this study, Culshaw *et al.* (1997) demonstrated that this protective effect was due to the production of IFNγ. Treatment with the hamster anti-IFNγ neutralizing antibody H22 reversed the ability of the adoptively transferred IELs to eliminate

C. muris from infected SCID mice. Further *in vitro* analysis indicated that the IELs isolated from *C. muris*-immune BALB/c mice secreted IFNγ following activation by either polyclonal stimuli (ConA or anti-CD3 treatment) or a soluble oocyst antigen preparation.

REFERENCES

Aguirre, S.A., Mason, P.H. and Perryman, L.E. (1994). Susceptibility of major histocompatibility complex (MHC) class I- and class II-deficient mice to *Cryptosporidium parvum* infection. *Infection and Immunity* **62**, 697–699.
Alak, J.I.B., Shahbazian, M., Huang, D.S., Wang, Y., Darban, H., Jenkins, E.M. and Watson, R.R. (1993). Alcohol and murine acquired immunodeficiency syndrome suppression of resistance to *Cryptosporidium parvum* infection during modulation of cytokine production. *Alcoholism: Clinical and Experimental Research* **17**, 539–544.
Baume, D.M., Robertson, M.J., Levine, H., Manley, T.J., Schow, P.W. and Ritz, J. (1992). Differential responses to interleukin 2 define functionally distinct subsets of human natural killer cells. *European Journal of Immunology* **22**, 1–6.
Bjorneby, J.M., Leach, D.R. and Perryman, L.E. (1991). Persistent cryptosporidiosis in horses with severe combined immunodeficiency. *Infection and Immunity* **59**, 3823–3826.
Blanshard, C., Jackson, A.M., Shanson, D.C., Francis, N. and Gazzard, B.G. (1992). Cryptosporidiosis in HIV-seropositive patients. *Quarterly Journal of Medicine* **85**, 813–823.
Boher, Y., Perez-Schael, I., Caceres-Dittmar, G., Urbina, G., Gonzalez, R., Kraal, G. and Tapia, F.J. (1994). Enumeration of selected leukocytes in the small intestine of BALB/c mice infected with *Cryptosporidium parvum*. *American Journal of Tropical Medicine and Hygiene* **50**, 145–151.
Brasseur, P., Lemeteil, D. and Ballet, J.J. (1988). Rat model for human cryptosporidiosis. *Journal of Clinical Microbiology* **26**, 1037–1039.
Caulada-Benedetti, Z., Al-Zamel, F., Sher, A. and James, S. (1991). Comparison of Th1- and Th2-associated immune reactivities stimulated by single versus multiple vaccination of mice with irradiated *Schistosoma mansoni* cercariae. *Journal of Immunology* **146**, 1655–1660.
Chardes, T., Velge-Roussel, F., Melvelec, P., Mevelec, M.-N., Buzoni-Gatel, D. and Bout, D. (1993). Mucosal and systemic cellular immune responses induced by *Toxoplasma gondii* antigens in cyst orally infected mice. *Immunology* **78**, 421–429.
Chehimi, J., Starr, S.E., Frank, I., D'Andrea, A., Ma, Xiaojing, MacGregor, R. R., Sennelier, J. and Trinchieri, G. (1994). Impaired Interleukin 12 production in human immunodeficiency virus-infected individuals. *Journal of Experimental Medicine* **179**, 1361–1366.
Chen, W., Harp, J.A., Harmsen, A.G. and Havell, E.A. (1993a). Gamma interferon functions in resistance to *Cryptosporidium parvum* infection in severe combined immunodeficient mice. *Infection and Immunity* **61**, 3548–3551.
Chen, W., Harp, J.A. and Harmsen, A. (1993b). Requirements for CD4+ cells and gamma interferon in resolution of established *Cryptosporidium parvum* infection in mice. *Infection and Immunity* **61**, 3928–3932.
Cheng, L., Rasmussen, K.R., Healey, M.C. and Yang, S. (1996). Primary and

secondary infections with *Cryptosporidium parvum* in immunosuppressed adult mice. *American Journal of Tropical Medicine and Hygiene* **55**, 324–329.

Chien, Y.-H., Jores, R. and Crowley, M.P. (1996). Recognition by γ/δ T cells. *Annual Review of Immunology* **14**, 511–532.

Culshaw, R.J., Bancroft, G.J. and McDonald, V. (1997). Gut intraepithelial lymphocytes induce immunity against *Cryptosporidium* infection through a mechanism involving gamma Interferon production. *Infection and Immunity* **65**, 3074–3079.

Darban, H., Enriquez, J., Sterling, C.R., Lopez, M.C., Chen, G., Abbaszadegan, M. and Watson, R.R. (1991). Cryptosporidiosis facilitated by murine retroviral infection with LP-BM5. *Journal of Infectious Diseases* **164**, 741–745.

Davis, C.P., McAllister, J.S. and Savage, D.C. (1973). Microbial colonization of the intestinal epithelium in suckling mice. *Infection and Immunity* **7**, 666–672.

Enriquez, F.J. and Sterling, C.R. (1991). *Cryptosporidium* infections in inbred strains of mice. *Journal of Protozoology* **38**, 100S–102S.

Enriquez, F.J. and Sterling, C.R. (1993). Role of CD4+ TH1- and TH2-cell secreted cytokines in cryptosporidiosis. *Folia Parasitologica* **40**, 307–311.

Ernest, J.A., Blagburn, B.L., Lindsay, D.S. and Current, W.L. (1986). Infection dynamics of *Cryptosporidium parvum* (Apicomplexa: *Cryptosporidiae*) in neonatal mice (*Mus musculus*). *Journal of Parasitology* **72**, 796–798.

Flannigan, T., Whalen, C., Turner, J., Soave, R., Toerner, J., Havlir, D. and Kotler, D. (1992). *Cryptosporidium* infection and CD4 counts. *Annals of Internal Medicine* **116**, 840–842.

Galli, S.J. and Wershil, B.K. (1995). Mouse mast cell cytokine production: role in cutaneous inflammatory and immunological responses. *Experimental Dermatology* **4**, 240–249.

Gardner, A.L., Roche, J.K., Weikel, C.S. and Guerrant, R.J. (1991). Intestinal cryptosporidiosis: pathophysiologic alterations and specific cellular and humoral immune responses in RNU/+ and RNU/RNU (athymic) rats. *American Journal of Tropical Medicine and Hygiene* **44**, 49–62.

Gazinelli, R.T., Hieny, S., Wynn, T.A., Wolf, S. and Sher A. (1993). Interleukin 12 is required for the T-lymphocyte-independent induction of interferon-γ by an intracellular parasite and induces resistance in T cell deficient hosts. *Proceedings of the National Academy of Sciences USA* **90**, 6115–6119.

Genta, R., Chappell, C., White, A., Kimball, K. and Goodgame, R. (1994). Duodenal morphology and intensity of infection in AIDS-related intestinal cryptosporidiosis. *Gastroenterology* **105**, 1769–1775.

Gomez Morales, M.A., Ausiello, C.M., Urbani, F. and Pozio, E. (1995). Crude extract and recombinant protein of *Cryptosporidium parvum* oocysts induce proliferation of human peripheral blood mononuclear cells *in vitro*. *Journal of Infectious Diseases* **172**, 211–216.

Gomez Morales, M.A., Ausiello, C.M., Guarino, A., Urhani, F., Spagnuolo, M.I., Pignata, C. and Pozio, E. (1996). Severe, protracted intestinal cryptosporidiosis associated with interferon γ deficiency: pediatric case report. *Clinical Infectious Diseases* **22**, 848–850.

Haas, W., Pereira, P. and Tonegawa, S. (1993). γδ cells. *Annual Review of Immunology* **11**, 637–685.

Harp, J.A. and Moon, H.W. (1991). Susceptibility of mast cell-deficient W/W^v mice to *Cryptosporidium parvum*. *Infection and Immunity* **59**, 718–720.

Harp, J.A. and Sacco, R.E. (1996). Development of cellular immune functions in neonatal to weanling mice: relationship to *Cryptosporidium parvum* infection. *Journal of Parasitology* **82**, 245–249.

Harp, J.A. and Whitmire, W.M. (1991). *Cryptosporidium parvum* infection in mice: Inability of lymphoid cells or culture supernatants to transfer protection from resistant adults to susceptible infants. *Journal of Parasitology* **77**, 170–172.

Harp, J.A., Wannemuehler, M.W., Woodmansee, D.B. and Moon, H.W. (1988). Susceptibility of germ-free or antibiotic-treated adult mice to *Cryptosporidium parvum*. *Infection and Immunity*. **56**, 2006–2010.

Harp, J.A., Chen, W. and Harmsen, A.G. (1992). Resistance of severe combined immunodeficient mice to infection with *Cryptosporidium parvum*: the importance of intestinal microflora. *Infection and Immunity* **60**, 3509–3512.

Harp, J.A., Whitmire, W.M. and Sacco, R. (1994). *In vitro* proliferation and production of gamma-interferon by murine CD4+ T cells in response to *Cryptosporidium parvum* antigen. *Journal of Parasitology* **80**, 67–72.

Harp, J.A., Franklin, S.T., Goff, J.P. and Nonnecke, B.J. (1995). Effects of *Cryptosporidium parvum* infection on lymphocyte phenotype and reactivity in calves. *Veterinary Immunology and Immunopathology* **44**, 197–207.

Heine, J., Moon, H.W. and Woodmansee, D.B. (1984). Persistent *Cryptosporidium* infection in congenitally athymic nude mice. *Infection and Immunity* **43**, 856–859.

Hunter, C.A. (1996). How are NK cell responses regulated during infection? *Experimental Parasitology* **84**, 444–448.

Hunter, C.A., Subauste, C.S., Van Cleave, V.H. and Remington, J.S. (1994). Production of γ-interferon by natural killer cells from *Toxoplasma gondii*-infected SCID mice: regulation by interleukin-10, interleukin-12, and tumor necrosis factor α. *Infection and Immunity* **62**, 2818–2824.

Johansen, G.A. and Sterling, C.R. (1994). Detection of prolonged *C. parvum* infection in immunocompetent adult C57BL/6 mice. *Journal of Eukaryotic Microbiology* **41**, 45S

Khan, I.A., Matsuura, T. and Kasper, L.H. (1994). Interleukin-12 enhances murine survival against acute toxoplasmosis. *Infection and Immunity* **62**, 1639–1642.

Kapel, N., Benhamou, Y., Buraud, M., Magne, D., Opolon, P. and Gobert, J.-G. (1996). Kinetics of mucosal ileal γ-interferon response during cryptosporidiosis in immunocompetent neonatal mice. *Parasitology Research* **82**, 664–667.

Kawase, I., Brooks, C.G., Kuribayashi, K., Olabuenaga, S., Newman, W., Gillis, S. and Henney, C. (1983). Interleukin 2 induces γ-interferon production: participation of macrophages and NK-like cells. *Journal of Immunology* **131**, 288–292.

Kelly, E.A.B., Cruz, E.S., Hauda, K.M. and Wassom, D.L. (1991). IFN-γ- and IL-5-producing cells compartmentalize to different lymphoid organs in *Trichinella spiralis*-infected mice. *Journal of Immunology* **147**, 306–311.

Kuhls, T.L., Greenfield, R.A., Mosier, D.A., Crawford, D.L. and Joyce, W. A. (1992). Cryptosporidiosis in adult and neonatal mice with severe combined immunodeficiency. *Journal of Comparative Pathology* **106**, 399–410.

Kuhls, T.L., Mosier, D.A., Abrams, V.L., Crawford, D.L. and Greenfield, R. A. (1994). Inability of interferon-γ and aminoguanidine to alter *Cryptosporidium parvum* infection in mice with severe combined immunodeficiency. *Journal of Parasitology* **80**, 480–485.

Kurzrock, R., Rosenblum, M.G., Sherwin, S.A., Rios, A., Talpaz, M., Quesada, J.R. and Gutterman, J.U. (1985). Pharmacokinetics, single-dose tolerance, and biological activity of recombinant γ-interferon in cancer patients. *Cancer Research* **45**, 2866–2872.

Leitch, G.J. and He, Q. (1994). Arginine-derived nitric oxide reduces fecal oocyst shedding in nude mice infected with *Cryptosporidium parvum*. *Infection and Immunity* **62**, 5173–5176.

Marcial, M.A. and Madara, J.L. (1986). *Cryptosporidium*: cellular localization, structural analysis of adsorptive cell–parasite membrane–membrane interactions in guinea pigs, and suggestion of protozoan transport by M cells. *Gastroenterology* **90**, 583–594.
Martinez, F., Mascaro, C., Rosales, M.J., Diaz, J., Cifuentes, J. and Osuna, A. 1992. *In vitro* multiplication of *Cryptosporidium parvum* in mouse peritoneal macrophages. *Veterinary Parasitology* **42**, 27–31.
McDonald, V. and Bancroft, G. (1994). Mechanisms of innate and acquired resistance to *Cryptosporidium parvum* infection in SCID mice. *Parasite Immunology* **16**, 315–320.
McDonald, V., Deer, R., Uni, S., Iseki, M. and Bancroft, G. (1992). Immune responses to *Cryptosporidium muris* and *Cryptosporidium parvum* in adult immunocompetent or immunocompromised (nude and SCID) mice. *Infection and Immunity* **60**, 3325–3331.
McDonald, V., Robinson, H.A., Kelly, J.P. and Bancroft, G.J. (1996). Immunity to *Cryptosporidium muris* infection in mice is expressed through gut CD4+ intraepithelial lymphocytes. *Infection and Immunity* **64**, 2556–2562.
McMeeking, A., Borkowsky, W., Klesius, P.H., Bonk, S., Holzman, R.S. and Lawrence, H.S. (1990). A controlled trial of bovine dialyzable leukocyte extract for cryptosporidiosis in patients with AIDS. *Journal of Infectious Diseases* **161**, 108–112.
Mead, J.R., Arrowood, M.J., Sidwell, R.W. and Healey, M.C. (1991a). Chronic *Cryptosporidium parvum* infections in congenitally immunodeficient SCID and nude mice. *Journal of Infectious Diseases* **163**, 1297–1304.
Mead, J.R., Arrowood, M.J., Healey, M.C. and Sidwell, R.W. (1991b). Cryptosporidial infections in SCID mice reconstituted with human or murine lymphocytes. *Journal of Protozoology* **38**, 59S–61S.
Mead, J.R., Ilksoy, N., You, X., Belenkaya, Y., Arrowood, M., Fallon, M. T. and Schinazi, R. (1994). Infection dynamics and clinical features of cryptosporidiosis in SCID mice. *Infection and Immunity* **62**, 1691–1695.
Molbak, K., Lisse, I.M., Hojlyng, N. and Aaby, P. (1994). Severe cryptosporidiosis in children with normal T-cell subsets. *Parasite Immunology* **16**, 275–277.
Moss, D.M. and Lammie, P.J. (1993). Proliferative responses of lymphocytes from *Cryptosporidium parvum*-exposed mice to two separate antigen fractions from oocysts. *American Journal of Tropical Medicine and Hygiene* **49**, 393–401.
Novak, S.M. and Sterling, C.R. (1991). Susceptibility dynamics in neonatal BALB/c mice infected with *Cryptosporidium parvum*. *Journal of Protozoology* **38**, 102S–104S.
Nussler, A. K. and Billiar, T. R. (1993). Inflammation, immunoregulation, and inducible nitric oxide synthase. *Journal of Leukocyte Biology* **54**, 171–178.
Perryman, L.E., Mason, P.A. and Chrisp, C.E. (1994). Effect of spleen cell populations on resolution of *Cryptosporidium parvum* infection in SCID mice. *Infection and Immunity* **62**, 1474–1477.
Petersen, C. (1992). Cryptosporidiosis in patients infected with the human immunodeficiency virus. *Clinical Infectious Diseases* **15**, 903–909.
Petry, F., Robinson, H.A. and McDonald, V. (1995). Murine infection model for maintenance and amplification of *Cryptosporidium parvum* oocysts. *Journal of Clinical Microbiology* **33**, 1922–1924.
Rasmussen, K.R. and Healey, M.C. (1992a). Experimental *Cryptosporidium parvum* infections in immunosuppressed adult mice. *Infection and Immunity* **60**, 1648–1652.
Rasmussen, K.R. and Healey, M.C. (1992b). *Cryptosporidium parvum*: experimental

infections in aged Syrian golden hamsters. *Journal of Infectious Diseases* **165**, 769–772.

Rasmussen, K.R. and Healey, M.C. (1992c). Dehydroepiandrosterone-induced reduction of *Cryptosporidium parvum* infections in aged Syrian golden hamsters. *Journal of Parasitology* **78**, 554–557.

Rasmussen, K.R., Arrowood, M.J. and Healey, M.C. (1992). Effectiveness of dehydroepiandrosterone in reduction of cryptosporidial activity in immunosuppressed rats. *Antimicrobial Agents and Chemotherapy* **36**, 220–222.

Rasmussen, K.R., Martin, E.G. and Healey, M.C. (1993). Effects of dehydroepiandrosterone in immunosuppressed rats infected with *Cryptosporidium parvum*. *Journal of Parasitology* **79**, 364–370.

Rehg, J.E. (1996a). Effect of Interferon-γ in experimental *Cryptosporidium parvum* infection. *Journal of Infectious Diseases* **174**, 229–232.

Rehg, J.E. (1996b). Effect of diethyldithiocarbamate on *Cryptosporidium parvum* infection in immunosuppressed rats. *Journal of Parasitology* **82**, 158–162.

Rehg, J.E., Hancock, M. L. and Woodmansee, D. B. (1987). Characterization of cyclophosphamide-rat model of cryptosporidiosis. *Infection and Immunity* **55**, 2669–2674.

Rehg, J.E., Hancock, M.L. and Woodmansee, D. B. (1988). Characterization of a dexamethasone-treated rat model of cryptosporidial infection. *Journal of Infectious Diseases* **158**, 1406–1407.

Roder, J.C., Lohmann-Matthes, M.-L., Domzig, W. and Wigzell, H. (1979). The *beige* mutation in the mouse. II. Selectivity of the natural killer (NK) cell defect. *Journal of Immunology* **123**, 2174–2181.

Rohlman, V.C., Kuhls, T.L., Mosier, D.A., Crawford, D.L. and Greenfield, R.A. (1993). *Cryptosporidium parvum* infection after abrogation of natural killer cell activity in normal and severe combined immunodeficiency mice. *Journal of Parasitology* **79**, 295–297.

Rossi, P., Pozio, E., Besse, M.G., Morales, M.A.G. and Larosa, G. (1990). Experimental cryptosporidiosis in hamsters. *Journal of Clinical Microbiology* **28**, 356–357.

Savage, D.C., Dubos, R. and Schaedler, R.W. (1968). The gastrointestinal epithelium and it autochthonous bacterial flora. *Journal of Experimental Medicine* **127**, 67–75.

Sherwood, D., Angus, K.W., Snodgrass, D.R. and Tzipori, S. (1982). Experimental cryptosporidiosis in laboratory mice. *Infection and Immunity* **38**, 471–475.

Sypek, J.P., Chung, C.L., Mayor, S.E., Subramanyam, J.M., Goldman, S. J., Sieburth, D.S., Wolf, S.F. and Schaub, R.G. (1993). Resolution of cutaneous leishmaniasis: interleukin-12 initiates a protective T helper 1 immune response. *Journal of Experimental Medicine* **177**, 1797–1802.

Tagliabue, A., Befus, A.D., Clark, D.A. and Bienenstock, J. (1982). Characteristics of natural killer cells in the murine intestinal epithelium and lamina propria. *Journal of Experimental Medicine* **155**, 1785–1796.

Tatalick, L.M. and Perryman, L.E. (1995). Effect of surface antigen-1 (SA-1) immune lymphocyte subsets and naive cell subsets in protecting *scid* mice from initial and persistent infection with *Cryptosporidium parvum*. *Veterinary Immunology and Immunopathology* **47**, 43–55.

Theodos, C.M., Sullivan, K.L., Griffiths, J.K. and Tzipori, S. (1997). Profiles of healing and nonhealing *Cryptosporidium parvum* infection in C57BL/6 mice with functional B and T lymphocytes: the extent of gamma interferon modulation determines the outcome of infection. *Infection and Immunity* **65**, 4761–4769.

Tilley, M., McDonald, V. and Bancroft, G.J. (1995). Resolution of cryptosporidial infection in mice correlates with parasite-specific lymphocyte proliferation associated with both Th1 and Th2 cytokine secretion. *Parasite Immunology* **17**, 459–464.

Tzipori, S., Rand, W. and Theodos, C. (1995). Evaluation of a two-phase scid mouse model preconditioned with anti-interferon-γ monoclonal antibody for drug testing against *Cryptosporidium parvum*. *Journal of Infectious Diseases* **172**, 1160–1164.

Ungar, B.L.P., Burris, J.A., Quinn, C.A. and Finkelman, F.D. (1990). New mouse models for chronic *Cryptosporidium* infection in immunodeficient hosts. *Infection and Immunity* **58**, 961–969.

Ungar, B.L.P., Kao, T.-Z., Burris, J.A. and Finkelman, F.D. (1991). *Cryptosporidium* infection in an adult mouse model. Independent roles for IFN-γ and CD4+ T lymphocytes in protective immunity. *Journal of Immunology* **147**, 1014–1022.

Urban, J.F., Fayer, R., Chen, S.-J., Gause, W.C. Gately, M.K. and Finkelman, F. (1996). IL-12 protects immunocompetent and immunodeficient neonatal mice against infection with *Cryptosporidium parvum*. *Journal of Immunology* **156**, 263–268.

Vakil, N.B., Schwartz, S.M., Buggy, B.P., Brummitt, C.F., Kherellah, M., Letzer, D.M., Gilson, I.H. and Jones, P.G. (1996). Biliary cryptosporidiosis in HIV-infected people after the waterborne outbreak of cryptosporidiosis in Milwaukee. *New England Journal of Medicine* **334**, 19–23.

Vassalli, P. (1992). The pathophysiology of tumor necrosis factors. *Annual Review of Immunology* **10**, 411–452.

Waters, W.R. and Harp, J.A. (1996). *Cryptosporidium parvum* infection in T-cell receptor (TCR)-α- and TCR-γ-deficient mice. *Infection and Immunity* **64**, 1854–1857.

Wershil, B.K., Theodos, C.M., Galli, S.J., and Titus, R.G. (1994). Mast cells augment lesion size and persistence during experimental *Leishmania major* infection in the mouse. *Journal of Immunology* **152**, 4563–4571.

Wherry, J.C., Schreiber, R.D. and Unanue, E.R. (1991). Regulation of γ-interferon by natural killer cells in SCID mice: Roles of tumor necrosis factor and bacterial stimuli. *Infection and Immunity* **59**, 1709–1715.

Whitmire, W. and Harp, J. (1990). *In vitro* murine lymphocyte blastogenic responses to *Cryptosporidium parvum*. *Journal of Parasitology* **76**, 450–452.

Whitmire, W. and Harp, J. (1993). Characterization of bovine cellular and serum antibody responses during infection by *Cryptosporidium parvum*. *Infection and Immunity* **59**, 990–995.

Wolf, S.F., Sieburth, D., and Sypek, J. (1994). Interleukin 12: a key modulator of immune function. *Stem Cells* **12**, 154–168.

Wyatt, C.R., Brackett, E.J., Perryman, L.E., Rice-Ficht, A.C., Brown, W.C. and O'Rourke, K.I. (1997). Activation of intestinal intraepithelial T lymphocytes in calves infected with *Cryptosporidium parvum*. *Infection and Immunity* **65**, 185–190.

Yang, S. and Healey, M.C. (1993). The immunosuppressive effects of dexamethasone administered in the drinking water to C57BL/6N mice infected with *Cryptosporidium parvum*. *Journal of Parasitology* **79**, 626–630.

Yang, S. and Healey, M.C. (1994). Patent gut infections in immunosuppressed adult C57BL/6N mice following intraperitoneal injection of *Cryptosporidium* oocysts. *Journal of Parasitology* **80**, 338–342.

Antibody-based Immunotherapy of Cryptosporidiosis

Joseph H. Crabb

*ImmuCell Corporation,
Portland, ME 04103, USA*

1. Introduction: Rationale for Therapy using Antibodies to
   Combat Cryptosporidiosis 122
   1.1. Role of humoral immunity in host resistance to cryptosporidiosis 123
2. Practical Considerations for Antibody Therapy 125
   2.1. Theoretical mechanisms of antibody action 126
   2.2. Site of infection and therapeutic strategy 128
   2.3. Fate and formulation of antibodies 128
   2.4. Therapy or prophylaxis 129
3. Early Studies of Immunotherapy for Cryptosporidiosis 130
   3.1. Case reports using bovine colostrum 130
   3.2. Passive transfer to offspring from convalescent or vaccinated animals 131
4. Laboratory Investigations of Anti-*Cryptosporidium* Antibodies 131
   4.1. Prophylaxis and use of antibodies *in vitro* 132
   4.2. Polyclonal antibodies 132
   4.3. Monoclonal antibodies 137
   4.4. Immunoglobulins G and M versus secretory immunoglobulin A 138
5. Studies in Humans using Antibodies 138
   5.1. Case reports 138
   5.2. Controlled studies 140
6. Future Prospects 141
   6.1. Monoclonal antibody 'cocktails' and polyclonal phage display 141
   6.2. Specific antigen-stimulated hyperimmune bovine
        colostrum immunoglobulin 142
   6.3. Immunotoxins 142
   6.4. Antibody prophylaxis 143
   6.5. Antibody–drug combination therapy 143
   References 143

Passive antibody immunotherapy (PAI) for cryptosporidiosis is a treatment strategy that has been actively pursued in laboratory studies and early-stage clinical studies for the last decade. Several experimental approaches have been initiated, including use of bovine colostrum and colostral antibodies (hyperimmune and natural), monoclonal antibodies, chicken egg yolk antibodies, and even orally administered human plasma antibodies. Most studies have employed oral administration to treat or prevent this intestinal infection. The interest in this treatment strategy has been sparked by the lack of an effective or approved therapy, increased awareness of the widespread nature of this parasite, epidemiological evidence that humoral immunity plays an important role in host resistance, and several early case reports of antibody therapy in which remarkable resolution of the disease was observed. Most studies using a variety of preparations of antibodies administered to animals and humans have shown some degree of efficacy, though the responses have been, for the most part, partial rather than complete resolution of the disease. This chapter examines critically the scientific rationale and the evidence for PAI for cryptosporidiosis, including practical considerations and future approaches.

1. INTRODUCTION: RATIONALE FOR THERAPY USING ANTIBODIES TO COMBAT CRYPTOSPORIDIOSIS

Passive antibody immunotherapy (PAI) for cryptosporidiosis has been pursued as both a research and product development strategy in humans and animals of veterinary significance for over a decade. During this period, over 40 publications have appeared describing the effects of various preparations of immune and nonimmune serum, hyperimmune bovine colostrum (HBC) and hyperimmune bovine colostral immunoglobulins (HBCIg), hen egg yolk antibodies (EYA), monoclonal antibodies (mAb), and orally administered pooled human immunoglobulins for intravenous administration (Igiv) against cryptosporidiosis in animals and humans (see Tables 1 and 2). In addition, no less than five commercial firms have anti-*Cryptosporidium* immunoglobulin products in various stages of human clinical trials. This approach has been encouraged by several factors:

(i) the increasing clinical significance of cryptosporidiosis in immunocompromised individuals (primarily those with acquired immune deficiency syndrome (AIDS)) and as a general public health threat due to waterborne disease outbreaks;
(ii) the lack of approved or recognized effective therapy against this parasite (after the failure of over 100 conventional and nonconventional compounds in various settings);
(iii) demonstrated success of PAI or prophylaxis using antibodies against other enteric infections;
(iv) epidemiological evidence of an important role for the humoral immune system in host control of the infection; and
(v) the description in several early case reports of spectacular reversal of

disease after treatment with *Cryptosporidium* HBC, which contains high levels of immunoglobulins (Igs). Whereas factors (i) and (ii) (above) have continued to increase in relevance to the *Cryptosporidium* problem, factors (iv) and (v) have decreased in relevance as rationales for the continued development of PAI against this disease. This report examines the foundations supporting continued development of PAI and recent results obtained using this strategy to combat cryptosporidiosis.

1.1. Role of Humoral Immunity in Host Resistance to Cryptosporidiosis

1.1.1. At-risk Populations Include Antibody-deficient Individuals

While it is well established that chronic or persistent *Cryptosporidium* infections are most prevalent in immunodeficient individuals, an oft cited rationale for pursuing PAI against cryptosporidiosis includes reports of increased incidence of chronic cryptosporidiosis in individuals with various antibody deficiency states. Persistent cryptosporidiosis has been described in patients with common variable hypogammaglobulinemia (Koch *et al.*, 1983), X-linked hyper-IgM syndrome (Notarangelo *et al.*, 1992), congenital hypogammaglobulinemia (Lasser *et al.*, 1979; Tzipori *et al.*, 1986), and IgA deficiency (Jacyna *et al.*, 1990; Melamed *et al.*, 1991). The overall incidence rates of chronic cryptosporidiosis in antibody immunodeficiency states has not been studied in any systematic fashion, and the literature reports are, for the most part, case reports. In some of these case reports, the antibody immunodeficiency status was associated with another underlying immunodeficiency (Koch *et al.*, 1983; Jacyna *et al.*, 1990), while in others the antibody defect appeared to be the only immune defect (Tzipori *et al.*, 1986). While the exact molecular etiologies remain to be elucidated in many of these immunodeficient, recent investigations suggest that most antibody immunodeficiencies are part of pleiotropic defect(s) affecting both humoral and cellular immune functions to some extent (Buckley, 1986; Zegers *et al.*, 1991). In certain cases, notably X-linked hyper-IgM syndrome and, to a lesser extent, common variable immunodeficiency, a specific defect in CD40 ligand is known to be involved (Banchereau *et al.*, 1994; Eibl and Wolf, 1995). This defect is now known to affect profoundly both B-cell maturation and T helper 1 and 2 (Th1–Th2) cellular responses (Banchereau *et al.*, 1994; Campbell *et al.*, 1996).

1.1.2. Epidemiologic Studies

Epidemiologic studies cited in support of a role for antibody in host resistance to cryptosporidiosis are mainly breast-fed versus non-breast-fed incidence rates. Specific studies report dramatically lower incidence rates in wholly breast-fed infants compared to those in bottle-fed babies (Mata et al., 1984; Hojlyng et al., 1986), while another study reported little or no difference (Mathan et al., 1985). It is not clear, however, that maternal secretory IgA (sIgA) in the breast milk was responsible for the lower incidence rates; it is reasonable to assume that contaminated water used in the formula preparation, consumption of contaminated foods, or simply poorer nutritional status led to the observed higher rates in non-breast-fed children, particularly when viewed in the light of documented higher rates in malnourished children and the finding of contaminated weaning foods in endemic areas (Current and Garcia, 1991). Further, conclusions drawn from retrospective analyses (such as the influence of breast milk Igs on incidence of cryptosporidiosis in these studies) must be viewed with caution, as the number of potential variables that could influence that particular outcome is large, and they would need to be identified and controlled in a prospective manner (for example, anti-*Cryptosporidium* antibody levels in the milk, associations with concurrent enteric infections, immune status of the subjects, etc.). Laboratory studies addressing the role of breast feeding in susceptibility to cryptosporidiosis appear to be a more appropriate mechanism to answer this question (see Section 3.2).

1.1.3. AIDS Patients: Antibody Status

The development of chronic cryptosporidiosis in AIDS patients has been correlated to CD4 status (Blanshard et al., 1992; Flanigan et al., 1992). In these studies, CD4 counts below 140 cells/mm^3 were associated with a higher proportion of persistent disease, while CD4 counts above 180 cells/mm^3 were associated with more spontaneous resolutions (Flanigan et al., 1992). Patients with CD4 counts below 50 cells/mm^3 were associated with fulminant cryptosporidiosis, having a very poor prognosis (Blanshard et al., 1992). Since CD4+ T cells orchestrate both cellular and humoral responses, this association alone cannot distinguish between humoral or cellular roles in host resistance to cryptosporidiosis. Recent studies on the role of cell-mediated immunity (CMI) have suggested strongly that it is sufficient to confer resistance to this infection (Ungar et al., 1991; Chen et al., 1993; Rehg, 1996; Urban et al., 1996). Further, ablation of antibody production in mice by administration of anti-μ antibodies had virtually no effect on neonatal mouse susceptibility to *Cryptosporidium* infection, or to resolution of the disease during maturation of these mice (Kilani et al., 1990). AIDS patients with cryptosporidiosis have been shown to mount humoral responses of IgG, IgM, and IgA class (Ungar and Nash, 1986; Ungar et al., 1986; Cozon et al., 1994). In Western blot studies, most sera from patients infected with *Cryptosporidium*, whether or not they had AIDS, recognized the 23kDa antigen, but non-AIDS patients' sera recognized a significantly wider array of antigens (Ungar and Nash, 1986). This may indicate a suboptimal functional antibody response in the AIDS patients; so, while antibody responses developed in AIDS patients, the lack of cell-mediated immunity (CMI) to help in the maturation of the antibody response could lead to a functionally ineffective antibody response. Indeed, antibody levels to the 23-kDa antigen in three infected AIDS patients declined significantly during the course of the infection, indicating impaired antibody response during the infection. With regard to vaccine responses in individuals infected with human immunodeficiency virus (HIV), responses to pneumococcal polysaccharide vaccines provide an enlightening illustration. The nonresponder group is associated with a lower CD4 cell count than the group that responded but, among those who responded, IgG levels were similar to those of normal control subjects (Rodriguez-Barradas et al., 1996). Interestingly, functional antibody responses to pneumococcal vaccines and to infection, as measured by an opsonophagocytosis assay, were significantly lower in HIV-infected individuals than in uninfected persons (Janoff et al., 1993). This was cited as a factor in the higher rates of pneumococcal disease in HIV-infected individuals found in the study, despite similar colonization rates compared to controls. This finding is relevant to generalizations of functional versus nonfunctional antibody responses in AIDS patients, particularly since host resistance to pneumococcal disease is considered to be determined primarily by humoral immunity. Relatively high levels of specific sIgA were found in secretions from chronic cryptosporidiosis patients with AIDS, and this was cited as evidence for the lack of a role of antibodies in host resistance against this parasite (Cozon et al., 1994). The requirement of some functional CMI in order to generate sIgA responses was cited to provide support for the idea that the antibody observed did not reflect mere polyclonal B-cell stimulation. However, without determining the functionality of the antibodies detected, it would be inappropriate to cite this as evidence for the lack of a role of antibody in host resistance to cryptosporidiosis.

2. PRACTICAL CONSIDERATIONS FOR ANTIBODY THERAPY

For PAI to be effective, antibody must be present at the site of infection long enough to interact with the appropriate stage or stages of the organism's life cycle, bind to the appropriate antigen(s), and activate the appropriate effector mechanisms (if any). Thus, for an intestinal parasite such as *Cryptosporidium*, considerations for an effective therapy include dosage schedule, gastrointestinal transit time and residence of the antibody in the small intestine (the primary site of infection), gastrointestinal degradation of the antibody and ways of protecting the antibody activity (formulation), and the antigenic specificity and antigenic relatedness of strains or isolates of *Cryptosporidium parvum*.

2.1. Theoretical Mechanisms of Antibody Action

2.1.1. Parasite Stages Susceptible to Antibodies

Cryptosporidium is an intracellular pathogen; therefore, effective PAI would generally require extracellular invasive stages (sporozoite or merozoite) to be present to permit antibody–parasite interaction and neutralization. Neutralization, in this case, is blocking the parasite's interaction with the host cell and preventing attachment and/or invasion. This mechanism is similar to the mechanism of antibody neutralization of a wide array of viruses and parasites, and is generally most effective in prophylaxis or very early infection (exemplified by the mode of action of several viral vaccines) (Whitton and Oldstone, 1990). It is feasible to have antibodies that do not neutralize attachment or invasion, but are brought into the host cell bound to the sporozoite and disrupt parasite development, resulting in an aborted infection or the production of defective oocysts. Several references speculate about this mechanism, but there are no objective data to support or refute it at the present time. A novel mechanism of intracellular antibody action has been recently described in rotavirus infections in mice, whereby non-neutralizing IgA mAbs were introduced to the mice via IgA-producing 'backpack' hybridoma tumors (Burns *et al.*, 1996). Interestingly, antibodies to neutralizing antigens VP7 and VP4 were not effective therapeutically, but antibodies to a structural antigen VP6 were effective both prophylactically and therapeutically. The data strongly supported the hypothesis that the effectiveness was due to anti-VP6 sIgA disrupting intracellular virus assembly during transcytosis through the infected epithelial cells. This finding provided a strong rationale for the use of antibodies for therapy of established infections with intracellular microbes. Such antibodies would have to be polymeric IgA, and delivered in a manner that retained the capacity for transcytosis (i.e. systemic administration, capable of binding the secretory component of IgA). In contrast, orally administered antibodies (which act primarily luminally) need to be targeted to surface neutralizing antigens, or antigens that could affect disruption of development upon entry into the host cell.

2.1.2. Role of Effector Mechanisms

Effector mechanisms (e.g. complement activation/lysis or opsonophagocytosis, binding to Fc receptors/antibody-dependent cell-mediated cytotoxicity), could greatly facilitate the action of antibodies acting luminally against the extracellular stages of *Cryptosporidium*. If such mechanisms were operative, potential targets could include all extracellular stages (oocyst, micro- and macrogametes, sporozoite, merozoite), and not be restricted to the invasive stages. Phagocytic and effector cells (lymphocytes, macrophages, mast cells, neutrophils) are present in the lamina propria and, to some extent, the epithelium, and are thought to be shed into the lumen of the gut at a low basal level (Brandtzaeg et al., 1987). Intraepithelial lymphocytes (IELs) are a unique example of effector cells present in the epithelium of the gut. The majority of IELs are γδ T cells, and appear to home directly to epithelial tissues from the circulation, rather than passing through a secondary lymphoid site (Picker and Butcher, 1992).

While effector cells are present to some extent at the relevant site of infection, it has not been shown that effector mechanisms are operative in PAI of cryptosporidiosis. Antibodies used for therapy against *Cryptosporidium* have been primarily mouse mAb, bovine polyclonal antibody, or egg antibodies; under these conditions, triggering of effector mechanisms would have to occur across species. Cross-species Fc receptor specificity for antibodies has been examined most extensively in the human and murine systems, and several Fcγ receptors of both species have binding affinities for both human and mouse IgG (Ravetch and Kinet, 1991).

2.1.3. Antibody Interactions with the Parasite

As described above, effective therapeutic antibodies should target appropriate epitopes on invasive parasites, extracellular stages, or intracellular targets of development. A hallmark of protozoan parasites is evasion of the immune system, and in many cases this is accomplished by stage-specific expression of unique antigens (thereby causing immune responses to stages that are subsequently lost during development). Well-characterized examples of such stage specific antigen expression are found among the apicomplexans *Plasmodium* spp. and *Toxoplasma gondii*, as well as the trypanosomes (Cross, 1990; Joiner and Dubremetz, 1993; Nardin and Nussenzweig, 1993). *Cryptosporidium* is unique among these protozoans in that neutralizing antigens are expressed on more than one stage (Bjorneby et al., 1990). Interestingly, HBCIg generated against oocyst antigens was found to label all stages in an immunoelectron microscopic study of the stage specificity of this antibody (Fayer et al., 1991).

Stage specificity appears to be less of a potential problem with PAI for cryptosporidiosis than with other protozoans, but the question of strain-to-strain antigenic variation still remains. Unfortunately, because of the inability to propagate *Cryptosporidium in vitro*, specific 'strains' do not exist; hence, references are made to 'isolates' of this parasite, which can be maintained indefinitely by passage through neonatal ruminants. To a limited extent, it has been shown that, while antigenic differences between isolates can be demonstrated (Mead et al., 1990), important protective epitopes are shared among diverse isolates (Uhl et al., 1992). This finding minimizes the fear that different isolates will be dramatically different in their response to anti-*Cryptosporidium* antibodies, but the possibility of antigenic change must be continuously monitored.

2.2. Site of Infection and Therapeutic Strategy

2.2.1 Routes of Administration

As cryptosporidiosis is most commonly a disease restricted to the enterocytes of the small intestine (in humans), many studies have focused on oral, rather than systemic, administration of antibodies. There are advantages in the oral dosage approach, the principal being that a large quantity of antibodies (resulting in high local concentrations) can be given without the serious safety concerns related to systemic administration of foreign proteins (as would be the case for xenogeneic, systemic administration). However, the drawbacks related to oral antibody therapy include the restriction of activity to the lumen of the gut, the possibility of gastrointestinal destruction of the antibody, and the limitation of the duration of antibody availability due to peristalsis. Compounding these concerns is the clinical significance of extraintestinal cryptosporidiosis. Indeed, the biliary tree is not an uncommon site of infection in many patients who have chronic cryptosporidiosis (Current and Garcia, 1991). While it is assumed that these extraintestinal infections result from dissemination from the primary intestinal site in immunodeficient hosts, the possibility has been raised that the biliary or gallbladder sites may be a cryptic reservoir of infectious parasites, disseminating to intestinal tissues during the progression of immunodeficiency (Soave, 1995). In the case of biliary infection, orally administered antibody has limited access. Biliary tract infection would be amenable only to secretory antibody therapy (i.e. sIgA) administered parenterally; alternatively, oral antibody could be administered directly into the bile duct via endoscopic retrograde cholangiopancreatoscopy.

2.3. Fate and Formulation of Antibodies

Orally administered antibodies would be expected to be degraded in the gastrointestinal tract, given the low gastric pH and the presence of gastric and intestinal proteases. However, HBC and HBCIg have been reported to be particularly resistant to gastrointestinal destruction, relative to other Igs. Early studies of administering HBCIg for other gastrointestinal infections cited remarkable 'resistance to proteolytic digestion in the intestine of a heterologous host' (Brussow et al., 1987). Indeed, there is evidence that bovine IgG1 is more resistant to proteolysis by trypsin, chymotrypsin and pepsin than other Igs (Butler, 1983; McClead and Gregory, 1984). Approximately 20% of immunoreactive bovine colostral Ig (as measured by radial immunodiffusion) was detectable in the ileal samples of human volunteers who consumed about 5 g of a ^{15}N-labeled bovine colostral Ig preparation after fasting (Roos et al., 1995). This recovery was cited as evidence of significant gastrointestinal resistance by comparison to the normal ileal digestibility of other milk proteins, reported in the literature to be 90% or more. Unfortunately, the 20% immunoreactive Igs detected in this study were not shown to be functional (i.e. antigen binding). Bovine Igs have also been cited as being particularly resistant to stomach acid (Rump *et al.*, 1992). Unfortunately, this assertion was based on exposure to 'artificial gastric juice (pH 2.4–4.0) for 24 h', but normal gastric juice is considered to have a pH$\leq$ of 1–2 (Kutchai, 1988). Indeed, exposure of HBCIg to pH 2.0 results in irreversible denaturation of antigen binding activity (J.H. Crabb and N.B. Turner, unpublished observations). Further, in studies of gnotobiotic piglets, even after pretreatment of the animals with cimetidine (which raised the mean gastric pH to 3.13), only 0.2–9% of the total activity of HBCIg was recovered in the stool (Tzipori *et al.*, 1994). These data indicate that the majority of antibody activity is destroyed during passage through the gastrointestinal tract even with modified gastric pH, and that issues of formulation need to be addressed to improve the proportion of active antibodies delivered via the oral route. The development of dosage formulations for oral immunotherapy of cryptosporidiosis must meet criteria of gastric resistance followed by rapid release in the proximal small intestine. Some of the formulations that could fulfill these criteria include co-administration with antacids, enteric coatings (pH dependent release), bioerodable coatings (delayed release, pH independent), or liposomal formulations (enzymatic induced release); in addition to protective coatings, formulations must be able to disperse rapidly when the coating is shed.

The transit time of orally administered antibody varies considerably, and is dependent on the individual subject's physiology, fed or fasted status, type of food ingested, etc. In the fasted state, antibodies have a transit time through the small intestine of the order of 1–2 hours (Roos *et al.*, 1995). In contrast, systemic administration of antibodies would be expected to have a much longer duration of activity, as IgA has a half-life in circulation of about 6 days. Therefore, oral antibody therapy requires several doses per day, whereas systemic or injectable antibody therapy would require doses of the order of once per week.

2.4 Therapy or Prophylaxis

With regard to the use of antibodies for the treatment of cryptosporidiosis, the vast majority of the studies reported to date have been concerned with treatment of established infection, due to the lack of any effective therapy for this infection. However, when studies of antibody therapies for other infections are reviewed (particularly gastrointestinal infections), prophylactic uses have proven to be much more effective than treatment of established infections. In this regard, oral antibodies have been used successfully in human studies of enterotoxigenic *Escherichia coli*-induced 'travelers diarrhea' (Tacket *et al.*, 1988; Freedman *et al.*, 1996), rotavirus infections in infants (Davidson *et al.*,

1989), *Shigella* infections (Tacket *et al.*, 1992), necrotizing enterocolitis (Eibl *et al.*, 1988), and *E. coli* infections in infants (Mietens *et al.*, 1979). With one exception (Mietens *et al.*, 1979), all these studies used antibodies prophylactically rather than as a treatment. Interestingly, in all the studies referred to above, clinical success was achieved, but sterilizing immunity was not established by the use of the passive antibodies. Animal studies of antibodies against *Cryptosporidium* using prevention rather than treatment models have corroborated this finding (Fayer *et al.*, 1989a,b; Perryman *et al.*, 1990; Doyle, *et al.*, 1993). Thus, while clinical benefit was observed and shedding of organisms was dramatically reduced, colonization still occurred, albeit at a dramatically reduced level (where studied). An exception to these results was the finding of sterilizing immunity when isolated sporozoites were preincubated with immune bovine serum before intrarectal inoculation (Riggs and Perryman, 1987).

These findings may be relevant to the use of antibodies for treating cryptosporidiosis (particularly in the AIDS patient population). Indeed, conventional wisdom in management of many opportunistic infections in AIDS patients has shifted towards prophylaxis rather than treatment. Prophylaxis against cryptosporidiosis with oral antibodies may not be practical from a cost standpoint, and the decision to conduct prospective clinical trials of prevention may not survive a cost–benefit analysis. In fact, most prophylactic uses of antimicrobial substances are implemented after demonstration of efficacy in therapeutic studies (due to the high cost of prospective prophylaxis studies in humans). Nevertheless, antibody prophylaxis against cryptosporidiosis may prove more effective than therapy.

3. EARLY STUDIES OF IMMUNOTHERAPY FOR CRYPTOSPORIDIOSIS

3.1. Case Reports using Bovine Colostrum

Early accounts of successful treatment of cryptosporidiosis described the use of bovine colostrum (BC) from cows hyperimmunized with oocyst–sporozoite mixtures; treatment of individual cases of cryptosporidiosis associated with AIDS, hypogammaglobulinemia, or chemotherapy resulted in dramatic responses to therapy (Tzipori *et al.*, 1986, 1987). These studies were initiated, in part, on the basis that cryptosporidiosis is a major disease in the newborn calf, that most cows have pre-existing immunity to the parasite, that immunization of cows would be expected to boost naturally acquired immunity, and that colostral feeding to calves had been shown to confer immunity against a variety of early calf infections. These early reports sparked a growing interest in this approach to immunotherapy, but numerous questions arose: (i) What component(s) in colostrum were responsible for the effect? (ii) Could normal bovine colostrum confer similar effects (given the immune status of natural cows? (iii) Could therapy be given orally (as opposed to via nasogastric tube, as used in these early studies)? (iv) If antibody was the important component, which immunoglobulin classes were primarily responsible for the effects? (v) Could mAb against specific sporozoite antigens confer similar efficacy? A case study similar to those referred to above, but using colostrum from cows with natural anti-*Cryptosporidium* antibodies, indicated that such preparations were ineffective in treating the disease in three patients (Saxon and Weinstein, 1987). These and other results have led to the belief that hyperimmunization is required to produce an efficacious preparation. The studies discussed below have provided answers to many of these questions.

3.2. Passive Transfer to Offspring from Convalescent or Vaccinated Animals

The early case reports coupled with intriguing epidemiological studies on the effects of breast feeding (Section 1) led to studies investigating the role of lactogenic immunity on susceptibility to cryptosporidiosis. Mice convalescent from experimental *Cryptosporidium* infections, and those which received booster inoculations, were unable to confer protection to their suckling offspring when the latter were challenged with oocysts (Moon et al., 1988). In a similar study, colostrum from cows inoculated with oocysts in complete Freund's adjuvant during 4–5 months before parturition was unable to affect the outcome in calves challenged with oocysts (Harp et al., 1989). Interestingly, in both these studies, colostral anti-*Cryptosporidium* titers were only moderately elevated compared with the control colostrum, and were dramatically lower than the titers reported after hyperimmunization regimens (albeit 'titers' are difficult to compare, as they reflect methods unique to the particular study being conducted). The lack of acquired resistance in breast-fed primates whose mothers had been inoculated with 100 oocysts was cited as evidence to refute the role of lactogenic immunity, but the mothers were not demonstrated to have significant anti-*Cryptosporidium* antibodies (Miller et al., 1991).

4. LABORATORY INVESTIGATIONS OF ANTI-*CRYPTOSPORIDIUM* ANTIBODIES

Investigations subsequent to these earlier reports have begun to address the above-raised questions regarding PAI, and have gone on to identify several potentially important antigens of *Cryptosporidium*, to begin defining stage-specific expression of relevant protective antigens, and to characterize isolate variability.

4.1. Prophylaxis and Uses of Antibodies *in vitro*

Important studies that established the role of antibodies in blocking infection by *Cryptosporidium* were neutralization studies *in vivo* and *in vitro*. By incubating isolated sporozoites with hyperimmune bovine serum, Riggs and Perryman (1987) neutralized their infectivity for neonatal mice. Using mAbs and HBC, it was later established that specific antigens on the sporozoite and the merozoite surface were involved in the neutralization of infectivity, and thus were potentially important targets for immunotherapy (Fayer *et al.*, 1989a; Riggs *et al.*, 1989; Bjorneby *et al.*, 1990; Perryman *et al.*, 1990). Using HBC, HBC whey, and HBCIg, studies *in vitro* confirmed that hyperimmune preparations could inhibit infection by *Cryptosporidium* in cell culture models significantly better than nonimmune preparations (Flanigan *et al.*, 1991; Doyle *et al.*, 1993). Inhibition was dependent on antibody dose, but not complete, even at the highest concentrations used (1000 µg/ml). This suggests that either the antibody had limited access to the sporozoites when oocysts were used for inoculation (which was done in both studies), or the antibodies reacted incompletely with the necessary neutralization epitopes. Interestingly, the isolates used for infection differed from those used for generation of the HBC. In addition, HBCIg antibodies eluted from Western blots of whole oocysts were significantly inhibitory and were about 10-fold more effective on a mass basis than unfractionated HBCIg (Doyle *et al.*, 1993).

4.2. Polyclonal Antibodies

4.2.1. Bovine Colostrum, Hyperimmune Bovine Colostrum and HBC Immunoglobulins

Given that over half of the laboratory studies presented in Table 1 and virtually all the human clinical studies to date have focused on BC and preparations derived therefrom, it is appropriate to review the properties of bovine colostral Ig. Unlike human colostrum, which contains sIgA as its major Ig class (>90% of Ig), BC antibodies are composed of >90% IgG1 (Butler, 1983). While human neonatal passive immunity is acquired through transplacental IgG transport and oral sIgA transfer from mother's milk to baby, no transplacental Ig transport occurs in bovines because of the multilayered placental tissue (five layers in ruminants = syndesmochorial) (Tizard, 1987). Thus, all passive immunity in calves is acquired through colostrum suckling in the first days of life. Unique mechanisms have evolved to accomplish efficient transfer of Ig in bovines. In a process called 'bagging-up', massive quantities (500–750 g) of circulating Ig are transported to the udder before parturition. BC contains both dimeric IgA (largely sIgA) and monomeric IgA, with the former derived primarily from local secretions and the latter derived directly from the circulation, together with IgG. The IgAs represent less than 5% of the total Ig in colostrum (Butler, 1983). It is not known precisely what the ratio of dimeric IgA to monomeric IgA is in colostrum, but it appears that the monomeric component is more prominent. Further, this ratio does not appear to differ after local or systemic immunization, as HBCIg generated by immunizations by intramuscular or subcutaneous injections contains the same sIgA/serum IgA ratio as colostrum from intramammary immunizations (J.H. Crabb and N.B. Turner, unpublished observations). Thus BC is a convenient source of antibodies, containing more than 50 times the concentration of antibodies found in conventional cow's milk. Most processes to extract the antibodies from BC retain all the Igs originally present in the colostrum, including IgM, IgA, and sIgA. With regard to which Ig class(es) are responsible for efficacy observed in studies employing BCIg or HBCIg, Fayer *et al.* (1990) demonstrated therapeutic efficacy in HBC-derived IgG, IgA, and IgM. This finding confirmed that antibodies are a primary active constituent in colostrum used for PAI.

As discussed above, because cryptosporidiosis is endemic in cattle herds (90% prevalence on US dairy farms) (USDA, 1993), pre-existing immunity is prevalent in cows. Cows can be relatively easily immunized during the 'dry' (i.e. nonlactating) period, when they receive most of their commercial vaccinations. Colostrum is collected for the first 2 days after calving, and can be frozen for pooling and eventual processing. Thus, production of HBC can be integrated relatively easily into commercial dairy practices. Since colostrum is merely early milk, and the antibody profile (relative proportion of Ig classes) is very similar to that found in commercial milk, there are few safety concerns related to the oral administration of BCIg (as long as no pathogen is present, and issues related to lactose intolerance and milk protein allergies are addressed). These are some of the features that have led to the attractiveness of BCIg as a commercially feasible therapeutic product.

BC contains, in addition to antibodies and nutrients, a plethora of serum and cellular constituents, including, but not limited to the following: lymphocytes, cytokines, enzymes, bioactive proteins such as lactoferrin and lactoglobulin, growth factors, peptides, transfer factor, protease inhibitors, bacteria, blood, minerals, and trace elements. Non-Ig substances in BC and serum have been shown to be inhibitory to *Cryptosporidium* (Flanigan *et al.*, 1991; Hill *et al.*, 1993; Watzl *et al.*, 1993). Thus, it became important to develop a reasonably well-characterized and standardized fraction of BC that consisted substantially of antibodies, for testing and eventual commercialization. While the multiple components found in BC could potentially have medical benefits, the most abundant proteins capable of reacting specifically with immunogens (and thus amenable to standardization) are antibodies.

Being a polyclonal antibody preparation, HBCIg and BCIg (in particular) contain *Cryptosporidium*-preparation specific antibodies as a relatively small proportion of the total antibody present. As discussed above, when HBCIg was subjected to binding and elution from Western blots, the neutralization capacity increased by an order of magnitude. Consistent with these findings, it is uncommon for even the best hyperimmune antibody preparations to contain more than 10% specific antibody. This could result in the requirement for administration of relatively large amounts of irrelevant antibodies during treatment. Alternatively, the irrelevant antibodies could 'buffer' the action of gastrointestinal degradative mechanisms, enabling a larger percentage of the specific antibody to reach the site of action than might otherwise have been possible. The practical considerations of therapy with BCIg are discussed in Section 2.

Laboratory studies with HBC and HBCIg have been instrumental in identifying several potentially important target antigens. A high molecular mass proteoglycan, termed GP900, that appears to be associated with the sporozoite, is a major antigen recognized by HBCIg (Petersen *et al.*, 1992). A circumsporozoite (CSP) like reaction, similar to that observed when *Plasmodium* sporozoites are incubated with neutralizing serum, is induced in sporozoites incubated with HBCIg (Riggs *et al.*, 1994). This preparation reacts with several high molecular mass antigens on both sporozoites and merozoites, and confirms the capacity of HBCIg to react with multiple *Cryptosporidium* stages discussed in Section 2 (Fayer *et al.*, 1991; Riggs *et al.*, 1994). Evidence has emerged linking the CSP-like reaction with a high molecular mass glycoprotein that could be the equivalent of GP900 (Riggs *et al.*, 1996). HBCIg has been shown to react with more than 30 specific antigens in Western blots; several of these antigens have apparent molecular masses similar to those of antigens recognized by efficacious mAbs (Tilley *et al.*, 1990; Riggs *et al.*, 1996).

In terms of efficacy of BC and its derivatives, Tables 1 and 2 list the major studies employing BC preparations. In review, the studies that demonstrated the highest efficacy against cryptosporidiosis were those using very high titered material; nonimmune and low titered material appears to be less effective. Further, studies employing prophylactic strategies or treatment of early stage infections appeared to have a higher degree of efficacy than studies engaged in treating established, chronic infection. In fact, the only laboratory studies demonstrating complete protection incubated isolated sporozoites with antibody preparations before intrarectal inoculation (Fayer *et al.*, 1989a; Riggs *et al.*, 1989; Bjorneby *et al.*, 1990; Perryman *et al.*, 1990). This relatively better outcome, associated with early treatment or prophylaxis, extends to virtually every laboratory study and antibody preparation reviewed here (Table 1). This finding is consistent with the luminal action of orally

Table 1 Antibody prophylaxis and treatment for cryptosporidiosis: experimental studies.

Reference	Study design			
	Preparation[a]	Species	Prevention/treatment[b]	Outcome[b,c]
Riggs and Perryman (1987)	HBS	Mice (neonates)	P	SS
Moon et al. (1988)	IMC	Mice (neonates)	P	NR
Harp et al. (1989)	HBC	Calves	P/T	NR
Arrowood et al. (1989)	IMC/mAb	Mice (neonates)	P/T	IMC: NR. mAb: SS
Fayer et al. (1989a)	HBC	Mice (neonates)	P	SS
Fayer et al. (1989b)	HBC	Calves	P	SS: diarrhea and oocyst shedding
Riggs et al. (1989)	mAbs	Mice (neonates)	P	SS
Fayer et al. (1990)	HBCIgG, A, M	Mice (neonates)	T	SS
Perryman et al. (1990)	HBCIg/mAb	Mice (neonates)	P	SS
Bjorneby et al. (1990)	mAb	Mice (neonates)	P	SS: sporozoites and merozoites
Cama and Sterling (1991)	EYA	Mice (neonates)	T	SS
Bjorneby et al. (1991)	mAb	Mice (adult nude)	T	SS
Tilley et al. (1991)	mAb	Mice (neonates)	T	SS
Flanigan et al. (1991)	HBC	In vitro	P	SS
Uhl et al. (1992)	mAb	Mice (neonates)	P	SS: diverse isolates
Doyle et al. (1993)	HBCIg	Calves and in vitro	P/T	SS
Perryman et al. (1993)	mAb	Mice (adult SCID)	T	SS
Watzl et al. (1993)	BC	Mice (retrovirus-infected)	P/T	SS
Tzipori et al. (1994)	HBCIg	Mice (SCID), piglets	T	Mice: SS. Piglets: NS
Riggs et al. (1994)	HBCIg	Mice (SCID)	T	SS
Kuhls et al. (1995)	Igiv	Mice (SCID)	P/T	P: SS. T: NS
Tatalik et al. (1995)	Immune calf serum	Mice (SCID and neonates)	P/T	NS

[a] BC, bovine colostrum (not specifically immunized); BCIg, BC immunoglobulin; HBC, hyperimmune bovine colostrum (immunized with Cryptosporidium); HBCIg, immunoglobulin fraction from HBC, generally > 50% bovine IgG; Igiv, (pooled normal human plasma immunoglobulin for intravenous administration; EYA, egg yolk antibodies (hyperimmune); IMC, immune mouse colostrum delivered via suckling; mAb, monoclonal antibody (murine)
[b] P, prevention; T, treatment.
[c] NR, no response; NS, response not statistically significant; SS, statistically significant response.

administered antibody. Tzipori *et al.* (1994) showed that efficacy of HBCIg and paromomycin in SCID and gnotobiotic piglets was related to severity of disease; the more severe the disease, the less effective the treatment. This effect was ascribed to rapid transit of the compounds in piglets with profound diarrhea, as well as the possibility of gastrointestinal degradation of the antibodies. The histopathology of the gastrointestinal architecture was also revealing; in severely infected animals, crypt infection and inflammation were pronounced. This severe inflammation would probably prevent access of luminally active compounds to the crypt region, resulting in an incomplete response.

4.2.2. Orally Administered Intravenous Pooled Human Immunoglobulins

Igiv was used as oral immunotherapy for cryptosporidiosis in one laboratory study and one human case. These preparations consist of plasma antibodies from pooled human donor plasma, and are suitable for intravenous administration to humans. Such preparations have been shown to have measurable titers against *Cryptosporidium* by enzyme-linked immunosorbent assay, but no detectable Western blot activity (Kuhls *et al.*, 1995), and this probably reflects natural exposure of a proportion of the donors. The fact that these preparations are approved for human use by the US Federal Drugs Administration has prompted some investigators to use them as oral therapy for cryptosporidiosis. An Igiv preparation was shown to reduce the intensity of infection in SCID mice when administered prophylactically, but not therapeutically (Kuhls *et al.*, 1995). A human case study reported resolution in a chemotherapy patient with cryptosporidiosis treated with oral Igiv (Borowitz and Saulsbury, 1991). This finding has not been confirmed in controlled studies, and the case report did not specify whether the resolution of disease coincided with cessation of chemotherapy.

4.2.3. Egg Yolk Antibodies

Hyperimmunization of laying hens results in high titered chicken antibodies (IgY) being transported to the yolk during egg production. This method of producing hyperimmune polyclonal antibodies provides a commercial opportunity similar to that described for HBCIg. In this case, the egg is the antibody source and is a foodstuff which can be produced in a manner consistent with normal egg production practices. EYA have been produced with very high anti-*Cryptosporidium* titers (similar magnitude to HBC) (Cama and Sterling, 1991). Treatment of neonatal mice infected with *Cryptosporidium* with a partially processed hyperimmune egg yolk preparation showed statistically significant improvement of gastrointestinal tract infection in a manner similar to HBC and HBCIg.

4.3. Monoclonal Antibodies

The use of mAbs against sporozoite antigens for immunotherapy has been the subject of numerous studies (Table 1). These studies have been invaluable in the identification of important antigens as immunotherapy or vaccine targets. Monoclonal antibodies of various isotypes reacting with cryptosporidial antigens of 15, 20, 23, 28/55/98 kDa, and high molecular mass glycolipid (mAb 18.44), have all been shown to reduce infectivity or ameliorate infection in various murine models (Arrowood et al., 1989; Riggs et al., 1989; Bjorneby et al., 1990, 1991; Perryman et al., 1990, 1993; Tilley et al., 1991; Perryman et al., 1993; Enriquez et al., 1996). Further, several of these mAbs have been shown to react with multiple developmental stages (Bjorneby et al., 1990), and their antigen targets are conserved among diverse isolates, as discussed by Uhl et al. (1992). The mAbs have been administered orally in most cases; in one study dimeric IgA against the 23-kDa antigen and a 200-kDa antigen generated via mucosal immunizations was transported to bile and intestinal mucosa when administered intravenously, intraperitoneally, or via 'backpack' tumors (Enriquez et al., 1996). These sIgA mAbs had antiparasitic effects in mouse models of infection. Interestingly, the sIgA mAbs did not appear to be significantly more potent when delivered transmucosally than orally administered mAbs. This is somewhat contrary to the thinking that mucosal delivery of antibodies should be more efficacious than oral, given a lower expected degree of degradation through not having to traverse the stomach, and being present theoretically at closer proximity to the infected sites of the intestine and biliary tract. This could be a function of the antigenic specificity of the particular mAbs tested; alternatively, combinations of mAbs may be required for maximal effect.

Monoclonal antibodies have advantages and disadvantages relative to polyclonal antibodies for immunotherapy of cryptosporidiosis. Advantages include: (i) the high level of specific activity (100% of the antibody administered is directed to the relevant epitope); (ii) the ability to 'humanize' the antibodies to take advantage of possible effector mechanisms that may require homologous species of antibody; (iii) the possibility of parenteral administration to achieve transmucosal delivery (discussed above); and (iv) the ability to construct specific 'cocktails' of mAbs, to exploit the relative importance of different antigens. Disadvantages of mAb therapy include: (i) the potential high cost associated with the manufacture of quantities that may be necessary for oral administration; (ii) the possibility that targeting single antigens may allow 'escape mutants' or 'antigenic drift' to render the mAb ineffective; (iii) maximum efficacy may require identification of antigens that are not currently known but which polyclonal antibodies may recognize; and (iv) the relatively low avidity mAbs possess relative to hyperimmune polyclonal antibodies.

4.4. Immunoglobulins G and M versus Secretory Immunoglobulin A

Intuitively, sIgA has a great deal of appeal for oral administration as it tends to survive the gastrointestinal tract better than IgG or IgM (J.H. Crabb, unpublished observations). However, it would be impractical to develop an sIgA antibody fraction from HBC, given the high levels of IgG relative to IgA (discussed above). While it would be possible to increase the specific anti-*Cryptosporidium* activity in the sIgA fraction through local immunizations (i.e. intramammary infusions or supramammary lymph node injections), these practices are not consistent with good dairy herd management (as they can predispose the animals to mastitis) they are not amenable to large-scale immunizations, and the total sIgA level is not significantly increased by using these practices. Further, a major attraction of the HBCIg approach to oral immunotherapy or prophylaxis is the relatively low cost, which makes the product economically feasible. The difficulties associated with purifying a bovine colostrum sIgA product coupled with the low abundance of this Ig class would diminish the economic advantages of this type of product. Generating sIgA from dimeric IgA can be accomplished *in vitro* by reconstituting with recombinant secretory component (M.R. Neutra, personal communication); it is unknown how robust this procedure is on a commercial scale, or how the sIgA will perform upon oral delivery. For reasons stated above, parenteral administration of dimeric IgA could be a very attractive option for immunotherapy, and would not require reconstitution *in vitro* with SC, as this would occur during transcytosis. Potential drawbacks of IgA include its inability to trigger most effector mechanisms.

5. STUDIES IN HUMANS USING ANTIBODIES.

Published studies in humans using anti-*Cryptosporidium* antibodies over the last decade are listed in Table 2. The vast majority of these studies used BC-derived antibodies, due, in part, to the inherently safe nature of this milk-based product.

5.1. Case Reports

Case studies have dominated this field, as small studies at an early stage of development can demonstrate proof of the principle and justify larger, more costly studies. Unfortunately, the predominance of these case reports can lead one to believe that antibodies are an effective therapy for cryptosporidiosis simply because there are more reports showing a response than not. For example, in Table 2, five case studies showed a therapeutic effect of anti-

Table 2 Antibody treatment for cryptosporidiosis: human studies.

Reference	Preparation[a]	Patient status and number	Study design[b]	Outcome[c] Clinical	Outcome[c] Parasitological
Tzipori et al. (1986)	HBC	Congenital hypogammaglobulinemia (1)	CS	R	R (temporary)
Tzipori et al. (1987)	HBC	AIDS (1); chemotherapy (1)	CS	R	R (one patient shed oocysts intermittently)
Saxon and Weinstein (1987)	BC	AIDS (2); hypogammaglobulinemia (1)	CS	NR	NR
Ungar et al. (1990)	HBC	AIDS (1)	CS	R	R (temporary)
Nord et al. (1990)	HBC	AIDS (5)	PC	NR	NS (reductions in 2 of 3)
Borowitz and Saulsbury (1991)	Igiv	Chemotherapy (1)	CS	R	–
Rump et al. (1992)	BCIg	AIDS (29); others (8); cryptosporidiosis (7)	OL	R	R (5/7)
Plettenberg et al. (1993)	BCIg	AIDS (25)	OL	R	R (64%)
Shield et al. (1993)	BCIg	AIDS (1)	CS	R	R
Fries et al. (1994)	HBCIg	AIDS (40; 20)	PC; OL	NR	SS
Greenberg and Cello (1996)	HBCIg	AIDS (24)	OL	R:SS	NS

[a] BC, bovine colostrum (not specifically immunized); BCIg, BC immunoglobulin; HBC, hyperimmune bovine colostrum (immunized with *Cryptosporidium*); HBCIg, immunoglobulin fraction from HBC, generally > 50% bovine IgG; Igiv, pooled normal human plasma immunoglobulin for intravenous administration, administered orally.
[b] CS, case study; OL, open label; PC, placebo-controlled, double-blind.
[c] NR, no response; NS, response not statistically significant; R = resolved; SS, statistically significant response.

bodies; however, the cumulative clinical experience in these studies was based on only six patients. Most of the controlled studies enrolled several times that number of patients in a single trial. Most of the case reports used whole colostrum, raw (when the microbiological levels were low), pasteurized, or irradiated. As discussed previously, because of the complexity of colostrum and the small number of patients used in each study, broad conclusions regarding the utility of antibodies in therapy for cryptosporidiosis simply cannot be made; it can only be stated that the results should encourage larger, well controlled studies.

5.2. Controlled Studies

As a follow-up to the early case reports a small, placebo-controlled study was conducted using HBC. Unfortunately, different lots of colostrum were used, and a significant baseline imbalance was observed in the stool volumes between the placebo and active groups (Nord *et al.*, 1990). These results did not demonstrate a significant response, but two of three patients in the active group experienced significant reductions in oocyst burden. An important lesson was also learned from this study that has guided clinical investigations over the years: improvement of an individual patient cannot necessarily be ascribed to the therapy. In this case, the two placebo-treated patients had the highest baseline stool volume, and also experienced the largest reduction in volume during treatment. Two moderately sized studies and a single case report tested the effects of BCIg in AIDS-related diarrhea or in cryptosporidial diarrhea in AIDS patients (Rump *et al.*, 1992; Plettenberg *et al.*, 1993; Shield *et al.*, 1993). These studies were open-label studies, and one of them analyzed the effect on cryptosporidiosis by retrospective analysis. The results suggested that oral administration of about 10 g of BCIg per day for 10 or more days caused clinical and microbiological improvement. Unfortunately, the lack of quantitative and systematic analysis of stool oocyst output precluded correlating the observed clinical benefits to a demonstrable reduction in oocyst output.

An open-label study of HBCIg was recently conducted in AIDS patients with chronic diarrhea and cryptosporidiosis (Greenberg and Cello, 1996). Patients were grouped into several cohorts based on their having cryptosporidiosis as the sole gastrointestinal disorder, cryptosporidiosis and another gastrointestinal pathogen, or AIDS enteropathy (diagnosed by endoscopic and colonoscopic examination and biopsy). Dosages of 40 g of HBCIg per day or 48 enteric-coated capsules per day were administered for a total of 21 days. Significant decreases in mean stool weights and marginal improvements in stool frequency were noted in the *Cryptosporidium*-alone group receiving the powder formulation ($n = 8$), and no improvement was noted in the other groups. With regard to stool oocyst output, while the percentage of stools with detectable oocysts decreased during and after treatment, stool oocyst levels were not determined in this study. It is interesting that the enteric-coated capsule treatment was ineffective. This may have been due to inefficient release of active product from the capsules in the proximal small intestine (see Section 2.3 for a discussion of formulation requirements). It is of interest that the presence of other gastrointestinal disorders in addition to cryptosporidiosis precluded effectiveness of the treatment.

A placebo-controlled, double-blind, one-way cross-over study of 40 patients tested the effects of 20 g/day HBCIg for 1–2 weeks in AIDS patients with cryptosporidiosis, followed by an open-label dose-escalation study, increasing the dosage to 80 g/day in some patients (Fries et al., 1994). These patients were pretreated with H_2-blockers to reduce stomach acid and improve gastric transit of active antibodies. No statistically significant clinical benefit was observed in either of these trials, but a consistent and significant reduction in stool oocyst shedding occurred in both studies. Whether the lack of clinical improvement was due to insufficient duration of treatment, the end-stage clinical status of the patients, concomitant gastrointestinal disorders in many of the patients (despite noninvasive efforts to exclude patients with other gastrointestinal pathogens), insufficient reductions in parasite burden to cause improvements in gastrointestinal function, or a combination of these possibilities, could not be determined.

6. FUTURE PROSPECTS

6.1. Monoclonal Antibody 'Cocktails' and Polyclonal Phage Display

As discussed in Section 4.3, mAb 'cocktails' against specific neutralizing antigens could prove to be a very effective approach to PAI against cryptosporidiosis, particularly if the mAbs were transformed into 'humanized' polymeric IgA and administered parenterally. These improvements should lead to increased duration of antibody at the site of infection and increased proximity to infected cells due to transcytosis, and they should enhance the possibility of intracellular disruption of parasite development (as discussed above). These studies are technologically feasible, but will probably have to await a better understanding of the appropriate antigenic specificities to be included in such a 'cocktail'. An approach that could combine this 'brute force' mAb approach with the advantages of the polyclonal antibody approach would be to use polyclonal phage display technology to generate libraries of Fv antibody fragments on the surface of filamentous phage (McCafferty et al., 1990). Relevant 'cocktails' could be selected by 'panning' with isolated sporozoites. Bulk transfer of V_L and V_H genes from the bacteriophage vector into eukaryotic expression systems can be accomplished by creating linked V_l–V_h gene pairs and transfection of appropriate eukaryotic host cells to express intact, glycosylated Fab fragments or intact antibodies (Sarantopoulos et al., 1994). This technology is in development, and has not yet been shown to result in polymeric IgA recombinants. However, this approach is an exciting combination of the attractive features of both monoclonal and polyclonal antibody strategies.

6.2. Specific Antigen-stimulated Hyperimmune Bovine Colostrum Immunoglobulin

Immunization with specific *Cryptosporidium* antigens could result in high titered colostrum against single, or a few, selected neutralizing antigens. In this manner, irrelevant specificities could be eliminated and the relative potency of the HBCIg resulting could be dramatically enhanced. Unfortunately, obtaining significant quantities of antigens purified from isolated parasites is very difficult, in part due to the inability to propagate this organism *in vitro*. However, several relevant *Cryptosporidium* antigen genes have been cloned, including genes encoding portions of GP900 and a 15kDa sporozoite protein (Petersen et al., 1992; Jenkins and Fayer, 1995). Interestingly, sheep anti-15kDa antibodies have been stimulated by DNA vaccination using the 15kDa construct, demonstrating the feasibility of using specific antigen (or DNA) stimulation of colostral antibodies (Jenkins et al., 1995). To elicit colostral antibodies to the 15kDa antigen, vaccination of mammary tissue with the DNA plasmid was required. These antibodies reacted with the native 15kDa antigen, but their therapeutic efficacy was not tested.

Specific antigen immunizations were performed on a calf using parasite-derived surface antigen-1 (SA-1) preparations (Tatalick and Perryman, 1995). These antigen preparations were derived from immunoaffinity chromatography using the protective mAb 17.41 (Riggs et al., 1989). Despite eliciting high titers against the native SA-1 antigen(s), PAI with immune serum did not protect or ameliorate infections in SCID mice. This study illustrates potential difficulties that may be encountered with single-antigen approaches.

6.3. Immunotoxins

Antibody–isotope, or antibody–toxin, conjugates could be developed that would target for destruction extracellular parasites or infected host cells, in a manner similar to the approach used for cancer therapy (Vitetta et al., 1987). This approach has not been reported for cryptosporidiosis.

6.4. Antibody Prophylaxis

As discussed in Section 2.4, the most effective use of PAI against cryptosporidiosis appears to be treatment of early stage disease, or even prophylaxis. The consensus of evidence to-date suggests that treatment of severe and chronic infections with the current antibody preparations is unlikely to be satisfactory. It is unclear if antibody prophylaxis will be economically feasible, as individuals at risk would need to take multiple daily doses of antibody, for as long as they remain at risk.

6.5. Antibody-Drug Combination Therapy

An attractive approach in this area would be to consolidate the favorable responses observed with antibodies and with certain drugs (in particular paromomycin or nitazoxanide) (see Tzipori, 1998). This approach would combine the (presumably) intracellular actions of the drugs with the extracellular neutralizing capacities of the antibodies, and would be likely to result in more favorable outcomes in severe infections than treatment with either agent alone. Combined use of antibody and drugs has not, however, been reported yet.

REFERENCES

Arrowood, M., Mead, J., Mahrt, J. and Sterling, C. (1989). Effects of immune colostrum and orally administered anti-sporozoite monoclonal antibodies on the outcome of *Cryptosporidium parvum* infection in neonatal mice. *Infection and Immunity* **57**, 2283-2288.

Banchereau, J., Bazan, F., Blanchard, D., Briere, F., Galizzi, J.P., Kooten, C. van, Liu, Y.J., Rousset, F. and Saeland, S. (1994). The CD40 antigen and its ligand. *Annual Review of Immunology* **12**, 881-922.

Bjorneby, J.M., Riggs, M.W. and Perryman, L.E. (1990). *Cryptosporidium parvum* merozoites share neutralization-sensitive epitopes with sporozoites. *Journal of Immunology* **145**, 298-304.

Bjorneby, J., Hunsaker, B., Riggs, M. and Perryman, L. (1991). Monoclonal antibody immunotherapy in nude mice persistently infected with *Cryptosporidium parvum*. *Infection and Immunity* **59**, 1172-1176.

Blanshard, C., Jackson, A.M., Shanson, D.C., Francis, N. and Gazzard, B.G. (1992). Cryptosporidiosis in HIV-seropositive patients. *Quarterly Journal of Medicine* **85**, 813-823.

Borowitz, S.M. and Saulsbury, F.T. (1991). Treatment of chronic cryptosporidial infection with orally administered human serum immune globulin. *Journal of Pediatrics* **119**, 593-595.

Brandtzaeg, P., Baklien, K., Bjerke, K., Rognum, T.O., Scott, H. and Valnes, K. (1987). Nature and properties of the human gastrointestinal immune system. In:

Immunology of the Gastrointestinal Tract (K. Miller and S. Nicklin, eds), Vol. 1, pp. 1–87. Boca Raton, FL: CRC Press.

Brussow, H., Hilpert, H., Walther, I., Sidotti, J., Meitens, C. and Bachmann, P. (1987). Bovine milk immunoglobulins for passive immunity to infantile rotavirus gastroenteritis. *Journal of Clinical Microbiology* **25**, 982–986.

Buckley, R.H. (1986). Humoral immunodeficiency. *Clinical Immunology and Immunopathology* **40**, 13–24.

Burns, J., Siadat-Pajouh, M., Krishnaney, A. and Greenberg, H. (1996). Protective effect of rotavirus VP6-specific IgA monoclonal antibodies that lack neutralizing activity. *Science* **272**, 104–107.

Butler, J.E. (1983). Bovine immunoglobulins: an augmented review. *Veterinary Immunology and Immunopathology* **4**, 43–152.

Cama, V.A. and Sterling, C.R. (1991). Hyperimmune hens as a novel source of anti-*Cryptosporidium* antibodies suitable for passive immune transfer. *Journal of Protozoology* **38**, 425–435.

Campbell, K.A., Ovendale, P.J., Kennedy, M.K., Fanslow, W.C., Reed, S.G. and Maliszewski, C.R. (1996). CD40 ligand is required for protective cell-mediated immunity to *Leishmania major*. *Immunity* **4**, 283–289

Chen, W., Harp, J., Harmsen, A. and Havell, E. (1993). γ-Interferon functions in resistance to *Cryptosporidium parvum* infection in severe combined immunodeficient mice. *Infection and Immunity* **61**, 3548–3551.

Cozon, G., Biron, F., Jeannin, M., Cannella, D. and Revillard J.P. (1994). Secretory IgA antibodies to *Cryptosporidium parvum* in AIDS patients with chronic cryptosporidiosis. *Journal of Infectious Diseases* **169**, 696–699.

Cross, G.A.M. (1990). Cellular and genetic aspects of antigenic variation in trypanosomes. *Annual Review of Immunology* **8**, 83–110.

Current W. and Garcia, L.S. (1991). Cryptosporidiosis. *Clinical Microbiology Reviews* **4**, 325–358.

Davidson, G.P., Daniels, E., Nunan, H., Moore, A.G., Whuyte, P.B.D., Franklin, K., McCloud, P.I. and Moore, D.J. (1989). Passive immunisation of children with bovine colostrum containing antibodies to human rotavirus. *Lancet* **ii**, 709–714.

Doyle, P.S., Crabb, J. and Petersen, C. (1993). Anti-*Cryptosporidium parvum* antibodies inhibit infectivity *in vitro* and *in vivo*. *Infection and Immunity* **61**, 4079–4084.

Eibl, M.M. and Wolf H.M. (1995). Common variable immunodeficiency: clinical aspects and recent progress in identifying the immunological defect(s). *Folia Microbiologica* **40**, 360–366.

Eibl, M.M., Wolf, H.M., Furnkranz, H. and Rosenkranz, M.D. (1988). Prevention of necrotizing entercolitis in low birth-weight infants by IgA–IgG feeding. *New England Journal of Medicine* **319**, 1–7

Enriquez, F.J., Palting, J., Hensel, J. and Riggs, M.W. (1996). Immunotherapy of cryptosporidiosis using IgA monoclonal antibodies to neutralizing-sensitive *C. parvum* antigens. *Joint Meeting of the American Society of Parasitologists and the Society of Protozoologists*: Abstracts, p. 123.

Fayer, R., Perryman, L.E. and Riggs, M.W. (1989a). Hyperimmune bovine colostrum neutralizes *Cryptosporidium* sporozoites and protects mice against oocyst challenge. *Journal of Parasitology* **75**, 151–153.

Fayer, R., Andrews, C., Ungar, B.L.P. and Blagburn, B. (1989b). Efficacy of hyperimmune bovine colostrum for prophylaxis of cryptosporidiosis in neonatal calves. *Journal of Parasitology* **75**, 393–397.

Fayer, R., Guidry, A and Blagburn, B. (1990). Immunotherapeutic efficacy of bovine

colostral immunoglobulins from a hyperimmunized cow against cryptosporidiosis in neonatal mice. *Infection and Immunity* **58**, 2962–2965.
Fayer, R., Barta, J.R., Guidry, A.J. and Blagburn, B.L. (1991). Immunogold labeling of stages of *Cryptosporidium parvum* recognized by immunoglobulins in hyperimmune bovine colostrum. *Journal of Parasitology* **77**, 487–490.
Flanigan, T., Marshall, R., Redman, D., Kaetzel, C. and Ungar, B. (1991). In vitro screening of therapeutic agents against *Cryptosporidium*: hyperimmune cow colostrum is highly inhibitory. *Journal of Protozoolology* **38**, 225S–227S.
Flanigan, T., Whalen, C., Turner, J., Soave, R., Toerner, J., Havlir, D. and Kotler, D. (1992). *Cryptosporidium* infection and CD4 counts. *Annals of Internal Medicine* **116**, 840–842.
Freedman, D.J., Tacket, C.O., Delehanty, A., Maneval, D.R. and Crabb, J.H. (1996). Milk antibodies against purified colonization factor antigens are protective in a human enterotoxigenic *Escherichia coli* challenge study. *96th General Meeting of the American Society of Microbiology*: Abstracts, p. 371.
Fries, L., Hillman, K., Crabb, J., Linberg, S., Hamer, D., Griffith, J., Keusch, G., Soave, R. and Petersen, C. (1994). Clinical and microbiologic effects of bovine anti-*Cryptosporidium* immunoglobulin (BACI) on cryptosporidial diarrhea in AIDS. *34th Interscience Conference on Antimicrobial Agents and Chemotherapy*: Abstracts, p. 198.
Greenberg, P. and Cello, J. (1996). Treatment of severe diarrhea caused by *Cryptosporidium parvum* with oral bovine immunoglobulin concentrate in patients with AIDS. *Journal of Acquired Immune Deficiency Syndrome and Human Retrovirology* **13**, 348–354.
Harp, J., Woodmansee, D. and Moon, H. (1989). Effects of colostral antibody on susceptibility of calves to *Cryptosporidium parvum* infection. *American Journal of Veterinary Research* **50**, 2117–2119.
Hill, B.D., Dawson, A.M. and Blewett, D.A. (1993). Neutralisation of *Cryptosporidium parvum* sporozoites by immunoglobulin and non-immunoglobulin components in serum. *Research in Veterinary Science* **54**, 356–360.
Hojlyng, N., Molback, K. and Jepsen, S. (1986). *Cryptosporidium* spp., a frequent cause of diarrhea in Liberian children. *Journal of Clinical Microbiology* **23**, 1109–1113.
Jacyna, M.R., Parkin J., Goldin, R. and Baron, J.H. (1990). Protracted enteric cryptosporidial infection in selective immunoglobulin A and *Saccharomyces* opsonin deficiencies. *Gut* **31**, 714–716.
Janoff, E., O'Brien, J., Thompson, P., Ehret, J., Meiklejohn, G., Duvall, G. and Douglas, J., jr (1993). *Streptococcus pneumoniae*: colonization, bacteremia and immune response among persons with human immunodeficiency virus infection. *Journal of Infectious Diseases* **167**, 49–56.
Jenkins, M. and Fayer, R. (1995). Cloning and expression of cDNA encoding an antigenic *Cryptosporidium parvum* protein. *Molecular and Biochemical Parasitology* **71**, 149–152.
Jenkins, M., Kerr, D., Fayer, R. and Wall, R. (1995). Serum and colostrum antibody responses induced by jet-injection of sheep with DNA encoding a *Cryptosporidium parvum* antigen. *Vaccine* **13**, 1658–1664.
Joiner, K. and Dubremetz, J. (1993). *Toxoplasma gondii*: a protozoan for the nineties. *Infection and Immunity* **61**, 1169–1172.
Kilani, R.T., Sekla, L. and Hayglass, K. (1990). The role of humoral immunity in *Cryptosporidium* spp. infection studies with B cell-depleted mice. *Journal of Immunology* **145**, 1571–1576.
Koch, K.L., Shankey, T.V., Weinstein, G.S., Dye, R.E., Abt, A.B., Current, W.L. and

Eyster, M.E. (1983). Cryptosporidiosis in a patient with hemophilia, common variable hypogammaglobulinemia, and the acquired immunodeficiency syndrome. *Annals of Internal Medicine* **99**, 337–340.

Kuhls, T.L., Orlicek, S.L., Mosier, D.A., Crawford, D.L., Abrams, V.L. and Greenfield, R.A. (1995). Enteral human serum immunoglobulin treatment of cryptosporidiosis in mice with severe combined immunodeficiency. *Infection and Immunity* **63**, 3582–3586.

Kutchai, H.C. (1988). The gastrointestinal system. In: *Physiology* (R.M. Berne and M.N. Levy, eds), edition 2, pp. 649–718. St Louis, MO: C.V. Mosby.

Lasser, K.H., Lewin, K.J. and Ryning, F.W. (1979). Cryptosporidial enteritis in a patient with congenital hypogammaglobulinema. *Human Pathology*, **10**, 234–240.

Mata, L., Bolanos, H., Pizarro, D. and Vives, M. (1984). Cryptosporidiosis in children from some highland Costa Rican rural and urban areas. *American Journal of Tropical Medicine and Hygiene* **33**, 24–29.

Mathan, M.M., Venkatesan, S., George, R., Mathew, M. and Mathan, V., I. (1985). *Cryptosporidium* and diarrhea in southern Indian children. *Lancet* **ii**, 1172–1175.

McCafferty, J., Griffiths, A.D., Winter, G. and Chiswell, D.J. (1990). Phage antibodies: filamentous phage displaying antibody variable domains. *Nature* **348**, 552–554.

McClead, R.E. and Gregory, S.A. (1984). Resistance of bovine colostral anti-cholera toxin antibody to *in vitro* and *in vivo* proteolysis. *Infection and Immunity* **44**, 474–478.

Mead, J.R., Humphreys, R.C., Sammons, D.W. and Sterling, C.R. (1990). Identification of isolate-specific sporozoite proteins of *Cryptosporidium parvum* by two-dimensional gel electrophoresis. *Infection and Immunity* **58**, 2071–2075.

Melamed, I., Griffiths, A. and Roifman, C. (1991). Benefit of oral immune globulin therapy in patients with immunodeficiency and chronic diarrhea. *Journal of Pediatrics* **119**, 486-489.

Mietens, C., Keinhorst, H., Hilpert, H., Gerber, H., Amster, H. and Pahud, J.J. (1979). Treatment of infantile gastroenteritis with specific bovine anti-*E. coli* milk immunoglobulins. *European Journal of Pediatrics* **132**, 239–252.

Miller, R.A., Bronsdon, M.A. and Morton, W.R. (1991). Failure of breast-feeding to prevent *Cryptosporidium* infection in a primate model. *Journal of Infectious Diseases* **164**, 826–827.

Moon, H.W., Woodmansee, D.B., Harp, J.A., Abel, S. and Ungar, B.L.P. (1988). Lacteal immunity to enteric cryptosporidiosis in mice: immune dams do not protect their suckling pups. *Infection and Immunity* **56**, 649–653.

Nardin, E.H. and Nussenzweig, R.S. (1993). T cell responses to pre-erythrocytic stages of malaria: role in protection and vaccine development against pre-erythrocytic stages. *Annual Review of Immunology* **11**, 687–727.

Nord, J., Ma, P., Dijohn, D., Tzipori, S. and Tacket, C. (1990). Treatment with bovine hyperimmune colostrum of cryptosporidial diarrhea in AIDS patients. *AIDS* **4**, 581–584.

Notarangelo, L.D., Duse, M. and Ugazio, A.G. (1992). Immunodeficiency with hyper-IgM (HIM). *Immunodeficiency Reviews* **3**, 101–121.

Perryman, L., Riggs, M., Mason, P. and Fayer, R. (1990). Kinetics of *Cryptosporidium parvum* sporozoite neutralization by monoclonal antibodies, immune bovine serum, and immune bovine colostrum. *Infection and Immunity* **58**, 257–259.

Perryman, L.E., Kegerreis, K.A. and Mason, P.H. (1993). Effect of orally administered monoclonal antibody on persistent *Cryptosporidium parvum* infection in scid mice. *Infection and Immunity* **61**, 4906–4908.

Petersen, C., Gut, J., Doyle, P.S., Crabb, J.H., Nelson, R.G. and Leech, J.H. (1992).

Characterization of a >900,000-M_r *Cryptosporidium parvum* sporozoite glycoprotein recognized by protective hyperimmune bovine colostral immunoglobulin. *Infection and Immunity* **60**, 5132–5138.

Picker, L.J. and Butcher, E.C. (1992). Physiological and molecular mechanisms of lymphocyte homing. *Annual Review of Immunology* **10**, 561–591.

Plettenberg, A., Stoehr, A., Stellbrink, H.-J., Albrecht. H. and Meigel, W. (1993). A preparation from bovine colostrum in the treatment of HIV-positive patients with chronic diarrhea. *Clinical Investigation* **71**, 42–45.

Ravetch, J.V. and Kinet, J.-P. (1991). Fc receptors. *Annual Review of Immunology* **9**, 457–92.

Rehg, J. (1996). Effect of interferon-γ in experimental *Cryptosporidium parvum* infection. *Journal of Infectious Diseases* **174**, 229–232.

Riggs, M. and Perryman, L. (1987). Infectivity and neutralization of *Cryptosporidium parvum* sporozoites. *Infection and Immunity* **55**, 2081–2087.

Riggs, M.W., McGuire, T. C., Mason, P.H. and Perryman, L.E. (1989). Neutralization-sensitive epitopes are expressed on the surface of infectious *Cryptosporidium parvum* sporozoites. *Journal of Immunology* **143**, 1340–1345.

Riggs, M.W., Cama, V.A., Leary, H.L., jr, and Sterling, C. R. (1994). Bovine antibody against *Cryptosporidium parvum* elicits a circumsporozoite precipitate-like reaction and has immunotherapeutic effect against persistent cryptosporidiosis in SCID mice. *Infection and Immunity* **62**, 1927–1939.

Riggs, M.W., Yount, P.A. and Stone, A.L. (1996). Protective monoclonal antibodies recognize a distinct epitope on a conserved apical complex exoantigen of *Cryptosporidium parvum* sporozoites and merozoites. *Joint Meeting of the American Society of Parasitologists and the Society of Protozoologists*: Abstracts, p. 87.

Rodriguez-Barradas, M., Groover, J., Lacke, C., Gump, D., Lahart, C. Pandey, J. and Musher, D. (1996). IgG antibody to pneumococcal capsular polysaccharide in human immunodeficiency virus-infected subjects: persistence of antibody in responders, revaccination in nonresponders, and relationship of immunoglobulin allotype to response. *Journal of Infectious Diseases* **173**, 1347–1353.

Roos, N., Mahe, S., Benamouzig, R., Sick, H., Rautureau, J. and Tome, D. (1995). ^{15}N-Labeled immunoglobulins from bovine colostrum are partially resistant to digestion in human intestine. *Journal of Nutrition* **125**, 1238–1244.

Rump, J.A., Arndt, R., Arnold, A., Bendick, C., Dichtelmuller, H., Franke, M., Helm, E.B., Jager, H., Kampmann, B., Kolb P., Kreuz, W., Lissner, R., Meigel, W., Ostendorf, P., Peter, H.H., Plettenberg, A., Schedel, I., Stelbrink, H.W. and Stephan, W. (1992). Treatment of diarrhoea in human immunodeficiency virus-infected patients with immunoglobulins from bovine colostrum. *Clinical Investigation* **70**, 588–594.

Sarantopoulos, S., Kao, C.Y.Y., Den, W. and Sharon, J. (1994). A method for linking V_L and V_H region genes that allows bulk transfer between vectors for use in generating polyclonal IgG libraries. *Journal of Immunology* **152**, 5344–5351.

Saxon, A. and Weinstein, W. (1987). Oral administration of bovine colostrum anti-*Cryptosporidium* antibody fails to alter the course of human cryptosporidiosis. *Journal of Parasitology* **73**, 413–415.

Shield, J., Melville, C., Novelli, V., Anderson, G., Scheimberg, I., Gibb, D. and Mila, P. (1993). Bovine colostrum immunoglobulin concentrate for cryptosporidiosis in AIDS. *Archives of Disease in Childhood* **69**, 451–453.

Soave, R. (1995). Editorial response: waterborne cryptosporidiosis – setting the stage for control of an emerging pathogen. *Clinical Infectious Diseases* **21**, 63–64.

Tacket, C.O., Losonsky, G., Link, H., Hoang, Y., Guesry, P., Hilpert, H. and Levine, M.M. (1988). Protection by milk immunoglobulin concentrate against oral challenge with enterotoxigenic *Escherichia coli*. *New England Journal of Medicine*. **318**, 1240–1243.

Tacket, C.O., Binion, S.B., Bostwick, E., Losonsky, G., Roy, M.J. and Edelman, R. (1992). Efficacy of bovine milk immunoglobulin concentrate in preventing illness after *Shigella flexneri* challenge. *American Journal of Tropical Medicine and Hygiene* **47**, 276–283.

Tatalick, L.M. and Perryman, L.E. (1995). Attempts to protect severe combined immunodeficient (SCID) mice with antibody enriched for reactivity to *Cryptosporidium parvum* surface antigen-1. *Veterinary Parasitology* **58**, 281–290.

Tilley, M., Fayer, R., Guidry, A., Upton, S. and Blagburn, B. (1990). *Cryptosporidium parvum* (Apicomplexa: Cryptosporidiidae) oocyst and sporozoite antigens recognized by bovine colostral antibodies. *Infection and Immunity* **58**, 2966–2971.

Tilley, M., Upton, S.J., Fayer, R., Barta, J.R., Chrisp, C.E., Freed, P.S., Blagburn, B.L., Anderson, B.C. and Barnard, S.M. (1991). Identification of a 15-kilodalton surface glycoprotein on sporozoites of *Cryptosporidium parvum*. *Infection and Immunity* **59**, 1002–1007.

Tizard, I. (1987). 'Immunity in the fetus and newborn'. In: *Veterinary Immunology* (D. Pederson, ed.), edition 3, Chap. 13, pp. 171–184. Philadelphia: W.B. Saunders.

Tzipori, S. (1998) Cryptosporidiosis: Laboratory investigations and chemotherapy. *Advances in Parasitology* **40**, 187–221.

Tzipori, S., Roberton, D. and Chapman, C. (1986). Remission of diarrhoea due to cryptosporidiosis in an immunodeficient child treated with hyperimmune bovine colostrum. *British Medical Journal* **293**, 1276–1277.

Tzipori, S., Roberton, D., Cooper, D. and White, L. (1987). Chronic cryptosporidial diarrhoea and hyperimmune cow colostrum. *Lancet* **ii**, 344–345.

Tzipori, S., Rand, W., Griffiths, J., Widmer, G. and Crabb, J. (1994). Evaluation of an animal model system for cryptosporidiosis: therapeutic efficacy of paromomycin and hyperimmune bovine colostrum-immunoglobulin. *Clinical and Diagnostic Laboratory Immunology* **1**, 450–463.

Uhl, E.W., O'Connor, R.M., Perryman, L.E. and Riggs, M.W. (1992). Neutralization-sensitive epitopes are conserved among geographically diverse isolates of *Cryptosporidium parvum*. *Infection and Immunity* **60**, 1703–1706.

Ungar, B. and Nash, T. (1986). Quantification of specific antibody response to *Cryptosporidium* antigens by laser densitometry. *Infection and Immunity*, **53**, 124–128.

Ungar, B.L.P., Soave, R., Fayer, R. and Nash, E. (1986). Detection of immunoglobulin M and G antibodies to *Cryptosporidium* in immunocompetent and immunocompromised person. *Journal of Infectious Diseases* **153**, 570–578.

Ungar, B.L.P., Ward, D.J., Fayer, R. and Quinn, C.A. (1990). Cessation of *Cryptosporidium*-associated diarrhea in an AIDS patient after treatment with hyperimmune bovine colostrum. *Gastroenterology* **98**, 486–489.

Ungar, B., Kao, T., Burris, J. and Finkelman, F. (1991). *Cryptosporidium* infection in an adult mouse model: independent roles for IFN-γ and CD4$^+$ T lymphocytes in protective immunity. *Journal of Immunology* **147**, 1014–1022.

Urban, J., Fayer, R., Chen, S., Gause, W., Gately, M. and Finkelman, F. (1996). IL-12 protects immunocompetent and immunodeficient neonatal mice against infection with *Cryptosporidium parvum*. *Journal of Immunology* **156**, 263–268.

USDA (1993). Cryptosporidium *is Common in Dairy Calves*. Fort Collins, CO: United States Department of Agriculture, National Dairy Heifer Evaluation Project, National Animal Health Monitoring System.

Vitetta, E.S., Fulton, R.J., May, R.D., Till, M. and Uhr, J.W. (1987). Redesigning nature's poisons to create anti-tumor reagents. *Science* **238**, 1098–1104.

Watzl, B., Huang, D., Alak, J., Hamid, D., Jenkins, E. and Watson, R. (1993). Enhancement of resistance to *Cryptosporidium parvum* by pooled bovine colostrum during murine retroviral infection. *American Journal of Tropical Medicine and Hygiene* **48**, 519–523.

Whitton, J. and Oldstone, M. (1990). Virus-induced immune response interactions principles of immunity and immunopathology. In: *Virology*, edition 2 (B.N. Fields, ed.) Vol. 1, pp. 369–381. New York: Raven Press.

Zegers, B.J., Rijkers, G.T. and Stoop, J.W. (1991). Humoral immunodeficiency: from description to the cellular and molecular basis of the defect. *Netherlands Journal of Medicine* **39**, 199–208.

Cryptosporidium: Molecular Basis of Host–Parasite Interaction

Honorine Ward and Ana Maria Cevallos

*Division of Geographic Medicine and Infectious Diseases,
New England Medical Center,
Tufts University School of Medicine,
Boston, MA 02111, USA*

1. Introduction 152
2. Oocysts and the Molecular Basis of Excystation and Oocyst Wall Formation 153
   2.1. Morphology and function 153
   2.2. Oocysts of other apicomplexans 154
   2.3. *Cryptosporidium* oocyst proteins 154
3. Invasive (Zoite) Stages and the Molecular Basis of Attachment, Invasion, and Parasitophorus Vacuole Formation 158
   3.1. Morphology and function 158
   3.2. Invasive stages of other apicomplexans 159
   3.3. Proteins of *Cryptosporidium* invasive stages 160
4. Intracellular Stages and the Molecular Basis of Stage Differentiation and Parasite Metabolism 173
   4.1. Morphology and function 173
   4.2. Intracellular stages of other apicomplexans 173
   4.3. Proteins of *Cryptosporidium* intracellular stages 174
5. Future Perspectives 177
6. Concluding Remarks 178
   References 178

Host–parasite interactions occur at a number of stages during the process of infection with Cryptosporidium. *Until recently, very little was known about the molecular basis of these interactions or of specific parasite and host molecules involved in them. Within the past decade significant advances have been made in our understanding of* Cryptosporidium *host–parasite interactions and in identifying molecules involved in them. However, with most interactions the story is far from complete and a number of gaps remain to be filled. This chapter reviews the existing knowledge of the molecular basis of various host–parasite interactions and of specific molecules that may be* involved in them, and identifies areas in which further investigation is necessary. Identification of these molecules and elucidation of molecular mechanisms underlying the host–parasite interaction are of vital importance in developing strategies to combat cryptosporidiosis by targeted chemo- and immunotherapy.

1. INTRODUCTION

Host–parasite interactions occur at a number of stages during the process of infection with *Cryptosporidium* (Figure 1). This process is initiated by ingestion of oocysts, which upon exposure to favorable conditions within the host undergo excystation. Released sporozoites attach to and invade host cells forming a parasitophorous vacuole (PV), where the parasite undergoes further

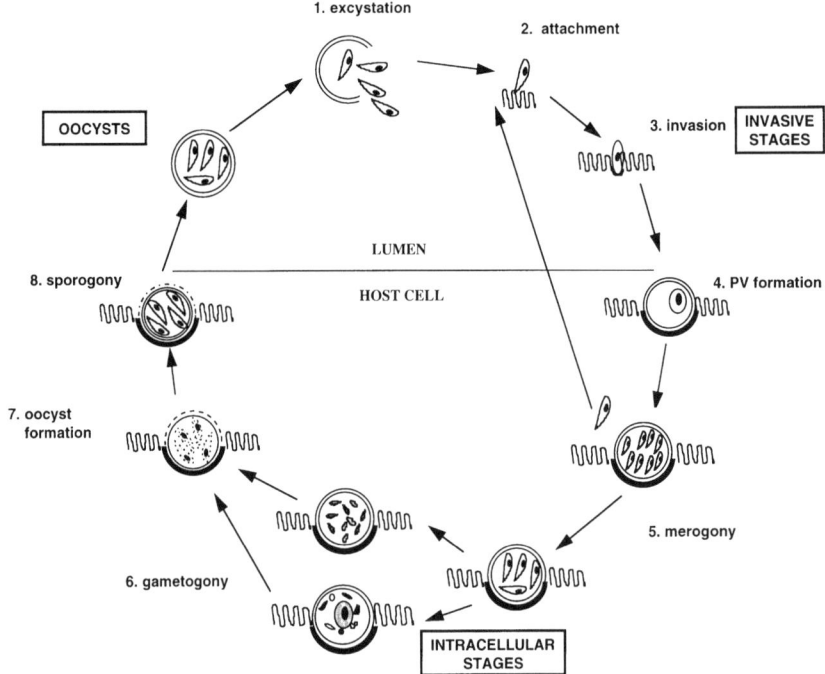

Figure 1 Host–parasite interactions of *C. parvum* developmental stages. Host–parasite interactions occur at a number of steps during *Cryptosporidium* infection. Oocysts, upon exposure to favorable conditions within the host undergo (1) excystation. Released sporozoites, and subsequently merozoites released from mature meronts (2) attach to and (3) invade host cells, forming a parasitophorous vacuole (PV) where the parasite undergoes further intracellular development, via (4) asexual (merogony) and (5) sexual (gametogony) cycles. Following fertilization, the oocyst wall is formed (6) and the oocyst undergoes sporulation (7) within the host, before being excreted into the environment.

intracellular development, through asexual as well as sexual cycles, eventually leading to formation of new oocysts that are capable of reinitiating the infectious cycle (this is discussed in detail by Tzipori and Griffiths in this volume). Although ultrastructural aspects of the processes of excystation (Reduker et al., 1985), attachment (Lumb et al., 1988b; Yoshikawa and Iseki, 1992), invasion, PV formation, and intracellular development of *Cryptosporidium* (Marcial and Madara, 1986; Tzipori, 1988; Flanigan et al., 1991; Rosales et al., 1993) have been characterized in some detail, until recently very little was known about the molecular basis of these host–parasite interactions or of specific parasite and host molecules involved in them. In part, this may have been due to the inability to cultivate large quantities of the organism *in vitro*, as well as the lack of standardized *in vitro* assay systems and animal models. However, increasing interest in *Cryptosporidium parvum* as an emerging human enteric pathogen has, in recent years, led to identification of a number of molecules which may be involved in pathogenesis. Since many of these molecules were initially identified and characterized using antibodies, they are generally referred to as 'antigens', and may be protein, carbohydrate or lipid or a combination thereof in nature. This chapter reviews the existing knowledge of the molecular basis of various host–parasite interactions and of specific molecules that may be involved in them, and identifies areas in which further investigation is necessary. In this regard, useful paradigms obtained from research on other apicomplexan parasites, such as *Toxoplasma*, *Plasmodium*, and *Eimeria*, which may be applicable to pathogenesis of cryptosporidial infection are also discussed. An attempt has been made to associate morphological and functional aspects of host–parasite interactions with specific parasite proteins, where known. Interactions of various developmental stages and proteins involved in them have been grouped into three sections, for convenience. However, this grouping is somewhat arbitrary, since the same protein may be expressed and function in several developmental stages. In addition, proteins expressed in one stage of the parasite may participate in interactions involving other stages.

2. OOCYSTS AND THE MOLECULAR BASIS OF EXCYSTATION AND OOCYST WALL FORMATION

2.1. Morphology and Function

Oocysts, the infective stage of the parasite, contain four naked sporozoites enclosed within a thick wall that serves a protective function by isolating the parasite from the external environment. The oocyst wall consists of two electron-dense layers separated by an electron-lucent space (Tzipori, 1988;

Fayer et al., 1990). Following ingestion by the host, oocysts survive the acid environment of the stomach and reach the small intestine where they undergo excystation by a process that is poorly understood (Hill et al., 1991). A number of conditions present in the intestinal milieu are capable of inducing excystation *in vitro*, including a temperature of 37°C, a slightly alkaline pH (7.6), bile salts, and trypsin (Fayer and Leek, 1984; Reduker and Speer, 1985; Robertson et al., 1993). However, unlike other coccidian parasites, none of these factors is essential for excystation *in vitro*. Recently, surface-associated protease activity in sporozoites (described in Section 3.3.1) has been implicated as playing a role in excystation.

Upon completion of the intracellular sexual cycle, oocyst formation occurs. Oocyst wall components are believed to be transported to the surface of developing oocysts in wall-forming bodies (Fayer et al., 1990). Sporulation occurs within the host, and mature oocysts are released into the intestinal lumen. Some oocysts, so-called 'single-walled' or 'thin-walled' oocysts are believed to contribute to an autoinfectious cycle (Tzipori, 1988; Current and Garcia, 1991). Again, the molecular mechanisms underlying the processes of oocyst wall formation, sporogony and release into the lumen, are poorly understood.

2.2. Oocysts of other Apicomplexans

Oocysts of other enteric apicomplexans, unlike those of *Cryptosporidium*, undergo sporulation outside the host and contain sporozoites within sporocysts (Current and Garcia, 1991). The oocyst wall of the closely related coccidian *Eimeria* spp. consists of an outer lipid layer and an inner glycoprotein layer, which is composed of a single 10- to 12-kDa protein (Stotish et al., 1978; Karim et al., 1996). Passive immunization with a monoclonal antibody (mAb) to this protein which is highly conserved among *Eimeria* spp. and which is also present in gametocytes resulted in partial protection against *E. tenella* infection (Karim et al., 1996).

2.3. *Cryptosporidium* Oocyst Proteins

The antigenic composition of oocysts has been studied using antibodies (obtained by immunization of animals with total oocyst preparations or sera from infected humans and animals) to identify specific proteins in oocysts. However, the majority of proteins recognized by these antibodies are present in sporozoites within the oocyst, with relatively few present in the oocyst wall. The former (i.e. sporozoite derived) are described below as 'oocyst/sporozoite' antigens, and those present in oocyst walls as 'oocyst wall' proteins.

2.3.1. Oocyst/Sporozoite Antigens

Polyclonal antibodies to oocysts have been shown to recognize a wide array of antigens, ranging from 14 to >200 kDa in molecular weight (Luft et al., 1987; Lumb et al., 1988a, 1989; McDonald et al., 1991; Ortega-Mora et al., 1992). Some of these antigens do not appear to be specific for *Cryptosporidium*. Thus, polyclonal antibodies raised against *C. parvum* oocysts were shown to cross-react with *Eimeria* spp. oocysts by immunofluorescence, and sera from *Eimeria*-infected lambs reacted with antigens ranging from 29 to 69 kDa in *C. parvum* oocysts by immunoblotting (Ortega-Mora et al., 1992). Studies have also shown differences in the antigenic composition of oocysts obtained from different patients and from different host species (McDonald et al., 1991; Nina et al., 1992b). As would be expected, the differences between species of *Cryptosporidium* are greater than within *C. parvum* (Nina et al., 1992b). Some of these differences are relatively minor, such as differences in the molecular size of antigens recognized by a monoclonal antibody. For example, mAb 182A6 recognizes a 47-kDa protein in human *C. parvum* isolates, whereas the equivalent antigen of bovine, ovine, or cervine isolates was consistently slightly larger at 48 kDa (Nina et al., 1992b). The significance of this size variation is unknown, but may indicate minor changes in the amino acid or carbohydrate composition.

Based on studies of lectin binding and glycosidase treatment, some oocyst/sporozoite proteins are thought to be glycosylated. Probing of blots of oocyst sonicates with lectins of varying specificity demonstrated the presence of 15 lectin-binding components (Luft et al., 1987). Eight of these bands (72, 81, 87, 92, 95, 99, and two >100 kDa) were specifically recognized by several lectins, including concanavalin A, *Dolichos biflorus* agglutinin, and wheat-germ agglutinin, suggesting that the carbohydrate residues in these proteins contain α-D-mannose (Man) and/or α-D-glucose (Glc), N-acetyl-α-D-galactosamine (GalNAc), and N-acetyl-β-D-glucosamine (GlcNAc) and/or sialic acid (Neu5Ac). The 72, 76, 98, and >100 kDa bands were recognized by immune sera obtained from experimentally infected mice. Treatment of the antigen preparation with a mixture of glycosidases, reduced immunoreactivity of these sera by enzyme-linked immunosorbent assay (ELISA), also suggesting that the epitopes recognized by them were glycosylated (Luft et al., 1987).

Other than their molecular weight, immunoreactivity with different *Cryptosporidium* species and isolates, and potential glycosylation status, very little else is known about the proteins described above, and their biological role during natural infection remains to be established.

2.3.2. Oocyst Wall Proteins

Cryptosporidium oocyst walls appear to contain a number of proteins (unlike those of *Eimeria* which, as described in Section 2.2, are composed of a single glycoprotein). Surface labeling of intact oocysts has been used by two groups of investigators to identify proteins localized to the surface of the oocyst wall. Lumb et al. (1988a) identified proteins of 15, 32, 47.5, 79, and 96 kDa by surface labeling of intact oocysts. However, sodium dodecylsulfate polyacrylamide gel electrophoresis (SDS PAGE) of Percoll-purified oocyst walls showed the presence of 21 bands with major bands of 14, 18, 19, 40, 65, 91, and 190 kDa molecular weight. Surface labeling of oocysts of three *Cryptosporidium* species by Tilley et al. (1990a) revealed differences in the banding pattern among them, with common bands of 32, 57, 120, 145–148, and 285–290 kDa. The discrepancies in the number and molecular weight of surface-labeled proteins in these studies are difficult to explain, but may reflect methodological or isolate differences.

Some oocyst wall surface proteins appear to be glycosylated since, as described below, reactivity of some mAbs to them are periodate sensitive. In addition, in a study of lectin binding to intact oocysts, lectins specific for GlcNAc were shown to agglutinate intact oocysts from four *C. parvum* isolates, suggesting that this sugar moiety is present on the surface of oocyst walls (Llovo et al., 1993). Whether these residues are present in glycoproteins or in polysaccharides remains to be determined.

Specific oocyst wall proteins (summarized in Table 1) have been identified using mAbs. A mAb which reacted with the oocyst walls of *C. parvum* and *C. muris* by immunofluorescence was found to recognize a band of 40 kDa molecular weight (Anusz et al., 1990). Another mAb, 1B5, which reacted with oocyst walls by immunofluorescence, recognized bands of 41 and 44 kDa molecular weight in immunoblots of *C. parvum* oocysts (McDonald et al., 1991; Nina et al., 1992a). Reactivity of this mAb was reduced by periodate treatment, suggesting that the epitope is glycosylated (McDonald et al., 1995). By immunoelectron microscopy (IEM), in addition to labeling the inner and outer layers of the oocyst wall, this mAb was also shown to react with the surface and dense bodies of macrogametocytes and with the PV of macro- and microgametocytes.

Bonnin et al. (1991a) characterized an oocyst wall antigen using mAb OW-IGO, which reacted with the oocyst wall by immunfluorescence and recognized major bands at 40 and 250 kDa (as well as several other bands) in immunoblots of oocyst extracts. The epitope recognized by OW-IGO was also sensitive to periodate treatment, suggesting that it is glycosylated. IEM revealed localization of the antigen to fibrillous material in electron-lucent vesicles at the periphery of developing macrogametes and in the PV surrounding them as well as in the oocyst wall. Another mAb, 2D7, obtained following multiple oral immunization of mice with oocysts, reacted with the interior of oocyst walls by immunofluorescence and recognized a band of 50 kDa by immunoblot analysis (Tilley et al., 1993). Unlike those of the other oocystwall proteins described above, the epitope recognized by mAb 2D7 did not appear to be glycosylated, since it was insensitive to periodate digestion.

Table 1 *Cryptosporidium parvum* oocyst-wall proteins.

Molecular weight (kDa)	Antibody	Immunolocalization	Comments	Reference
40	mAb	IF: oocyst wall	Reacts with *C. muris* oocysts also	Anusz et al. (1990)
40, 250	mAb OW1-IGO (IgM)	IF: oocyst wall IEM: macrogametes, PV; oocyst wall	Epitope glycosylated (periodate sensitive)	Bonnin et al. (1991a)
41, 44	mAb 1B5 (IgM)	IF: oocyst wall IEM: macro- and microgametocytes, PV; oocyst wall	Epitope glycosylated (periodate sensitive)	McDonald et al. (1991, 1995), Nina et al. (1992b)
190	mAb 11B2 (IgG1)	IF: oocyst wall residual body	Deduced amino acid sequence of genomic DNA clones shows cysteine-rich repeats	Lally et al. (1992), Ranucci et al. (1993)
50	mAb 2D7 (IgM)	IF: oocyst wall	Epitope not glycosylated (periodate insensitive)	Tilley et al. (1993)
70	mAb OW3	IF: oocyst wall, intracellular stages in infected MDCK cells	Sequence of cDNA clone homologous to *C. parvum* hsp 70; OW3 mRNA expressed in infected MDCK cells	Mead et al. (1994) Bonafonte et al. (1996) Khramstsov et al. (1995)

hsp, heat shock protein; IgG1, immunoglobulin G1; IgM, immunoglobulin M; IEM, immunoelectron microscopy; IF, immunofluorescence; mAb, monoclonal antibody; MDCK, Madin-Darby canine kidney cells; PV, parasitophorous vacuole.

A 190-kDa protein present in oocyst walls was described by two different groups. Genomic DNA clones encoding this protein were identified, initially by Lally *et al.* (1992) and subsequently by Ranucci *et al.* (1993) by screening λgt11 genomic *C. parvum* expression libraries with antisera to oocysts. The deduced amino acid sequence of these clones revealed unusually high contents of cysteine, proline, and histidine. In addition, there are two repeat motifs, both characterized by the presence of cysteine residues at conserved positions. The function of the repeat units in this protein is unknown, although it has been speculated that because of their cysteine content they may have a role in stabilizing protein structure through disulfide bonds. This protein appears to be conserved among various human and bovine *C. parvum* isolates. Polyclonal and monoclonal antibodies raised to the recombinant protein recognized a band of 190 kDa in oocyst lysates by immunoblotting and reacted with the oocyst wall as well as with the residual body within oocysts by confocal immunofluorescence microscopy (Ranucci *et al.*, 1993).

Another oocyst wall protein was identified using mAb OW3 to an oocyst-wall antigen (Mead *et al.*, 1994; Bonafonte *et al.*, 1996). Analysis of the nucleotide sequence of a cDNA clone which was isolated using this mAb showed identity to a *C. parvum* gene previously described by Khramtsov *et al.* (1995). This gene encodes a protein with homology to the cytoplasmic form of the heat shock protein hsp70. Monospecific polyclonal antibodies to the recombinant hsp70 protein recognized a band of 70 kDa in *C. parvum* oocysts. A recent study showed that mRNA coding for the OW3 protein was detected in infected MDCK cells from 4 to 96 hours after infection, with peak expression at 24 hours (Mead *et al.*, 1996). By confocal immunofluorescence microscopy OW3-reactive protein could be visualized at 48, 72, and 96 hours after infection, but not at 4 or 24 hours. Initially the immunofluorescence pattern of the protein was punctate, suggesting the presence of wall-forming bodies. At later stages these structures appeared to migrate to the periphery of the parasite and then to coalesce into wall-like plates.

3. INVASIVE (ZOITE) STAGES AND THE MOLECULAR BASIS OF ATTACHMENT, INVASION AND PARASITOPHOROUS VACUOLE FORMATION

3.1. Morphology and Function

Invasive, or zoite, stages of *C. parvum* include sporozoites and merozoites. Sporozoites initiate infection by attaching to and invading host cells. Merozoites are released into the intestinal lumen, following rupture of infected cells, and invade adjacent epithelial cells. Morphologically, sporozoites and merozoites are very similar. They are surrounded by a pellicle composed of three membranes (one outer and two inner) and contain apical complex organelles such as rhoptries, micronemes, conoid, and dense granules (Marcial and Madara, 1986; Tzipori, 1988).

Attachment of sporozoites to the microvillus membrane of intestinal epithelial cells is a primary event in the initial host–parasite interaction. Ultrastructural studies have demonstrated that sporozoites and merozoites attach to host cells by their anterior pole (Lumb *et al.*, 1988b; Tzipori, 1988). Using an *in vitro* model of attachment to fixed MDCK (Madin-Darby canine kidney) cells, studies have shown that sporozoite attachment is dose and time dependent and is optimal at a temperature of 37°C, at pH 7.2–7.6, and in the presence of the divalent cations Ca^{2+} and Mn^{2+} (Hamer *et al.*, 1994).

Following attachment, zoite invasion occurs by invagination of the host cell plasma membrane, which engulfs and eventually completely surrounds the sporozoite to form the parasitophorous vacuole. Electron dense bands are formed in the adjoining host cell cytoplasm. During formation of the parasitophorus vacuole, a membrane-bound vacuole develops in the anterior third of the sporozoite in the region where rhoptries and micronemes had previously been visualized. Subsequently, the outer membrane of the sporozoite and the membrane around the anterior vacuole both appear to fuse with the host cell membrane adjacent to the conoid (Lumb *et al.*, 1988b; Tzipori, 1988).

3.2. Invasive Stages of other Apicomplexans

Host cell attachment and invasion by zoites of other apicomplexans such as *Toxoplasma* and *Plasmodium* has been well characterized (Sam-Yellowe, 1996). Following recognition and attachment to target cells by surface molecules, a tight junction is formed that is drawn backwards along the parasite, shedding membrane and associated surface molecules as the organism is internalized. During invasion, the organelles of the apical complex, including rhoptries, micronemes, and dense granules, discharge their contents to facilitate parasite entry, to form the parasitophorous vacuole membrane, and to reseal the host cell membrane (Sam-Yellowe, 1996). In *Toxoplasma* (Smith, 1995; Grimwood and Smith, 1996) and *Plasmodium* (Sam-Yellowe, 1992; Pasloske and Howard, 1994) a number of surface as well as apical complex (including rhoptry, microneme, and dense granule) proteins which mediate these interactions have been described, and the genes encoding them have been cloned and sequenced. Many of these proteins are shed from the surface of the parasite during invasion, and some of them have been demonstrated to be deposited in trails during gliding movement of zoites *in vitro* (Stewart and Vanderberg, 1988; Entzeroth *et al.*, 1989). Another common characteristic of some of these proteins is the finding that they are anchored in the membrane via a glycophosphatidylinositol (GPI) linkage (Tomavo et al., 1989; Hoessli et al., 1996). In addition to surface and apical complex proteins, cytoskeletal proteins such as actin have been shown to facilitate parasite entry by mediating gliding motility of zoites (Dobrowolski and Sibley, 1996).

3.3. Proteins of *Cryptosporidium* Invasive Stages

3.3.1. Surface and Apical Complex Proteins

Unlike the sporozoites of other apicomplexans, which express a paucity of surface proteins, *C. parvum* sporozoites have been shown to have at least 20 surface proteins ranging from 11 to >200 kDa molecular weight, as determined by radio-iodination of intact parasites (Tilley and Upton, 1990; Doyle et al., 1993; Petersen, 1993). Some of these have been characterized and implicated as playing a role in attachment and/or invasion. Some proteins have been localized to apical complex organelles and, by analogy to those from other apicomplexans, are also believed to function in these initial host–parasite interactions. These surface and apical complex proteins, which represent the majority of those identified in *C. parvum*, are described in this section. Many of these proteins share common characteristics. For example, many of them appear to be glycosylated based on lectin binding or loss of antibody reactivity following periodate or glycosidase treatment. In addition, many of them are 'shed', 'released', or 'exocytosed' from the surface of invasive stages and may be deposited in trails during gliding movement. Furthermore, many of these proteins appear to be expressed in sporozoites as well as merozoites.

As will be evident from the following descriptions, different investigators, using a variety of antibodies, have described at least three groups of *C. parvum* sporozoite surface proteins of similar molecular weight. Some of these proteins have been shown to have similar characteristics, whereas others appear to be quite different. In many instances these may indeed be the same protein with the minor differences in reported molecular weight being explicable by methodological or possibly isolate variations. On the other hand, they may represent different proteins of similar molecular mass. Some of these discrepancies could be resolved if oocyst isolates and reagents such as mAbs are exchanged among investigators. Comparison of N-terminal amino acid sequence or of nucleotide sequence of genes encoding these proteins, when they become available, will also help to confirm their identity.

(a) *Low molecular weight proteins*
(i) *15- to 20-kDa proteins.* A number of investigators have described sporozoite antigens in the 15–20 kDa range. These antigens are recognized by sera from infected humans (Lumb et al., 1988a; Mead et al., 1988) and

animals (Luft *et al.*, 1987; Lumb *et al.*, 1988a; Mead *et al.*, 1988; Hill *et al.*, 1990; Whitmire and Harp, 1991; Reperant *et al.*, 1992; Peeters *et al.*, 1992), as well as by hyperimmune bovine colostrum (Tilley *et al.*, 1990b) and mAbs (Tilley *et al.*, 1991, 1993; Gut and Nelson, 1994; Tilley and Upton, 1994). Some of these have been characterized in some detail (summarized in Table 2) and appear to represent the same molecule, but others are clearly different proteins.

A protein named GP15 (and subsequently CP15) was identified by Tilley *et al.* (1991) using an immunoglobulin A (IgA) MAb 5C3, which reacted with both sporozoites and merozoites by immunofluorescence. Surface localization of the protein in both stages was shown by IEM. By immunoblot analysis the mAb reacted with bands of 14–16 kDa molecular weight in *C. parvum*. The epitope recognized by the mAb was presumed to be carbohydrate in nature based on competition for binding with GlcNAc. Further evidence for the presence of carbohydrate (Man/Glc) residues was suggested by binding of the lectin concanavalin A to a protein of similar molecular weight. The mAb 5C3 was reported to reduce oocyst excretion as well as histological evidence of infection in the suckling mouse model. Another series of five mAbs (two of which were directed against carbohydrate epitopes) to a 15-kDa protein obtained following oral immunization of mice was described by the same investigators (Tilley *et al.*, 1993). These mAbs reacted with the surface of sporozoites by immunofluorescence. By IEM, in addition to labeling the surface of sporozoites and merozoites, the mAbs reacted with electron-lucent oocyst residual granules. One of these mAbs 2B3, reacted with antigen shed from the surface of sporozoites during gliding movement on glass slides (Tilley and Upton, 1994).

Jenkins *et al.* (1993) used a rat antiserum to gel-isolated CP15 to isolate a clone from a sporozoite cDNA library. The predicted amino acid sequence from this clone contained two potential sites for N-linked glycosylation, a signal sequence, and a transmembrane region. Antisera to the recombinant protein recognized bands of 15 and 60 kDa (among other bands) in immunoblots of native sporozoite proteins, as did the antiserum used to screen the libraries, whereas mAb 5C3 recognized only the 15kDa band. Subsequently, a genomic clone was isolated by hybridization with an oligonucleotide probe corresponding to a portion of the cDNA sequence (Khramtsov *et al.*, 1997). Monospecific antibodies (purified from rat polyclonal antisera using the recombinant protein) recognized a 15kDa band by immunoblot analysis of *C. parvum* proteins.

Jenkins and Fayer (1995) also identified a cDNA clone using antisera to unfractionated parasite proteins to screen cDNA libraries. Analysis of the DNA sequence of this clone showed an open-reading frame which could encode a 14kDa protein. No putative *N*-glycosylation sites were found. However, several potential sites for *O*-glycosylation were detected. Antisera to the recombinant protein recognized a band of 15 kDa in *C. parvum*

Table 2 *Cryptosporidium parvum* zoite low molecular weight surface proteins.

Molecular weight (kDa)	Antibody	Immunolocalization	Comments	Reference
15 (14–16)	mAb 5C3 (IgA)	IF: sporozoite and merozoite IEM: sporozoite and merozoite surface	Epitope glycosylated (periodate sensitive); mAb 5C3 reduced infection in mice	Tilley *et al.* (1991)
15	mAbs: 2B3 (IgG2b), 4G2 (IgG2a), 4B10 (IgG2a), 5E3 (IgG1), 4C3 (IgG3)	IF: sporozoite surface IEM: sporozoite and merozoite surface	2B3 and 4C3 Ags glycosylated (periodate-sensitive); 4G2, 4B10, and 5E3 Ags not glycosylated (periodate insensitive); 2B3 Ag shed in trails during gliding movement	Tilley *et al.* (1993), Tilley and Upton (1994)
15/60	pAb to gel-isolated 15-kDa band	IF: surface and apical end of sporozoite	cDNA and genomic clones isolated	Jenkins *et al.* (1993), Khramtsov *et al.* (1997)
15	pAb to unfractionated *C. parvum* proteins	IF: sporozoite surface, interior structures, oocyst surface	Sequence of cDNA clone different from that described by Jenkins *et al.* (1993) and Khramtsov *et al.* (1997)	Jenkins and Fayer (1995)
15	mAb 11A5	IF: sporozoite and merozoite surface	11A5 Ag shed in trails during gliding movement and sporozoite invasion of MDCK cells; epitope glycosylated (periodate sensitive, lectin binding)	Gut and Nelson (1994)
15/35	pAb to recombinant protein	IF: sporozoite (diffuse)	cDNA clones isolated	Petersen *et al.* (1992a)

Table 2 Continued.

Molecular weight (kDa)	Antibody	Immunolocalization	Comments	Reference
17	pAb to gel-isolated 17-kDa band	IF: sporozoite (interior)	Immunodominant Ag during infection of mice	Reperant et al. (1992)
18–20	pAb to gel-isolated 18–20 kDa band	IF: apical end of sporozoite	Immunodominant copro Ag	El-Shewy et al. (1994)
20 (23)	mAbs: C3B4 (IgG1); C8C5 (IgG3); C6B6 (IgG1); C1D3 (IgG1); 7D10	IF: sporozoite surface	C6B6 and 7D10 epitopes not glycosylated (periodate insensitive); C3B4Ag, C8C5Ag, and C6B6Ag shed in trails during gliding movement; mAbs C6B6 and 7D10 reduced infection in mice; cDNA clones isolated	Mead et al. (1988) Arrowood et al. (1989, 1991), Perryman et al. (1996)
23	pAb	IEM: anterior end of pellicles of sporozoite and merozoite	pAb recognizes multiple bands by immunblot (? shared epitopes)	Lumb et al. (1989)
23	pAb to recombinant protein	IF: sporozoite (diffuse)	cDNA clone isolated	Petersen et al. (1992a)
25	mAb 3E3 (IgG2b)	IF: sporozoite surface	3E3 Ag shed in trails during gliding movement; epitope not glycosylated (periodate insensitive)	Tilley et al. (1993), Tilley and Upton (1994)

Ag, antigen; IEM, immunoelectron microscopy; IF, immunofluorescence; Ig, Immunoglobulin; mAb, monoclonal antibody; MDCK, Madin-Darby canine kidney; pAb, polyclonal antibody.

proteins. In this report the antigen is stated to be present on the surface as well as interior structures of sporozoites, as well as on oocysts. The relationship of the protein encoded by this gene to that described earlier (Jenkins et al., 1993) is not discussed. However, the differences in the sequences as well as in the recombinant proteins suggest that they are different proteins with similar molecular masses.

Petersen et al. (1992a) identified a number of clones using polyclonal anti-C. parvum antisera to screen genomic DNA expression libraries. Affinity purified antibodies to recombinant proteins expressed by five of these clones recognized doublets of 15 and 35 kDa molecular weight by immunoblot analysis and reacted diffusely with fixed sporozoites by immunfluorescence. The DNA sequence of these clones has not yet been reported.

Another report described a protein of similar molecular weight, named p17 (Reperant et al., 1992). In this study IgG and IgA antibodies in sera and intestinal secretions of infected mice were shown by immunoblot analysis to recognize predominant bands of 16–19 kDa as early as 8 days and persisting for a month after infection. Antibodies to this protein were raised by immunizing mice with a 17kDa band isolated from SDS polyacrylamide gels of oocyst proteins. This antisera recognized bands of 16–18 kDa in immunoblots and reacted with sporozoites but not oocysts by immunfluorescence.

A 15-kDa sporozoite protein recognized by mAb 11A5 was described by Gut and Nelson (1994). The antigen recognized by this mAb was present in trails deposited by sporozoites gliding on poly-L-lysine coated slides, was shed by them during invasion of MDCK cells in vitro and was also present in meronts. The epitope recognized by mAb 11A5 was periodate sensitive, suggesting that it was carbohydrate in nature. This was confirmed by specific binding of galactose- (Gal) and GalNAc-specific lectins such as Jacalin and Helix pomatia to the antigen. The protein was Triton-soluble, suggesting that it was membrane bound.

Antigens 18–20 kDa in molecular weight were detected in the stools of calves and humans with cryptosporidiosis (El-Shewy et al., 1994). Monospecific antibodies to gel-purified 18–20 kDa C. parvum proteins reacted with the apical end of sporozoites by immunofluorescence and recognized a 20kDa band by immunoblotting.

(ii) *20- to 25kDa proteins.* A second group of low molecular weight proteins (20–25 kDa) has also been described by several groups. Ungar and Nash (1986) described an antibody response to a 23kDa protein measured quantitatively by laser densitometry in 93% of patients with active cryptosporidiosis. A 20kDa protein was reported to be recognized by sera from infected humans as well as animals (Mead et al., 1988; Hill et al., 1990). Proteins in this molecular weight range have been characterized further using polyclonal or monoclonal antibodies (summarized in Table 2). A rabbit polyclonal antibody made to a gel-eluted 23kDa protein, presumed to be the same as that described by Ungar and Nash, was shown to label the inner and outer membranes of the pellicle of sporozoites and merozoites by IEM (Lumb et al., 1989). However, this antibody was not specific for the 23kDa protein, since it recognized a number of other bands by immunoblot analysis.

Affinity purified antibodies to recombinant protein expressed by a genomic DNA clone (identified by Petersen et al. (1992a) in the same study described in Section 3.3.1.a.i) recognized a band of 23 kDa molecular weight by immunoblot analysis and reacted diffusely with fixed sporozoites by immunofluorescence. The DNA sequence of this clone has not yet been reported (Petersen et al., 1992a).

Various mAbs have also been shown to recognize proteins 20–25 kDa in molecular weight. mAbs C6B6, C8C5, C3B4, and C1D3 recognized a 20-kDa (presumed to be the same as p23) protein on immunoblots and reacted uniformly with the surface of sporozoites and merozoites by immunfluorescence (Mead et al., 1988; Arrowood et al., 1989; Perryman et al., 1996). mAb 3E3 reacted with a 25kDa antigen, and again was presumed to be the same as p23 (Tilley et al., 1993; Tilley and Upton, 1994). This antigen (recognized by mAbs C6B6, C8C5, and 3E3) also appears to be shed from the surface of sporozoites during gliding movement (Arrowood et al., 1991; Tilley and Upton, 1994). mAbs C6B6 as well as a mAb 7D10 raised to affinity purified 23-kDa antigen, decreased infection in neonatal mice orally challenged with oocysts (Perryman et al., 1996). Epitopes of the 23kDa antigen recognized by these two mAbs are conserved among a number of different geographic isolates, and do not appear to be glycosylated (Perryman et al., 1996). A number of cDNA clones encoding epitopes recognized by C6B6 and 7D10 were isolated using these mAbs to screen a cDNA library (Perryman et al., 1996). One clone, contained the longest open-reading frame which would be capable of encoding a protein of 11.3 kDa and showed a single potential N-glycosylation site.

(b) *High molecular weight glycoproteins* High molecular weight surface and/or apical complex proteins of *Cryptosporidium* with a number of common characteristics have also been described by many investigators (summarized in Table 3).

A high molecular weight glycoprotein named GP900 was first identified by probing λgt11 genomic expression libraries with polyclonal anti-*Cryptosporidium* antibodies known to react with sporozoites and oocysts (Petersen et al., 1992a). Affinity purified antibodies to S34, one of the clones obtained, reacted with the anterior half of sporozoites by immunofluorescence and reacted with a protein >900 kDa by immunoblotting (Petersen et al., 1992a, b). mAbs 10C6, 7B3, and E6 also recognized the same >900 kDa protein by immunoblotting (Petersen et al., 1992b). These mAbs reacted with the anterior region of sporozoites, but not with oocysts. mAb 10C6 also reacted with intracellular meronts in infected MDCK monolayers by immunofluorescence. Treatment of parasite lysates with peptide N-glycosidase F (PNGaseF) abolished reactivity of all three mAbs, suggesting that the epitope they recognized

Table 3 *Cryptosporidium parvum* zoite high molecular weight surface and apical complex proteins.

Molecular weight (kDa)	Antibody	Immunolocalization	Comments	Reference
>900 (GP900) 38	mAbs: 7B3 (IgG1); 10C6 (IgG1); E6 (IgG1); pAbs to recombinant protein	IF: apical pole of sporozoites, surface of merozoites IEM: micronemes of zoites	Ag shed in trails during gliding movement. mAbs 7B3, 10C6, E6 epitopes glycosylated (PNGaseF sensitive); GP900 gene has five structural domains including 2 cysteine rich domains. pAbs to recombinant protein inhibit invasion *in vitro*	Petersen *et al.* (1992a,b, 1996, 1997)
>200	mAbs: 8B2 (IgG2b); 8D7 (IgG2b)	IF: apical pole of sporozoites	Epitopes glycosylated (periodate sensitive)	Tilley *et al.* (1993)
>500, multiple bands	mAb AT7G10	IEM: micronemes, dense granules of sporozoites, merozoites, PV macrogametes, oocyst walls	Epitope glycosylated (periodate sensitive)	Robert *et al.* (1994)
>200 190 40	mAb 2B2 (IgM)	IF: sporozoite surface. IEM: surface of invasive stages, all intracellular stages, dense granules of macrogametocytes, PV	Epitope glycosylated (periodate sensitive)	McDonald *et al.* (1995)
>900 136–150 46	mAbs: 4G12 (IgG1) 4E9 (IgM)	IF: apical pole of sporozoites. surface of merozoites, intracellular stages	Ag shed during excystation; epitopes glycosylated (both periodate sensitive, 4G12 epitope PNGaseF sensitive, 4E9 epitope PNGaseF insensitive); mAb 4E9 inhibited sporozoite attachment to and decreased infection of Caco-2A cells.	Ward *et al.* (1997)

Table 3 Continued.

Molecular weight (kDa)	Antibody	Immunolocalization	Comments	Reference
12–1400 (CSL) 46–230	mAbs: 3E2 (IgM); 3B12 (IgM); 3E6 (IgM); 3A12 (IgM); 3A11 (IgM)	IF: apical pole of sporozoites. IEM: apical complex organelles including dense granules	Epitopes glycosylated (periodate sensitive); Ag shed into culture supernatant; mAbs induce a circumsporozoite like reaction mAbs; 3E2, 3B12, and 3A11 protected mice against oocyst challenge; mAb 3E2 reduced invasion of Caco-2 cells by sporozoites	Langer and Riggs (1996), Riggs et al. (1996)
63–210	mAbs: TOU, HAD	IF: apical pole of sporozoites, merozoites. IEM: micronemes of zoites, PV of trophozoites and macrogametes	TOU epitope glycosylated (periodate sensitive); HAD epitope not glycosylated (periodate insensitive)	Bonnin et al. (1991b)
100	mAbs: ABD (IgG); BAX (IgG); SPO (IgG)	IF: apical pole of sporozoites. IEM: micronemes of zoites dense granules of macrogametes, PV of macrogametes and oocysts	Epitopes glycosylated (periodate sensitive)	Bonnin et al. (1993)
110 >200	mAbs: LOI; BKE	IF: apical pole of sporozoites, intracellular stages. IEM: dense granules of merozoites, macrogamonts, PV	Epitopes glycosylated (periodate-sensitive); Ag is Triton-X-100 insoluble	Bonnin et al. (1995)

Ag, antigen; IEM, immunoelectron microscopy; IF, immunofluorescence; Ig, immunoglobulin; mAb, monoclonal antibody; pAb, polyclonal antibody; PNGaseF, peptide-N-glycosidase F; PV, parasitophorus vacuole.

was N-glycosylated. Affinity purified antibodies to the recombinant S34 clone reacted with a series of bands <190 kDa in N-deglycosylated lysates, suggesting that the protein backbone has a molecular weight of <190 kDa (Petersen *et al.*, 1992b). This glycoprotein, appears to be membrane bound, since it is soluble in Triton X-100. It is encoded by a single-copy gene that resides on the largest *Cryptosporidium* chromosome and is present in both bovine and human isolates (Petersen *et al.*, 1992b). Antibodies to GP900 are present in abundance in partially purified Ig from hyperimmune bovine colostrum, which was shown to inhibit infectivity of *C. parvum* in MDCK cells *in vitro* as well as in infected calves *in vivo* (Petersen *et al.*, 1992b; Doyle *et al.*, 1993). GP900 has also been shown to be present in trails deposited by sporozoites displaying gliding motility and was localized to micronemes of invasive stages by IEM (Petersen *et al.*, 1996, 1997). The gene encoding GP900 has been cloned and sequenced. GP900 was shown to be the product of a 7-kB message, encoded by a gene containing five structurally distinct domains, including two cysteine-rich domains, two mucin-like, threonine-rich, trinucleotide repeat domains, and a large domain containing degenerate 8-mer amino acid repeats. Consensus sequences for transmembrane association, a putative cytoplasmic domain and N- and O-linked glycosylation sites are also present (Gousset *et al.*, 1997; Petersen *et al.*, 1997).

A high molecular weight glycoprotein was also identified using two mAbs 4G12 and 4E9 raised to sporozoite surface proteins (Ward, 1996; Ward *et al.*, 1997). mAb 4G12 reacted with a band that co-migrated with the >900kDa band as well as with bands of 136–150 kDa in sporozoite lysates. By immunofluorescence this mAb reacted with the anterior pole of sporozoites, with the surface of purified merozoites and with intracellular stages of the parasite in infected Caco-2A cells, but not with oocysts. mAb 4E9, in addition to recognizing the >900kDa band also reacted with bands of >200 and 46 kDa molecular weight. By immunofluorescence this mAb reacted with the anterior pole of sporozoites, with merozoites and intracellular stages as well as with a subset of oocysts that appeared to be undergoing excystation. The proteins recognized by both mAbs are 'shed' into the supernatant culture medium of live sporozoites as well as of excysting oocysts. Probing of these shed proteins with lectins of varying sugar specificity revealed binding of GalNAc-, Gal-, and Man/Glc-specific lectins, suggesting the presence of these residues. Periodate treatment reduced binding of both mAbs, suggesting that they both recognized carbohydrate epitopes. PNGase F treatment abolished reactivity of mAb 4G12 but not of 4E9, suggesting that the epitope recognized by the former but not the latter was dependent on N-glycosylation. In order to determine the role of the protein in adhesion and invasion, ELISA-based *in vitro* attachment and infection assays were used (see the chapter by Tzipori in this volume). Purified IgM from mAb 4E9 decreased attachment of sporozoites to fixed Caco-2A cells in a dose-dependent fashion when compared to an isotype-matched control mAb. In addition, purified IgM from this mAb decreased infection of these cells by *C. parvum* oocysts, also in a dose-dependent fashion. These results suggest that this protein may be involved in attachment to and invasion of host cells by the parasite. This was confirmed further by the finding that the purified 'shed' proteins themselves bound to Caco-2A cells using an ELISA-based binding assay.

A group of five mAbs (3E2, 3B12, 3E6, 3A12, and 3A11) raised to affinity-purified *C. parvum* antigens were also shown to recognize very high molecular weight proteins (Langer and Riggs, 1996; Riggs *et al.*, 1996). These mAbs elicit a 'circumsporozoite-like' reaction, which was first reported to occur with polyclonal bovine colostral antibody preparations (Riggs *et al.*, 1994) and is characterized by the progressive formation and release of antigen–antibody complexes from the posterior end of sporozoites, reminiscent of the circumsporozoite reaction initially described in malarial parasites (Cochrane *et al.*, 1976). These mAbs protected mice from oocyst challenge in the neonatal mouse model. The antigen recognized by them was localized to apical complex organelles (including dense granules) of sporozoites by IEM. Immunoblot analysis of sporozoites or merozoites probed with the mAbs showed reactivity with multiple bands ranging from 46 to 1400 kDa. The high molecular weight bands (1200–1400 kDa), designated CSL, were shown to contain epitopes which were carbohydrate in nature and which were conserved among two human and three bovine isolates from geographically diverse areas. One of the mAbs, 3E2, inhibited infectivity of sporozoites for Caco-2 cells *in vitro* (Langer and Riggs, 1996). CSL, purified by isoelectric focusing, was shown to bind to Caco-2 cells using an immunofluorescence assay employing mAb 3E2. Incubation of Caco-2 cells with CSL decreased subsequent infection of these cells by sporozoites. These results suggest that CSL may mediate attachment and invasion.

McDonald *et al.* (1995) described mAb 2B2 which reacted with the surface of sporozoites by immunofluorescence. This mAb was reported to react with a very high molecular weight band, which was not resolved by 10% SDS PAGE. Additional bands of 190 and 40 kDa were also seen. Reactivity of the mAb was abolished by periodate treatment, suggesting that the epitope is glycosylated. By IEM, mAb 2B2 reacted with invasive and endogenous (including sexual) stages in intestinal tissue of infected SCID mice as well as with the PV.

Two other mAbs, 8B2 and 8D7, obtained after oral immunization of mice, reacted with the apical end of sporozoites by immunofluorescence and recognized very high molecular weight bands (>200 kDa) by immunoblot analysis (Tilley *et al.*, 1993). The epitopes recognized by these mAbs were also periodate sensitive, suggesting that they were carbohydrate in nature.

Robert *et al.* (1994) also described a high molecular weight band of >500 kDa (in addition to multiple other bands) recognized by mAb AT7G10 in immunoblots of oocyst/sporozoite proteins. The epitope recognized by this mAb was also periodate sensitive, suggesting its carbohydrate nature. By IEM, the epitope was localized to micronemes as well as some dense granules of sporozoites and merozoites. In addition, labeling was present in the PV membrane of all intracellular stages, the wall-forming bodies of macrogametes, and the outer oocyst wall.

Microneme antigens of invasive stages of *C. parvum* were identified using mAbs TOU and HAD (Bonnin et al., 1991b). These mAbs reacted with the anterior pole of sporozoites by immunofluorescence. Subsequently, three other mAbs, ABD, BAX, and SPO, to microneme antigens were identified (Bonnin et al., 1993). The antigen recognized by these mAbs had a molecular weight of 100 kDa, as assessed by immunoblotting. Periodate treatment abolished reactivity of the mAbs, suggesting that the antigen contained carbohydrate epitopes. By immunofluorescence these mAbs reacted with the apical region of sporozoites from five different *C. parvum* isolates, suggesting that the antigen is conserved among them. By IEM, the mAbs reacted with micronemes of sporozoites, merozoites, and early trophozoites. They also labeled electron-dense granules located in the peripheral cytoplasm of early and late macrogametes, as well as in the PV of gametes and of sporulating oocysts, but not oocyst walls.

An antigen present in dense granules of zoites and macrogamonts was identified using mAbs LO1 and BKE (Bonnin et al., 1995). By immunofluorescence these mAbs reacted with the anterior half of sporozoites from nine different *Cryptosporidium* isolates as well as with intracellular stages. These mAbs also reacted with Triton-X100-insoluble structures in infected MDCK monolayers, suggesting that the antigen may be present in structural components of the PV. A major protein of 110 kDa as well as a weakly reactive band of much higher molecular weight were recognized by the mAbs as determined by SDS PAGE and immunoblotting of oocysts. The antigen recognized by the mAbs was presumed to contain carbohydrate residues based on periodate sensitivity. IEM of ileal tissue from infected lambs revealed that the antigen was present in electron-dense granules in the middle region of merozoites, but not in dense granules of the apical complex. Reactivity of the mAbs was also seen in dense granules in the cytoplasm of macrogamonts as well as in the PV of various intracellular stages.

A number of the antigens described above appear to be similar in many respects, including high molecular weight, glycosylation, surface and/or apical complex localization, and presence in more than one developmental stage of the parasite. Their localization to the surface and to apical complex organelles, together with the findings that mAbs to some of them inhibit attachment and invasion *in vitro* and decrease infection *in vivo* in animal models suggest that they play an important role in the host-parasite interaction. However, their relationship to each other remains to be determined.

(c) *Gal/GalNAc-specific Lectin* A number of cell–cell interactions including those between host and parasite are mediated by lectins, a class of carbohydrate-binding proteins. The abundance of glycoconjugates in the glycocalyx coat and microvillus membrane of intestinal cells supports the involvement of lectin–carbohydrate interactions in the recognition and adhesion of sporozoite to host cells. A surface-associated lectin which may mediate attachment of sporozoites to host cells has been identified in *C. parvum* sporozoites (Thea *et al.*, 1992; Joe *et al.*, 1994; Ward, 1996; Ward *et al.*, 1996). Lectin activity in intact and lysed sporozoites was identified using a hemagglutination assay and found to be optimal at pH 7.5, and in the presence of the divalent cations Ca^{2+} and Mn^{2+} (Joe *et al.*, 1994). The sugar specificity of this lectin was determined by a hemagglutination inhibition assay using a wide range of simple and substituted mono- and disaccharides as well as glycoproteins. The results indicated that the lectin was most specific for the monosaccharides Gal and GalNAc, which inhibited lectin activity at minimum inhibitory concentrations (MICs) of 4 and 14 mM, respectively (Joe *et al.*, 1994). Gal- and GalNAc-containing disaccharides were shown to be potent inhibitors of hemagglutination, the best being Gal(β1-3)GalNAc and Gal(α1-4)Gal (Joe *et al.*, 1994). In addition, the lectin bound avidly to mucins such as bovine submaxillary mucin and hog gastric mucin, which is of interest since the parasite exists in a mucin-rich environment. The role of this lectin in mediating attachment was studied using lectin-specific saccharides and glycoproteins to inhibit adherence of sporozoites to intestinal epithelial cells. The monosaccharides Gal and GalNAc, the disaccharides, lactose, and melibiose as well as the mucins bovine submaxillary mucin, hog gastric mucin, and hydatid cyst P1 glycoprotein inhibited attachment in a dose-dependent fashion, further suggesting the possibility that the lectin is involved in these events. The role of Gal residues in mediating attachment was further studied using attachment of sporozoites to CHO cell glycosylation mutants (Ward, 1996; Ward *et al.*, 1996, 1997). Attachment of sporozoites to the Lec-2 mutant, which expresses increased terminal Gal residues, was increased compared to the parent strain. In addition, sialidase treatment of intestinal epithelial cells leading to exposure of terminal Gal residues resulted in increased attachment of sporozoites to these cells. The lectin has been partially purified by galactose-affinity chromatography.

(d) *Proteases* Parasitic protozoa including apicomplexans such as *Plasmodium* and *Eimeria* have been shown to express surface proteases, some of which are believed to function in attachment and invasion of host cells (Hadley *et al.*, 1983; Fuller and McDougald, 1990). The presence of proteases in *C. parvum* has recently been shown by a number of investigators.

(i) *Arginine aminopeptidase.* An aminopeptidase activity that preferentially cleaves amino-terminal arginine from a synthetic substrate was described by Okhuysen *et al.* (1994). This enzyme appears to be an integral membrane protein of sporozoites that is expressed during and after excystation but is not present in intact oocysts or on oocyst walls. Enzyme activity could be inhibited by chelators of divalent cations, suggesting that it required metal ions (Okhuysen et al., 1996). Activity of the enzyme was shown to increase during excystation of oocysts in vitro. Specific inhibitors of the enzyme activity also inhibited excystation in a dose-dependent fashion, suggesting that the enzyme may have a functional role in this process (Okhuysen et al., 1996). The protein itself has not been identified. However, a DNA clone encoding it has been identified and partially sequenced. Analysis of this sequence revealed a high degree of homology to the enzyme from other eukaryotes and a single zinc-binding motif characteristic of other aminopeptidases (Okhuysen and Chappell, 1997).

(ii) *Cysteine protease.* A metallo-dependent cysteine protease associated with the surface of sporozoites was reported by Nesterenko et al. (1995). Activity of this protease was optimal at pH 6.5–7.0 and could be inhibited by chelators of divalent cations and cysteine protease inhibitors. The enzyme, partially purified by preparative isoelectric focusing, was shown to have a molecular weight of 24 kDa. Antibodies to the enzyme were reported to react with the surface of sporozoites by immunofluorescence.

Petersen et al. (1996) identified a cathepsin-L like cysteine protease of sporozoites, named cryptopain, and cloned and sequenced the gene encoding it (Petersen et al., 1996; Huang and Peterson 1997). The gene was reported to have an open-reading frame of 1203 bp, encoding a predicted 45 kDa pre-proprotein and a 25 kDa mature protein. Biotinylated phe-ala, a cysteine protease inhibitor which is not taken up by cells, inhibited *C. parvum* infection *in vitro*, suggesting a surface location for the enzyme and a possible role in invasion.

(iii) *Serine protease.* The presence of a serine protease activity associated with excystation of oocysts was shown by Forney et al. (1996c). Hydrolysis of azocasein at pH 7.0 was not detected in intact oocysts, but was shown to increase progressively upon excystation at 37°C. Excystation could be inhibited by serine protease inhibitors, suggesting a role for the enzyme in excystation. The same authors also showed, by immunofluorescence as well as IEM, that the human serine protease inhibitor α-1-antitrypsin bound to sporozoites, but not oocysts (Forney et al., 1996a). Serine protease inhibitors were also shown to decrease infection in a bovine fallopian tube epithelial cell culture system (Forney et al., 1996b).

3.3.2. Cytoskeletal and Structural Proteins

In other apicomplexans such as *Toxoplasma*, cytoskeletal proteins have been shown to play an important role in attachment and invasion by mediating gliding and propulsive movements of zoites (Dobrowolski and Sibley, 1996). The role of cytoskeletal proteins in *C. parvum* has not been characterized in any detail, although genes encoding some of them have been isolated.

(a) *Actin* A single copy actin gene located on a 1200kbp chromosome was identified in *C. parvum* (Nelson et al., 1991; Kim et al., 1992). The deduced amino acid sequence predicted a protein of 42 kDa and showed homology to other protozoal actins. Inhibition of shedding of surface proteins during gliding motility of sporozoites by cytochalasin D suggests a role for microfilaments in mediating this motility (Gut and Nelson, 1994).

(b) *Tubulin* DNA sequences encoding *C. parvum* α- and β-tubulin have been identified and analyzed (Nelson *et al.*, 1991; Edlind *et al.*, 1994). Recently, β-tubulin sequences expressed in yeast were used as a model for screening antimicrotubule drugs *in vitro* (Edlind *et al.*, 1996). The role of microtubules in infection has been evaluated *in vitro* (Wiest *et al.*, 1993). Treatment with colchicine and vinblastine reduced infection of HT29 cells by sporozoites, in a dose-dependent manner, suggesting a role for microtubules in invasion.

4. INTRACELLULAR STAGES AND THE MOLECULAR BASIS OF STAGE DIFFERENTIATION AND PARASITE METABOLISM

4.1. Morphology and Function

Intracellular stages of *C. parvum* (including type I and II meronts, and micro- and macrogamonts, which develop into oocysts) are located inside the PV in a unique intracellular but extracytoplasmic niche. The parasitophorus membranes are composed of host as well as parasite components. Although, as described in the previous sections, some antigens of invasive stages have been localized to the parasitophorous membranes, very little is known about parasite molecules present in them. The base of intracellular stages of the parasite is in direct contact with the host cytoplasm at the site of the 'feeder organelle'. It is presumed that the parasite obtains essential nutrients from the host cell via this structure (Tzipori, 1988; see also the chapter by Tzipori in this volume); however, virtually nothing is known about this process or of the parasite molecules involved in it. In addition, little is known about the metabolic pathways utilized by the parasite to survive within the host cell.

Within the host cell the parasite undergoes stage differentiation via asexual (merogony) and sexual (gametogony, fertilization, oocyst formation, and sporogony) cycles. Once again, very little, if anything, is known about the signals that trigger stage differentiation or of the developmentally regulated genes involved in the process.

4.2. Intracellular Stages of other Apicomplexans

Like those of *Cryptosporidium*, the intracellular stages of other Apicomplexans such as *Toxoplasma*, *Plasmodium*, and *Eimeria* also exist within a PV

(Sam-Yellowe, 1996). However, unlike those of *Cryptosporidium*, which have an extracytoplasmic location, the intracellular stages of these parasites are located within the host cell cytoplasm. Formation of the parasitophorous vacuolar membrane as well as an associated intravacuolar tubovesicular network in these parasites has been well characterized at the ultrastructural level. In addition, specific proteins (particularly those of rhoptries and dense granules) present in these membranes have been identified and characterized (Joiner and Dubremetz, 1993; Smith, 1995; Sam-Yellowe, 1996). Some of these proteins are postulated to have a structural role or to function in acquisition of nutrients from the host (Joiner and Dubremetz, 1993).

Although stage-specific proteins have been identified in some apicomplexans, very little is known about stage differentiation (Joiner and Dubremetz, 1993), or of the developmentally regulated genes and their products in these parasites.

4.3. Proteins of *Cryptosporidium* Intracellular Stages

4.3.1. Antigens Expressed in Multiple Stages

As described in Sections 1 and 2, a number of antigens have been shown to be immunolocalized to more than one stage of the parasite. For example, some oocyst wall proteins such as those recognized by the mAbs 1B5 (McDonald *et al.*, 1995), OW-IGO (Bonnin *et al.*, 1991a), and OW-3 (Mead *et al.*, 1996), have also been shown to be present in developing sexual stages and in the PV surrounding them. These findings probably represent expression of the protein during synthesis, assembly, and transport to the oocyst wall. A number of invasive stage proteins have also been immunolocalized to intracellular stages. Some of these proteins may be expressed in multiple stages and may mediate the same functions in these stages. However, many of them have been characterized using mAbs to carbohydrate epitopes that are frequently present in multiple protein bands by immunoblot analysis. It is therefore difficult to determine whether these represent the same protein expressed in many developmental stages or cross-reactive carbohydrate epitopes present in other proteins in these stages.

4.3.2. Enzymes Involved in Synthetic and Metabolic Pathways

As mentioned above, very little is known about the synthetic and metabolic pathways utilized by the parasite to survive within host cells. Recently, a few enzymes involved in some of these pathways have been identified and are being investigated as potential drug targets. Although some of them have been characterized in oocysts or sporozoites, they are described in this section because it is likely that they function in metabolism of intracellular stages as well.

(a) *Folate metabolism and dihydrofolate reductase–thymidylate synthase* A number of apicomplexans express a bifunctional enzyme, dihydrofolate reductase–thymidylate synthase (DHFR-TS), which is involved in folate biosynthesis and is the target of antifolate chemotherapeutic agents. These drugs have been found to be ineffective in cryptosporidiosis. Recently, a gene encoding DHFR-TS was isolated from genomic DNA libraries by hybridization with a probe amplified from *C. parvum* genomic DNA by a polymerase chain reaction (PCR) using generic thymidylate synthase primers (Vasquez *et al.*, 1996). The *C. parvum* DHFR-TS is a single copy gene on a 1200 kbp chromosome. Analysis of the DNA sequence of this gene revealed a single 1563 bp open-reading frame which encodes a 179 residue amino-terminal DHFR domain connected by a 55 amino acid junction peptide to a 287 residue C-terminal TS (thymidylate synthase) domain. DNA sequence polymorphisms in this gene were identified in human and bovine isolates. The protein itself has not been identified as yet.

(b) *Enzymes involved in polyamine biosynthesis* A recent study showed that, unlike that in other apicomplexans, polyamine biosynthesis in *C. parvum* is initiated from agmantine which is formed from arginine via arginine decarboxylase (ADC) and that polyamine back-conversion occurs via spermidine:spermine *N*-acetyl transferase (SSAT) (Yarlett *et al.*, 1996). Substrate analogues of ADC and SSAT were found to be effective inhibitors of *C. parvum* growth *in vitro*, suggesting that polyamine biosynthesis may be a potential drug target.

(c) *Enzymes involved in carbohydrate metabolism* Enzymes such as glucose phosphate isomerase, phosphoglucomutase, carboxyl esterase, hexokinase, lacatate dehydrogenase, phosphoenolpyruvate carboxykinase, malate dehydrogenase, phosphofructokinase, and pyruvate kinase have been identified in *C. parvum* (Ogunkolade *et al.*, 1993; Denton *et al.*, 1996; Malek *et al.*, 1996). Lysosomal hydrolases such as β-*N*-acetylglucosaminidase, β-galactosidase, β-*N*-acetylgalactosaminidase, and acid phosphatase have also been identified (Malek *et al.*, 1996).

A putative gene encoding acetyl-CoA synthase of *C. parvum* has been cloned and sequenced (Khramtsov *et al.*, 1996). The gene is present in a single copy, on a chromosome 1.08 Mbp in size. Analysis of the deduced amino acid sequence revealed a 694 amino acid protein with a predicted molecular weight of 78 kDa. Further characterization and subcellular localization of these enzymes remains to be determined.

(d) *Protein disulfide isomerase* Protein disulfide isomerase catalyzes native disulfide bond formation and proper protein folding. A 1876 genomic DNA clone encoding a *C. parvum* PDI has been isolated and sequenced (Blunt *et al.*, 1996). Analysis of the sequence predicted a molecular weight of 54 kDa for the protein and showed strong homology to other PDIs.

(e) *Topoisomerase II* Topoisomerases represent a potential drug target in *cryptosporidium*. A genomic DNA clone encoding a type II topoisomerase in *C. parvum* has been identified and partially sequenced (Christopher and Dykstra, 1994).

(f) *Antioxidant enzymes* The antioxidant enzyme superoxide dismutase, which was found to have a pI of 4.1 and an apparent molecular weight of 35 kDa, has been reported to be present in low levels in *C. parvum* (Ogunkolade *et al.*, 1993; Entrala *et al.*, 1997).

Mitochondrial enzymes have been assayed for but could not be detected in *C. parvum* (Malek *et al.*, 1996), which may be consistent with the reported lack of mitochondria in this organism (Current and Garcia, 1991). Also, the apparent absence of peroxidase activity as well as glutathione- and trypanathione dependent enzyme systems suggest that there is no classical respiratory transport chain in the parasite (Entrala *et al.*, 1997).

4.3.3. Other C. parvum Genes and Proteins

A few other *C. parvum* genes (some encoding proteins of uncertain function) have been cloned and are included in this section for the sake of completion.

(a) *HemA* A *C. parvum* genomic region *hemA*, encoding a putative hemolysin, was identified by screening a genomic expression library in λ zap II for hemolytic activity. The protein predicted by the nucleotide sequence is predominantly hydrophilic and has a calculated molecular weight of 26 kDa (Steele *et al.*, 1995). It contains one potential *N*-glycosylation site and there are numerous serine and threonine residues that could serve as sites of O-linked glycosylation. A tetrapeptide that is repeated four consecutive times in the carboxy terminus of the predicted protein was found which may be a membrane-spanning amphipathic helix. *hemA* appears to be actively transcribed during cryptosporidial infection *in vivo*. The protein itself has not been identified or characterized. However, based on its hemolytic activity, it has been suggested that it may function in disruption of host and/or parasite membranes during formation of the parasitophorus vacuole, release of intracellular stages into the intestinal lumen, or pore formation (for nutrient transport) in the feeder organelle.

(b) *Elongation factor 1α* Proteins of 15 and 45 kDa molecular weight were identified using mAb C6C1 (Mead *et al.*, 1994, 1996). cDNA clones were isolated using this mAb to screen a sporozoite cDNA library (Bonafonte *et al.*, 1996). These clones were expressed in *Escherichia coli* and shown to encode a 47kDa fusion protein which was recognized by mAb C6C1. Analysis of the DNA sequences of these clones revealed homology to elongation factor 1α (EF-1α).

(c) *Elongation factor 2* Elongation factor 2 (EF-2) is a ubiquitous eukaryotic protein which is essential for protein synthesis. The *C. parvum* EF-2 gene has been cloned and sequenced (Jones *et al.*, 1995). Southern and

Northern blot analyses showed that EF-2 is a single copy gene expressed in oocysts.

5. FUTURE PERSPECTIVES

As described in this chapter, a number of *C. parvum* proteins have been identified that may play a role in various host–parasite interactions. However, in most cases these proteins have not been fully characterized at the biochemical or molecular levels, and their functional role in various host–parasite interactions has not been clearly elucidated. These are obviously areas of active research being pursued by a number of investigators.

A number of other questions remain to be answered. What is the biochemical composition of the oocyst wall? How are its components synthesized, assembled, and transported, and how is its formation regulated? A number of *C. parvum* molecules (unlike those of other Apicomplexa) appear to be glycosylated. Are these carbohydrate components synthesized by the parasite or are they host derived? What is their functional role in parasite glycoconjugates? Many *C. parvum* proteins are shed or exocytosed from the surface of the parasite. Are these proteins linked to the membrane by GPI anchors and cleaved by endogenous phospholipases? Or are they released by proteolytic cleavage? Which host-derived nutrients are essential for intracellular survival of the parasite, and what parasite molecules are involved in their acquisition? What synthetic and metabolic pathways are utilized by the parasite, and what are the parasite enzymes involved in these pathways? What signals (host, parasite, or environmental) trigger differentiation of one developmental stage of the parasite to the next, and what genes are involved in the regulation of these processes? Thus far, most research has focused on parasite molecules. What are the host molecules that serve as receptors for recognition, attachment, and invasion? Do host molecules regulate expression of parasite proteins? Which parasite and host molecules are involved in signaling? These are a few of the many issues that need to be addressed.

One of the major drawbacks to investigating the molecular basis of host–parasite interactions of *C. parvum* is the lack of a method for continuous propagation of the parasite *in vitro*, which should be an area of active investigation. Another avenue of research which would be invaluable is the development of a system for DNA transformation and stable transfection of parasites as has been achieved for *Toxoplasma* and *Plasmodium* (Kim *et al.*, 1993; van Djik *et al.*, 1994).

CONCLUDING REMARKS

Within the past decade significant advances have been made in our understanding of *Cryptosporidium* host–parasite interactions and in identifying molecules involved in them. However, these advances represent the tip of the iceberg, since with most interactions the story is far from complete and a number of gaps remain to be filled. These areas of investigation should, however, be greatly facilitated by the increasingly rapid technological advances in molecular and cellular biology. Identification of specific molecules involved in, and elucidation of molecular mechanisms underlying, the host–parasite interaction are of vital importance in developing strategies to combat cryptosporidiosis by targeted chemo- and immunotherapy.

ADDENDUM

Since this chapter was written the following relevant information has been published.
The entire gene encoding COWP, a previously described oocyst wall protein (Ranucci *et al.*, 1993) was cloned and shown to encode a protein containing 1622 amino acids. Ultrastructural studies localized the protein to cytoplasmic inclusions and wall-forming bodies of macrogametes and to the inner layer of the wall of mature oocysts (Spano *et al.*, 1997).
The presence of glycolytic enzymes involved in the Embden Meyerhoff pathway in *Cryptosporidium parvum* was confirmed in a recent study. Enzyme activities were localized to the cytoplasmic fraction by subcellular fractionation (Entrala and Mascaro, 1997).

REFERENCES

Anusz, K.Z., Manson, P.H., Riggs, M.W. and Perryman, L.E. (1990). Detection of *Cryptosporidium parvum* oocysts in bovine feces by monoclonal antibody capture enzyme-linked immunosorbent assay. *Journal of Clinical Microbiology* **28**, 2770–2774.

Arrowood, M.J., Mead, J.R., Mahrt, J.L. and Sterling, C.R. (1989). Effects of immune colostrum and orally administered antisporozoite antibodies on the outcome of *Cryptosporidium parvum* infections in neonatal mice. *Infection and Immunity* **57**, 2283–2288.

Arrowood, M.J., Sterling, C.R. and Healy, M.C. (1991). Immunofluorescent microscopical visualization of trails left by gliding *Cryptosporidium parvum* sporozoites. *Journal of Parasitology* **77**, 315–317.

Blunt, D.S., Montelone, B.A., Upton, S.J. and Khramtsov, N.V. (1996). Sequence of the parasitic protozoan, *Cryptosporidium parvum*, putative protein disulfide isomerase-encoding DNA. *Gene* **181**, 221–223.

Bonafonte, M.T., Lloyd, R.M., Garmon, K.D., You, X., Arrowood, M.J., Schinazi, R.F. and Mead J.R. (1996). Cloning and expression of sporozoite and oocyst *Cryptosporidium parvum* recombinant proteins. *Journal of Eukaryotic Microbiology* **41**, 83S (Abstract). Bonafonte, M.T., Priest, J.W., Garman, D., Arrowood, M.J. and Mead, J. (1997) Isolation of gene coding for elongation factor-1 alpha in *Cryptosporidium parvum*. *Biochimica Biophysica Acta* **1351**, 256–260.

Bonnin, A., Dubremetz, J.F. and Camerlynck, P. (1991a). Characterization and immunolocalization of an oocyst wall antigen of *Cryptosporidium parvum* (Protozoa: Apicomplexa). *Parasitology* **103**, 171–177.

Bonnin, A., Dubremetz, J.F. and Camerlynck, P. (1991b). Characterization of microneme antigens of *Cryptosporidium parvum* (Protozoa: Apicomplexa). *Infection and Immunity* **59**, 1703–1708.

Bonnin, A., Dubremetz, J.F. and Camerlynck, P. (1993). A new antigen of *Cryptosporidium parvum* micronemes possessing epitopes cross-reactive with macrogamete granules. *Parasitology Research* **79**, 8–14.

Bonnin, A., Gut, J., Dubremetz, J.F., Nelson, R.G. and Camerlynck, P. (1995). Monoclonal antibodies identify a subset of dense granules in *Cryptosporidium parvum* zoites and gamonts. *Journal of Eukaryotic Microbiology* **42**, 395–401.

Christopher, L.J. and Dykstra, C.C. (1994). Identification of a Type II topoisomerase gene from *Cryptosporidium parvum*. *Journal of Eukaryotic Microbiology* **41**, 28S.

Cochrane, A.H., Aikawa, M., Jeng, M. and Nussenzweig, R.S. (1976). Antibody-induced ultrastructural changes of malarial sporozoites. *Journal of Immunology* **116**, 859–867.

Current, W.L. and Garcia, L.S. (1991) Cryptosporidiosis. *Microbiology Reviews* **4**, 325–358.

Denton, H., Brown, S.M.A., Roberts, C.W., Alexander, J., McDonald, V., Thong, K.W. and Coombs, G.H. (1996). Comparison of the phosphofructokinase and pyruvate kinase activities of *Cryptosporidium parvum*, *Eimeria tenella* and *Toxoplasma gondii*. *Molecular and Biochemical Parasitology* **76**, 23–29.

Dobrowolski, J.M. and Sibley, L.D. (1996). *Toxoplasma* invasion of mammalian cells is powered by the actin cytoskeleton of the parasite. *Cell* **84**, 933–939.

Doyle, P.S., Crabb, J. and Petersen, C. (1993). Anti-*Cryptosporidium parvum* antibodies inhibit infectivity *in vitro* and *in vivo*. *Infection and Immunity* **61**, 4079–4084.

Edlind, T., Visvesvara, G., Li, J. and Katiyar, S. (1994). *Cryptosporidium* and microsporidial β-tubulin sequences: predictions of benzimidazole sensitivity and phylogeny. *Journal of Eukaryotic Microbiology* **41**, 38S.

Edlind, T., Li, J. and Katiyar, S. (1996). Expression of *Cryptosporidium parvum* β-tubulin sequences in yeast: potential model for drug development. *Journal of Eukaryotic Microbiology* **43**, 86S.

El-Shewy, K., Kinlani, R.T., Hegazy, M.M., Makhlouf, L.M. and Wenman, W.M. (1994). Identification of low-molecular mass coproantigens of *Cryptosporidium parvum*. *Journal of Infectious Diseases* **169**, 460–463.

Entrala, E. and Mascaro, C. (1997). Glycolytic enzymes in *Cryptosporidium parvum* oocysts. *FEMS Microbiology Letters* **151**, 51–57.

Entrala, E., Mascaro, C. and Barret, J. (1997). Anti-oxidant enzymes in *Cryptosporidium parvum* oocysts. *Parasitology* **114**, 13–17.

Entzeroth, R., Zgrzebski, G. and Dubremetz, J.F. (1989). Secretion of trails during gliding motility of *Eimeria nieschulzi* (Apicomplexa, Coccidia) sporozoites visualized

by a monoclonal antibody and immuno-gold silver enhancement. *Parasitology Research* **76**, 174–175.
Fayer, R. and Leek, R.G. (1984). The effects of reducing conditions, medium, pH, temperature, and time on *in vitro* excystation of *Cryptosporidium. Journal of Protozoology* **31**, 567–569.
Fayer, R., Speer, C.A. and Dubey, J.P. (1990). In: *Cryptosporidiosis of Man and Animals* (J.P. Dubey, C.A. Speer, and R. Fayer, eds). Boca Raton, FL: CRC Press.
Flanigan, T.P., Aji, T., Marshal, R., Soave, R., Aikawa, M. and Kaetzel, C. (1991). Asexual development of *Cryptosporidium parvum* within a differentiated human enterocyte cell line. *Infection and Immunity* **59**, 234–239.
Forney, J.R., Yang, S. and Healey, M.C. (1996a). Interaction of the human serine protease inhibitor α-1-antitrypsin with *Cryptosporidium parvum* 2. *Journal of Parasitology* **82**, 496–502.
Forney, J.R., Yang, S., Du, C. and Healey, M.C. (1996b). Efficacy of serine protease inhibitors against *Cryptosporidium parvum* infection in a bovine Fallopian tube epithelial cell culture system. *Journal of Parasitology* **82**, 638–640.
Forney, J.R., Yang, S. and Healey, M.C. (1996c). Protease activity associated with excystation of *Cryptosporidium parvum* oocysts. *Journal of Parasitology* **82**, 889–892.
Fuller, A. and McDougald, L. (1990). Reduction in cell entry of *Eimeria tenella* (coccidia) sporozoites by protease inhibitors, and partial characterization of proteolytic activity associated with intact sporozoites and merozoites. *Journal of Parasitology* **76**, 464–467.
Gousset, L., Wu, J., Huang, J.X., Barnes, D. and Petersen, C. (1997). *C. parvum* GP900, a surface glycoprotein of invasive stages, contains cysteine rich and variant threonine rich domains. *Abstracts, Keystone symposium on Molecular and Cellular Biology of Apicomplexan Protozoa, Park City, UT.*
Grimwood, J. and Smith, J.E. (1996). *Toxoplasma gondii*: the role of parasite surface and secreted proteins in host cell invasion. *International Journal of Parasitology* **26**, 169–173.
Gut, J. and Nelson, R.G. (1994). *Cryptosporidium parvum* sporozoites deposit trails of 11A5 antigen during gliding locomotion and shed 11A5 antigen during invasion of MDCK cells *in vitro. Journal of Eukaryotic Microbiology* **41**, 42S-43S.
Hadley, T.J., Aikawa, M. and Miller, L. (1983). *Plasmodium knowlesi*: studies on the invasion of rhesus erythrocytes by merozoites in the presence of protease inhibitors. *Experimental Parasitology* **5**, 306–311.
Hamer, D.H., Ward, H., Tzipori, S., Pereira, M.E.A., Alory, J.P. and Keusch, G.T. (1994). Attachment of *Cryptosporidium parvum* sporozoites to MDCK cells *in vitro. Infection and Immunity* **62**, 2208–2213.
Hill, B.D., Blewett, D.A., Dawson, A.M. and Wright, S. (1990). Analysis of the kinetics, isotype and specificity of serum and coproantibody in lambs infected with *Cryptosporidium. Research in Veterinary Science* **48**, 76–81.
Hill, B.D., Blewett, D.A., Dawson, A.M. and Wright, S. (1991). *Cryptosporidium parvum*: investigation of sporozoite excystation *in vivo* and association of merozoites with intestinal mucus. *Research in Veterinary Science* **51**, 264–267.
Hoessli, D.C., Davidson, E.A., Schwarz, R.T, and Nasir-Ud-Din (1996). Glycobiology of *Plasmodium falciparum*: an emerging area of research. *Glycoconjugate Journal* **13**, 1–3.
Huang, J.X. and Petersen, C. (1997). Soluble expression of cryptopain, a cysteine proteinase of C. *parvum. Abstract, Keystone Symposium on Molecular and Cellular Biology of Apicomplexan Protozoa, Park City, UT.*
Jenkins, M.C. and Fayer, R. (1995). Cloning and expression of cDNA encoding an

antigenic *Cryptosporidium parvum* protein. *Molecular and Biochemical Parasitology* **71**, 149–152.

Jenkins, M.C., Fayer, R., Tilley, M. and Upton, S.J. (1993). Cloning and expression of a cDNA encoding epitopes shared by 15- and 60-kilodalton proteins of *Cryptosporidium parvum* sporozoites. *Infection and Immunity* **61**, 2377–2382.

Joe, A., Hamer, D.H., Kelley, M.A., Pereira, M.E.A., Keusch, G.T., Tzipori, S. and Ward, H.D. (1994). Role of a Gal/GalNAc-specific sporozoite surface lectin in *C. parvum*–host cell interaction. *Journal of Eukaryotic Microbiology* **41**, 44S-45S.

Joiner, K.A. and Dubremetz, J.F. (1993). *Toxoplasma gondii*: a protozoan for the nineties. *Infection and Immunity* **61**, 1169–1172

Jones, D.E., Tu, T.D., Mathur, S., Sweenney, R.W. and Clark, D.P. (1995). Molecular cloning and characterization of a *Cryptosporidium parvum* elongation factor-2 gene. *Molecular and Biochemical Parasitology* **71**, 143–147.

Karim, M.J., Basak, S.C. and Trees, A.J. (1996). Characterization and immunoprotective properties of a monoclonal antibody against the major oocyst wall protein of *Eimeria tenella*. *Infection and Immunity* **64**, 1227–1232.

Keithly, J.S., Zhu, G., Upton, S.J., Woods, K., Martnez, M.P. and Yarlett, N. (1997) The polyamines of *Cryptosporidium parvum*: Implications for chemotherapy. *Molecular and Biochemical Parasitology* (In press).

Khramtsov, N.V., Tilley, M., Blunt, D.S., Montelone, B.A. and Upton S.J. (1995). Cloning and analysis of a *Cryptosporidium parvum* gene encoding a protein with homology to cytoplasmic form of Hsp70. *Journal of Eukaryotic Microbiology* **42**, 416–422.

Khramtsov, N.V., Blunt, D.S., Montelone, B.A. and Upton S.J. (1996). The putative acetyl-CoA synthetase gene of *Cryptosporidum parvum* and a new conserved protein motif in acetyl-CoA synthetases. *Journal of Parasitology* **82**, 423–427.

Khramtsov, N.V., Oppert, B., Montelone, B.A. and Upton S.J. (1997). Sequencing, analysis and expression in *Escherichia coli* of a gene encoding a 15 kDa *Cryptosporidium parvum* protein. *Biochemical and Biophysical Research Communications* **230**, 164–166.

Kim, K., Gooze, L., Petersen. C., Gut, J. and Nelson, R.G. (1992). Isolation, sequence and molecular karyotype analysis of the actin gene of *Cryptosporidium parvum*. *Molecular and Biochemical Parasitology* **50**, 105–114.

Kim, K., Soldati, D. and Boothroyd, J.C. (1993). Gene replacement in *Toxoplasma gondii* with chloramphenicol acetyl transferase as a selectable marker. *Science* **262**, 911–914.

Lally, N.C., Baird, G.D., McQuay, S.J., Wright, F. and Oliver, J.J. (1992). A 2359-base pair DNA fragment from *Cryptosporidium parvum* encoding a repetitive oocyst protein. *Molecular and Biochemical Parasitology* **56**, 69–78.

Langer, R.C. and Riggs, M.W. (1996). Neutralizing monoclonal antibody protects against *Cryptosporidium parvum* infection by inhibiting sporozoite attachment and invasion. *Journal of Eukaryotic Microbiology* **43**, 76S–77S

Llovo, J., Lopez, A., Fabregas, J. and Muñoz, A. (1993). Interaction of lectins with *Cryptosporidium parvum*. *Journal of Infectious Diseases* **167**, 1477–1480.

Luft, B.J., Payne, D., Woodmansee, D. and Kim, C.W. (1987). Characterization of the *Cryptosporidium* antigens from sporulated oocysts of *Cryptosporidium parvum*. *Infection and Immunity* **55**, 2436–2441.

Lumb, B.J., Lanser, J.A. and O'Donoghue, P.J.. (1988a). Electrophorectic and immunoblot analysis of *Cryptosporidium* oocysts. *Immunology and Cell Biology* **66**, 369–376.

Lumb, R., Smith, K., O'Donoghue, P.J. and Lanser, J.A. (1988b). Ultrastructure of the attachment of *Cryptosporidium* sporozoites to tissue culture cells. *Parasitology Research* **74**, 531–536.

Lumb, B.J., Smith, P.S., Davies, R., O'Donoghue, P.J., Atkinson, H.M. and Lanser, J.A. (1989). Localization of a 23 000 MW antigen of *Cryptosporidium* by IEM. *Immunology and Cell Biology* **67**, 267–270.

Malek, S., Lindmark, D.G., Jarroll, E.L., Wade, S. and Schaaf, S. (1996). Detection of selected enzyme activities in *Cryptosporidium parvum*. *Journal of Eukaryotic Microbiology* **43**, 82S.

Marcial, M.A. and Madara, J.L. (1986). *Cryptosporidium*: Cellular localization, structural analysis of absorptive cell-parasite membrane–membrane interactions in guinea pigs, and suggestion of protozoan transport by M cells. *Gastroenterology* **90**, 583–594.

McDonald, V., Deer, R.M.A., Nina, J.M.S., Wright, S., Chiodini, P.L. and McAdam, K.P.W.J. (1991). Characteristics and specificity of hybridoma antibodies against oocyst antigens of *Cryptosporidium parvum* from man. *Parasite Immunology* **13**, 251–259.

McDonald, V., McCrossan, M.V. and Petry, F. (1995). Localization of parasite antigens in *Cryptosporidium parvum* infected epithelial cells using monoclonal antibodies. *Parasitology* **110**, 259–268.

Mead, J.R., Arrowood, M.J. and Sterling, C.R. (1988). Antigens of *Cryptosporidium* sporozoites recognized by immune sera of infected animals and humans. *Journal of Parasitology* **74**, 135–143.

Mead, J.R., Lloyd, R.M., You, X., Tucker-Burden, C., Arrowood, M.J. and Schinazi, R.F. (1994). Isolation and partial characterization of *Cryptosporidium* sporozoite and oocyst wall recombinant proteins. *Journal of Eukaryotic Microbiology* **41**, 51S.

Mead, J.R., Bonafonte, M.T., Arrowood, M.J. and Schinzai, R.F. (1996). *In vitro* expression of mRNA coding for a *Cryptosporidium parvum* oocyst wall protein. *Journal of Eukaryotic Microbiology* **43**, 84S–85S.

Nelson, R.G., Kim,K., Gooze, L., Petersen, C. and Gut, J. (1991). Identification and isolation of *Cryptosporidium parvum* genes encoding microtubule and microfilament proteins. *Journal of Protozoology* **38**, 52S–55S.

Nesterenko, M.V., Tilley, M. and Upton, S.J. (1995). A metallo-dependent cysteine proteinase of *Cryptosporidium parvum* associated with the surface of sporozoites. *Microbios* **83**, 77–88.

Nina, J.M.S., McDonald, V., Deer, R.M.A., Wright, S.E., Dyson, D.A., Chiodini, P.L. and McAdam, K.P.W.J. (1992a). Comparative study of the antigenic composition of oocyst isolates of *Cryptosporidium parvum* from different hosts. *Parasite Immunology* **14**, 227–232.

Nina, J.M.S., McDonald, V., Dyson, D.A., Catchpole, J., Uni, S., Iseki, M., Chiodini, P.L. and McAdam, K.P.W.J. (1992b). Analysis of oocyst wall and sporozoite antigens from three *Cryptosporidium* species. *Infection and Immunity* **60**, 1509–1513.

Ogunkolade, B.W., Robinson, H.A., McDonald V., Webster, K. and Evans, D.A. (1993). Isoenzyme variation within the genus *Cryptosporidium*. *Parasitology Research* **79**, 385–388.

Okhuysen, P.C. and Chappell, C.L. (1997). Cloning and partial sequence of *Cryptosporidium parvum* sporozoite arginine aminopeptidase. *Abstracts, Keystone Symposium on Molecular and Cellular Biology of Apicomplexan Protozoa. Park City, UT*.

Okhuysen, P.C., Dupont, H.L., Sterling, C.R. and Chappell, C.L. (1994). Arginine aminopeptidase, an integral membrane protein of the *Cryptosporidium parvum* sporozoite. *Infection and Immunity* **62**, 4667–4670.

Okhuysen, P.C., Chappell, C.L., Kettner, C. and Sterling, C.R. (1996). *Cryptosporidium parvum* metalloaminopeptidase inhibitors prevent *in vitro* excystation. *Antimicrobial Agents and Chemotherapy* **40**, 2781–2784.
Ortega-Mora, L.M., Troncoso, J.M., Rojo-Vazquez, F.A. and Gomez-Bautista, M. (1992). Cross-reactivity of polyclonal serum antibodies generated against *Cryptosporidium parvum* oocysts. *Infection and Immunity* **60**, 3442–3445.
Pasloske B.L. and Howard R.J. (1994). Malaria, the red cell, and the endothelium. *Annual Review of Medicine* **45**, 283–295.
Peeters, J.E., Villacorta, I., Vanopdenbosch, E., Vandergheynst, D., Naciri, M., Ares-Mazás, E. and Yvoré, P. (1992). *Cryptosporidium parvum* in calves: kinetics and immunoblot analysis of specific serum and local antibody responses (immunoglobulin A [IgA], IgG, and IgM) after natural and experimental infections. *Infection and Immunity* **60**, 2309–2316.
Perryman, L.E., Jasmer, D.P., Riggs, M.W., Bohnet, S.G., McGuire, T.C. and Arrowood, M.J. (1996). A cloned gene of *Cryptosporidium parvum* encodes neutralization-sensitive epitopes. *Molecular and Biochemical Parasitology* **80**, 137–147.
Petersen, C. (1993). Cellular biology of *Cryptosporidium parvum*. *Parasitology Today* **9**, 87–91.
Petersen, C., Gut, J., Leech, J.H. and Nelson, R.G. (1992a). Identification and initial characterization of five *Cryptosporidium parvum* sporozoite antigen genes. *Infection and Immunity* **60**, 2343–2348.
Petersen, C., Gut, J., Doyle, P.S., Crabb, J.H., Nelson, R.G. and Leech, J.H. (1992b). Characterization of a >900,000-M_r *Cryptosporidium parvum* sporozoite glycoprotein recognized by protective hyperimmune bovine colostral immunoglobulin. *Infection and Immunity* **60**, 5132–5138.
Petersen, C., Huang, J.X., Barnes, D., Bonnin, A. and Gousett, L. (1996). *Cryptosporidium parvum* invasion. *Abstract, Joint Meeting of the American Society of Parasitologists and the Society of Protozoology*, Tucson, AZ.
Petersen, C., Barnes, D., Gousett, L., Wu, J. and Huang, J.X. (1997). The structure and function of *C. parvum* GP900. *Abstracts, Keystone Symposium on Molecular and Cellular Biology of Apicomplexan Protozoa. Park City, UT*.
Ranucci, L., Muller, H.M., La Rosa, G., Reckmann, I., Gomez Morales, M.A., Spano, F., Pozio, E. and Crisanti, A. (1993). Characterization and immunolocalization of a *Cryptosporidium* protein containing repeated aminoacid motifs. *Infection and Immunity* **61**, 2437–2356.
Reduker, D.W. and Speer, C.A. (1985). Factors influencing excystation in *Cryptosporidium* oocysts in cattle. *Journal of Parasitology* **71**, 112–115.
Reduker, D.W., Speer, C.A. and Blixt, J.A. (1985). Ultrastructure of *Cryptosporidium parvum* oocysts and excysting sporozoites as revealed by high resolution scanning electron microscopy. *Journal of Protozoology* **32**, 708–711.
Reperant, J.M., Naciri, M., Chardes, T. and Bout, D.T. (1992). Immunological characterization of a 17 kDa antigen from *Cryptosporidium parvum* recognized early by mucosal IgA antibodies. *FEMS Microbiology Letters* **99**, 7–14.
Riggs, M.W., Cama, V.A., Leary, H.L. and Sterling, R.C. (1994). Bovine antibody against *Cryptosporidium parvum* elicits a circumsporozoite precipitate-like reaction and has immunotherapeutic effect against persistent cryptosoporidiosis in SCID mice. *Infection and Immunity* **62**, 1927–1939.
Riggs, M.W., Yount, P.A., Stone A.L. and Langer, R.C. (1996). Protective monoclonal antibodies define a distinct, conserved epitope on an apical complex exoantigen of *Cryptosporidium parvum* sporozoites. *Journal of Eukaryotic Microbiology* **43**, 74S–75S (Abstract). Riggs, M.W., Stone, A.L., Yount, P.A., Langer, R.C., Arrowood, M.J. and Bently, D.I. (1997) Protective monoclonal antibody defines a

circumsporozoite-like glycoprotein exoantigen of *Cryptosporidium parvum* sporozoites and merozoites. *Journal of Immunology* **158**, 1787–1795.

Robert, B., Antoine, H., Dreze, F., Coppe, P. and Collard, A. (1994). Characterization of a high molecular weight antigen of *Cryptosporidium parvum* micronemes possessing epitopes that are cross-reactive with all parasitic life cycle stages. *Veterinary Research* **25**: 384–398

Robertson, L.J., Campbell, A.T. and Smith, H.V. (1993). *In vitro* excystation of *Cryptosporidium parvum*. *Parasitology* **106**, 13–19.

Rosales, M.J., Mascaro, C. and Osuna, A. (1993). Iltrastructural study of *Cryptosporidium* development in Madin–Darby canine kidney cells. *Veterinary Parasitology* **45**, 267–273.

Sam-Yellowe, T.Y. (1992). Molecular factors responsible for host cell recognition and invasion in *Plasmodium falciparum*. *Journal of Protozoology* **39**, 181–189.

Sam-Yellowe, T.Y. (1996). Rhoptry organelles of the Apicomplexa: their role in host cell invasion and intracellular survival. *Parasitology Today* **12**, 308–316.

Smith, J.E. (1995). An ubiquitous intracellular parasite: the cellular biology of *Toxoplasma gondii*. *International Journal of Parasitology* **25**, 1301–1309.

Spano, F., Puri, C., Ranucci, L., Putignani, L. and Crisanti, A. (1997). Cloning of the entire COWP gene of *Cryptosporidium parvum* and ultrastructural localization of the protein during sexual parasite development. *Parasitology* **114**, 427–437.

Steele, M.I., Kuhls, T.L., Nida, K., Reddy Meka, C.S., Halabi, I.M., Mosier, D.A., Elliott, W., Crawford, D.L. and Greenfield, R.A. (1995). A *Cryptosporidium parvum* genomic region encoding hemolytic activity. *Infection and Immunity* **53**, 3840–3845.

Stewart, M.J. and Vanderberg, J.P. (1988). Malaria sporozoites leave behind trails of circumsporozoite protein during gliding motility. *Journal of Protozoology* **35**, 389–393.

Stotish, R.L., Wang, C.C. and Meyenhofer, M. (1978). Structure and composition of the oocyst wall of *Eimeria tenella*. *Parasitology* **64**, 1074–1081.

Thea, D.M., Pereira, M.E.A., Kotler, D., Sterling, C.R. and Keusch, G.T. (1992). Identification and partial purification of a lectin on the surface of the sporozoite of *Cryptosporidium parvum*. *Journal of Parasitology* **78**, 886–893.

Tilley, M. and Upton, S.J. (1990). Electrophoretic characterization of *Cryptosporidium parvum* (KSU-1 isolate) (Apicomplexa: Cryptosporidiidae). *Canadian Journal of Zoology* **68**, 1513–1519.

Tilley, M. and Upton, S.J. (1994). Both CP15 and CP25 are left as trails behind gliding sporozoites of *Cryptosporidium parvum* (Apicomplexa). *FEMS Microbiology Letters* **120**, 275–278.

Tilley, M., Upton, S.J., Blagburn, B. and Anderson, B.C. (1990a). Identification of outer oocyst wall proteins of three *Cryptosporidium* (Apicomplexa: Cryptosporidiidae) species by ^{125}I surface labeling. *Infection and Immunity* **58**, 252–253.

Tilley, M., Fayer, R., Guidry, A., Upton, S.J. and Blagburn, B.L. (1990b). *Cryptosporidium parvum* (Apicomplexa: Cryptosporidiidae) oocyst and sporozoite antigens recognized by bovine colostral antibodies. *Infection and Immunity* **58**, 2966–2971.

Tilley, M., Upton, S.J., Fayer, R., Barta, J.R., Chrisp, C.E., Freed, P.S., Blagburn, B.L., Anderson, B.C. and Barnard, S.M. (1991). Identification of a 15–kilodalton surface glycoprotein on sporozoites of *Cryptosporidium parvum*. *Infection and Immunity* **59**, 1002–1007.

Tilley, M., Eggleston, M.T. and Upton, S.J. (1993). Multiple oral inoculations with *Cryptosporidium parvum* as means of immunization for production of monoclonal antibodies. *FEMS Microbiology Letters* **113**, 235–240.

Tomavo, S., Schwarz, R.T. and Dubremetz, J.F. (1989). Evidence for glycosyl-

phosphatidylinositol anchoring of *Toxoplasma gondii* major surface antigens. *Molecular and Cell Biology* **9**, 4576–4580
Tzipori, S. (1988). Cryptosporidiosis in perspective. *Advances in Parasitology* **27**, 63–129.
Ungar, B.L.P. and Nash, T.E. (1986). Quantification of specific antibody response to *Cryptosporidium* antigens by laser densitometry. *Infection and Immunity* **53**, 124–128.
van Djik, M.R., Waters, A.P. and Janse, C.J. (1994). Stable transfection of malarial parasite blood stages. *Science* **268**, 1358–1362.
Vasquez, J.R., Gooze, L., Kim, K. Gut, J., Petersen, C. and Nelson, R. (1996). Potential antifolate resistance determinants and genotypic variation in the bifunctional dihydrofolate reductase–thymidylate synthase gene from human and bovine isolates of *Cryptosporidium parvum*. *Molecular and Biochemical Parasitology* **79**, 153–165.
Ward, H. D. (1996). Glycobiology of parasite infection: role of carbohydrate-binding proteins and their ligands in the host–parasite interaction. In: *Glycosciences: Status and Perspectives* (Gabius, H.-J. and Gabius, S. eds), pp. 399–409. Weinheim: Chapman & Hall.
Ward, H., Joe, A., Tzipori, S. Pereira, M.E.A and Keusch, G.T. (1996). Role of a carbohydrate-binding protein in attachment of *Cryptosporidium parvum* sporozoites to host cells. *Glycobiology* **6**, 412.
Ward, H., Joe, A., Verdon, R., Hamer, D., Tzipori, S., Pereira, M.E.A. and Keusch, G.T. (1997). Role of carbohydrates and carbohydrate-binding proteins in *Cryptosporidium* host–parasite interaction. *Abstract, Keystone Symposium on Molecular and Cellular Biology of Apicomplexan Protozoa*. Park City, UT.
Whitmire, W.M. and Harp, J.A. (1991). Characterization of bovine cellular and serum antibody responses during infection by *Cryptosporidium parvum*. *Infection and Immunity* **59**, 990–995.
Wiest, P.M., Johnson, J.H. and Flanigan, T.P. (1993). Microtubule inhibitors block *Cryptosporidium parvum* infection of a human cell line. *Infection and Immunity* **61**, 4888–4890.
Yarlett, N., Martinez, M.P., Zhu, G., Keithly, J.S., Woods, K. and Upton, S.J. (1996). *Cryptosporidium parvum*: polyamine biosynthesis from agmatine. *Journal of Eukaryotic Microbiology* **43**, 73S (Abstract).
Yoshikawa, H. and Iseki, M. (1992). Freeze–fracture study of the site of attachment of *Cryptosporidium muris* in gastric glands. *Journal of Protozoology* **39**, 539–544.

Cryptosporidiosis: Laboratory Investigations and Chemotherapy

Saul Tzipori

Division of Infectious Diseases, Tufts University School of Veterinary Medicine, North Grafton, MA 01536, USA

1. Introduction 188
2. Production of Oocysts for Biomedical Research 188
   2.1. Oocysts from calves 189
   2.2. Detection of oocysts 190
3. Propagation in Cell Culture 190
   3.1. *In vitro* assays for drug screening 191
   3.2. Quantitation of parasites *in vitro* 192
   3.3. Oocyst and sporozoite viability 193
   3.4. Cell culture systems for studying the host cell–pathogen interface .. 194
4. Animal Models 195
   4.1. Models for evaluating agents for treatment or prevention 196
   4.2. Pathophysiologic studies in animals 203
5. New Approaches to Drug Design 205
6. Laboratory Investigations on Drugs Presently Used in Humans 208
   6.1. Paromomycin 209
   6.2. Nitazoxanide (NTZ) 210
7. Concluding Remarks 212
   References 212

Much progress has been achieved in the last decade in terms of development of laboratory techniques, reagents and in vivo *models. They have undoubtedly contributed to better and more accurate investigations. Despite a concerted effort by many investigators, however, breakthroughs have been minimal. The development of adequate* in vitro *and* in vivo *techniques for drug screening, and the intensified and systematic screening, has so far not resulted in the discovery of an effective therapy. The reason for the failure may well be due to the unique biological niche the parasite occupies (discussed at length in the first chapter in this volume). Its location beneath the cell membrane, but outside the cell cytoplasm, may prove a crucial element that needs to be considered when designing new therapeutic approaches. Laboratory* investigations on two drugs currently used against chronic *Cryptosporidium parvum* in acquired immune deficiency syndrome (AIDS) are discussed. This chapter also provides information and the rationale for work in progress in our laboratory that relates to the development of novel approaches for control of the disease. This includes the identification of molecular targets of parasite origin for drug design, and studies on the structure–activity relationships of partially effective drugs with a view to synthesize more effective derivatives. Other investigations attempt to establish the role of secretory antibody, and the merit of repeated mucosal immunizations as a means of providing protection to individuals with AIDS who are at risk of developing chronic *C. parvum* infection.

1. INTRODUCTION

The inability to continuously propagate *Cryptosporidium parvum in vitro* remains the most important obstacle to performing the kind of studies possible with so many other pathogens of major interest. This limitation is reflected in the lack of access to stages of the life cycle other than sporozoites and oocysts, and the need to use animals to generate oocysts for laboratory investigations. Furthermore, oocysts generated in animals have a limited viability span. For instance, infectivity is markedly diminished within 6–8 weeks for *in vitro* and *in vivo* infectivity studies. They may, however, still be useful for animal propagation, and as a source of antigen for serology, biochemistry, and genetic work. At present there are no methods that allow an indefinite storage of infectious material, and thus isolates have to be continuously passaged through animals, usually calves. The lack of recognized and characterized molecular, antigenic, or virulence markers makes it impossible to conduct meaningful comparative studies among isolates obtained from different hosts, geographic locations, or propagated in different host species. This makes the use of standard isolates over time for comparative studies, impossible. A limited cell culture propagation, largely confined to the asexual stages, is available but provides only a restricted scope.

This chapter reviews the procedures and tools available for laboratory investigations relevant to biomedical research, with an emphasis on defining aspects of the host–parasite relationship, and the development and evaluation of effective therapy. The chapter also highlights new research directions currently in progress in our laboratory.

2. PRODUCTION OF OOCYSTS FOR BIOMEDICAL RESEARCH

Oocysts and sporozoites are the only stages of the life cycle which can be produced in large quantities in experimentally infected calves. Oocysts are readily purified and concentrated from calf feces. Oocysts can also be obtained from experimentally infected small ruminants immediately after birth (Blewett, 1988; Tilley and Upton, 1991). Some investigators use rodents to generate small quantities of oocysts (Hill *et al.*, 1991; Suresh and Rehg, 1996). Another major source of material is from chronically, *C. parvum*-infected human patients with AIDS (Giacometti *et al.*, 1996). However, these sources tend to provide small and variable quantities of material which makes referenced, comparative, studies on genetic variation, virulence, and host susceptibility impossible.

2.1. Oocysts from Calves

Calves are infected immediately after birth, preferably before feeding of colostrum. They should be housed in complete isolation to protect the calves from exposure to other pathogens, including *C. parvum* from other sources, and to protect the environment from the infected animals. Infected calves constitute the greatest risk to humans because of the large number of oocysts generated by them. There are several published reports of animal handlers (Current *et al.*, 1983) and veterinary students (Anderson *et al.*, 1982) who became infected through contact with infected calves. Infected calves are kept in metabolic cages once diarrhea begins, to facilitate the collection of all fecal material. During the course of oocyst shedding, calves are fed an enriched calf electrolyte formula, which facilitates the purification of oocysts by decreasing the fecal fat and roughage contents.

There are several purification, concentration, and sterilization procedures, some of which were reviewed by Current (1990). The method we have adopted in our laboratory combines several of them. Briefly, homogenized and filtered calf feces are mixed with saturated NaCl solution and centrifuged at 1000 g for 15 minutes at 4°C. The supernatant containing the oocysts is removed and is diluted 1 : 3 in distilled water and centrifuged at 4000 g for 15 minutes. The pellet is resuspended and washed in 0.1% thiosulfate solution before pelleting and resuspending in Alseveri's solution for the Percoll gradient centrifugation (Waldman *et al.*, 1986, Arrowood, and Sterling, 1987). After counting in a hemocytometer, oocysts are stored at 4°C in antibiotic solution and are used within 6–8 weeks for infectivity studies. Oocysts are treated with 1.75% hypochlorite solution for 7 minutes before using them for *in vitro* or *in vivo* studies. Using these and similar methods it is possible to obtain up to 10^{10} oocysts from a single calf.

2.2. Detection of Oocysts

The presence of *C. parvum* oocysts in the infected host is identified by either direct staining of fecal smears (Ma *et al.*, 1984), by indirect immunofluorescence (Garcia *et al.*, 1987), by polymerase chain reaction (PCR) (Laxer *et al.*, 1991; Webster *et al.*, 1993; Awad-El-Kariem *et al.*, 1994), by enzyme-linked immunosorbent assay (ELISA) (Dagan *et al.*, 1995; Graczyk *et al.*, 1996), or by flow cytometry (Arrowood *et al.*, 1995). Of these methods the modified acid-fast staining technique is the cheapest and simplest to perform, and in our hands is as sensitive as detection by immunofluorescence (unpublished data). For diagnostic purposes, ELISA can be more sensitive in formalin-fixed samples or in poorly preserved specimens (Dagan *et al.*, 1995), in which residual antigens derived from oocysts as well as shared antigens from other stages of the life cycle are present in feces. PCR is more sensitive for detecting *C. parvum* in small numbers, in fixed, in frozen, or often in poorly preserved samples (see the chapter by Widmer, 1998, in this volume). Detection of water-borne oocysts is discussed at length in the chapter by Fricker and Crabb in this volume.

3. PROPAGATION IN CELL CULTURE

In the decade since the first report of *C. parvum* growth in cell culture (Current and Haynes, 1984), there have been more than 25 publications describing *in vitro* cultivation techniques (Eggleston *et al.*, 1994). Although there has been some improvement, the overall extent of parasite growth remains limited largely to the asexual phase. Cells from a variety of sources can be infected with *C. parvum*, in which growth peaks at 48–72 hours and then gradually declines. The failure to maintain the intensity of infection *in vitro* has been attributed to the lack of the thin-shell oocyst stage that is thought to be a key factor in autoinfection (Current and Garcia 1991). In some cell lines production of a small number of oocysts has been reported. They include Caco-2 cells (Buraud *et al.*, 1991), bovine fallopian tube epithelial cells (Yang *et al.*, 1996b), and MDBK cell line (Villacorta *et al.*, 1996). However the amount of oocysts produced is smaller than the inoculum. Upton and coworkers have compared the growth of *C. parvum* in various cell lines, improved some of the parameters required to enhance *in vitro* excystation, and identified a supplemental medium formulation that is said to enhance parasite growth further (Upton *et al.*, 1994a, b, 1995). The growth cycle of *C. parvum* in cell culture has been described in detail by Current and Haynes (1984).

3.1. In vitro Assays for Drug Screening

With all the limitations of cell culture propagation, the two merogeny stages of the life cycle have been useful for developing *in vitro* assays for screening compounds and immune reagents for inhibitory activity against *C. parvum*. The system developed in our laboratory for routine *in vitro* screening is simple, inexpensive, quick, and reproducible. Confluent monolayers of MDBK cells selectively cloned for susceptibility to *C. parvum* infection are grown in 96-well microtiter plates. At confluency they are inoculated simultaneously with the test drugs and 5.0×10^4 oocysts per well. Compounds are tested at four different concentrations, in quadruplicate. Depending on solubility, drugs are either dissolved in media or in 0.2% dimethyl sulfoxide (DMSO). Paromomycin is included in every assay as positive control at 2 mg/ml in quadruplicate. Paromomycin is dissolved in water initially and, depending on the solubility of the drugs to be tested, is diluted either in medium or in 0.2% DMSO. Again depending on drug solubility, DMSO or medium are added to four infected wells containing no drug (negative control), and against which the percentage inhibitory drug activity is measured and calculated. Wells are fixed 48 hours after inoculation and then reacted in the indirect immunofluorescence assay to determine intensity of infection. The extent of infection in the four infected negative control wells is quantitated by a computer video image device, specifically designed for this purpose. The percentage inhibition of the test compounds at four different concentrations, and the positive control drug paromomycin, are then calculated against the infected negative control wells to determine the percentage inhibitory activity of each concentration of the test compounds. In parallel, the cytotoxicity for each drug concentration is determined by one or two assays, measuring the release cell products such as formazan (Cory *et al.*, 1991) or lactate dehydrogenase (LDH) (Griffiths *et al.*, 1994). The percentage toxicity is calculated and scored to reflect levels of cell toxicity in relation to each specific drug concentration. The results are then tabulated, and charted with 95% confidence intervals (CI) to reflect statistically significant drug inhibitory activities against *C. parvum* in relation to paromomycin and the negative medium control. If a drug shows inhibitory activity of 50% or higher, coupled with low cytotoxicity, the tests are repeated, often with a different range of drug concentrations. Drugs that repeatedly show inhibitory activity equal or greater than paromomycin (75–85%) in this assay are considered for the mouse assay described below.

The *in vitro* system described above has been used extensively over the last 3 years, well over 200 compounds having been screened for activity against *C. parvum*. The system is highly standardized, reproducible, and consistent at the top 50% of the scale, as judged by the positive control drug paromomycin, which consistently inhibits the rate of growth at 2 mg/ml by 75–85% against infected control media. The assay is less sensitive at the lower end of the scale, with inhibitory activity below 40% tending to have greater variation. Drugs showing below 40–50% inhibition normally are not pursued further. The poor sensitivity and reproducibility of the assay below 40–50% inhibition is independent of the type of cell, growth conditions, and infectious dose (unpublished data).

Other *in vitro* systems for drug screening have been described, using different cell lines and procedures, including A-549 (Giacometti *et al.*, 1996), T84 (Flanigan *et al.*, 1991), Caco-2 cells (McDonald *et al.*, 1990), and HCT-8 (Woods *et al.*, 1996). There are many variations regarding time of onset of drug treatment in relation to infection with oocysts or sporozoites, days of incubation and methods of fixation, quantitation, analysis, and presentation of the data. These variations often make it difficult to compare results generated by different laboratories. Reproducibility among *in vitro* systems is a major problem, which in part is attributed to the nature of the oocysts used. Oocyst infectivity and/or excystation rate vary with age, with the method of storage and purification including presence of chemical and fecal residues, time of incubation, and origin. Unfortunately, these variables, which profoundly affect the results of *in vitro* assays, are difficult to control or predict. Therefore the outcome of *in vitro* drug screening should only be reported when the results are repeatedly reproducible.

3.2. Quantitation of Parasites *In Vitro*

The method used for quantitation varies. Some investigators manually count the number of parasites seen under high power, in fixed monolayers stained with either Giemsa (Giacometti *et al.*, 1996) or hematoxylin and eosin (Flanigan *et al.*, 1991), by immunoflurescent labeling (Tilley *et al.*, 1991; Griffiths *et al.*, 1994), or visualized unfixed using interference contrast microscopy (Current and Haynes, 1984; Upton *et al.*, 1994a). Others have used ELISA (Ward *et al.*, 1995; Woods et al., 1995; You *et al.*, 1996), or a computer videoimage device to qunatitate parasites in fixed monolayers (unpublished data). The latter two techniques provide a rapid and more comprehensive evaluation of the rate of infection in wells compared with random selection of fields when counting is performed manually. These techniques tend to be bias free and the data are readily analyzed statistically.

3.2.1. ELISA technique

Cell monolayers are fixed 24–48 hours after infection, and ELISA is performed to quantitate the rate of infection. Three slightly different ELISA methods have been described (Ward *et al.*, 1995; Woods *et al.*, 1995; You *et al.*, 1996; Verdon *et al.*, 1997). The ELISA described by Ward *et al.* (1995), which uses biotinylated second antibody with an avidin–biotin–alkaline phosphatase system, appears to increase the sensitivity of the assay, and is independent of the cell line used (Verdon et al., 1997). This method was found to be as sensitive as an immunofluorescence-based detection method. The ELISA method of quantitating infection is rapid and reproducible and is useful for screening a large number of compounds. However, the effect of various agents on parasites or host cells, morphology cannot be ascertained without microscopy. In addition, cytotoxicity assays should also be included in parallel, to make sure that high inhibitory activity is not due to cell death resulting from direct toxicity of the agents to the cells. Even mildly toxic agents will have an impact on infected cells first, which are the first to die. When this occurs the percentage inhibition of the drug may be overestimated.

3.2.2. Computer-based Video-image Technique

This computer-based video image analysis system, which was modified for the purpose of quantitation of immunofluorescent-labeled parasites, provides an accurate method for quantitating the rate of infection in fixed monolayers. It is rapid, simple, sensitive, reproducible, and free of operator bias. In addition, it provides for microscopic assessment of the morphological integrity of the cell monolayers. We use MDBK cells because this cell line lacks background fluorescence when reacted in the immunofluorescence technique, an essential characteristic for accurate quantitation of fluorescing parasites without background interference (Gunzer and Tzipori, unpublished data). Although HCT-8 cells were somewhat more sensitive to *C. parvum* than MDBK (Upton et al., 1994a), the background fluorescence of HCT-8 cells interfered with the image analysis. Briefly, once a microscopic field is captured, the camera is activated and a live image appears on the monitor. The software adapts the system to the fluorescence light intensity of the sample. Once the parasite forms are judged to be satisfactory against the background of the MDBK cells, the image is stored and the analyzer is activated to quantitate the entire assay under the same settings.

3.3. Oocyst and Sporozoite Viability

As mentioned earlier, oocysts can survive for up to 18 months at 4°C (Yang et al., 1996a). However, their infectivity diminishes rapidly at first, then more gradually. Determining the viability of oocysts of unknown history, e.g. from water sources, or of sporozoites under laboratory conditions remains a major problem. Vital dyes, e.g. 6–diamidino–2–phenylindole (DAPI) and propidium iodide, or the rate of *in vitro* excystation are used by some investigators to estimate viability. However, neither of these is as reliable as infectivity (Black et al., 1996). Infectivity of oocysts remains the gold standard assay, although neither the cell culture technique nor the infant mouse, in which infectivities are normally tested, are particularly sensitive when measuring small infectious doses. In contrast to the infant mouse, in which the minimal infective dose was reported to be 312 oocysts (Finch *et al.*, 1993), the median infective dose for human volunteers in one study was 132 oocysts (DuPont *et al.*, 1995), and in infant primates 10–30 oocysts (Miller *et al.*, 1990), illustrating the poor sensitivity of the infant mouse. Furthermore, as pointed out in the first chapter in this volume, some human *C. parvum* isolates do not infect mice (unpublished data). Sporozoite viability is normally assayed in cell culture, or in infant mice inoculated rectally (Riggs and Perryman, 1987) or intraperitoneally (unpublished data).

3.4. Cell Culture Systems for Studying the Host Cell–Pathogen Interface

Cell culture models are not confined to screening of drugs only. Sophisticated cell culture systems, developed by some, can be used to examine the impact of *C. parvum* on cell function, resistance to infection, and cell death (apoptosis). Using monolayers of mature polarized intestinal epithelial cells (T-84), Adams *et al.* (1994) have shown that infection with *C. parvum* leads to a time and dose dependent reduction in epithelial barrier function. This was manifested by a drop in resistance, as measured in Ussing chambers, and it reflects an increase in monolayer permeability. They have shown that T-84 monolayers became permeable to horseradish peroxidase (molecular mass 44 kDa) at 48 hours after infection, coinciding with the release of LDH, both of which indicated the occurrence of modest cellular injury. In contrast, Griffiths *et al.* (1994) found that, despite the high rate of infection, Caco-2 monolayers remained impermeable to proteins for up to 72 hours. While both studies demonstrate a time- and dose-dependent host cell dysfunction and death due to *C. parvum* infection, it appears that Caco-2 cell monolayers are better able to maintain a relatively impermeable barrier to macromolecules, despite widespread cell death. However, Caco-2 cells did become permeable to small molecules <1 kDa, but remained impermeable to exogenously added, or endogeneously produced proteins of >1881 Da. This is because infected dying Caco-2 cells, as evidenced by LDH release into the apical surface, are extruded from the monolayer much the same as *in vivo*, and is consistent with an acquired defect in paracellular channel integrity (Griffiths *et al.*, 1994).

Another study utilizing the same system of Caco-2 monolayer cells grown on filters and mounted in Ussing chambers revealed a significant increase in short circuit current when the cells were exposed to fecal supernatant obtained from patients with diarrhea attributed to *C. parvum*. The effect was time and dose dependent, saturable, and Cl^- and Ca^{2+} dependent, suggesting a secretory diarrhea presumably mediated by an enterotoxin-like activity (Guarino et al., 1995), which could have been either of host or parasite origin. There is evidence that death of extruded C. parvum-infected cells is mediated by apoptosis (G. Widmer, G.J. Griffiths, and S. Tzipori, unpublished data), as infection of Caco-2 cells with C. parvum causes disruption of monolayers and cell death. The possibility that apoptosis occurs in cells extruded from monolayers was investigated by monitoring the extent of DNA fragmentation in monolayers infected with C. parvum. Using the TUNEL technique and by labeling parasites with immunofluorescence, individual infected cells were found to undergo apoptosis (Griffiths et al., 1994). DNA laddering indicative of apoptotic cell death was observed following infection with C. parvum oocysts as well as in positive controls induced with tumor necrosis factor α (TNFα) and cycloheximide. The extent of apoptosis correlated with the number of cells shed from the monolayer, regardless of the stimulus causing cell detachment. We speculate that cell detachment and apoptosis might interfere with in vitro maturation, and possibly the completion of the life cycle, of C. parvum. We are therefore investigating the impact of inhibitors of apoptosis on cell death, as a means of prolonging survival of infected cells, and possibly the completion of the parasite's life cycle. To this aim, a Caco-2 line overexpressing Bcl-2, a known inhibitor of certain pathways leading to apoptosis (Tsujimoto, 1989; Hockenberry et al., 1990), was generated. However, the outcome suggests that apoptosis in C. parvum-infected cells proceeds along a pathway only partially inhibited by Bcl-2, as the cell line did become resistant to TNF α-mediated apoptosis (Widmer et al., 1996). To investigate further the role of apoptosis in cell death induced by C. parvum infection in vitro, Caco-2 lines stably transformed with crmA (Tewari and Dixit, 1995) and p35 (Xue and Horvitz, 1995) were generated. crmA from cowpox virus and p35 from baculovirus have been shown to inhibit apoptosis. The advantage of these inhibitors, as compared to Bcl-2, is that they act at a step in the pathway leading to cell death which is further downstream than Bcl-2. Transformed cell lines are currently being tested by monitoring LDH release and cell detachment resulting from infection with C. parvum.

4. ANIMAL MODELS

Animal models for C. parvum are required to understand, (a) the clinical course of the infection and the disease it induces in humans, (b) the pathology and pathophysiology of digestion, (c) immunology and immunopathology, and (d) for drug testing and evaluation. For each one of those aspects of the host–pathogen interaction, different animal models were used. For instance, clinically affected species such as lambs, calves, and piglets, were used to acquire insight not only into the specific effect of infection in the respective species, but also into the clinical course and pathogenesis of acute cryptosporidiosis in neonates in general. Because of the significance of cryptosporidiosis in other mammals, namely calves and other small ruminants, since 1980 studies have been conducted to investigate the infection and disease that are unique to these economically important domestic animals (Tzipori, 1983; O'Donoghue, 1995).

The pathophysiologic impact of cryptosporidiosis on the entire host (Argenzio *et al.*, 1990, 1993), as well as at specific gastrointestinal sites (Moore *et al.*, 1995; J.K. Griffiths and S. Tzipori, unpublished data), were investigated using piglets or calves. For immunological and immunopathological studies, the mouse (Ungar *et al.*, 1990, 1991; Chen *et al.*, 1993a, b; Theodos *et al.*, 1996), the simian immunodeficiency virus (SIV) infected monkey (A. Carville and S. Tzipori, unpublished data), and other neonates including lambs (Hill *et al.*, 1990) and calves (Fayer *et al.*, 1989; Peeters *et al.*, 1992) were used. These studies provided as much information about cryptosporidiosis in general as in relation to the specific species (Tzipori, 1988; Current and Garcia, 1991; O'Donoghue, 1995).

Since the use of animal models for immunological studies and pathogenesis of *C. parvum* infection are described elsewhere in this volume (see the chapter by Theodos, 1998), the next part of this chapter describes animal models used to evaluate control measures (see section 4.1), and highlights pertinent aspects of recent studies on the pathophysiology of the disease investigated in piglets and calves (see section 4.2).

4.1. Models for Evaluating Agents for Treatment or Prevention

Since the search for an effective therapy against persistent *C. parvum* infection remains the major objective of any drug development program, the use of an immunodeficient host is highly desirable. The ideal immunocompromised animal model should (a) allow a rapid and consistent establishment of persistent infection in adult animals, (b) require a low to moderate infectious dose, and (c) generate a profound watery diarrhea, dehydration, malabsorption, weight loss, and emaciation. Moreover, it is critical that the infection should colonize a major portion of the gastrointestinal tract with the ability to spread to the hepatobiliary system, resulting in morphological changes similar to those seen in humans. There are several reported rodent models that are being used to screen drugs for their efficacy against *C. parvum* infection, none of which meet all the above criteria. These include (a) the normal neonatal mouse (Tzipori *et al.*, 1982a; Fayer and Ellis, 1993a) or the neonatal SCID mouse (Rohlman *et al.*, 1993; Tzipori *et al.*, 1994), (b) the dexamethasone-suppressed adult mouse model (Yang and Healey, 1993), (c) the immunosuppressed rat model (Rehg *et al.*, 1987; Brasseur *et al.*, 1988), (d) the adult SCID mouse model (Mead et al., 1995), and (e) the anti-interferon-γ (IFNγ) conditioned weaned SCID model (Tzipori et al., 1995).

Each model offers its own advantages and limitations. A disadvantage to the immunosuppressed adult mouse and rat models is that these animals will resolve their *C. parvum* infection unless immunosuppression is maintained. This could create potential toxicity problems, since administration of an immunosuppressive drug could possibly result in untoward interactions between the experimental and the immunosuppressive drugs. An added disadvantage to the rat model is that more drug is required than with a murine model, due to differences in size. While the adult SCID mouse offers an immunocompromised model that can be easily randomized and manipulated, it takes a minimum time of 2–3 weeks before these mice start to shed detectable levels of oocysts (Mead et al., 1995; Tzipori et al., 1995). In addition, while the hepatobiliary system becomes infected in adult SCID mice, it generally takes 12–13 weeks before extraintestinal infections are consistently observed in all mice. This lengthens the time for which mice need to be maintained for a drug trial, and thereby increases the cost of the trial.

4.1.1. The Anti-IFNγ Conditioned SCID Mouse

In our experience, the anti-IFNγ conditioned weaned SCID mouse offers several advantages over all the above-mentioned models, including: (a) weaned mice are easier to obtain, manipulate, and randomize than infants; (b) the small intestine is more heavily infected; (c) mice demonstrate consistent shedding of oocysts by day 6 of infection, which allows for completion of a drug efficacy trial against acute infection within 16–18 days; and (d) the hepatobiliary system is involved before 5 weeks of infection (Tzipori et al., 1995). This rapid involvement of the hepatobiliary system offers a model in which drug efficacy can be tested against both gastrointestinal and extraintestinal forms of *C. parvum* infection. This model has now been adopted by several other laboratories for screening agents against *C. parvum*.

The acute model. The IFNγ-conditioned SCID mouse can be used as a model of the acute phase of infection, which occurs during the first 20 days after *C. parvum* challenge when infection is limited to the gastrointestinal tract. This model is used to analyze orally administered compounds that are not systemically absorbed from the gastrointestinal tract (e.g. hyperimmune bovine colostrum or paromomycin) for efficacy against *C. parvum* forms that colonize accessible epithelial cell surfaces of the gastrointestinal lumen.

The chronic model. The chronic phase of *C. parvum* infection occurs between 32 and 47 days after *C. parvum* challenge of anti-IFNγ treated mice, in which extraintestinal infections can be observed in all mice. This phase may be used to analyze the effects of systemically or orally administered compounds for their efficacy against *C. parvum* forms colonizing inaccessible epithelial cell surfaces such as the hepatobiliary tract, abscessed crypts, occluded stomach pits, and pancreatic ducts. There would be little point in testing drugs that are not absorbed from the gastrointestinal tract against hepatobiliary tract infections in the chronic infection model. The acute model has been used extensively in the last 3 years to assay many compounds that showed inhibitory activities against *C. parvum* infection in the *in vitro* system. It has proved to be sensitive and reproducible, as judged by the consistent inhibitory activity by paromomycin, the positive control drug, whether given orally twice daily or through medicated water.

Other models. Other models of *C. parvum* that are presently being used include the neonatal mouse (Tzipori *et al.*, 1982a, 1994; Fayer *et al.*, 1990; Blagburn *et al.*, 1991), the immunosupressed mouse (Yang and Healey, 1993), the immunosuppressed rat (Rehg *et al.*, 1987, 1988; Brasseur *et al.*, 1988), the piglet diarrhea model (Tzipori, 1985; Tzipori *et al.*, 1994), and the SIV-infected rhesus monkey model (unpublished data).

4.1.2. The Infant Mouse Model

The infant mouse, aged 4–6 days, is highly susceptible to infection and can be challenged with a dose as low as 310 oocysts (Finch *et al.*, 1993). The mice remain clinically healthy and continue to shed oocysts for 10–14 days, after which they recover fully and oocyst shedding ceases. Neonates invariably are used for assessing preventive treatment rather than therapy because of the short window of infection. ICR litters are randomized into the desired groups on day 5, one day prior to challenge. Normally treatment commences when mice are challenged. The following is a brief outline of the model used in our laboratory, which differs somewhat from those described by others (Fayer *et al.*, 1990; Blagburn *et al.*, 1991). The course of treatment is for 5–6 days, with two oral doses of 30–50 µl volume per dose by gavage. Mice are euthanized 18 hours after the end of treatment, and the entire gastrointestinal track is removed, weighed, and homogenized (1 : 40, weight/volume) in phosphate buffered saline (PBS). The number of oocysts present in a given volume is counted in a hemocytometer chamber. The enumeration is repeated five times per mouse and averaged. The mean ($\pm$ standard deviation (SD)) per treatment group is calculated and expressed as the percentage inhibition as compared with placebo. Paromomycin is normally used as the positive control drug. The limitations of the infant mouse are that only a small amount of drugs can be given for only a short period, they are susceptible to excessive manipulation, weight variation, and maternal acceptance after randomization, and the model is limited to drug prophylaxis.

4.1.3. The Immunosuppressed Rodent Models

Several methods of inducing immunosuppression in rodents have been described, of which the administration of dexamethasone (DEX) in drinking water to either adult C57/BL mice (Yang and Healey, 1993) or to adult female rats (Rehg et al., 1988), is the most convenient. Immunosuppressed C57/BL mice given 10–100 µg/ml DEX in drinking water were challenged with 10^6 oocysts/mouse on day 14 after immunosuppression. With a dose of 10 µg/ml mice began to shed oocysts 3 days after infection, and this continued for 45 days before succumbing to drug toxicity, which provided a sufficient window for drug testing. It was not clear whether hepatobiliary tract infection occurred. The rat model only requires 0.25 µg/ml DEX for 10 days to become immunosuppressed (Rehg et al., 1988). Rats were challenged with 10^6 oocysts, and by day 21 oocysts were detected in the feces of 75% of infected rats, and bile duct infection was evident in 68% by day 30. Another method of immunosuppression in rats was achieved by a combined twice-weekly injections of 25 mg hydrocortisone acetate and a low protein diet. A drawback regarding all the methods of immunosupression, however, is the associated levels of mortality and other health complications.

4.1.4. The Piglet Diarrhea Model

This model is used to evaluate therapeutic agents that have shown considerable activity in cell culture and in the anti-IFNγ conditioned SCID mouse model described earlier. The obvious advantage of the piglet model is the development of acute watery diarrhea and other gastrointestinal symptoms, which all rodent models lack (Tzipori, 1985). The manifestation of diarrhea, a key symptom in human patients, is essential for the evaluation of oral therapeutic agents. Severe diarrhea affects the drug transit time through the gastriontestinal tract and can markedly reduce the therapeutic action of the drug (Tzipori et al., 1994).

Colostrum deprived newborn piglets are highly susceptible to cryptosporidiosis, which leads to watery diarrhea, dehydration, malabsorption, wasting, and often sudden death within 2–5 days of onset of symptoms (Tzipori et al., 1982b, 1994; Vitovec and Koudela, 1992). In contrast, piglets raised under conventional conditions rarely experience diarrhea due to this infection, although cryptosporidiosis in swine is extremely prevalent, as judged by serological surveys (Quilez et al., 1996) and isolation studies (Sanford, 1987). Swine, like cattle (Lorenzo-Lorenzo et al., 1993; Scott et al., 1995), continue to excrete oocysts intermittently in feces for long periods of time. It is not clear whether they regularly become reinfected or if they maintain a low-grade but persistent infection. Even colostrum deprived piglets become rapidly resistant to *C. parvum*, and infection becomes mostly asymptomatic 7–10 days after birth. It is thought that because of the widespread and high prevalence of *C. parvum* (Quilez *et al.*, 1996), good maternal protection is afforded to piglets during the first few days after birth. Studies have shown that the majority of pigs become subclinincally infected shortly after weaning when aged 4–5 weeks (Vanopdenbosch and Wellemans, 1985; Quilez *et al.*, 1996).

Because colostrum deprived newborn piglets are highly susceptible to infection with many other pathogens, and to avoid cross-infection among experimental groups, cesarean derived gnotobiotic piglets maintained in complete isolation are used for drug evaluation. They are divided into groups and challenged 24 hours after birth. Drug treatment normally begins on day 3, with the onset of watery diarrhea and oocyst excretion in feces. The drug is either mixed in the milk diet or administered orally over 10–11 days. Parameters used to assess therapy include gastrointestinal symptoms, including the severity of diarrhea and the extent of illness, weight change, oocyst shedding, and the degree of mucosal damage and infection at the end of treatment (Tzipori *et al.*, 1994).

The piglet diarrhea model helps identify realistic interactions between diarrhea, effective therapy, and drug toxicity. It highlights the impact of the severity of diarrhea on the outcome of treatment; the more severe the diarrheal illness, which accelerates gut transit time by the drug, the poorer the clinical response to treatment with paromomycin (Tzipori *et al.*, 1994). These studies also showed the detrimental effect of stomach acidity on the efficacy of hyperimmune bovine colostrum and paromomycin. Finally, individual variation in response to treatment can, as in human patients, be assessed in the context of the extent of infection and severity of illness. The limitations of this model include: (a) litter variation in terms of litter size (8–12), and piglet size (0.8–2.0 kg); (b) the infection and illness, although possibly severe, are self-limiting and animals recover within 14 days; (c) the overall susceptibility of piglets can vary greatly from highly susceptible, resulting in death, to mild infection with rapid recovery, even when using the same inoculum; and (d) the model is labor intensive and expensive as compared with rodent models. However, the piglet diarrhea model is the only standard diarrhea model available that can be used under controlled laboratory conditions.

4.1.5. *The SIV-infected Rhesus Monkey Model*

The ultimate model that mimics many of the disease aspects observed in humans with AIDS is a primate with AIDS (Blanchard *et al.*, 1987; Baskerville *et al.*, 1991; Kaup *et al.*, 1994). Immunocompetent primates develop a self-limiting acute diarrhea, which can be life-threatening in the young (reviewed by Miller *et al.*, 1990; Riggs, 1990). Progressive, fatal illness was described in macaques infected with SIV and with cryptosporidiosis involving the gastrointestinal tract, complicated by hepatobiliary and respiratory tract infections (King, 1986; Osborn et al., 1989; Riggs, 1990). While the expense, limited availability, and many other issues, including ethical considerations, limit the extensive use of this model, there are some aspects of the disease and intervention that can only truly be investigated in such animals. Among them is the delineation of the various immunological defects in the gastrointestinal tract mucosa, associated with progression of SIV infection, which contribute to persistence of infection and to mucosal damage during cryptosporidiosis.

Potential for vaccine development. The premise that HIV-infected individuals can be protected against cryptosporidiosis through regular mucosal immunizations with *C. parvum* antigen is also more appropriately investigated in primates. The ability of HIV-infected individuals to mount an antibody response depends on the level of circulating CD4 T lymphocytes, and diminishes considerably at very low CD4 cell counts (Carson et al., 1995). Patients with counts below 180 are at risk of developing chronic *C. parvum* infection (Blanshard et al., 1992; Flanigan et al., 1992). With low CD4 T cell counts, *C. parvum*-specific secretory immunoglobulin A (IgA), which is elevated in patients with chronic cryptosporidiosis (Cozon et al., 1994; Benhamou et al., 1995), appears to play no role in eliminating the infection. Elevated IgA in these patients results from chronic *C. parvum* infection. We believe, however, that elevated *C. parvum*-specific secretory IgA *before* infection could protect patients against moderate exposure to *C. parvum* present in contaminated food or water. It is conceivable, therefore, that regular and frequent mucosal immunization with *C. parvum* antigen could protect patients with CD4 T cell counts well below 150 cells/mm^3 against chronic disease. In the absence of other control measures against the disease in this subpopulation, the development of a *C. parvum* vaccine to stimulate secretory antibodies as a means of protection has merit and deserves serious consideration.

As most people are primed early in life through infection, intranasal boosting with an appropriate mucosal adjuvant might be sufficient to stimulate protective mucosal levels of secretory IgA. We initially used mice to determine the best source of parasite antigen for mucosal stimulation and resistance to challenge, and the choice of a noninvasive route of administration. We found that formalin-inactivated excysted sporozoites, mixed with cholera toxin as the mucosal adjuvant, provided good stimulation of secretory IgA production in mice and in rhesus monkeys (unpublished data). We have recently immunized SIV-infected monkeys, which are being regularly boosted. They are being monitored for secretory IgA levels in saliva and feces in order to establish the effect of declining CD4 T cell numbers on secretory IgA levels and, ultimately, protection against challenge with *C. parvum*. Protection of newborn calves by oral immunization with inactivated oocysts (Harp and Goff, 1995), and antibody stimulation resulting from immunizing sheep with recombinant plasmid DNA encoding two surface sporozoite epitopes (Jenkins et al., 1995), suggest further that immunization as a form of protection against cryptosporidiosis is worth considering.

4.1.6. Secretory Antibodies and the Backpack Mouse Hybridoma Model

The potential therapeutic efficacy of hyperimmune bovine colostrum and specific monoclonal IgG and IgM antibodies is discussed at length in the chapter by Crabb. Here, the rationale and techniques used for the production of secretory IgA antibodies against C. parvum as a potential immunotherapeutic approach and their evaluation in the backpack hybridoma mouse model (Neutra et al., 1991) will be outlined briefly.

Production of secretory IgA mAb. Secretory antibodies in the intestine provide a first line of defense against mucosal infections. Previous studies at Neutra's laboratory have demonstrated that secretory IgA monoclonal antibodies (mAbs) directed against surface antigens of bacteria and viruses can protect against mucosal infections by preventing contact of the corresponding microorganisms with epithelial surfaces (Kraehenbuhl and Neutra, 1994). Our hypothesis is that dimeric IgA directed against surface epitopes shared by sporozoites and merozoites can provide immunoprotection against cryptosporidiosis, and at the same time help identify the epitopes that are protective in the mucosal system. Effective secretory IgA molecules can potentially be humanized in the future. However, unlike bacteria and viruses, which can be neutralized by a single mAb, protozoa have complex surface structures and the process of attachment and invasion of host cells may involve multiple steps and molecules. Therefore, inhibition of attachment and invasion may require a combination of several mAbs. Initially, six mucosally derived dimeric IgA-producing hybridomas from Peyer's patch lymphocytes, were obtained from six separate fusions. These were produced by repeated oral immunizations of mice with oocysts. In contrast, 50 new IgA hybridomas were recently derived from a single fusion of lymphocytes of a mouse infected at infancy and then twice immunized orally with oocysts. The mAbs are presently being characterized by ELISA, Western blots, dot blot, deglycosylation, and immunofluorescence using intact oocysts and infected tissue sections.

Preliminary screening in the *in vitro* assay demonstrated varying degrees of inhibitory activitiy, but none was above 50%. While the inhibitory activitiy of single sIgA mAbs against C. parvum is low to moderate, a combination of several mAbs selected for their unique epitope and other antigenic specificities may yet prove much more effective. This optimism is based on the fact that rabbit polyclonal antibodies produced specifically against sporozoites are highly neutralizing *in vitro* and *in vivo*.

The backpack hybridoma model. In this model secretory IgA-producing tumor cells are injected subcutaneously onto the back of young adult BALB/c mice. The tumor cells secrete secretory IgA into the mouse tissue, which becomes systemic, and are transported onto mucosal membranes in much the same way as are endogenous IgA molecules. This includes the surfaces of the hepatobiliary system as well as the gastriontestinal tract. Mice which received secretory IgA tumor-secreting cells against cholera toxin or salmonella were highly protected against subsequent oral challenge with the corresponding agents (Winner *et al.*, 1991; Michetti *et al.*, 1992). To investigate the impact of secretory IgA on intestinal colonization in mice, three hybridomas were tested in the backpack tumor-bearing mouse model (Apter *et al.*, 1993). In our studies mice were injected subcutaneously with tumor cells and, once the tumors had reached 1 cm in diameter 6–8 days later, the mice were conditioned with anti-IFNγ mAb and then challenged with *C. parvum* oocysts. In this assay mice with the highest levels of secretory IgA in two of the three mAb groups, had much reduced levels of parasites in feces, indicating a good correlation between IgA fecal and serum levels and protection (X. Zhou, S. Tzipori and M.R. Neutra, unpublished data). Similar results were reported by others who had followed the same procedure for IgA production and testing them in the mouse backpack hybridoma technique (Enriquez *et al.*, 1996). The protective effect of several combinations of mAbs will be investigated next.

4.2. Pathophysiologic Studies in Animals

Studies on intestinal pathophysiology include investigations of absorption in intact animals, and electrophysiology in Ussing chambers. Coupled with structural information these functional data can offer a powerful insight into how *C. parvum* alters the host intestine. Described below are studies performed in piglets and in calves, both of which develop acute watery diarrhea, as do humans.

4.2.1. Studies in Piglets

The aim of these studies was to shed light on the structural and functional basis of diarrhea due to *C. parvum* during evolving host cell–parasite interactions in the small intestine over 12–48 hours after infection, before major mucosal reorganization had occurred. Segments of intestine were directly inoculated *in vivo*, harvested, and studied in Ussing chambers (Moore *et al.*, 1995). As expected, villus, but not crypt, architectural changes corresponded to extent of infection. While solute and macromolecular permeability did not increase, the glucose responsive, short-circuit current was diminished 48 hours after inoculation. These findings parallel structural observations that include loss of the Na^+/glucose transporting villus epithelium, without loss of crypt epithelium. Na^+ absorption by the impaired villi due to *C. parvum* infection can, however, be stimulated by glutamine, and to a lesser extent by glucose (Argenzio et al., 1994). This glutamine-stimulated Na^+ absorption by impaired villi is thought to be mediated by prostaglandin-sensitive apical Na^+/H^+ exchange mechanisms. With increased intensity of infection leading to profound architectural changes, NaCl absorption is markedly suppressed, while net Cl^- secretion into the lumen is somewhat increased. On balance, it appears that the watery diarrhea seen first in infected piglets is largely due to the abnormal intestinal function leading to impaired absorptive processes rather than to increased intestinal secretion. This is also consistent with the observation (unpublished) that piglets ultimately succumb to starvation due to poor absorption rather that to dehydration *per se*.

4.2.2. Studies in calves

We speculated that different parts of the small intestine, the major site of a clinically significant *C. parvum* infection, respond differently to infection. Indeed, studies conducted in calves, showed that the proximal and distal small intestine do not suffer the same physiological fate after site-specific inoculation with *C. parvum* (J.K. Griffiths and S. Tzipori, unpublished). Neonatal calves deprived of colostrum underwent site-specific surgery, in which 10^9 *C. parvum* oocysts were deposited in the proximal or distal small intestine (Moore et al., 1995). A removable tie was used to keep the oocysts within the specific region, and was removed 2 hours after surgery, once excystation and infection had occurred. The animals were fed intravenously to decrease gut motility, and were sacrificed 48 hours after surgery. The infected sites, plus control sites from the same animals located proximal to the intentionally infected sites, were mounted in Ussing chambers and studied for physiological changes. As an additional control, sham-infected calves underwent identical surgical procedures.

In the distal ileum, infection led to a fall in baseline short-circuit current (I_{sc}) at 48 hours. Short-circuit current is generally associated with Cl^- secretion and NaCl absorption. At the site of infection, the baseline I_{sc} fell from 19.7 ± 0.7 to 14.2 ± 0.4 $\mu A/cm^2$ ($P < 0.001$). This result is consistent with what has been found previously in *C. parvum*-infected piglets (Argenzio et al., 1990, 1993; Moore et al., 1995). In contrast, at a proximal duodenal site, the I_{sc} increased with infection (60.9 ± 1.6 rising to $67.6 + 1.7$ $\mu A/cm^2$, $P = 0.005$). All other prior published Ussing chamber studies of infected animals have focused on the distal small intestine, and this result was unexpected. Thus different sites in the intestine may respond differently to the same inoculum of parasites. Cryptosporidial diarrhea in people with AIDS is related to the extent of infection (Genta et al., 1993), and it may prove to be true that infection in one region of the intestinal tract is more serious than infection in another with different physiological implications concerning diarrhea.

In addition, there is evidence that there are quantifiable physiological effects in the intestine that occur at distant sites from the actual site of infection. As outlined above, the I_{sc} fell from 19.7 ± 0.7 to 14.2 ± 0.4 µA/cm² in the distal ileum (ΔI_{sc} = 5.2 µA/cm², $P < 0.001$). Surprisingly, uninfected control ileal site proximal to the infected site also exhibited a significantly decreased baseline I_{sc}, falling from 18.1 ± 0.9 to 11.0 ± 0.4 µA/cm² (ΔI_{sc} = 7.1 µA/cm², $P < 0.001$). When examined for evidence of infection, the proximal control ileal site showed little morphological change ($P = 0.0028$), a smaller change in the crypt/villus ratio ($P < 0.0001$), and minimal infection ($P < 0.0001$), indicating that only scanty numbers of parasites had traveled counter to the luminal flow over the 48-hour period. Thus this equally large change in the I_{sc} of the internal control site could not be explained by the extent of infection or mucosal changes observed. Conjecturally, this physiological effect, which clearly cannot be attributed directly to parasite presence or mucosal damage, is due to hormonal, neuronal, or second messenger alterations induced by an infection occurring elsewhere in the gastrointestinal tract (J.K. Griffiths and S. Tzipori, unpublished). Argenzio and colleagues (1996) have recently published evidence implicating the enteric nervous system in the diarrhea of piglet cryptosporidiosis. Their data suggest that prostaglandins alter NaCl transport by stimulating cholinergic interneurons that innervate VIPergic and cholinergic motor nerves. Our data independently lead to a similar conclusion, i.e. that infection at one site in the intestine may lead to distant effects that are not directly related to the actual presence of the parasite at all sites.

5. NEW APPROACHES TO DRUG DESIGN

C. parvum continues to be a serious and life-threatening disease in patients with AIDS, in the elderly, and in children with chronic diarrhea and malnutrition that is directly linked to *C. parvum*, and against which there is no consistently effective treatment. In 1980, with the recognition of *C. parvum* as a serious contagious disease of newborn calves (Tzipori *et al.*, 1980; Anderson, 1981), the need for effective treatment was obvious. It was assumed at the time that *C. parvum*, which taxonomically is closely related to coccidia, would be highly susceptible to anticoccidial agents because of its peculiar and presumed vulnerable extracytoplasmic location. This, however, was far from the truth, as 16 highly effective anticoccidial agents proved ineffective against *C. parvum* (Tzipori *et al.*, 1982a).

The search for effective chemotherapeutic and immunotherapeutic agents began in earnest in the early 1990s. Clearly the impetus is largely due to the need to control the chronic diarrhea and wasting that *C. parvum* inflicts on patients with AIDS (Griffths, 1998). Although by 1990 as many as 100 therapeutic modalities had been tested (Ungar, 1990), and many others have been tested since (Brasseur *et al.*, 1991, 1994; Giacometti *et al.*, 1996; Woods *et al.*, 1996), these were developed against other infections.

Unfortunately, very few in the private sector are actively pursing the development of compounds that specifically target *C. parvum*. There are three major reasons for this lack of interest: (a) there is a legitimate risk involved with undertaking the development of drugs against an infectious agent which is notoriously unresponsive to any drug treatment; (b) many technical limitations exist, which are reflected in a poor understanding of many aspects of the disease, and makes for an expensive and a risky investment; and (c) a somewhat misguided perception of a relatively small market. While the last point may be true with regard to the number of patients with AIDS and cryptosporidiosis who will benefit the most from such a treatment, it would be wrong to ignore other potential beneficiaries. Benefits include treatment and prophylaxis in the face of outbreaks in day-care centers (Alpert *et al.*, 1984; Cordell and Addiss, 1994), water-borne outbreaks (MacKenzie *et al.*, 1994; Fricker and Crabb, 1998, this volume), and elderly individuals suffering a severe bout of diarrhea due to *C. parvum* (Bannister and Mountford, 1989; Neill *et al.*, 1996). *C. parvum*-related chronic diarrhea and malnutrition in children in developing countries, a long ignored condition, will also benefit from effective treatment or prevention. Last, but not least, cryptosporidiosis in young calves is a disease with serious economic consequences, and a reasonably low cost treatment will undoubtedly be attractive to producers. Treatment of infected susceptible animals may be another way of reducing the enormous environmental contamination, which is hazardous to public health.

The resistance of *C. parvum* to chemotherapeutic agents remains the parasite's greatest mystery and also contributes to its seriousness as a pathogen. It may be associated with the unique intracellular niche that *C. parvum* occupies, which makes it less accessible to drugs entering epithelial cells, as compared with other intracellular organisms (see Tzipori and Griffiths, 1998). Elucidation of the function and molecular trafficking through the feeder organelle and through the parasitophorous membrane may hold the key to this question. The ubiquitous nature and the ease with which the parasite moves among humans and their domestic animals may have over the last 50 years exposed *C. parvum* to all the major groups of antimicrobial agents, and might in part have also contributed to resistance to drugs.

The acute phase of cryptosporidiosis, which involves parts of or the entire gastrointestinal tract, normally occurs in the immunocompetent host. The chronic phase develops over a prolonged period of persistent infection and includes the involvement of extragastrointestinal tract sites, namely the hepatobiliary and pancreatic duct systems. Chronic infection also leads to profound mucosal changes, which often result in abscessation and obliteration of crypts and stomach pits making them inaccessible to luminal therapy. One may postulate that, if an effective therapy were to become available, early luminal treatment of acute infection will prevent the development of chronic disease. Certainly, treatments against infections confined to the gastrointestinal tract are easier and simpler to develop and can be administered orally. Chronic infection presumably can only be eliminated by systemically active agents.

The search for new molecular targets for future drug design and production is in progress in a number of laboratories. Many metabolic enzymes have served as useful targets for drug development against other pathogens. Essential metabolic enzymes of parasite origin that are structurally or functionally sufficiently different from that of the mammalian host cell are, theoretically, suitable targets for drug design. Understanding the biochemistry and molecular biology of the Apicomplexa, and of *C. parvum* in particular, will help shift the emphasis of drug development from random screening towards a more rational approach. Distinct features of parasite metabolism and their metabolic enzymes can be exploited for the design of specific antiparasitic agents. For example, dihydrofolate reductase of *Plasmodium falciparum*, *Toxoplasma gondii*, and *Eimeria tenella*, ionosine monophosphate dehydrogenase of *E. tenella*, and ubiquinol–cytochrome c reductase of *E. tenella* are some of the metabolic enzymes that have been targeted for the design of specific antiparasitic drugs. In addition, it has been found that *E. tenella* has a unique mannitol metabolic pathway that is absent in host cells (Schmatz *et al.*, 1989), which makes a potential target for drug development.

At present, little is known about the metabolic enzymes of *C. parvum*. The bifunctional dihydrofolate reductase–thymidylate synthase (DHFR-TS) gene from *C. parvum* has been cloned and characterized only recently (Vazquez *et al.*,1996). This enzyme is present as a single-copy gene and maps to a 1.2 Mbp chromosome. DHFR-TS is essential for folate biosynthesis, and many antifolate drugs that target this enzyme have been used successfully against other apicomplexan parasites but not *C. parvum*. Sequence analysis has identified many potential antifolate resistance determinants in *C. parvum* DHFR-TS. Ile51 and Thr108, which are shown to confer antifolate resistance in *P. falciparum* DHFR-TS, are present in analogous positions in *C. parvum* DHFR-TS. It has been speculated that Cys113 in *C. parvum* DHFR-TS might also be responsible for conferring antifolate resistance, based on analysis of comparable position in *P. falciparum* DHFR-TS.

The ribonucleotide reductase (RNR) of *C. parvum*, a two-subunit (R1R2) enzyme essential for the *de novo* biosynthesis of deoxyribonucleotides, was identified in our laboratory as a potential target for drug design. Anti-RNR drugs are used for treatment of cancer (Stubbe, 1990) and against certain herpes infections (Dutia *et al.*, 1986; Liuzzi *et al.*, 1994). The R1 subunit of *C. parvum* is 2331 nucleotides long and codes for a predicted protein of 777 amino acids. Though the RNR activity was not detected in the extracts of oocysts or sporozoites, *C. parvum* R1-specific mRNA was identified in gut tissues of infected mice (R. Balakrishnan and S. Tzipori, unpublished data), indicating that the gene is transcribed. Other metabolic enzymes, including respiratory enzymes, that have been identified and characterized in other members of the Apicomplexa would potentially make good targets for drug development. The presence of some of the genes responsible for respiratory enzymes in *C. parvum* have recently been confirmed (G. Carraway, L.K. Griffiths, and S. Tzipori, unpublished data), even though *C. parvum* has no morphologically recognizable mitochondria-like structures (Current and Reese, 1986), and there are suggestions that it may even be an obligate fermentor. Denton *et al.* (1996) have shown the presence of pyrophosphate dependent phosphofructokinase (PPi-PFK) and adenosine diphosphate (ADP) dependent pyruvate kinase (ADP-PK) in the soluble extracts of *C. parvum* oocysts. Since PPi-PFK increases the energy efficiency of glycolysis, it is likely that PPi-PFK in *C. parvum* represents an evolutionary adaptation towards anerobiosis (Mertens, 1991), and it could represent a stage-specific process. There are no data to date that suggest *C. parvum* is solely fermentative. If it is, selective inhibition of pyrophosphate dependent, fermentative enzymes could be targeted for selective drug development against cryptosporidiosis.

C. parvum appears to have unique biological and metabolic characteristics which set it apart from other Apicomplexa. It has recently been shown that extracts of *C. parvum* show arginine decarboxylase (ADC), rather than ornithine decarboxylase (ODC) activity, which is more consistent with plant metabolic pathways. Furthermore, inhibitor of ADC has also been shown to inhibit the growth of *C. parvum* meronts and gametes *in vitro* (Yarlet *et al.*, 1995).

Other potential molecular targets for drug design include parasite surface lectins and other molecules that are responsible for parasite attachment and invasion of host cells (Joe *et al.*, 1995). The identification of naturally occurring analogs, or the design of such molecules, could provide a means of interfering with the early parasite attachment and prevent cell penetration.

6. LABORATORY INVESTIGATIONS ON DRUGS PRESENTLY USED IN HUMANS

The effort to identify drugs effective against *C. parvum* through *in vitro* and *in vivo* screening continues in several laboratories. A considerable number of compounds have been reported to have inhibitory activity against *C. parvum in vitro* (Marshall and Flanigan, 1992; Egraz-Bernard *et al.*, 1996; Giacommetti *et al.*, 1996; Woods *et al.*, 1996) or *in vivo* (Fayer and Ellis, 1993a, b; Brassuer *et al.*, 1991, 1994; Healey *et al.*, 1995; Mancassola *et al.*, 1995; Mead *et al.*, 1995; Rehg, 1991a, b, 1993, 1994, 1995). Of these compounds that have been tested in either animals or in patients, none were 100%

effective in eliminating or preventing *C. parvum* infection. In this section, laboratory investigations relating to two drugs that are currently being used in human patients will be presented. There are other drugs that are prescribed by clinicians, of which azythromycin is one that is also considered beneficial (Hicks *et al.*, 1996).

6.1. Paromomycin

Paromomycin was reported to be partially effective in several small trials, including a 70% response rate in one placebo-controlled study (White *et al.*, 1994). Benefit from paromomycin treatment in patients with AIDS was also reported by Scaglia *et al.* (1994), Mohri *et al.* (1995), and Flanigan *et al.* (1996). Most patients, however, eventually relapsed while still on medication. In other trials, paromomycin treatment of patients with AIDS and cryptosporidiosis was even less effective (unpublished data). The dose used in these patients was 1–2 g/day. Higher doses were avoided because of the risk of toxicity due to presumed systemic drug absorption from an injured gastrointestinal tract. However, in one study a dose of 1.4 g/day given over several weeks showed neither evidence of side-effects nor appreciable levels of paromomycin in the serum (Bissuel *et al.*, 1994).

Of the many compounds tested under experimental conditions in our laboratory, paromomycin is still the only drug to show consistently good inhibitory activity at high doses in all three models: in cell culture at 2 mg/ml, in the anti-IFNγ conditioned SCID mouse at a daily dose of 2000 mg/kg (Tzipori *et al.*, 1995), and in the piglet diarrhea model at a daily dose of 500 mg/kg (Tzipori *et al.*, 1994). No signs of toxicity in these animals were apparent with these doses. Others have reported similar results using different *in vitro* (Marshall and Flanigan, 1992; Egraz-Bernard *et al.*, 1996) or *in vivo* (Fayer and Ellis, 1993a, b; Rehg, 1994; Verdon *et al.*, 1994; Healey *et al.*, 1995) models.

Aminoglycosides, such as paromomycin, possess activity against prokaryotic, eukaryotic, and mammalian cells, depending upon their chemical substitution and pharmacophore pattern at varying positions within their respective molecular scaffolds. Paromomycin was synthesized in the mid-1960s but, because of potential renal toxicity in humans, it is not used in therapy. Paromomycin, which is not significantly absorbed after ingestion (Bissuel *et al.*, 1994), manifests its major activity against intracellular developmental *C. parvum* forms, with no effect on extracellular forms. Data generated recently in our laboratory demonstrate that, given the structure of the host apical surface after infection, paromomycin *in vitro* probably reaches the parasite via the parasitophorous vacuolar membrane, as well as through vesicular uptake and accumulation within the cell cytoplasm and transportation via the feeder organelle (J. Griffiths, R. Balakrishnan and S. Tzipori, unpublished).

The effective cellular uptake of paromomycin is less than 1%, which accounts for the high dose required for effective inhibitory activities in all three models, and may explain the partial activity in human patients when used in moderate doses. As seen in the piglet diarrhea model (Tzipori et al., 1994), severe watery diarrhea, which leads to a rapid transit time through the gastrointestinal tract, must also be a factor in further reducing the drug's activity in patients.

Based on these observations we have initiated a study to synthesize paromomycin derivatives that are more effective against *C. parvum* by applying medicinal chemistry and drug design principles (M. Nelson and S. Tzipori, unpublished data), with a view to synthesize derivatives that are more readily taken up by the host cell. An increased cellular uptake will much reduce the dose required for effective treatment, increase efficacy, and hopefully reduce drug toxicity considerably, a major limiting factor in the use of paromomycin in human patients (Tan et al., 1995). However, it is possible that the limited inhibitory activity of paromomycin against *C. parvum* is unrelated to cellular drug uptake. The correlation between resistance to paromomycin and a particular sequence on the ribosomal RNA of *Escherichia coli*, recently determined by nuclear resonance spectroscopy (Fourmy et al., 1996), places *C. parvum* rRNA among those that are partially resistant (R. Balakrishnan, unpublished data).

6.2. Nitazoxanide (NTZ)

NTZ is a nitrothiazole benzamide compound, first synthesized in 1976, with a wide range of antimicrobial activity against protozoa, helminths, and bacterial pathogens. A single dose (50 mg/kg) was used against *Taenia saginata* and *Hymenolepis nana* in humans (Rossignol and Maisonneuve, 1985). In an open-label study of 15 subjects in Mexico, parasite clearance was said to have been almost 100% in patients with AIDS and cryptosporidiosis (Feregrino et al., 1996). The recommended treatment by these investigators was 1 or 2 g/day over a period of 10–30 days. In contrast, Davis et al. (1996) reported different results from a clinical trial in the USA involving 22 patients with AIDS and *C. parvum* who were assigned to receive between 500–2000 mg/day orally for 4 weeks. Preliminary analysis of the data revealed that 15 (68%) had a reduction in bowel-movement frequency at doses of 1000 mg/day and above, four had complete resolution of diarrhea, and four others lacked detectable oocysts in the stool. There was no apparent toxicity associated with doses used. These investigators concluded that NTZ appears to have a favorable clinical but less pronounced parasitologic effect on *C. parvum*, and suggested that greater benefit might result with longer duration and/or higher NTZ dose. NTZ received US Federal Drugs Administration approval for limited compassionate use in patients, and controlled clinical trials are underway to determine its true efficacy and benefit in the treatment of chronic *C. parvum* infection.

Because of the current use of NTZ in humans, we performed extensive preclinical studies to determine NTZ efficacy in our *in vitro* system, and in the two animal models described earlier (S. Tzipori, C.M. Theodos, J.K. Griffiths, and A. Fairfield, unpublished data). NTZ had a high inhibitory activity against *C. parvum* in cell culture, being 85–94% at concentrations of 2.5–10 µg/ml in 0.2% DMSO solution. Little cytotoxicity was seen.

NTZ performed less well *in vivo*. In the anti-IFNγ conditioned SCID mouse model NTZ showed little to no effect at daily doses of 100 and 200 mg/kg orally in two divided doses for 10 days. There was no appreciable reduction in oocyst shedding or extent of mucosal parasite infection as compared with the placebo treated mice. In the piglet diarrhea model there was no significant effect when piglets were treated with 125 mg/kg orally in two divided doses for 11 days, as determined by extent and duration of diarrhea, oocyst shedding, and extent of mucosal infection at the end of the trial. However, when piglets received a dose of 250 mg/kg NTZ, the outcome was different. There was a statistically significant reduction in oocyst shedding in feces in the last 4–5 days of treatment in the NTZ group as compared with the placebo group. There was also a considerable reduction in the extent of mucosal infection at the end of the treatment. NTZ treatment was, however, less effective than paromomycin given at 500 mg/kg in this model. Only paromomycin caused a marked reduction in diarrhea in these experiments. NTZ induced some diarrhea in NTZ-treated uninfected piglets, indicating that the drug can be diarrhaegenic at moderate to high doses, observations which were also noted in human patients (unpublished observations). It has been shown previously in the piglet diarrhea model that diarrhea has a profound impact on drug treatment, presumably because of a rapid transit time through the gastrointestinal tract (Tzipori *et al.*, 1994). It is therefore conceivable that NTZ could have been more effective in piglets had it been less diarrheagenic at 250 mg/kg. The NTZ preclinical trial in pigs mimics much more closely the observations made in humans treated with NTZ against chronic cryptosporidiosis than any of the other models. Cell culture showed very high efficacy, higher than paromomycin at a much lower dose, while in the SCID mouse there was no appreciable effect.

While the models for drug testing, including the more sensitive pig diarrhea model, are far from optimal, they are hardly responsible for the failure to identify truly effective agents. Truly effective drugs will be readily detected by most of the currently described models. However, the limitation of the available models lies in their inability to reflect accurately poorly or partially effective drugs, a problem which is also encountered when assessing marginally effective drugs in patients. Under such circumstances there is often a tendency to rely too heavily on cryptic statistical analyses to demonstrate significance, when in reality only marginal, and ultimately therapeutically insignificant, effects are observed.

7. CONCLUDING REMARKS

Much progress has been achieved in the last decade in terms of the development of laboratory techniques, reagents, and *in vivo* models. This has undoubtedly contributed to better and more accurate investigations. Despite a concerted effort by many investigators, however, breakthroughs have been minimal. The development of adequate *in vitro* and *in vivo* techniques for drug screening, and intensified and systematic screening, has so far not resulted in the discovery of an effective therapy. The reason for the failure may well be due to the unique biological niche that *C. parvum* parasite occupies, as discussed at length in the first chapter in this volume. Its location beneath the cell membrane, but outside the cell cytoplasm, may prove a crucial element that needs to be considered when designing new therapeutic approaches.

Laboratory investigations on two drugs currently used against chronic *C. parvum* infection in AIDS patients are discussed. This chapter also provides information and the rationale for work in progress in our laboratory that relates to the development of novel approaches to the control of the disease. This includes the identification of molecular targets of parasite origin for drug design, and studies on the structure–activity relationships of partially effective drugs with a view to synthesizing more effective derivatives. Other investigations are aimed at attempting to establish the role of secretory antibody, and the merit of repeated mucosal immunizations as a means of providing protection to individuals with AIDS who are at risk of developing chronic *C. parvum* infection.

REFERENCES

Adams, R.B., Guerrant, R.L., Zu, S., Fang, G. and Roche, J.K. (1994). *Cryptosporidium parvum* infection of intestinal epithelium: morphologic and functional studies in an *in vitro* model. *The Journal of Infectious Diseases* **169**, 170–177.
Alpert, G., Bell, L.M., Kirkpatrick, C.E., Budnick, L.D., Campos, J.M., Friedman, H.M. and Plotkin, S.A. (1984). Cryptosporiodiosis in a day-care center. *New England Medical Journal* **311**, 860–861.
Anderson, B.C. (1981). Pattern of shedding of cryptosporidial oocysts in Idaho calves. *Journal of the American Medical Association* **178**, 865–867.
Anderson, B.C., Donndelinger, T., Wilkins, R.M. and Smith, J. (1982). Cryptosporidiosis in a veterinary student. *Journal of the American Veterinary Medical Association* **180**, 408–409.
Apter, F.M., Michetti, P., Winner, L.S., III, Mack, J.A., Mekalanos, J.J. and Neutra,

M.R. (1993). Analysis of the roles of anti-lipopolysaccharide and anti-cholera toxin IgA antibodies in protection against *Vibrio cholerae* and cholera toxin using monoclonal IgA antibodies *in vivo*. *Infection and Immunity* **61**, 5279–5285.

Argenzio, R.A., Liacos, J.A., Levy, M.L., Meuten, D.J., Lecce, J.G. and Powell, D.W. (1990). Villous atrophy, crypt hyperplasia, cellular infiltration, and impaired glucose–Na absorption in enteric cryptosporidiosis of pigs. *Gastroenterology* **98**, 1129–1140.

Argenzio, R.A., Lecce, J. and Powell, D.W. (1993). Prostanoids inhibit intestinal NaCl absorption in experimental porcine *cryptosporidiosis*. *Gastroenterology* **104**, 440–447.

Argenzio, R.A., Rhoads, J.M., Armstrong, M. and Gomez, G. (1994). Glutamine stimulates prostaglandin-sensitive Na(+)–H+ exhange in experimental porcine cryptosporidiosis. *Gastroenterology* **106**, 1418–1428.

Argenzio, R.A., Armstrong, M. and Rhoads, J.M. (1996). Role of the enteric nervous system in piglet Cryptosporidiosis. *The Journal of Pharmacology and Experimental Therapeutics* **279**, 1109–1115.

Arrowood, M.J. and Sterling, C.R. (1987). Isolation of *Cryptosporidium* oocysts and sporozoites using discontinuous sucrose and isopycnic Percoll gradients. *Journal of Parasitology* **73**, 314–319.

Arrowood, M.J., Hurd, M.R. and Mead, J.R. (1995). A new method for evaluating experimental cryptosporidial parasite loads using immunofluorescent flow cytometry. *Journal of Parasitology* **81**, 404–409.

Awad-El-Kariem, F.M., Wahurst, D.C. and McDonald, V. (1994). Detection and species identification of *Cryptosporidium* oocysts using a system based on PCS and endonuclease restriction. *Parasitology* **109**, 19–22.

Bannister, P. and Mountford, R.A. (1989). *Cryptosporidium* in the elderly: a cause of life-threatening diarrhoea. *American Journal of Medicine* **86**, 507–508.

Baskerville, A., Ramsay, A.D., Millward-Sadler, G.H., Cook, R.W. and Cranage, M.P. (1991). Chronic pancreatitis and biliary fibrosis associated with cryptosporidiosis in simian AIDS. *Journal of Comparative Pathology* **105**, 415–421.

Benhamou, Y., Kapel, N., Hoang, C., Matta, H., Meillet, D., Magne, D., Raphael, M., Gentilini, M., Opolon, P. and Gobert, J-G. (1995). Inefficacy of intestinal secretory immune response to *Cryptosporidium* in aquired immunodefiency syndrome. *Gastroenterology* **108**, 627–635.

Bissuel, F., Cotte, L., Rabodonirina, M., Rougier, P., Piens, M.A. and Trepo, C. (1994). Paromomycin: an effective treatment for cryptosporidial diarrhea in patients with AIDS. *Clinical Infectious Diseases* **18**, 447–449.

Black, E.K., Finch, G.R., Taghi-Kilani, R. and Belosevic, M. (1996). Comparison of assays for *Cryptosporidium parvum* oocysts viability after chemical disinfection. *FEMS Microbiology Letters* **135**, 187–189.

Blagburn, B.L., Sundermann, C.A., Lindsay, D.S., Hall, J.E. and Tidwell, R.R. 1991. Inhibition of *Cryptosporidium parvum* in neonatal Hsd: (ICR) BR Swiss mice by polyether ionophores and aromatic amidines. *Antimicrobial Agents and Chemotherapy* **35**, 1520–1523.

Blanchard, J.L., Baskin, G.B., Murphey-Corb, M. and Martin, L.N. (1987). Disseminated cryptosporidiosis in simian immunodeficiency virus/delta-infected rhesus monkeys. *Veterinary Pathology* **24**, 454–456.

Blanshard, C., Jackson, A.M., Shanson, D.C., Francis, N. and Gazzard, B.G. (1992). Cryptosporidiosis in HIV-seropositive patients. *Quarterly Journal of Medicine, New Series* **85**, 813–823.

Blewett, D.A. (1988). Quantitative techniques in *Cryptosporium* research. In:

Proceedings of the 1st International Workshop on Cryptosporidiosis (Angus and Blewett, eds). Edinburgh.
Brasseur, P., Lemeteil, D. and Ballet, J.J. (1988). Rat model for human cryptosporidiosis. *Journal of Clinical Microbiology* **26**, 1037–1039.
Brasseur, P., Lemeteil, D. and Ballet, J.J. (1991). Anticryptosporidiosis drug activity screened with an immunosuppressed rat model. *Journal of Protozoology* **38**, 230S–231S.
Brasseur, P., Favennec, L., Lemeteil, D., Roussel, F. and Ballet, J.J. (1994). An immunosuppressed rate model for evaluation of anti-*Cryptosporidium* activity in sinefungin. *Folia Parasitologica* **41**, 13–16.
Buraud, M., Forget, E., Favennec, L., Bizet, J., Gobert, J. G. and Deluol, A. M. (1991). Sexual stage development of cryptosporidia in the Caco-2 cell line. *Infection and Immunity* **59**, 4610–4613.
Carsen, P.J., Schut, R.L., Simpson, M.L., O'Brien, J. and Janoff, E.N. (1995). Antibody class and subclass responses to pneumococcal polysaccharides following immunization of human immunodeficiency virus-infected patients. *The Journal of Infectious Diseases* **172**, 340–345.
Chen, W., Harp, J.A., Harmsen, A.G. and Havell, E.A. (1993a). Gamma interferon functions in resistance to *Cryptosporidium parvum* infection in severe combined immunodeficient mice. *Infection and Immunity* **61**, 3548–3551.
Chen, W., Harp, J.A. and Harmsen, A.G. (1993b). Requirements for CD4+ cells and γ-interferon in resolution of established *Cryptosporidium parvum* infection in mice. *Infection and Immunity* **61**, 3928–3932.
Cordell, R. and Addiss, D. (1994). Cryptosporidiosis in child care settings: a review of the literature and recommendations for prevention and control. *Pediatric Infectious Disease Journal* **13**, 310–317.
Cory, A.H., Owen, T.C., Barltrop, J.A. and Cory, J.G. (1991). Use of aqueous soluble tetrazolium/formazan assay for cell growth assays in culture. *Cancer Communications* **3**, 207–212.
Cozon, G., Biron, F., Jeannin, M., Cannella, D. and Revillard, J-P. (1994). Secretory IgA antibodies to *Cryptosporidium parvum* in AIDS patients with chronic cryptosporidiosis. *The Journal of Infectious Diseases* **169**, 696–699.
Current, W.L. (1990). Techniques and laboratory maintenance of *Cryptosporidium*. In: *Cryptosporidiosis of Man and Animals* (J.P. Dubey, C.A. Speer and R. Fayer, eds), pp. 31–49. Boston: CRC Press.
Current, L. and Garcia, L.S. (1991). Cryptosporidiosis. *Clinical Microbiology Reviews* **4**, 325–358.
Current, W.L. and Haynes, T.B. (1984). Complete development of *Cryptosporidium* in cell cultures. *Science* **224**, 603–605.
Current, W.L. and Reese, N.C. (1986). A comparison of endogenous development of three isolates of *Cryptosporidium* in suckling mice. *Journal of Protozoology* **33**, 98–108.
Current, W.L., Reese, N.C., Ernst, J.V., Bailey, W.S., Heyman, M.B. and Weinstein, W.M. (1983). Human cryptosporidiosis in immunocompetent and immunodeficient persons. Studies of an outbreak and experimental transmission. *New England Journal of Medicine* **308**, 1252–1257.
Dagan, R., Fraser, D., El-On, Y., Kassis, I., Deckelbaum, R.J. and Turner, S. (1995). Evaluation of an enzyme-immunoassay for detection of *Cryptosporidium* spp. in stool specimens from infants and young children. *American Journal of Tropical Medicine and Hygiene* **52**, 134–138.
Davis, L.J., Soave, R.E., Dudley, R.E., Fessel, J.W., Faulkner, S. and Mamaxos, J.P. (1996). Nitazoxanide (NTZ) for AIDS-related cryptosporidial diarrhea (CD): an

open-label safety, efficacy and pharmacologic study. *36th Interscience Conference on Antimicrobial Agents and Chemotherapy*, New Orleans, September 1996, abstract LM50, p. 289.
Denton, H., Brown, S.M.A., Roberts, C.W., Alexander, J., McDonald, V., Thong, K. and Coombs, G.H. (1996). Comparison of the phosphofructokinase and pyruvate kinase activities of *Cryptosporidium parvum*, *Eimeria tenella* and *Toxoplasma gondii*. *Molecular and Biochemical Parasitology* **76**, 23–29.
DuPont, H.L., Chappel, C.L., Sterling, C.R., Okhuysen, P.O., Rose, J.B. and Jakubowski, W. (1995). The infectivity of *Cryptosporidium parvum* in healthy volunteers. *New England Journal of Medicine* **332**, 855–859.
Dutia, B.M., Frame, M.C., Subak-Sharpe, J.H., Clark, W.N. and Marsden, H.S. (1986). Specific inhibition of herpesvirus ribonucleotide reductase by synthetic peptides. *Nature* **321**, 439–441.
Eggleston, M.T., Tilley, M. and Upton, S.J. (1994). Enhanced development of *Cryptosporidium parvum in vitro* by removal of oocyst toxins from infected cell monolayers. *Proceedings of the Helminthology Society, Washington* **61**, 118–121.
Egraz-Bernard, M., Favennec, L., Agnamey, P., Ballet, J.J. and Brasseur, P. (1996). Inhibition of complete development of *Cryptosporidium parvum* in Caco-2 cells. *Letters* **15**, 897–899.
Enriquez, F.J., Palting, J., Hensel, J. and Riggs, M.W. (1996). Immunotherapy of cryptosporidiosis using IgA monoclonal antibodies to neutralizing-sensitive *C. parvum* antigens. In: *49th Meeting of the Society of Protozoologists*. Tuscon, AZ, abstract 162, p. 123.
Fayer, R. and Ellis, W. (1993a). Glycoside antibiotics alone and combined with tetracyclines for prophylaxis of experimental cryptosporidiosis in neonatal BALB/c mice. *Journal of Parasitology* **79**, 553–558.
Fayer, R. and Ellis, W. (1993b). Paromomycin is effective as prophylaxis for cryptosporidiosis in dairy calves. *Journal of Parasitology* **79**, 771–774.
Fayer, R., Andrews, C., Ungar, B.L.P. and Blagburn, B.L. (1989). Efficacy of hyperimmune bovine colostrum for prophylaxis of cryptosporidiosis in neonatal calves. *Journal of Parasitology* **75**, 393–397.
Fayer, R., Guidry, A. and Blagburn, B.L. (1990). Immunotherapeutic efficacy of bovine colostral immunoglobulins from a hyperimmunized cow against cryptosporidiosis in neonatal mice. *Infection and Immunity* **58**, 2962–2965.
Feregrino, G.M., Higuera, R.F., Rossignol, J.F., Cavier, R., Villarreal, C., Padierna, O.J. and Hidalgo, H. (1996). Extraordinary potency of the Nitozoxanida, a new antiparasitary against the *Cryptosporium* infections in advanced AIDS. In: *XI International Conference on AIDS*, Vancouver.
Finch, G.R., Daniels, C.W., Black, E.K., Schaefer, F.W. and Belosevic, M. (1993). Dose response of *Cryptosporidium parvum* in outbred neonatal CD-1 mice. *Applied and Environmental Microbiology* **59**, 3661–3665.
Flanigan, T., Marshall, R., Redman, D., Kaetzel, C. and Ungar, B. (1991). In vitro screening of therepeutic agents against *Cryptosporidium*: hyperimmune cow colostrum is highly inhibitory. *Journal of Protozoology* **38**, 225S-227S.
Flanigan, T., Whalen, C., Turner, J., Soave, R., Toerner, J., Havlir, D. and Kotler, D. (1992). *Cryptosporidium* infection and CD4 counts. *Annals of Internal Medicine* **116**, 840–842.
Flanigan, T.P., Ramratnam, B., Graeber, C., Hellinger, J., Smith, D., Wheeler, D., Hawley, P., Heath-Chiozzi, M., Ward, D.J., Brummitt, C. and Turner, J. (1996). Prospective trial of paromomycin for cryptosporidiosis in AIDS. *American Medical Journal* **100**, 370–372.
Fourmy, D., Recht, M.I., Blanchard, S.C. and Puglisi, J.D. (1996). Structure of the A

site of *Escherichia coli* 16S ribosomal RNA complexed with an aminoglycoside antibiotic. *Science* **274**, 1367–1371.
Fricker, C.R and Crabb, J.H. (1998) Water-borne cryptosporidiosis: detection methods and treatment. *Advances in Parasitology* **40**, 237–274.
Garcia, L.S., Brewer, T.C. and Bruckner, D.A. (1987). Fluorescence detection of *Cryptosporidium* oocysts in human fecal specimans using monoclonal antibodies. *Journal of Clinical Microbiology* **25**, 119–121.
Genta, R.M., Chappell, C.L., White, A.C., j., Kimball, K.T. and Goodgame, R.W. (1993). Duodenal morphology and intensity of infection in AIDS-related intestinal cryptosporidiosis. *Gastroenterology* **105**, 1769–1775.
Giacometti, A., Cirioni O. and Scalise, G. (1996). *In vitro* activity of macrolides alone and in combination with artemisin, atovaquone, dapsone, minocycline or pryimethamine against *Cryptosporidium parvum*. *Journal of Antimicrobial Chemotherapy* **38**, 399–408.
Graczyk, T.K., Cranfield, M.R. and Fayer, R. (1996). Evaluation of commercial enzyme immunoassay (EIA) and immunofluorescent antibody (FA) test kits for detection of *Cryptosporidium* oocysts of species other than *Cryptosporidium parvum*. *American Journal of Tropical Medicine and Hygiene* **54**, 274–279.
Griffiths, J.K. (1998). Human cryptosporidiosis: epidemiology, transmission, clinical disease, treatment and diagnosis. *Advances in Parasitology* **40**, 37–81.
Griffiths, J.K., Moore, R., Dooley, S., Keusch, G.T. and Tzipori, S. (1994). *Cryptosporidium parvum* infection of Caco-2 cell monolayers induces an apical monolayer defect, selectively increases transmonolayer permeability, and causes epithelial cell death. *Infection and Immunity* **62**, 4506–4514.
Guarino, A., Canani, R.B., Casola, A., Pozio, E., Russo, R., Bruzzese, E., Fontana, M. and Rubino, A. (1995). Human intestinal cryptosporidiosis: secretory diarrhea and enterotoxic activity in Caco-2 cells. *Journal of Infectious Diseases* **171**, 976–983.
Harp, J.A. and Goff, J.P. (1995). Protection of calves with a vaccine against *Cryptosporium parvum*. *Journal of Parasitology* **81**, 54–57.
Healy, M.C., Yang, S., Rasmussen, K.R., Jackson, M.K. and Du, C. (1995). Therapeutic efficacy of paromomycin in immunosuppressed adult mice infected with *Cryptosporidium parvum*. *Journal of Parasitology* **81**, 114–116.
Hicks, P., Zwiener, R.J., Squires J. and Savell, V. (1996). Azithromycin therapy for *Cryptosporidium parvum* infection in four children infected with human immunodeficiency virus. *Journal of Pediatrics* **129**, 297–300.
Hill, B.D., Blewett, D.A., Dawson, A.M. and Wright, S. (1990). Analysis of the kinetics, isotype and specificity of serum and coproantibody in lambs infected with *Cryptosporidium parvum*. *Research in Veterinary Science* **48**, 76–81.
Hill, B.D., Blewett, D.A., Dawson, A.M. and Wright, S. (1991). *Cryptosporidium parvum*: investigation of sporozite excystation *in vivo* and the association of merozoites with intestinal mucus. *Research in Veterinary Science* **51**, 264–267.
Hockenberry, D., Nunez, G., Milliman, C., Schreiber, R.D. and Korsmeyer, S.J. (1990). Bcl-2 is an inner mitochondrial membrane protein that blocks programmed cell death. *Nature* **348**, 334–336.
Jenkins, M., Kerr, D., Fayer, R. and Wall, R. (1995). Serum and colostrum antibody responses induced by jet-injection of sheep with DNA encoding a *Cryptosporidium parvum* antigen. *Vaccine* **13**, 1658–1664.
Joe, A., Pereira, M.E.A., Keusch, G.T., Tzipori, S. and Ward, H.D. (1995). Role of a galactose-specific *Cryptosporidium parvum* lectin in attachment of sporozoites to host cells. In: *Proceedings of the 95th Annual Meeting of the American Society of Microbiology*, Washington, DC, abstract B-472, p. 247.
Kaup, F.J., Kuhn, E.M., Makoschey, B. and Hunsmann, G. (1994). Cryptosporidiosis

of liver and pancreas in rhesus monkeys with experimental SIV infection. *Journal of Medical Primatology* **23**, 304–308.
King, N.W. (1986). Simian models of acquired immunodeficiency syndrome (AIDS): a review. *Veterinary Pathology* **23**, 345.
Kraehenbuhl, J.P. and Neutra, M.R. (1994). Monoclonal secretory IgA for protection of the intestinal mucosa against viral and bacterial pathogens. In: *Mucosal Immunology* (R. Ogra, J. Mestecky, J. McGhee, J. Bienenstock, M. Lamm and W. Strober, eds), pp. 403–410. New York: Academic Press.
Laxer, M.A., Timblin, B.K. and Patel, R.J. (1991). DNA sequences for the specific detection of *Cryptosporidium parvum* by the polymerase chain reaction. *American Journal of Tropical Medicine and Hygiene* **45**, 688–694.
Liuzzi, M., Deziel, R., Moss, N., Beaulieu, P., Bonneau, A., Bousquet, C., Chafouleas, J.G., Garneau, M., Jaramillo, J., Krogsrud, R.L., Lagace, L., McCollum, R.S., Nawoot, S. and Guindon, Y. (1994). HSV ribonucleotide reductase with antiviral activity *in vivo*. *Nature* **372**, 695–698.
Lorenzo-Lorenzo, M.J., Ares-Mazas, E. and Villacorta Martinez de Maturana, I. (1993). Detection of oocysts and IgG antibodies to *Cryptosporidium parvum* in asymptomatic adult cattle. *Veterinary Parasitology* **47**, 9–15.
Ma, P., Villanueva, T.G., Kaufman, D. and Gillooley, J.F. (1984). Respiratory cryptosporidiosis in the acquired immune deficiency syndrome. Use of modified cold Kinyoun and Hemacolor stains for rapid diagnoses. *Journal of the American Medical Association* **252**, 1298–1301.
MacKenzie, W.R., Hoxie, N.J., Prctor, M.E., Gradus, M.S., Blair, K.A., Peterson, D.E., Kazmierczak, J.J., Addiss, D.G., Fox, K.R. and Rose, J.B. (1994). A massive outbreak in Milwaukee of *Cryptosporidium* infection transmitted through the public water supply. *New England Journal of Medicine* **331**, 161–167.
Mancassola, R., Reperant, J.M., Naciri, M. and Chartier, C. (1995). Chemoprophylaxis of *Cryptosporidium parvum* infection with paromomycin in kids and immunological study. *Antimicrobial Agents and Chemotherapy* **39**, 75–78.
Marshall, R.J. and Flanigan, T.P. (1992). Paromomycin inhibits *Cryptosporidium* infection of a human enterocyte cell line. *Journal of Infectious Diseases* **165**, 772–774.
McDonald, V., Stables, R., Warhurst, D.C., Barer, M.R., Blewett, D.A., Chapman, H.D., Connolly, G.M., Chiodini, P.l. and McAdam, K.P.W.J. (1990). *In vitro* cultivation of *Cryptosporium parvum* and screening for anticryptosporidial drugs. *Antimicrobial Agents and Chemotherapy* **34**, 1498–1500.
Mead, J.R., You, X., Pharr, J.E., Belenkaya, Y., Arrowood, M.J., Fallon, M.T. and Schinazi, R.F. (1995). Evaluation of maduramicin and alborixin in a SCID mouse model of chronic cryptosporidiosis. *Antimicrobial Agents and Chemotherapy* **39**, 854–858.
Mertens, E. (1991). Pyrophosphate dependent phosphofructokinase, an anaerobic glycolytic enzyme. *FEBs Letters* **288**, 1–5.
Michetti, P., Mahan, M.J., Slauch, J.M., Mekalanos, J.J. and Neutra, M.R. (1992). Monoclonal secretory IgA protects mice against oral challenge with the invasive pathogen *Salmonella typhimurium*. *Infection and Immunity* **60**, 1786–1792.
Miller, R.A., Brondson, M.A. and Morton, W.R. (1990). Experimental cryptosporidiosis in a primate model. *Journal of Infectious Diseases* **161**, 312–315.
Mohri, H., Fujita, H., Asakura, Y., Katoh, K., Okamoto, R., Tanabe, J., Harano, J., Noguchi, T., Inayama, Y., Amano, T. and Okubo, T. (1995). Case report: therapy of paromomycin in effective for respiratory infection and hypoxia by *Cryptosporidium* with AIDS. *American Journal of Medical Science* **309**, 60–62.
Moore, R., Tzipori, S., Griffiths, J., Johnson K. and Lomakina, I. (1995). Temporal

changes in permeability of piglet ileum after site-specific infection by *Cryptosporidium parvum*. *Gastroenterology* **108**, 1030–1039.
Neill, M.A., Rice, S.K., Ahmad, N.V. and Flanigan, T.P. (1996). Cryptosporidiosis: an unrecognized cause of diarrhea in elderly hospitalized patients. *Clinical Infectious Diseases* **22**, 167–170.
Neutra, M., Weltzin, R., Winner, L., Mack, J., Michetti, P., Morrison, L., Fields, B., Mekalanos, J. and Kraehenbuhl, J.P. (1991). Identification and use of protective monoclonal IgA antibodies against viral and bacterial pathogens. In: *Immunology of Milk and the Neonate, Advanced Experimental Biology Medicine* (J. Mestecky and R. Ogra, eds), pp. 179–182. New York: Plenum Press.
O'Donoghue, P.J. (1995). *Cryptosporidium* and cryptosporidiosis in man and animals. *International Journal of Parasitology* **25**, 139–195.
Osborn, K.G., Prahalada, S., Lowenstine, L.J., Gardner, M.B., Maul, D.H. and Henrickson, R.V. (1989). The pathology of an epizootic of acquired immunodeficiency in rhesus macaques. *American Journal of Pathology* **114**, 94–99.
Peeters, J.E., Villacorta, I., Vanopdenbosch, E., Vandergheynst, D., Naciri, M., Ares-Mazas, E. and Yvore, P. (1992). *Cryptosporidium parvum* in calves: kinetics and immunoblot analysis of specific serum and local antibody responses (immunglobulin A [IgA], IgG, and IgM) after natural and experimental infections. *Infection and Immunity* **60**, 2309–2316.
Quilez, J., Ares-Mazas, E., Sanchez-Acedo, C., del Cacho, E., Clavel, A. and Causape, A.C. (1996). Comparison of oocyst shedding and the serum immune response to *Cryptosporidium parvum* in cattle and pigs. *Parasitol Resources* **82**, 529–234.
Rehg, J.E. (1991a). Anticryptosporidial activity is associated with specific sulfonamides in immunosuppressed rats. *Journal of Parasitology* **77**, 238–240.
Rehg, J.E. (1991b). Anti-cryptosporidial activity of macrolides in immunosuppressed rats. *Journal of Protozoology* **38**, 228s-229s.
Rehg, J.E. (1993). Anticryptosporidial activity of lasalocid and other ionophorous antibodies in immunosuppressed rats. *Journal of Infectious Diseases* **168**, 1566–1569.
Rehg, J.E. (1994). A comparison of anticryptosporidial activity of paramomycin with that of other aminoglycosides and azithromycin in immunosuppressed rats. *Journal of Infectious Diseases* **170**, 934–938.
Rehg, J.E. (1995). The activity of halofuginone in immunosuppressed rats infected with *Cryptosporidium parvum*. *Journal of Antimicrobial Chemotherapy* **35**, 391–397.
Rehg, J.E., Hancock, M.L. and Woodmansee, D.B. (1987). Characterization of cyclophosphamide–rat model of cryptosporidiosis. *Infection and Immunity* **55**, 2669–2774.
Rehg, J.E., Hancok, M.L. and Woodmansee, D.B. (1988). Characterization of a dexamethasone-treated rat model of cryptosporidial infection. *Journal of Infectious Diseases* **158**, 1406–1407.
Riggs, M.W. (1990). Cryptosporidiosis in cats, dogs, ferrets, raccoons, opossums, rabbits and non-human primates. In: *Cryptosporidiosis of Man and Animals* (J.P. Dubey, C.A. Speer and R. Fayer, eds), pp. 113–120. Boston: CRC Press.
Riggs, M.W. and Perryman, L.E. (1987). Infectivity and neutralization of *Cryptosporidium* sporozoites. *Infection and Immunity* **55**, 2081–2087.
Rohlman, S., Kuhls, V.C., Mosier, D.A., Crawford, D.L., Hawkins, D.R. and Abrams, V.L. (1993). Therapy with atovaquone for *Cryptosporium parvum* infection in neonatal severe combined immunodeficiency mice. *Journal of Infectious Diseases* **168**, 258–260.
Rossignol, J.F. and Maisonneuve, H. (1985). Nitazoxanide in the treatment of *Taenia*

saginata and *Hymenolepis nana* infections. *American Journal of Tropical Medicine and Hygiene* **33**, 511–512.
Sanford, S.E. (1987). Enteric cryptosporidial infection in pigs: 184 cases (1981–1985). *Journal of American Veterinary Medical Association* **190**, 695–698.
Scaglia, M., Atzori, C., Marchetti, G., Orso, M., Maserati, R., Orani, A., Novati, S. and Olliaro, P. (1994). Effectiveness of aminosidine (paromomycin) sulfate in chronic *Cryptosporidium* diarrhea in AIDS patients: an open, uncontrolled, prospective clinical trial. *Journal of Infectious Diseases* **170**, 1349–1350.
Schmatz, D.M., Baginsky, W.F. and Turner, M.J. (1989). Evidence for and characterization of a mannitol cycle in *Eimeria tenella*. *Molecular and Biochemical Parasitology* **32**, 263–270.
Scott, C.A., Smith, H.V., Mtambo, M.M. and Gibbs, H.A. (1995). An epidemiological study of *Cryptosporidium parvum* in two herds of adult beef cattle. *Veterinary Parasitology* **57**, 277–288.
Stubbe, J. (1990). Ribonucleotide reductases. *Advanced Enzymolecular Related Areas Molecular Biology* **63**, 349–419.
Suresh, P. and Rehg, J.E. (1996). Comparative evaluation of several techniques for purification of *Cryptosporidium parvum* oocysts from rat feces. *Journal of Clinical Microbiology* **34**, 38.
Tan, W.W., Chapnick, E. K., Abter, E. I., Haddad, S., Zimbalist, E. H. and Lutwick, L.I. (1995). Paromomycin-associated pancreatitis in HIV-related cryptosporidiosis. *Annals of Pharmacotherapy* **29**, 22–24.
Tewari, M. and Dixit, V.M. (1995). Fas- and tumor necrosis factor-induced apoptosis is inhibited by the poxvirus *crmA* gene product. *Journal of Biology Chemistry* **270**, 3255–3260.
Theodos, C.M. (1998). Innate and cell-mediated immune responses to *Cryptosporidium parvum*. *Advances in Parasitology* **40**, 83–115.
Theodos, C.M., Sullivan, K., Gull, T. and Tzipori, S. (1996). Analysis of the *Cryptosporidium parvum*-specific immune response generated in healing and non-healing murine models of infection. *FASEB Journal* **10**, A1182.
Tilley, M. and Upton, S. (1991). Sporozoites and merozoites of *Cryptosporidium parvum* share a common epitope recognized by a monoclonal antibody and two-dimensional electrophoresis. *Journal of Parasitology* **38**, 48S-49S.
Tilley, M., Upton, S.J., Fayer, R., Barta, J.R., Chrisp, C.E., Freed, P.S., Blagburn, B.L., Anderson, B.C. and Bernard, S.M. (1991). Identification of a 15–kilodalton surface glycoprotein on sprozoites of *Cryptosporium parvum*. *Infection and Immunity* **59**, 1002–1007.
Tsujimoto, Y. (1989). Overexpression of the human *bcl*-2 gene products results in growth enhancement of Epstein–Barr virus-immortalized B cells. *Proceedings of the National Academic of Sciences USA* **86**, 1958–1962.
Tzipori, S. (1983). Cryptosporidiosis in humans and in animals. *Microbiological Reviews* **47**, 84–96.
Tzipori, S. (1985). The relative importance of enteric pathogens affecting neonates of domestic animals. *Advances in Veterinary Science and Comparative Medicine* **29**, 103–203.
Tzipori, S. (1988). Cryptosporidiosis in perspective. *Advances in Parasitology* **27**, 63–129.
Tzipori, S. and Griffiths, J.K. (1998) Natural history and biology of *Cryptosporidium parvum*. *Advances in Parasitology* **40**, 5–36.
Tzipori, S., Campbell, I., Sherwood, D., Snodgrass, D.R. and Whitelaw, A. (1980). An outbreak of calf diarrhoea attributed to cryptosporidial infection. *Veterinary Record* **107**, 579–580.

Tzipori, S., Campbell, I. and Angus, T. (1982a). The theraputic effect of 16 antimicrobiol agents on *Cryptosporidium* infection in mice. *Aust. J. Experimental Biology Medical Science* **60**, 187–190.

Tzipori, S., Smith, M., Makin, T. and Halpin, C. (1982b). Enterocolitis in piglets caused by *Cryptosporidium* sp. purified from calf faeces. *Veterinary Parasitology* **11**, 121–126.

Tzipori, S., Rand, W., Griffiths, J.K., Widmer, G. and Crabb, J. (1994). Evaluation of an animal model system for cryptosporidiosis: therapeutic efficacy of paromomycin and hyperimmune bovine colostrum-immunoglobulin. *Clinical and Diagnostic Laboratory Immunology* **1**, 450–463.

Tzipori, S., Rand, W. and Theodos, C. (1995). Evaluation of a two-phase *scid* mouse model preconditioned with anti-interferon-γ monoclonal antibody for drug testing against *Cryptosporidium parvum*. *Journal of Infectious Diseases* **172**, 1160–1164.

Ungar, B.L.P. (1990). Cryptosporidiosis in humans (*Homo saienus*). In: *Cryptosporidiosis of Man and Animals* (J.P. Dubey, C.A. Speer and R. Fayer, eds), pp. 60–82. Boston: CRC Press.

Ungar, B.L.P., Burris, J.A., Quinn, C.A. and Finkelman, F.D. (1990). New mouse models for chronic *Cryptosporidium* infection in immunodeficient hosts. *Infection and Immunity* **58**, 961–969.

Ungar, B.L.P., Kao, T-C., Burris, J.A. and Finkelman, F.D. (1991). *Cryptosporidium* infection in an adult mouse model: independent roles for IFN and CD4+ T lymphocytes in protective immunity. *Journal of Immunology* **147**, 1014–1022.

Upton, S.J., Tilley, M. and Brillhart, D.B. (1994a). Comparative development of *Cryptosporidium parvum* (Apicomplexa) in 11 continuous host cell lines. *FEMS Microbiology Letters* **118**, 233–236.

Upton, S.J., Tilley, M., Nesterenko, M.V. and Brillhart, D.B. (1994b). A simple and reliable method of producing *in vitro* infections of *Cryptosporidium parvum* (Apicomplexa). *FEMS Microbiology Letters* **118**, 45–50.

Upton, S.J., Tilley, M. and Brillhart, D.B. (1995). Effects of select medium supplements on *in vitro* development of *Cryptosporidium parvum* in HCT-8 cells. *Journal of Clinical Microbiology* **33**, 371–375.

Vanopdenbosch, E. and Wellemans, G. (1985). Detection of antibodies to *Cryptosporidium* by indirect immunofluorecensce (drop method). Prevalence of antibodies in different animal species. *Vlaams Diergeneeskd Tijdschr* **54**, 49–54.

Vazques, J.R., Gooze, L., Kim, K., Gut, J., Peterson, C. and Nelson, R.G. (1996). Potential antifolate resistance determinants and genotypic variation in the bifunctional dihydrofolate reductase–thymidylate synthase gene from human and bovine isolates of *Cryptosporidium parvum*. *Molecular and Biochemical Parasitology* **79**, 153–165.

Verdon, R., Polianski, J., Gaudebout, C., Marche, C., Garry, L. and Pocidalo, J.J. (1994). Evaluation of curative anticryptosporidial activity of paromomycin in a dexamethasone-treated rat model. *Antimicrobial Agents and Chemotherapy* **38**, 1681–1682.

Verdon, R., Keusch, G.T., Tzipori, S., Grubman, S.M., Jefferson, D.M. and Ward, H.D. (1997). An *in vitro* model of infection of human biliary epithelial cells by *Cryptosporidium parvum*. *Journal of Infectious Diseases* (in press).

Villacorta, I., de Graaf, D., Charleir, G. and Peeters, J.E. (1996). Complete development of *Cryptosporium parvum* in MDBK cells. *FEMS Microbiology Letters* **142**, 129–132.

Vitovec, J. and Koudela, B. (1992). Pathogenesis of intestinal cryptosporidiosis in conventional and gnotobioti piglets. *Veterinary Parasitology* **43**, 25–36.

Waldman, E., Tzipori, S. and Forsyth, J.R.L. (1986). Separation of *Cryptosporidium*

species oocysts from feces by using a percoll discontinuous density gradient. *Journal of Clinical Microbiology* **23**, 199.
Ward, H., Joe, A., Kelly, M., Verdon, R., Tzipori, S., Percira, M. and Keusch, G.T. (1995). Use of ELISA-based assays to study *C. parvum*-host cell interaction. In: *Abstracts of the American Society of Tropical Medicine and Hygiene Annual Meeting*, San Antonio.
Webster, K.A., Pow, J.D.E., Giles, M., Catchpole, J. and Wooward, M.J. (1993). Detection of *Cryptosporidium parvum* using a specific polymerase chain reaction. *Veterinary Parasitology* **50**, 35–44.
White, A.C., Chappell, C.L., Hayat, C.S., Kimball, K.T., Flanigan, T.P. and Goodgame, R.W. (1994). Paromomycin for cryptosporidiosis in AIDS: a prospective, double-blind trial. *Journal of Infectious Diseases* **170**, 419–424.
Widmer, G. (1998). Genetic heterogeneity and PCR detection of *Cryptosporidium parvum*. *Advances in Parasitology* **40**, 219–235.
Widmer, G., Tzipori, S. and Griffiths, G.J. (1996). *Cryptosporidium parvum* induces apoptosis in Caco-2 and MDBK monolayers. In: *49th Meeting of the Society of Protozoologists*, Tuscon, AZ, abstract 228, p. 143.
Winner, L., Mack, Weltzin, R., Mekalanos, J., Kraehenbuhl, J.P. and Neutra, M.R. (1991). New model for analysis of muscosal immunity: intestinal secretion of specific monoclonal IgA from hybridoma tumors protects against *V. Cholerae*. *Infection and Immunity* **59**, 977–982.
Woods, K.M., Nesterenko, M.V. and Upton, S.J. (1995). Development of a microtitre ELISA to quantify development of *Cryptosporidium parvum in vitro*. *FEMS Microbiology Letters* **128**, 89–93.
Woods, K.M., Nesterenko, M.V. and Upton, S.J. (1996). Efficacy of 101 antimicrobials and other agents on the development of *Cryptosporidium parvum in vitro*. *Annals of Tropical Medicine and Parasitology* **90**, 603–615.
Xue, D. and Horvitz, H.R. (1995). Inhibition of the *Caenorhabditis elegans* cell-death protease CED-3 by a CED-3 cleavage site in baculovirus p35 protein. *Nature* **377**, 248–251.
Yang, S. and Healey, M.C. (1993). The immunosuppressive effects of dexamethasone administered in drinking water to C57BL/6N mice infected with *Cryptosporidium parvum*. *Journal of Parasitology* **79**, 626–630.
Yang, S., Healey, M.C. and Du, C. (1996a). Infectivity of preserved *Cryptosporidium parvum* oocysts for immunosuppressed adult mice. *FEMS Immunology and Medical Microbiology* **13**, 141–145.
Yang, S., Healey, M.C., Du, C. and Zhang, J. (1996b). Complete development of *Cryptosporidium parvum* in bovine fallopian tube epithelial cells. *Infection and Immunity* **64**, 349–354.
Yarlet, N., Upton, S.J., Zhu, G. and Keithly, J. (1995). Polyamine metabolism: a rational drug target for *Cryptosporium parvum*. *Molecular Parasitology meeting*. September. MBL, Woods Hole, MA.
You, X., Arrowood, M.J., Lejkowski, M., Xie, L., Schinazi, R.F. and Mead, J.R. (1996). A chemiluminescence immunoassay for evaluation of *Cryptosporidium parvum* growth *in vitro*. *FEMS Microbiology Letters* **136**, 251–256.

Genetic Heterogeneity and PCR Detection of *Cryptosporidium parvum*

Giovanni Widmer

*Tufts University School of Veterinary Medicine,
North Grafton,
MA 01536, USA*

1. Relevance of Studying Genetic Polymorphism in *Cryptosporidium Parvum* 224
2. Heterogeneity in *Cryptosporidium* 225
   2.1. Phenotypic variability 226
   2.2. Genetic variability 227
3. PCR Detection of *Cryptosporidium Parvum* 233
   References 236

A variety of methods have been applied to the study of genotypic and phenotypic polymorphism in Cryptosporidium parvum. *Results from these studies have consistently shown the existence of different genotypes and phenotypes within the species. A long-term goal of this work is the identification of markers for virulence in humans and animals and the elucidation of transmission cycles of* C. parvum. *Achievement of these goals will depend on the identification of highly polymorphic loci. Of particular interest are polymorphisms amenable to typing by polymerase chain reaction (PCR), as* C. parvum *cannot be expanded* in vitro. *Fingerprinting of isolates by restriction of PCR fragments or allele-specific PCR has given promising results. As originally observed by isoenzyme analysis, genetic fingerprinting has confirmed the occurrence in humans of unique* C. parvum *genotypes which are not found among calf isolates. This observation remains to be reconciled with the cross-infectivity of* C. parvum *to ruminant and nonruminant hosts and the important role that bovines play in the epidemiology of* C. parvum *and human cryptosporidiosis.*

Although PCR detection of C. parvum *DNA from individual oocysts has been reported, the sensitivity of PCR detection when working with environmental or fecal samples is significantly reduced. Therefore, PCR is currently not used for routine diagnosis or environmental monitoring for* C. parvum. *Inhibitors present in environmental samples, mainly in water and soil, which can negatively affect PCR recoveries, have been identified, and several methods have been proposed to circumvent these problems. The further refinement of detection and genetic fingerprinting protocols will provide essential tools for indentifying environmental sources of oocysts and elucidating transmission cycles.*

1. RELEVANCE OF STUDYING GENETIC POLYMORPHISM IN *CRYPTOSPORIDIUM PARVUM.*

Understanding genetic heterogeneity in *Cryptosporidium parvum* is a first step towards the rational management of this pathogen, on both a clinical and an environmental level. Clinical observations (Saltzberg *et al.*, 1991; Günthard *et al.*, 1996) and experimental infections (Pozio *et al.*, 1992) have revealed a significant range in symptoms and the response to drug treatment. As in all interactions between parasites and host, it is likely that this variability results from a combination of factors. Information on the genetic make-up of *C. parvum* poulations will be relevant for predicting the course of an infection, but will have to be evaluated in the context of a dynamic interaction between host and parasite. Identification of genetic markers for virulence and drug sensitivity are one of the common goals driving the study of *C. parvum* heterogeneity. This information will facilitate the management of cryptosporidiosis, particularly in immuncompromised individuals at risk of developing chronic infections. The identification of genes responsible for specific phenotypic properties is a difficult task. The absence of an *in vitro* system for parasite maintenance and the lack of genetic transformation tools as used in related protozoa (Donald and Roos, 1993; Goonewardene *et al.*, 1993; Van Dijk *et al.*, 1996) are a significant obstacle to the achievement of this task.

From an environmental point of view there is considerable interest in studying genetic polymorphism in *C. parvum*. Specifically, regulatory agencies and the water industry are concerned about oocyst contamination in water storage and distribution systems (Solo-Gabriele and Neumeister, 1996). Concerns about watershed contamination with *C. parvum* and other water-borne pathogens are likely to increase as expanding urbanization and agricultural land use introduce new sources of human and zoonotic pathogens in watersheds. In view of the limitations of current water treatment methods (McKenzie *et al.*, 1994, Widmer *et al.*, 1996), preventive measures aimed at reducing contamination of source water are a realistic alternative to costly water filtration systems. Impementation of preventive measures depends on the identification of potential oocyst sources, such as dairy farms and septic systems. Genetic fingerprinting of water-borne oocysts can assist in this task. Although fingerprinting by PCR assays to distinguish between human and animal sources of oocysts is already a technical reality (Bonnin *et al.*, 1996; Carraway *et al.*, 1996, 1997), problems in the application of PCR to environmental samples persist (Tsai and Olson, 1992; Johnson *et al.*, 1995; Sluter *et al.*, 1997).

In contrast to recently emerged pathogens, studies on more 'established' Apicomplexa has progressed much further. In particular, research on malaria has been driven by the high mortality and morbidity caused by this parasite, in particular by *Plasmodium falciparum*. The need for effective vaccines and antimalarial drugs has produced a wealth of genetic information on this parasite. For example, the combined GenEMBL sequence database listed 1478 entries for *P. falciparum*, as compared to 15 for *C. parvum*. The taxonomic proximity between *Cryptosporidium* and *Plasmodium* (Johnson *et al.*, 1990) will facilitate the transfer of knowledge from other Apicomplexa to less studied parasites. For instance, the isolation of new *C. parvum* genes is greatly facilitated by the availability of homologous sequences from related species.

A potentially useful lead in the search for new polymorphic genetic markers in *C. parvum* is the description of numerous microsatellites in *P. falciparum* (Su and Wellems, 1996). These sequences, also known as simple sequence repeats, are frequently polymorphic and used for genotyping in numerous organisms. Length polymorphism at such loci are often detected by PCR using primers flanking the microsatellite (Hearne *et al.*, 1992). The identification of such loci in *C. parvum* will open new possibilities for fingerprinting and for the identification of clinically relevant markers.

2. HETEROGENEITY IN *CRYPTOSPORIDIUM*

Genotypic and phenotypic markers for the identification of different species of *Cryptosporidium* have been identified (Nina *et al.*, 1992; Ogunkolade *et al.*, 1993; Awad-El-Kariem *et al.*, 1994). The taxonomic status of *C. parvum*, *C. baileyi*, and *C. muris* as individual species is supported by isoenzyme analysis, antigenic profiles, and restriction fragment length polymorphism (RFLP). A similar conclusion can be drawn from the analysis of sequence variability between *C. parvum* and *C. wrairi* at a locus identified by Laxer *et al.* (1991) and used for the development of a widely used *C. parvum*-specific PCR assay (Chrisp and LeGendre, 1994). Within a 895–bp DNA fragment 22 point mutations were identified between the two species, whereas no differences between two unrelated *C. parvum* isolates were found.

Within *C. parvum*, both phenotypic and genotypic studies have identified polymorphism among isolates. In these studies the taxonomic unit is commonly referred to as an 'isolate', loosely defined as one or multiple oocyst samples originating from an individual host. Some isolates are maintained by serial animal passage, usually through newborn calves, and assigned an isolate code such as Iowa, MD or GCH1. Isolate propagation by this method is not only expensive, but also carries the risk of isolate contamination by endogenous *C. parvum*. Single-cell clones, which for many years have been the standard in other protozoa (Rosario, 1981; Dvorak, 1984) are not available in *C. parvum*, as *in vitro* cultivation systems capable of expanding individual sporozoites do not exist. Individual oocysts can be propagated in neonatal mice (K. Johnson, S. Tzipori, and G. Widmer, unpublished data), but lines obtained from such infections are not clonal as they originate from four sporozoites. Although genetic fingerprints obtained from natural *C. parvum* populations provide valuable information, interpretation of results has to take into consideration the possibility of mixed parasite populations. Genetically mixed isolates of *C. parvum* were recently identified using PCR-RFLP analysis (Carraway *et al.*, 1997).

2.1. Phenotypic Variability

Phenotypic differences between *C. parvum* isolates have been identified using isoenzyme analysis and Western blotting. The former technique was applied to characterize 10 *C. parvum* isolates, four of human origin and six of calf origin or maintained in calves (Awad-El-Kariem *et al.*, 1995). With the exception of one human isolate which displayed a calf-like phosphoglucomutase (PGM) profile, human and animal isolates generated different PGM and hexokinase profiles. The phenotypic differences among human isolates appeared unrelated to their geographical origin, as the only sample originating from Africa was indistinguishable from the other human samples from the UK. Similar observations were made in an earlier study by the same group (Ogunkolade *et al.*, 1993). Four *C. parvum* isolates, two of human and one each of bovine and ovine origin, were analysed for their PGM and glucose phosphate isomerase (GPI) profiles. Again, PGM profiles were identical among animal isolates and different from the two human profiles, whereas GPI was monomorphic. The application of isoenzyme analysis to *Cryptosporidium* species is limited by the considerable number of oocysts required. Lysates prepared from 10^7 oocysts were needed to obtain detectable enzyme levels. In addition, from the 12 enzymes assayed, activity from two enzymes only were clearly detectable, limiting the number of loci that can be typed. The deficiency of oocyst lysates in a number of enzymes is consistent with the fact that oocysts are nondividing forms and most likely metabolically inactive. Enzyme polymorphism such as seen with PGM could, however, be exploited for the development of allele-specific PCR assays by identifying specific nucleotide changes responsible for altered enzyme mobility.

Antigenic differences among *C. parvum* isolates and different *Cryptosporidium* species were investigated using monoclonal and polyclonal antibodies raised against *C. parvum*. In one study (Nina *et al.*, 1992), three different phenotypes were detected with two monoclonal antibodies. Similarly, as with the isoenzyme analysis, polymorphism among human isolates was observed, but no shared profiles between human and calf isolates were detected. Nichols *et al.* (1991) also observed antigenic differences between human and calf isolates, as well as differences among human isolates and among bovine isolates. In comparison with the isoenzyme technique, Western blotting was more sensitive, requiring 2×10^4 oocysts (Nichols *et al.*, 1991), which is 500-fold less than for isoenzyme detection. Antigens appear also to be more polymorphic than enzymes, revealing in some cases three or more phenotypes. A direct comparison of the extent of polymorphism by the two techniques has, however, not been reported.

2.2. Genetic Variability

Molecular techniques have the advantage over phenotypic methods in that they require significantly less oocysts. This is particularly relevant when working with a parasite like *C. parvum*, which cannot be amplified *in vitro*. In addition, noncoding sequences and silent mutations can be examined. Techniques capable of directly fingerprinting environmental samples are of particular interest to the water industry, and PCR is likely to play an important role in future screening and typing methods. The resistance of *C. parvum* oocysts to many chemical disinfectants, a challenge for the water industry, is a significant advantage for molecular biology research, as exogenous nucleic acid can easily be removed without affecting oocyst integrity or viability. For molecular fingerprinting protocols in particular, the feasibility of completely removing contaminating nucleic acids with bleach is an advantage. Techniques based on random amplification of DNA, also known as RAPD (Welsh and McClelland, 1990) depend on uncontaminated DNA, since foreign DNA will generate artifactual bands.

RAPD analysis was used to compare 23 *C. parvum* isolates, comprising 14 human samples and 8 bovine samples from Australia (Morgan *et al.*, 1995). Included in this analysis was also the MD isolate from Scotland, which was also used in some of the isoenzyme studies described above. Because of its wide distribution as a laboratory isolate, this isolate is a good candidate for an international benchmark for studies on *C. parvum* polymorphism. Inclusion of such a standard in future studies would facilitate the comparison of data obtained in different laboratories. RAPD profiles from the Australian study were evaluated using Jaccard's distance index (Jaccard, 1908) and two tightly linked clusters, one comprising the majority of human samples, the other including most calf samples, were identified. In addition to a clear correlation with the host, profiles also seemed to be affected by the geographical origin of the isolate. In a second study using RAPD analysis (Carraway *et al.*, 1996), five calf-passaged isolates, including MD, were compared using two arbitrary primers. Differences were found between an isolate from Holland and the remaining samples from different locations within the USA. A human isolate originally obtained from a patient with acquired immune deficiency syndrome

(AIDS), and propagated for 4 years in calves grouped with the other isolates from the USA.

In both studies RAPD analysis detected minor differences between fingerprints from calf isolates, but the significance of these changes is difficult to assess, as the RAPD procedure is prone to generating artifactual differences due to minor variation in experimental parameters (Ellsworth et al., 1993). Polymorphic bands generated by random amplification can, however, be sequenced in order to determine whether they originate from genuine polymorphic sites or whether they are the product of spurious amplification. Authentic polymorphic bands could then be used to generate species or isolate-specific primers. In addition to genetic fingerprinting applications, RAPD methods offer a quick way of identifying new species-specific sequences without the need for screening DNA libraries. This approach is particularly relevant for emerging pathogens, for which only limited sequence information is available, and was applied by Morgan et al. (1996) to develop new primers diagnostic for *C. parvum*.

Defined genetic polymorphisms were identified by different laboratories and have generated valuable information on genetic variability in *C. parvum*. This information is now making the study of the molecular epidemiology of cryptosporidiosis feasible. Sequence heterogeneity within *C. parvum* was initially demonstrated by RFLP analysis (Ortega et al., 1991). This study was based on the use of a nonspecific human DNA probe (Longmire et al., 1990), and was consistent with phenotypic studies in showing differences between human and calf isolates. Specific sequence polymorphism was first demonstrated by Kilani and Wenman (1994), who compared three fragments of the ribosomal small subunit (SSU) gene with previously published homologous sequences from *C. parvum* isolated from naturally infected Australian goat kids (Johnson et al., 1990). Homologies in this comparison were in the range 96–99%, depending on the region. Also included in this study was the sequence published by Cai et al. (1992), with 91% homology. This lower value is consistent with the fact that the sequence published by Cai et al. was erroneously identified as a *C. parvum* sequence, but was in fact from *C. muris* (V. McDonald, personal communication).

Carraway et al. (1996) compared the SSU and the adjacent intergenic spacer 1 (ITS1) from five calf isolates of *C. parvum*, including GenBank entry L16996 submitted by Blagburn et al. Heterogeneity identified in this study was comparable to the range observed by Kilani and Wenman (1994), showing within a 300-bp segment a total of 13 single-base changes as well as one insertion and deletion. From a 136-bp fragment of isolate MD originating from Scotland, it appeared that the geographical origin was not reflected in the ribosomal sequence. In the same study the heterogeneity within one isolate was investigated by examining different DNA clones of a SSU PCR fragment obtained from calf isolate UCP (Tzipori et al., 1994). Interestingly, heterogeneity within this isolate was comparable to interisolate differences.

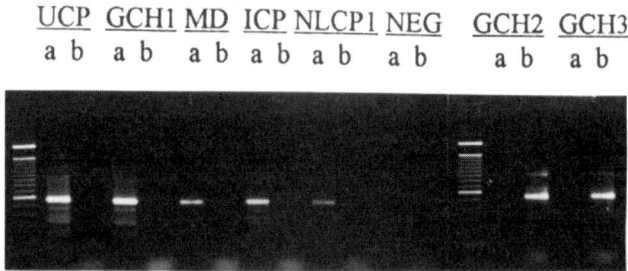

Figure 1 Ribosomal PCR fingerprints of five bovine and two human *C. parvum* isolates. A conserved upstream primer (cry7) in the SSU rRNA gene was used in combination with one of two sequence-specific downstream primers (cryITS1 or cry21) to amplify genomic DNA recovered from oocysts. PCR reactions identified two genotypes at this locus. (a) PCR products with primers cry7 and cryITS1; (b) PCR products with cry7 and cry21. Isolates showing a PCR product in lane (a) are of bovine origin (UCP, GCH1, MD, ICP, NLCP1); isolates positive in the (b) lanes were isolated directly from two HIV-positive individuals (GCH2, GCH3). NEG, negative PCR control; markers are 100-bp DNA ladders.

Confirming the observation on the SSU gene, differences between two ITS1 DNA sequences from isolates MD and UCP showed only few single-base changes and one deletion. In contrast, ITS1 polymorphism between calf isolates and one human isolate were more pronounced. This extensive polymorphism in the ITS1 regions was targeted for the development of PCR primers capable of selectively amplifying one genotype only. As expected from the sequence analysis, PCR analysis of five calf and two isolates from AIDS patients grouped the isolates according to their host origin (Figure 1).

Consistent with the RAPD analysis, human isolate GCH1 maintained in calves over a period of 4 years had the same genotype as found in the other bovine isolates but, due to the limitations of the isolate propagation method, it cannot be excluded that the orginal GCH1 has been overgrown by endogenous *C. parvum*. An interesting difference between the SSU and ITS1 polymorphism was that two SSU genotypes were detected in each calf isolate, whereas only one of two known genotypes was found in the ITS1 region. The detection of multiple SSU ribosomal sequences within the same isolate could be explained by the presence of multiple ribosomal gene copies within the genome (Zamani *et al.*, 1996) or, alternatively, by the presence of genetically mixed populations. The analysis of polymorphisms in multicopy genes is less amenable to PCR fingerprinting, as discriminating between a mixed population and heterogeneous gene copies within the genome is difficult to accomplish without clonal parasite lines. Distinct ribosomal genes were identified in the genome of *P. falciparum*, which led to the discovery of differentially expressed ribosomal genes during the life cycle of malaria (McCutchan *et al.*, 1995).

Currently no evidence for such a phenomenon has been found in *Cryptosporidium*.

An new sequence polymorphism in the dihydrofolate reductase–thymidylate synthase gene was recently described by Vasquez et al. (1996). In this study the DHFR-TS sequences of two *C. parvum* isolates, one from an AIDS patient and one from a calf, were compared, and 38 point mutations were identified within the 1562–bp coding sequence. The authors emphasize the high degree of divergence within this sequence relative to the homologous locus in *P. falciparum*. Comparison of the *C. parvum* α-tubulin and β-tubulin genes showed a comparable similarity value as in DHFR. Future sequence information from coding and noncoding regions will reveal whether this level of heterogeneity is representative of the entire genome.

Bonnin et al. (1996) examined genetic polymorphism within an unidentified 1500-bp *C. parvum* repetitive sequence using PCR-RFLP. Amplification and subsequent digestion with two restriction enzymes identified two variants of this sequence. In agreement with observations on other loci, all calf isolates were monomorphic at this locus, whereas both sequences were present among 13 human samples.

PCR-RFLP was also used by Carraway and coworkers to differentiate between *C. parvum* isolates (Carraway et al,., 1997). A polymorphic RsaI site was identified within a threonine-rich open-reading frame. Digestion of PCR products spanning this site allowed the identification of three profiles: a RsaI-positive profile was found in all calf isolates examined, as well as in some human isolates; a RsaI-negative profile in some human isolates; and a mixed profile in other human isolates. To examine further the nature of this polymorphism, PCR products originating from three isolates (GCH1, GCH2, and GCH3, see Figure 1) were cloned and sequenced. An alignment of these sequences (Figure 2) identified multiple alleles, including two with point mutations within the RsaI site and three with an intact site. Consistent with the RFLP fingerprints, two clones from the calf isolate (pGC91, GCH1) had no mutations within the restriction site, whereas clones originating from a human isolate with mixed profile (GCH3) had both, a RsaI-positive and RsaI-negative sequence. The sequence obtained from GCH2 had a point mutation within the restriction site, which is consistent with the RsaI-negative fingerprint for this isolate. Most of the additional point mutations located within this 143-bp region co-segregated with the RsaI genotype, but the occurrence of three additional mutations suggest that multiple alleles are present at this locus.

The issue of stability of genetic markers in *C. parvum* is critically important for our understanding of the epidemiology of this parasite, and has been addressed in two recent studies. Bonnin et al. (1996) assessed the stability of RFLP markers by repeated sampling and fingerprinting of oocyst samples from individuals positive for the human immunodeficiency virus (HIV). Isolates from two patients showed no genotypic changes, but unfortunately the time period examined was not indicated. Also, sampling from AIDS patients

```
       1
pGc91  AACAACAACA    GTGCCAACGA    CAACTACTAC    TACCAAGAGA    GACGAAATGA
GCH1   nnnnnnnn..... ............  ............  ............  ............
GCH3   ............  ............  ............  ............  ............
GCH2   ............  ..........A.  .......A..A.  ...T........  ..........G...
GCH3   ............  ..........A.  .......A..A.  ...T........  ............
       51
pGc91  CAACAACAAC    GACACCA —     TTACCTGATA    TCGGTGACAT    TGAAATTACA
GCH1   ............  ..........—   ............  ............  ............
GCH3   ............  ..........—   ............  ............  ............
GCH2   ............  A.......CCA   ............  ............  ............
GCH3   ............  A.......CCA   ............  ............  ............
       101
pGc91  CCAATCCCAA    TTGAAAAGAT    GTTGGATAA G  TAC ACAAGAA    TGATTTATGA
GCH1   ............  ............  ..........                ......nnnnnnn
GCH3   ............  ............  .C........                ......nnnnnnn
GCH2   ............  ............  ..........  ...T.         ......nnnnnnn
GCH3   ............  ............  ..........  . C.T         ......nnnnnnn
                                                RsaI
```

Figure 2 Multiple sequence analysis of a 143–bp fragment from the *C. parvum* polythreonine open-reading frame identifies point mutations within the RsaI restriction site as well as additional deletions and point mutations. pGC91 originates from calf isolate GCH1. See text and the legend to Figure 1 for details of each isolate.

with chronic cryptosporidiosis might reduce the likelihood of identifying changes in the genetic profile as compared to acute infections. The stability of *C. parvum* genotypes following host change was also addressed by RAPD analysis of a bovine isolate serially passaged in mice (Morgan *et al.*, 1995), but this experiment failed to identify genetic changes.

After considering the information on heterogeneity in *C. parvum*, we are presented with the challenge of reconciling this data with what is known on the transmission of this parasite, in particular regarding human cryptosporidiosis. Specifically, the occurrence of human infections with parasite populations genotypically and phenotypically different from those seen in calves remains to be explained. Human infections originate either from oocysts shed by calves, often via contaminated water, or by direct human-to-human transmission. Calf-to-human transmission has long been postulated to be the main source of human infection, and the infectivity of calf-derived oocysts has been documented in human volunteer studies (DuPont *et al.*, 1995; Chappell *et al.*, 1996). The reverse transmission route has also been confirmed experimentally (Pozio *et al.*, 1992), although failure of propagating human-derived oocysts in calves has been observed (S. Tzipori and G. Widmer, unpublished data). Secondary human infections during outbreaks attributed to human-to-human transmission have also been reported (Miron *et al.*, 1991; Millard *et al.*, 1994). Together, these observations have confirmed the infectivity of bovine *C. parvum* oocysts for humans, whereas the reverse might not always be the case. In light of the genetic differences observed among human isolates, it is tempting to speculate that parasites with the unique human genotype might have reduced infectivity for bovines. Although scientific evidence for this model is lacking, calves experimentally infected with four *C. parvum* isolates, one of calf origin, and three from AIDS patients, showed differences in the severity of infections between groups (Pozio *et al.*, 1992). Interestingly, the two groups of calves showing the mildest infections, as defined by the number of oocyst shed and the duration of watery diarrhea, were those inoculated with oocysts originating from AIDS patients who ultimately died as a consequence of chronic cryptosporidiosis, whereas calves infected with oocysts of bovine origin or from an AIDS patient who suffered a mild, transient diarrhea presented more severe infections. It is tempting to speculate from this study that different symptoms are caused by parasites of different genotype. In contrast to the study of calf infections, observations with mice infected with three different *C. parvum* isolates, one of bovine and two of human origin, failed to show differences in infectivity (Current and Reese, 1986). The identification in future experimental infection studies of the genotypic profile of the oocysts used as inoculum will no doubt facilitate the interpretation of such experiments.

Together, genetic studies and experimental infections suggest that a selective mechanism triggered by a change in the intestinal environment might be involved in shaping the genetic make-up of *C. parvum* populations. A simple selection model is, however, not compatible with the apparent absence of human-type markers in calf isolates examined to date (Bonnin *et al.*, 1996; Carraway *et al.*, 1996). In the absence of such profiles among the bovine isolates, it is unclear how the genetic diversity among *C. parvum* isolates could arise from selection, unless human-type strains are present but have remained undetected due to their small number.

Studies with human volunteers infected with *C. parvum* are being conducted in order to study human cryptosporidiosis (DuPont *et al.*, 1995; Chappell *et al.*, 1996). The genotype of *C. parvum* in DNA samples from selected volunteers are currently being characterized (G. Widmer and C.L. Chappell, unpublished) with the aim of monitoring the genetic profile of *C. parvum* following bovine-to-human transmission, and gaining a better understanding of changes in population make-up in response to a change in host. Genetic markers associated with certain patterns of infecton might emerge, as fingerprinting of *C. parvum* will be expanded to new polymorphic loci. These studies will lead to the identification of virulence markers, genetic markers associated with certain clinical properties, and response to drugs.

Our understanding of the epidemiology of *C. parvum* will progress with the identification of new polymorphic markers, in particular genetic loci with multiple alleles, which can be typed by PCR. Recent work on microsatellites in *P. falciparum* (Su and Wellems, 1996) suggests that such polymorphisms might provide useful markers in *C. parvum*. In contrast to the conservation of microsatellites between species of higher eukaryotes (Schlötterer et al., 1991), alignment of the subunit 1 the of ribonucleotide reductase sequence from C. parvum (R. Balakrishnan, unpublished data) and P. falciparum showed no conservation of a microsatellite sequence present in P. falciparum. Identifying such loci in C. parvum will therefore require screening of genomic libraries or PCR-based protocols for the identification of such sequences (Lench et al., 1996).

3. PCR DETECTION OF C. PARVUM

Numerous PCR methods for the detection of C. parvum have been published (Table 1). Three studies have compared the sensistivity of different PCR protocols (Laberge et al., 1996; Rochelle et al., 1997; Sluter et al., 1997), providing essential information for identifying detection methods with the highest sensitivity and specificity. In order to provide generally applicable information on specific PCR protocols, optimization of each primer pair with respect to a number of experimental parameters is required, the foremost among these being the primer annealing temperature and the Mg^{2+} concentration. The large number of possible combinations between experimental variable makes comparison of PCR protocols labor intensive and limits the number of primers that can be included. Sluter et al. (1997) compared four primer pairs targeting unrelated loci, whereas Rochelle et al. (1997) tested three and Laberge et al. (1996) two. In two comparisons, one pair of primers targeting the ribosomal SSU gene and the primers published by Laxer et al. (1991) were included. Laxer's method, which uses primers specific for an unidentified region, was among the first to be published and continues to be widely used. Laxer's primers gave equal level of sensitivity as primers targeted to the ribosomal SSU (Rochelle et al., 1997; Sluter et al., 1997). Considering that several copies of the ribosomal genes are present in *Cryptosporidium* (Zamani et al., 1996), this observation raises the possibility that the sequence amplified by Laxer's primers is also present in multiple copies. Sluter et al. (1997) observed an approximately 10-fold higher sensitivity with Laxer's and the ribosomal primers as compared to a primer pair targeting a fragment of an oocyst wall protein (Lally et al., 1992; Ranucci et al., 1993), whereas Laberge et al. (1996) detected no difference using primer sets targeting the same loci. Rochelle et al. (1997) and Sluter et al. (1997) also assessed the sensitivity of a primer pair described by Webster et al. (1993), and reported either poor sensitivity or no detection.

Detection limits in the single oocyst range were reported with several PCR protocols and mostly with purified oocysts or DNA (Johnson et al., 1995; Laberge et al., 1996; Leng et al., 1996; Morgan et al., 1996; Stinear et al., 1996; Rochelle et al. 1997), although some protocols achieved similar

Table 1 Summary of published PCR detection methods for *C. parvum* oocysts.

Reference	Target	Visualization	cycles[a] PCR	Sensitivity[b]
Laxer et al. (1991)	Unknown	Ethidium bromide, chemiluminescence	35	30 fg DNA
Ranucci et al. (1993)	Oocyst wall protein	Ethidium bromide	35	40/ND
Webster et al. (1993)	Unknown	Ethidium bromide, ^{32}P	40	200/ND
Awad-El-Kariem et al. (1994)	SSU rRNA	Ethidium bromide	38	ND
Johnson et al. (1995)	SSU rRNA	Ethidium bromide, chemiluminescence	39	< 10/ND
Leng et al. (1996), Laxer et al. (1991)		Ethidium bromide, alkaline phosphatase	35	1/< 100
Mayer and Palmer (1996), Ranucci et al. (1993)		Digoxigenin	40/40	ND/370
Laberge et al. (1996), Laxer et al. (1991), Lally et al. (1992)		Ethidium bromide, chemiluminescence	40	1 (in milk)
Morgan (1996)	RAPD fragment	Ethidium bromide	?	ND/1
Benigno Balabat et al. (1996), Laxer et al. (1991)		Ethidium bromide, chemiluminescence	35/35	500/500 per g stool
Stinear et al. (1996)	Heat shock protein	Ethidium bromide, silver staining	Touchdown RT-PCR	ND/1
Gobet et al. (1997), Laxer et al. (1991)		Ethidium bromide,	50	ND/100
Rochelle et al. (1997), Awad-El-Kariem et al. (1994), Laxer et al. (1991), Webster et al. (1993)		Ethidium bromide, chemiluminescence	40	1/5
Sluter et al. (1997), Laxer et al. (1991), Ranucci et al. (1993), Webster et al. (1993)	SSU rRNA	Ethidium bromide	40	10/40

ND, not determined.
[a] Two numbers indicate nested PCR.

sensitivities with oocysts in stool or water concentrates (Morgan et al., 1996; Stinear et al., 1996; Rochelle et al., 1997). Visualization of PCR products using Southern blotting are typically required to detect less than 10 oocysts in crude samples. Detection of PCR products using radioactivity, chemiluminescence, or alkaline phosphatase have been used to improve sensitivity. Nested PCR protocols have also been described (Benigno Balabat et al., 1996; Mayer and Palmer, 1996; Rochelle et al., 1997) as well as a reverse transcriptase (RT) PCR method (Stinear et al., 1996).

The detection of oocysts in stool samples and water concentrates remains a challenge, and several studies have addressed this problem. Methods for detection of water-borne oocysts are discussed by Fricker and Crabb (1998). PCR detection of oocysts in stool typically does not achieve the same sensitivity as with purified oocysts (Table 1). Several methods for the removal of substances inhibitory to PCR have been described, and fall into three groups. (1) conventional oocyst purification using isopycnic or density gradients followed by DNA extraction; (2) physical separation of oocysts from fecal material by magnetic technology or cell sorting; and (3) direct DNA extraction from diluted stool or washed stool pellets. As no systematic comparison of methods has been published, the choice of approach is probably determined as much by personal preference and experience with certain procedures as by scientific reasoning.

Methods suitable for processing fixed or frozen stool sample are particularly valuable for retrospective and epidemiological studies. Formalin fixation has been shown to reduce PCR sensitivity (Greer et al., 1995; Johnson et al., 1995), whereas potassium dichromate has not been reported to reduce sensitivity (Johnson et al., 1995). For long-term storage freezing is the preferred method. According to our observations, oocysts subject to freezing are, however, not efficiently recovered by gradient purification, requiring instead extraction of fecal DNA. In our laboratory the method of Bukhari and Smith (1995) has given satisfactory results with a majority of human and pig samples. Nevertheless, a certain proportion of stool samples fail to generate PCR products, and require a supplementary DNA purification step.

Although several PCR methods for the detection of oocysts in stool have been described, no comparative study has been published. Such a study would greatly facilitate the choice of methodology. Of particular interest are comparisons performed on multiple stool samples from different species, fresh and preserved by different methods. The absence of *in vitro* propagation methods and the potential for genotypic changes in *C. parvum* emphasize the need for direct analysis of stool and environmental samples. Direct PCR of oocysts in stool will, therefore, continue to be highly relevant for clinicians and investigators interested in monitoring and genotyping *C. parvum*.

REFERENCES

Awad-El-Kariem, F.M., Warhurst, D.C. and McDonald, V. (1994). Detection and species identification of *Cryptosporidium* oocysts using a system based on PCR and endonuclease restricition. *Parasitology* **109**, 19–22.
Awad-El-Kariem, F.M., Robinson, H.A., Dyson, D.A., Evans, D., Wright, S., Fox, T.M. and McDonald, V. (1995). Differentiation between human and animal strains of *Cryptosporidium parvum* using isoenzyme typing. *Parasitology* **110**, 129–132.
Benigno Balabat, A., Jordan, G.W., Tang, Y.J. and Silva, J. jr. (1996). Detection of *Cryptosporidium parvum* DNA in human feces by nested PCR. *Journal of Clinical Microbiology* **34**, 1769–1772.
Bonnin, A., Fourmaux, M.N., Dubremetz, J.F., Nelson, R.G., Gobet, P., Harly, G., Buisson, M., Puygauthier-Toubas, D., Gabriel-Pospisil, F., Naciri, M. and Camerlynck, P. (1996). Genotyping human and bovine isolates of *Cryptosporidium parvum* by polymerase chain reaction–restriction fragment length polymorphism analysis of a repetitive DNA sequence. *FEMS Microbiology Letters* **137**, 207–211.
Bukhari, Z. and Smith, H.V. (1995). Effect of three concentration techniques on viability of *Cryptosporidium parvum* oocysts recovered from bovine feces. *Journal of Clinical Microbiology* **33**, 2592–2595.
Cai, J., Collins, M.D., McDonald, V. and Thompson, D.E. (1992). PCR cloning and nucleotide sequence determination of the 18S rRNA genes and internal transcribed spacer 1 of the protozoan parasites *Cryptosporidium parvum* and *Cryptosporidium muris*. *Biochimica et Biophysica Acta* **1131**, 317–320.
Carraway, M., Tzipori, S. and Widmer G. (1996). Identification of genetic heterogeneity in the *Cryptosporidium parvum* ribosomal repeat. *Applied and Environmental Microbiology* **62**, 712–716.
Carraway, M., Tzipori, S. and Widmer, G. (1997). A new restriction fragment length polymorphism from *Cryptosporidium parvum* identifies genetically heferogeneous parasite populations and genotypic changes following transmission from bovine to human host. *Infection and Immunity* **65**, 3958–3960.
Chappell, C.L., Okhuysen, P.C., Sterling, C.R. and DuPont, H.L. (1996). *Cryptosporidium parvum*: intensity of infection and oocyst excretion patterns in healthy volunteers. *Journal of Infectious Diseases* **173**, 232–236.
Chrisp, C.E. and LeGendre, M. (1994). Similarities and differences between DNA of *Cryptosporidium parvum* and *C. wrairi* detected by the polymerase chain reaction. *Folia Parasitologica* **41**, 97–100.
Current, W.L. and Reese, N.C. (1986). A comparison of the endogenous development of three isolates of *Cryptosporidium* in suckling mice. *Journal of Protozoology* **33**, 98–108.
Donald, R.K.G. and Roos, D.S. (1993). Stable molecular transformation of *Toxoplasma gondii*: a selectable dihydrofolate reductase–thymidylate synthase marker based on drug-resistance mutations in malaria. *Proceedings of the National Academy of Science, USA* **90**, 11 703–11 707.
DuPont, H.L., Chappell C.L., Sterling, C.R., Okhuysen, P.C., Rose, J.B. and Jakubowski, W. (1995). Infectivity of *Cryptosporidium parvum* in healthy volunteers. *New England Journal of Medicine* **332**, 855–859.
Dvorak, J.A. (1984). The natural heterogeneity of *Trypanosoma cruzi*: biological and medical implications. *Journal of Cellular Biochemistry* **24**, 357–371.
Ellsworth, D.L., Rittenhouse, K.D. and Honeycutt, R.L. (1993). Artifactual variation in randomly amplified polymorphic DNA banding patterns. *Biotechniques* **14**, 214–218.

Fricker, C.R. and Crabb, J.H. (1998). Water-borne cryptosporidiosis: detection methods and treatment options. *Advances in Parasitology* **40**, 241–278.
Gobet, P., Buisson, J.C., Vagner, O., Naciri, M., Grappin, M., Comparot, S., Harly, G., Aubert, D., Varga, I., Camerlynk, P. and Bonnin, A. (1997). Detection of *Cryptosporidium parvum* DNA in formed human feces by a sensitive PCR-based assay including uracyl-*N*-glycosidase inactivation. *Journal of Clinical Microbiology* **35**, 254–256.
Goonewardene, R., Daily, J., Kaslow, D., Sullivan, T.J., Duffy, P., Carter, R., Mendis, K. and Wirth, D. (1993). Transfection of the malaria parasite and expression of firefly luciferase. *Proceedings of the National Academy of Science, USA* **90**, 5234–5236.
Greer, C.E., Wheeler, C.M. and Manos M.M. (1996). PCR amplification from parafin-embedded tissue: sample preparation and the effects of fixation. In: *PCR Primer* (C.W. Dieffenbach and G.S. Dveksler, eds), pp. 99–112. Cold Spring Harbor: Cold Spring Harbor Laboratory Press.
Günthard, M., Meister, T., Liithy, R. and Weber, R. (1996). Intestinale Kryptosporidiose bei HIV-Infektion. *Deutsche Medizinische Wochenschrift* **121**, 686–692.
Hearne, C.M., Ghosh, S. and Todd, J.A. (1992) Microsatellites for linkage analysis of genetic traits. *Trends in Genetics* **8**, 288–294.
Jaccard, P. (1908) *Bulletin Societe Vaudoise Sciences Naturelles* **44**, 223–270.
Johnson, A.M., Fielke, R., Lumb, R. and Baverstock, P.R. (1990) Phylogenetic relationship of *Cryptosporidium* determined by ribosomal RNA sequence comparison. *International Journal for Parasitology* **20**, 141–147.
Johnson, D.W., Pieniazek, N.J., Griffin, D.W., Misener, L. and Rose, J.B. (1995). Development of a PCR protocol for sensitive detection of *Cryptosporidium* oocysts in water samples. *Applied and Environmental Microbiology* **61**, 3849–3855.
Kilani, R.T. and Wenman, W.M. (1994). Geographical variation in the 18S rRNA gene sequence of *Cryptosporidium parvum*. *International Journal for Parasitology* **24**, 303–306.
Laberge, I., Ibrahim, A., Barta, J.R. and Griffiths, M.W. (1996). Detection of *Cryptosporidium parvum* in raw milk by PCR and oligonucleotide probe hybridization. *Applied and Environmental Microbiology* **62**, 3259–3264.
Lally, N.C., Baird, G.D., McQuai, S.J., Wright, F. and Olivar, J.J. (1992). A 2359-base pair DNA fragment from *Cryptosporidium parvum* encoding a repetitive oocyst protein. *Molecular and Biochemical Parasitology* **56**, 69–78.
Laxer, M.L., Timblin, B.K. and Patel, R.J. (1991). DNA sequence for the specific detection of *Cryptosporidium parvum* by the polymerase chain reaction. *American Journal of Tropical Medicine and Hygiene* **45**, 688–694.
Lench, N.J., Norris, A., Bailey, A., Booth, A. and Markham, A.F. (1996). Vectorette PCR isolation of microsatellite repeat sequences using anchored dinucleotide repeat primers. *Nucleic Acids Research* **24**, 2190–2191.
Leng, X., Mosier, D.A. and Oberst, R.D. (1996). Simplified method for recovery and PCR detection of *Cryptosporidium* DNA from bovine feces. *Applied and Environmental Microbiology* **62**, 643–647.
Longmire, J.L., Kraemer, P.M., Brown, N.C., Hardekopf, L.C. and Deaven, L.L. (1990). A new multi-locus DNA fingerprinting probe: pV47-2. *Nucleic Acids Research* **18**, 1658.
Mayer, C.L. and Palmer, C.J. (1996). Evaluation of PCR, nested PCR, and fluorescent antibodies for detection of *Giardia* and *Cryptosporidium* species in wastewater. *Applied and Environmental Microbiology* **62**, 2081–2085.
McCutchan, T.F., Li, J., McConkey, G.A., Rogers, M.J. and Waters, A.P. (1995).

The cytoplasmic ribosomal RNAs of *Plasmodium* spp. *Parasitology Today* **11**, 134–138.
McKenzie, W.R., Hoxie, N.J., Proctor, M.E., Gradus, M.S., Blair, K.A., Peterson, D.E., Kazmierzak, J.J., Addiss, D.G., Fox, K.R., Rose, J.B. and Davies, J.P. (1994). A massive outbreak in Milwaukee of *Cryptosporidium* infection transmitted through the public water supply. *New England Journal of Medicine* **331**, 161–167.
Millard, P.S., Gensheimer, K.F., Addiss, D.G., Sosin, D.M., Beckett, G.A., Houck-Jankoski, A. and Hudson, A. (1994). An outbreak of cryptosporidiosis from fresh-pressed apple cider. *Journal of the American Medical Association* **272**, 1592–1596.
Miron, D., Joram, K. and Dagan, R. (1991). Calves as a source of an outbreak of cryptosporidiosis among young children in an agricultural closed community. *Pediatric Infectious Diseases* **10**, 438–441.
Morgan, U.M., Constantine, C.C., O'Donoghue, P., Meloni, B.P., O'Brien, P.A. and Thompson, R.C.A. (1995) Molecular characterization of *Cryptosporidium* isolates from human and other animals using random amplified polymorphic DNA analysis. *American Journal of Tropical Medicine and Hygiene* **52**, 559–564.
Morgan, U.M., O'Brien, P.A. and Thompson, R.C.A. (1996). The development of diagnostic PCR primers for *Cryptosporidium* using RAPD-PCR. *Molecular and Biochemical Parasitology* **77**, 103–108.
Nichols, G. L., McLauchlin, J. and Samuel, D. (1991). A technique for typing *Cryptosporidium* isolates. *Journal of Protozoology* **38**, 237S–240S.
Nina, J.M.S., McDonald, V., Deer, R.M.A., Wright, S.E., and Dyson, D.A., Chiodini, P.L. and McAdam, K.P.W.J. (1992). Comparative study of the antigenic composition of oocyst isolates of *Cryptosporidium parvum* from different hosts. *Parasite Immunology* **14**, 227–232.
Ogunkolade, B.W., Robinson, H.A., McDonald, V., Webster, K. and Evans, D.A. (1993). Isoenzyme variation within the genus *Cryptosporidium*. *Parasitology Research* **79**, 385–388.
Ortega, Y., Sheehy, R.R., Cama, V.A., Oishi, K.K. and Sterling, C.R. (1991) Restriction length polymorphism analysis of *Cryptosporidium parvum* isolates of bovine and human origin. *Journal of Protozoology* **38**, 40S–41S.
Pozio, E., Angeles Gomez, M.A., Mancini Barbieri, F. and La Rosa, G. (1992). *Cryptosporidium*: different behaviour in calves of isolates of human origin. *Transactions of the Royal Society of Tropical Medicine and Hygiene* **86**, 636–638.
Ranucci, L., Müller, H.M., La Rosa, G., Reckmann, I., Gomez Morales, M.A., Spano, F., Pozio, E. and Crisanti, A. (1993). Characterization and immunolocalization of a *Cryptosporidium* protein containing repeated amino acid motifs. *Infection and Immunity* **61**, 2347–2356.
Rochelle, P.A., De Leon, R., Stewart, M.H. and Wolfe, L.R. (1997). Comparison of primers and optimization of PCR conditions for detection of *Cryptosporidium parvum* and *Giardia lamblia* in water. *Applied and Environmental Microbiology* **63**, 106–114.
Rosario, V. (1981). Cloning of naturally occuring mixed infections of malaria parasites. *Science* **212**, 1037–1038.
Saltzberg, D.F., Kotloff K.L., Newman, J.L. and Fastiggi, R. (1991). *Cryptosporidium* infection in acquired immunodeficiency syndrome: not always a poor prognosis. *Journal of Clinical Gastroenterology* **13**, 94–97.
Schlötterer, C., Amos, B. and Tautz, D. (1991). Conservation of polymorphic simple sequence loci in cetacean species. *Nature* **354**, 63–65.
Sluter, S.D., Tzipori, S. and Widmer, G. (1997). Parameters affecting PCR detection of waterborne *Cryptosporidium parvum* oocysts. *Applied Microbiology and Biotechnology* (in press).

Solo-Gabriele, H. and Neumeister, S. (1996). US outbreaks of cryptosporidiosis. *Journal of the American Waterworks Association* **88**, 76–86.
Stinear, T., Matusan, A., Hines, K. and Sandery, M. (1996). Detection of a single viable *Cryptosporidium parvum* oocyst in environmental water concentrates by reverse transcription–PCR. *Applied and Environmental Microbiology* **62**, 3385–3390.
Su, X. and Wellems, T. (1996). Towards a high resolution *Plasmodium falciparum* linkage map: polymorphic markers from hundreds of simple sequence repeats. *Genomics* **33**, 430–444.
Tsai, Y.L. and Olson, B.H. (1992). Detection of low numbers of bacterial cells in soils and sediments by polymerase chain reaction. *Applied and Environmental Microbiology* **58**, 754–757.
Tzipori, S., Rand, W., Griffiths, J., Widmer, G. and Crabb, J. (1994). Evaluation of an animal model system for cryptosporidiosis: therapeutic efficacy of paromomycin and hyperimmune bovine colostrum-immunoglobulin. *Clincal and Diagnostic Laboratory Immunology* **1**, 450–463.
Van Dijk, M.R., Janse, C.J. and Waters, A.P. (1996). Expression of a *Plasmodium* gene introduced into subtelomeric regions of *Plasmodium berghei* chromosomes. *Science* **271**, 662–665.
Vasquez, J.R., Gooze, L., Kim, K., Gut, J., Peterson, C. and Nelson R.G. (1996). Potential antifolate resistance determinants and genotypic variation in the bifunctional dihydrofolate reductase–thymidilate synthase gene from human and bovine isolates of *Cryptosporidium parvum*. *Molecular and Biochemical Parasitology* **79**, 153–156.
Webster, K.A., Pow, J.D.E., Giles, M., Catchpole, J. and Woodward, M.J. (1993). Detection of *Cryptosporidium parvum* using a specific polymerase chain reaction. *Veterinary Parasitology* **50**, 35–44.
Welsh, J. and McClelland, M. (1990). Fingerprinting genomes using PCR with arbitrary primers. *Nucleic Acids Research* **18**, 7213–7218.
Widmer, G., Carraway, M. and Tzipori, S. (1996). Water-borne *Cryptosporidium*: a perspective from the USA. *Parasitology Today* **12**, 286–289.
Zamani, F., Upton, S.J. and LeBlancq, S.M. (1996). Ribosomal RNA gene organization in *Cryptosporidium parvum*. In: *Program and Abstracts of the 45th Annual Meeting of the American Society of Tropical Medicine and Hygiene*, Baltimore, MD, abstract 455.

Water-borne Cryptosporidiosis: Detection Methods and Treatment Options

Colin R. Fricker[1] and Joseph H. Crabb[2]

*[1]Thames Water Utilities, Reading, RG2 0JN, UK,
and [2]ImmuCell Corporation, Portland, ME, 04103, USA*

1. Water-borne Outbreaks: A Brief History....................................242
   1.1. Outbreaks as a function of type of source water......................244
   1.2. Treatment or system failures 245
   1.3. Sources of infectious parasites......................................246
   1.4. Infectious dose of *Cryptosporidium* and the 'recovery–infectivity paradox' ..247
2. Occurrence of *Cryptosporidium* in Source Water 247
3. Detection of *Cryptosporidium* in Water 249
   3.1. Collection methods..250
   3.2. Methods for separation of oocysts from environmental debris..........253
   3.3. Methods for detection...256
   3.4. Methods for determining oocyst viability 260
4. Regulatory Status and Environmental Laws 264
   4.1. The USA..264
   4.2. The UK...266
5. Water Treatment Options 267
   5.1. Current municipal water treatment practices 267
   5.2. Disinfection options 269
   5.3. Watershed management...270
   5.4. Theoretical considerations 271
   References...272

Since the infamous outbreak in Milwaukee, WI, USA, of water-borne cryptosporidiosis affecting over 400 000 people, there have been at least 20 smaller outbreaks associated with this parasite in the UK and North America. These events have led to an explosion of interest in and research on the nature of cryptosporidiosis as a dangerous water-borne pathogen, particularly patients with acquired immune deficiency syndrome (AIDS). In addition, several major enviromental laws and proposed regulations specifically address the control of this parasite. The possible ramifications of these laws include billions of dollars of modifications to water-treatment facilities in the

USA. Unfortunately, the methods used to gather the information on which thses laws are based have serious deficiencies that could lead to gross underestimation of the magnitude of this problem. The present review considers gaps in our understanding of water-borne cryptosporidiosis, new methods under investigation that could improve our ability to monitor water for the presence of this organism, and treatment and control strategies to limit the threat to our water supplies.

1. WATER-BORNE OUTBREAKS: A BRIEF HISTORY

The magnitude of growth of *Cryptosporidium* as a serious water-borne pathogen can be illustrated by reviewing the appearance of this pathogen in human disease since its first description in 1907 (Tyzzer, 1907). Sixty-five years transpired from that description until the first human case was described (Nime *et al.*, 1976). About 6 years passed from the first human case descriptions until the first outbreak was recorded (Current *et al.*, 1983). About that time, the Centers for Disease Control and Prevention (CDC, 1982) described 21 cases of cryptosporidiosis in AIDS patients in the USA, occurring over a 3-year period. Over the ensuing 10 years, the threat of this organism to the lives and livelihood of the immunocompromised began to be recognized, and at least 12 well-documented water-borne outbreaks occurred in the USA and UK (Lisle and Rose, 1995). Then, in 1993, the largest outbreak of water-borne disease in recorded history occurred in Milwaukee, WI, affecting over 400 000 individuals (MacKenzie *et al.*, 1994). This massive outbreak has raised the level of public awareness, as well as increasing the level of surveillance of the public water supplies. This increased recognition of the role of *Cryptosporidium* as a water-borne disease threat has prompted retrospective reviews of water-associated infections. Prior to the elevated awareness and investigations since the mid-1980s, 502 water-borne outbreaks of disease in the USA (affecting 111 000 individuals) were reported to the CDC from 1971 (when the surveillance system was begun) through 1985, peaking in 1980 (CDC, 1996). In only about 50% of the cases was the responsible agent identified. Given the large number of water-borne outbreaks associated with *Cryptosporidium* in recent years, combined with the high occurrence of the parasite in surface waters (see Section 2), and the relatively low recognition of the parasite as a major water-borne threat in previous years, it is intriguing to speculate about the contribution of this organism to the previous outbreaks.

Since the Milwaukee outbreak, there have been at least 23 reported water-borne outbreaks of cryptosporidiosis in the UK and North America (where surveillance and epidemiology for this parasite has been reasonably comprehensive) (Table 1). The increase in number of outbreaks compared to previous years is likely due to the increased awareness since the infamous Milwaukee outbreak, including enhanced surveillance and more familiarity of hospital

Table 1 *Crytosporidium* outbreaks in North America and the UK, 1993–1996

Year	Location	No. of confirmed cases	Source	Reference
1996	WoodRiver Valley, ID	3	?	Anon. (1996k)
1996	Cortland Co., NY	20	Apple cider	Anon. (1996j)
1996	Clay Co., FL	16	? (dual outbreak *Cryptosporidium/ Giardia*)	Anon. (1996i)
1996	Fresno Co., CA	22	Water park/ recreation	Anon. (1996h)
1996	Kelowna, BC	> 200	? Lake/drinking	Anon. (1996g)
1996	Cranbrook, BC	29	? Lake/drinking	Anon. (1996f)
1996	Waitsfield, VT	2	Cattle contact	Anon. (1996d)
1996	Collingwood, Ont.	30	? Lake/drinking	Anon. (1996a)
1995	Alachua, FL	72	Drinking	Solo-Gabriele and Neumeister (1996)
1995	Blue Earth Co., MN	15	Food	Besser-Wiek *et al.* (1996)
1995	Worcester Co., MA	54	? Drinking water	Anon. (1996c)
1995	Cobb Co., GA	58	Water park/ recreation	Anon. (1996e)
1995	Devon, UK	500[a]	Drinking water	Anon. (1995)
1994	Walla Walla Co., WA	134	Well	Solo-Gabriele and Neumeister (1996)
1994	Missouri	101	Motel pool	Anon. (1996b)
1994	NJ	418[a]	Lake/recreation	Anon. (1996b)
1993–1994	Clark Co., NV	103[a]	Lake/drinking	Golstein *et al.* (1996)
1993	UK	64	Well	Morgan *et al.* (1995)
1993	Yakima Co., WA	7	Well	Solo-Gabriele and Neumeister (1996)
1993	Newport, ME	160[a]	Apple cider	Millard *et al.* (1995)
1993	Oshkosh, WI	51	Pool	MacKenzie *et al.* (1995)
1993	Cook Co., MN	27	Lake/drinking	Solo-Gabriele and Neumeister (1996)
1993	Milwaukee, WI	403 000[a]	Lake/drinking	MacKenzie *et al.* (1994[a])

[a]Estimated.

laboratories in detection of the parasite in stool specimens. Alternatively, the increased incidence of outbreaks could be due to a general decline in the quality of drinking water supplies and/or inappropriate or inaccurate methods for monitoring water for the presence of this parasite. While the methods for monitoring water for the presence of this parasite are, indeed, inaccurate (see Section 3), the notion of general water decline seems less likely, particularly given the very high and steady seroprevalence of anti-*Cryptosporidium* antibodies in the general population. A compilation of seroprevalence studies in small groups in the period 1981–1990 yielded a mean seropositivity rate of 54% (range 25–91%) (O'Donoghue, 1995). A large study of over 800 individuals seen in an Oklahoma Children's Hospital reported seroprevalence rates of 38% in children and 58% in adolescents (Kuhls et al., 1994). These data indicate that exposure to *Cryptosporidium* has remained fairly constant and widespread over the last decade.

1.1. Outbreaks as a Function of Type of Source Water.

It is noteworthy that recent outbreaks have been attributed to drinking water, recreational water, food, and contact with infected individuals and cattle (see Table 1). The bulk of the outbreak-related incidence (and certainly the overwhelming majority of cases) can be ascribed to primary infection from consumption of contaminated drinking water. Cases associated with recreational water and contact with infected individuals can be considered secondary transmission (although the source in recreational water incidents cannot be clearly differentiated between accidental defecation incidents and contaminated source water). Similarly, one of the three described food-borne outbreaks can be considered secondary transmission (the other two appeared to be directly attributable to apple cider that had been contaminated). That secondary transmission (in the general population) can be considered a minor contributor to the overall incidence of outbreak-related disease is confirmed by follow-up epidemiology of the Milwaukee outbreak. Among adults having diarrhea during the outbreak in Milwaukee, there was only a 4.2% rate of secondary transmission to household contacts (MacKenzie et al., 1995b). Secondary transmission is probably higher in the AIDS population, given the increased susceptibility and the higher likelihood of encountering infectious oocysts in this group (Sorvillo et al., 1994).

Among drinking water outbreaks, similar numbers of incidents have occurred in association with groundwater (wells) as with surface water sources; these groundwater-associated outbreaks have almost always been demonstrated to be under the direct influence of a contamination source (sewage, or surface water influx) (Solo-Gabriele and Neumeister, 1996). In contrast, surface-water associated outbreaks are less well characterized as to the source of the infectious parasites. A few outbreaks have been traced to contamination of the distribution system rather than the source water itself (Solo-Gabriele and Neumeister, 1996). However, given the high prevalence of parasites in surface waters, and the aging distribution systems in most large municipal water supplies, it is currently difficult to prove whether a given-surface water-related outbreak is associated with contaminated source water or breaches in the distribution system. When molecular epidemiological tools (probes) become available which can accurately distinguish between specific *Cryptosporidium* isolates, outbreaks can be more accurately traced back to their sources.

1.2. Treatment or System Failures

It is well established that *Cryptosporidium* oocysts are refractory to chlorine treatment of drinking water (Fayer, 1995) and coarse filtration methods normally performed on surface drinking waters are not adequate for efficient oocyst removal (due to the small size and chemically resistant oocyst wall). Since groundwater sources are typically not treated, except by chlorination in some cases, these outbreaks cannot be characterized as treatment failures. However, groundwater-associated outbreaks were found in virtually every case to be associated with either contamination with raw sewage, or runoff from oocyst-contaminated surface water sources (Solo-Gabriele and Neumeister, 1996). Therefore, most groundwater outbreaks can be characterized as system failures.

In all well-documented cases of surface-source-related outbreaks, the treatment facility was in compliance with federal regulations during the outbreak (<1.0 nephelometric turbidity units (ntu), and <1 colony forming units (cfu) coliform per 100 ml finished water), and there were no documented interruptions in filtration or treatment of the water (Lisle and Rose, 1995; Solo-Gabriele and Neumeister, 1996; MacKenzie *et al.*, 1994a; Goldstein *et al.*, 1996). However, in the Milwaukee outbreak and several outbreaks that occurred prior to 1993, at least one of several treatment deficiencies were noted (Lisle and Rose, 1995). In particular, the turbidity of the treated water from the Milwaukee treatment facility implicated in the outbreak rose dramatically, with peaks of > 1.5 ntu on two separate days that were coincident with the outbreak. Continuous turbidity was not monitored; 8-hour readings were taken during this period (MacKenzie *et al.*, 1994a). In addition to inadequate monitoring of turbidity of finished water, filter backwash was recycled through the treatment facility (which could concentrate oocysts), treatments were not adjusted to compensate for higher runoff from a rainy spring, watersheds were not monitored for oocysts, and no barriers were present between the water facility intake and potential sources of runoff contamination (Lisle and Rose, 1995). These cases could be considered to be treatment or system deficiencies, if not overt failures. In contrast to the correlation of outbreaks with documented deficiencies, there has been at least one outbreak associated with a modern treatment facility; at Clark Co., Nevada (Goldstein et al., 1996). In this instance, approximately 100 cases of water-borne infections occurred during a period where water treatment was well documented, and unremarkable. In addition, monthly monitoring of source water and twice-monthly monitoring of finished water was negative for *Cryptosporidium*. These data underscore the need for water treatment practices that focus on important *Cryptosporidium* control points and oocyst monitoring methods that yield accurate and actionable results.

1.3. Sources of Infectious Parasites

While *Cryptosporidium parvum* is capable of infecting virtually every mammal, oocyst shedding of long duration or high magnitude does not occur in immunocompetent adults of any species. In contrast, the calf reservoir of *Cryptosporidium* is likely to be the most significant as it relates to the widespread prevalence and quantity of infectious oocysts. The USDA National Dairy Heifer Evaluation Project provided some startling data bearing on water-borne cryptosporidiosis (USDA, 1993). In 28 states tested in the north-east, midwest and west in 1991 and 1992, more than 90% of farms had animals that excreted *Cryptosporidium*. This very large incidence was similar in farms of various sizes, although very small farms (< 100 cows) had a slightly lower incidence rate. The lack of effective treatments coupled with a surveyed 2% awareness rate among dairy farmers indicates that the potential for human exposure from the cattle reservoir is great (USDA, 1993). An infected, newborn calf can shed oocysts for up to 2 weeks, with oocyst concentrations in diarrheic stools ranging from 10^5 to 10^7/ml, producing 1–5 × 10^{10} total oocysts excreted (J.H. Crabb, unpublished observations). This incidence rate in calves coupled with approximately 9 million dairy cows in the USA, giving birth to one calf per year (Halladay, 1994), creates the potential for staggering numbers of oocysts excreted per year (9 × 10^6 dairy calves per year × 90% potential incidence × 10^{10} shed per calf = 8 × 10^{16} oocysts per year shed in the USA). The calf shedding levels are similar in magnitude to levels of oocysts excreted by AIDS patients with chronic cryptosporidiosis; up to 10^8 per gram of stool during heavy shedding (Fries et al., 1994). Thus, both agricultural runoff in virtually all dairy sheds, and raw sewage in regions with significant AIDS patient populations have the potential for high infectivity based on the numbers of infectious oocysts likely to be present.

Several studies have been conducted to investigate the potential for commonly found non-mammal species to harbor and/or shed *C. parvum* oocysts. It was found that waterfowl have the capacity to harbor infectious oocysts of *C. parvum* after experimental oral inoculation (Graczyk et al., 1996a). While these ducks did not support growth and development of the parasite in their intestinal tracts, they could shed a significant percentage of the oral inoculum that was infectious for neonatal mice. In similar studies, *C. parvum* oocysts were not infectious for various reptiles or amphibians tested, but detectable inoculum-derived oocysts were found in stools for up to 14 days after inoculation (Graczyk *et al.*, 1996b). It would appear that, given the large populations of waterfowl in and around public water supplies, that this mode of dissemination could be a significant contributor to water-borne cryptosporidiosis.

1.4. Infectious Dose of *Cryptosporidium* and the 'Recovery–Infectivity Paradox'

Using a calf isolate of *C. parvum* (the 'Iowa isolate'), DuPont *et al.* (1995) determined the median infective dose to be 132 oocysts in human volunteer studies. Oocyst concentrations recovered from water frozen during the Milwaukee outbreak were reported to be 0.7–13 oocysts per 100 l using the 'standard technique' or a modification using more efficient 'collection' membranes (MacKenzie *et al.*, 1994). Given estimated human consumption of drinking water at about 1–2 l/day, these data raise a paradox; if 50–100 oocysts/l of the 'Iowa' isolate had to be consumed to account for such a high attack rate (up to 50% in the primary outbreak region), either the samples of water taken during the outbreak contained anomalously low numbers of oocysts (sampling error) or the detection method(s) used in this instance were about 0.1–1% efficient. Recent data suggest that oocyst distributions in water supplies follow essentially a normal distribution (Haas and Rose, 1996), and the >200 l samples that were taken from drinking water in the affected Milwaukee region during the peak of the outbreak appeared to be representative samples. This information suggests that dramatic inefficiency of detection is responsible for this paradox. Essentially confirming this interpretation are results of an analysis of the current 'standard method' (see Section 3) demonstrating a mean recovery of 2.7% and a mode of 0% reported from several commercial laboratories evaluating cartridge filters seeded with *Cryptosporidium* oocysts (Clancy *et al.*, 1994). Subsequent sections of this article review the status of methods for detection of oocysts in water supplies.

2. OCCURRENCE OF *CRYPTOSPORIDIUM* IN SOURCE WATER

The prevalence of *Cryptosporidium* oocysts in drinking water has been the subject of numerous studies over the last 10 years, and has been one of the issues that has driven the efforts to develop a reliable method of detecting the presence of this parasite in both raw and finished waters (Madore *et al.*, 1987; Ongerth and Stibbs, 1987; LeChevallier *et al.*, 1991a,b; Rose *et al.*, 1991; The National *Cryptosporidium* Survey Group, 1992; McTigue *et al.*, 1996). The survey work in the USA and UK has been carefully reviewed as well (Lisle and Rose, 1995). Most of these surveys used the 'standard method' or a modification thereof. This method (referred to as the 'standard method', 'ASTM method', or 'ICR method') uses a 10-inch wound cartridge filter of 1 μm nominal porosity for collection of 100–1000 l followed by filter media dispersal, separation, centrifugation, density gradient sedimentation, filter manifold collection, and detection via immunofluorescence microscopy (reviewed in Section 3). Using this method, oocysts were detected in 10–100% of samples tested from surface source water in the USA, and approximately 5–50% of all surface sources in the UK in a review of all studies referenced above. In studies with larger sample sizes, occurrence rates were significantly higher. Interestingly, some studies found dramatically higher oocyst occurrence in sources known to be under the influence of domestic or agricultural wastes (Rose *et al.*, 1991; McTigue *et. al.*, 1996), while another study found no correlation between occurrence and protected versus 'impacted' source water (LeChevallier *et al.*, 1991a). A recent study utilizing a mobile laboratory equipped with the ICR method sampled raw and finished water from 82 utilities in 39 states and found *Cryptosporidium* in 73% of the raw water and 17% of finished water (McTigue *et al.*, 1996). This prevalence is similar in magnitude to that described in other studies. Source water was ranked according to the prevalence of pathogens (*Cryptosporidium* and *Giardia*) found, and developed the following rank (from high to low prevalence): creek > river > reservoir > lake (McTigue *et al.*, 1996).

A very important finding to emerge from this study was that there was a poor correlation between particle counts and oocyst occurrence. Therefore, particle counting (and, by analogy, turbidimetry) are poor predictors of the likelihood of oocyst contamination. Thus, while particle counting and turbidity can be used to monitor the overall functioning of a treatment plant, such data should not be used to predict the presence or absence of oocyst contamination, or as a replacement for a reliable and accurate oocyst monitoring method and program. The finding of significant occurrence in finished waters employing conventional filtration confirms the notion that conventional filtration does not provide an adequate barrier to the passage of oocysts. Given the acknowledged inefficiencies in recovery of oocysts using the methods employed in these surveys (see Sections 3 and 4), the occurrence of oocysts in both raw and finished water would be expected to be far greater.

In a study examining 85 raw-water sources in 1988–1990, which found 87% of these sources to be harboring oocysts, reference was made to the observation that the majority of the 'oocysts observed in the samples were not viable', based on morphological criteria (LeChevallier *et al.*, 1991a). The viability issue is addressed below (see Section 3), but it must be noted that viability assessments, whether accurate or not, reflect not only the effects of residence in the water supply and/or treatment effect, but the effects of the sampling and downstream processing of the sample. Thus, it may be inappropriate to comment on the potential viability of a sample that has been subjected to a multi-step isolation procedure exposing the oocysts to high shear forces, large changes in osmotic balance, and process steps that may be otherwise detrimental.

3. DETECTION OF *CRYPTOSPORIDIUM* IN WATER

The ability to detect oocysts of *Cryptosporidium* has become a necessity for water utilities in recent years due to the large number of outbreaks of disease which have been documented. Until the late 1980s, water microbiologists were unaware of the transmission of this organism via water, but there is little doubt that in recent years detection of *Cryptosporidium* has become the water microbiologist's most pressing problem. Several characteristics of the organism have meant that the traditional methods used for assessing the microbiological quality of water involving the detection of indicator organisms are not suitable for determining the level of contamination of water with *Cryptosporidium*. The organism cannot be easily cultured *in vitro*, and therefore methods for detection have relied principally on a microscopic examination of water samples. The infectious dose of the organism is thought to be low (see Section 1.4), and therefore large volumes of water need to be concentrated before examination takes place. The traditional indicator organisms are used on the basis that they are likely to be more numerous in fecal material than pathogens, have similar resistance to removal through water-treatment processes and have similar survival characteristics in the environment. *Cryptosporidium* oocysts are extremely robust and survive in the environment for much longer periods of time than bacterial indicators such as *Escherichia coli*. Furthermore, the oocyst is resistant to chlorine disinfection and thus, in fully treated waters, the absence of bacterial indicators has no relationship to the possible presence of viable, potentially infectious oocysts. Thus, the methodology required for the detection of *Cryptosporidium* oocysts is completely different from that traditionally used in the water industry and the methods which are currently available are at best tentative.

A plethora of methods or adaptations of methods for the detection of *Cryptosporidium* in water have been described and much attention has been given to the specific detection of oocysts. However, the overall procedure can be broken down in several stages (all of which require optimization), namely, sample collection and concentration, separation of oocysts from contaminating debris and detection of oocysts. In addition, there is considerable interest in determining if oocysts are viable and therefore potentially infectious. Thus, methods have been and are currently being developed to assess the viability of oocysts found in the environment. The majority of information in the literature addresses a single aspect of the whole process and there is relatively little information on the efficiency of the overall method. One particular problem is that recovery efficiencies are often quoted, but there is scant information on the water quality used, the age of the oocysts and the length of time that oocysts have been in contact with water, before the recovery efficiency experiments are performed. Each of these factors can have a significant effect on the overall recovery efficiency and thus it is almost impossible to compare the effectiveness of two methods that have been performed in different laboratories. Furthermore, the criteria used for determining that a particle in a sample preparation is in fact a *Cryptosporidium* oocyst differs from method to method. Some workers use only the fact that oocysts fluoresce when labeled with a fluorescein isothiocyanate tagged anti *Cryptosporidium* monoclonal antibody and that it is in the size range 4–6 µm to determine that a particle is an oocyst, while others will use differential interference microscopy or nucleic acid stains to confirm internal morphology. This more detailed examination may appear to be more thorough, but may be difficult to accomplish if debris is present. Care must therefore be taken to ensure that the particles being counted are oocysts, whether or not they appear to contain sporozoites, and that algal and yeast cells are excluded from any counts that are made. A further problem with published data on the recovery efficiencies of different methods is that the method of seeding organisms into water varies between laboratories and is seldom standardized. In order to determine accurately the number of organisms being seeded into a sample, oocysts must either be counted directly on a microscope slide and washed into the sample or a number of replicates must be counted and the range of counts obtained used to calculate confidence limits which should be quoted. Sadly, this is seldom done and, if it is, the data are generally not presented in published articles.

3.1. Collection Methods

3.1.1. *Current Methods and Problems*

Initial attempts to detect *Cryptosporidium* in water used methodology based on that previously developed for the detection of *Giardia* (Jakubowski and Ericksen, 1979) and these methods became adopted as 'standards' for the water industry ('standard method') (Le Chevallier *et al.*, 1991a,b; Musial *et al.*, 1987; Rose *et al.*, 1988). Large volumes of water (100–1000 l) are passed through polypropylene cartridge filters of nominal 1-µm pore size at a flow rate of 1–5 l/min. Trapped material is then eluted by cutting the filter open and washing, either by hand or by stomaching using a dilute detergent solution. Several washes of the filter material may be required, and some workers use a preliminary backflush of the filter prior to cutting open the filter. The resulting washings from these cartridges sometimes totals 3 or 4 l and they must then be further concentrated by centrifugation. The ability to recover *Cryptosporidium* oocysts using this concentration was originally reported to be in the range 14–44% (Musial *et al.*, 1987), although much lower recovery efficiencies (<1–30%) have been reported (Ongerth and Stibbs, 1987; Vesey and Slade, 1991; Shepherd and Wyn-Jones, 1996), which may be due to differences in water quality, laboratory efficiency, or any of the other reasons discussed above. In addition, analysis by commercial laboratories of oocyst-seeded cartridges produced even poorer recoveries, as discussed in Section 1.4 (Clancy *et al.*, 1994)

Ongerth and Stibbs (1987) suggested that large flat-bed membranes could be used for the concentration of oocysts from water samples, and indeed many workers have now adopted this procedure. Membranes are held in stainless-steel supports (e.g. the 'tripod'), water pumped through, and the concentrated material recovered by 'scraping' the surface of the membrane, together with washing with dilute detergent. The material is then concentrated further by centrifugation. However, as with all membrane filtration techniques, the ability to filter the sample is dependent on the water quality. With low-turbidity water, it is relatively easy to filter 10–20 l, but with some high-turbidity waters it is only possible to filter 1–2 l. Thus, while the use of absolute filters ensures that no oocysts pass through the filter, in some cases only small volumes of water can be concentrated. Furthermore, the use of absolute filters often means that the amount of material collected is large, and therefore examination of the concentrate may be difficult. As with cartridge filtration, a range of recovery efficiencies has been reported for flat-bed membranes. These differences are probably due to the many factors described above for cartridge filtration and also to the different membranes used. Ongerth and Stibbs (1987) described the use of a 293- or 142-mm diameter, 2.0-µm pore size polycarbonate membrane and this has been used in further studies (Nieminski *et al.*, 1995). In a study of the efficiencies of several different membranes for recovering both *Cryptosporidium* oocysts and *Giardia* cysts, Shepherd and Wyn-Jones (1996) suggested that 1.2-µm cellulose acetate membranes gave a higher recovery than the 2.0-µm polycarbonate membranes preferred by Ongerth and Stibbs (1987). However, in their study, Shepherd and Wyn-Jones produced some results which were somewhat unusual, in that for many of their experiments, they reported higher recovery efficiencies for raw water than for tap water, which is contrary to what most workers find. However, the reported difference between the 1.2- and 2.0-µm membranes was not large and no statistical difference was reported for the two membranes.

The other well-established procedure for concentrating *Cryptosporidium* oocysts and *Giardia* cysts is the calcium carbonate flocculation procedure developed by Vesey et al. (1993a). In this procedure, a fine precipitate of calcium carbonate is formed in a water sample by the addition of calcium chloride and sodium bicarbonate, followed by adjusting the pH to 10.0 with sodium hydroxide. After allowing the precipitate to settle, the supernatant fluid is aspirated off and the sedimented material resuspended after dissolving the calcium carbonate using sulfamic acid. Recovery efficiencies using this method have been reported to be as high as 70% for both *Cryptosporidium* and *Giardia* (Vesey et al., 1993a, 1994; Campbell et al., 1994; Shepherd and Wyn-Jones, 1996). More recent work has demonstrated that this is the upper limit of the detection efficiency and that lower recoveries are usually encountered. Use of 'environmentally challenged' organisms for seeding experiments, together with leaving the oocysts in contact with water for a few days prior to analysis, showed that recovery rates of 30–40% were more normally seen.

These three concentration techniques are most often used for routine sampling, although there continues to be much debate over which method is the most appropriate. Realistically, there is no one single method which is most suitable for all situations. The choice of method should be made with due regard to a number of parameters including the purpose of sampling, the water quality and the facilities in the laboratory which will perform the analysis. Ideally, the method chosen should efficiently concentrate as large a sample as possible and yield a concentrate which can be examined easily. While large volume samples are preferred for most purposes, the problem arises when the sample concentrate is to be examined. For medium- to high-turbidity waters, the volume of the final concentrate is often far too large to be examined using currently available methods, and this leads to only a portion of the material being examined. Thus many workers prefer to concentrate only a small volume of water initially and examine the entire concentrate, while others take large samples and examine only a fraction of the final concentrate. Either approach is defensible, but the methods used to concentrate small volumes (i.e. 10–20 l) tend to be easier to perform and generally have a higher recovery efficiency, and so it is often preferable to take a larger number of small volume samples and to examine all the concentrated material. However, if the quality of the water to be examined is high, then it may well be preferable to take larger volume samples if all of the concentrate can be examined, and Dawson et al. (1993) suggest that it may be possible to use membrane filters rather than spiral-wound cartridge filters. Other factors which may affect the choice of concentration method include the site of sample collection and the distance which samples must be transported. Cartridge filtration can be used in the field using simple equipment and a small battery or petroleum operated pump, whereas membrane filtration requires somewhat more equipment which is not as easy to transport, and in any case is not suitable for high-turbidity waters. Calcium carbonate flocculation, while being suitable for both low- and high-turbidity waters, requires that 10–20 l samples be sent to the laboratory and this may be limiting for field studies. However, when sampling from the water distribution system in customer's premises it may be preferable to take a 10 l 'grab' sample, since it is unlikely that customers would be prepared for large-volume samples to be taken from their kitchen taps.

3.1.2. *New and Emerging Collection Methods*

The search for new methods for concentrating water samples to detect the presence of protozoan parasites continues and many methods have been evaluated, including cross-flow filtration, continuous-flow centrifugation, and vortex-flow filtration (Whitmore, 1995). Of these, the one which shows the most promise and has therefore received the most attention is vortex-flow filtration (Fricker *et al.*, 1995; Jonas *et al.*, in press a). Vortex-flow filtration utilizes a cylindrical filter with a defined pore size, which is contained within a housing and which rotates during sampling. The rotation of the filter causes 'Taylor vortices' to be created within the chamber, and these vortices serve continually to 'scrub' the membrane clean during sample concentration, while maintaining a low shear zone that results in minimized cell damage (Rolchigo, 1995). The procedure has been shown to be capable of concentrating raw-water samples of up to 100 l, but can also be used for smaller sample volumes. Importantly, the unit is configured in such a way as to permit the recovery of very small concentrate volumes (5–25 ml). The recovery efficiency using this procedure has been shown to be in excess of 70% for both *Cryptosporidium* and *Giardia* (Jonas *et al.*, in press) and the equipment can also be 'plumbed in' at treatment works, and can be used in the field or in the laboratory. This system appears to be the most versatile available for difficult to process feed stocks, such as raw water. In addition, the use of small-pore membranes create the possibility for simultaneous recovery, and thus detection, of multiple and diverse pathogens, such as *Cryptosporidium* and viruses (Jonas *et al.*, in press b).

3.2. Methods for Separation of Oocysts from Environmental Debris

3.2.1. *Flotation*

Since the concentration of *Cryptosporidium* oocysts is based almost exclusively on particle size, the techniques are nonspecific, and consequently concentrate a large amount of extraneous material that may be present in the water, which includes organic and inorganic particulates and other organisms including bacteria, yeasts, and algae. This material interferes with the successful detection of oocysts, either by increasing the total volume that needs to be examined, by obscuring oocysts, or by mimicking them. Thus some form of separation technology is normally required in order to reduce the time taken to examine the sample and to prevent oocysts being missed. The majority of workers use some form of density centrifugation to separate oocysts from background debris and thus reduce the amount of material to be examined. Sheather (1923) first described the use of sucrose density centrifugation to separate parasites from fecal material in clinical samples, and this basic technique has been adopted for use with environmental samples, although some workers prefer to use Percoll–sucrose or Percoll–Percoll gradients (Arrowood and Sterling, 1987; Musial et al., 1987; LeChevalier et al., 1991a). Whatever method is used, several groups have demonstrated that this is an extremely inefficient procedure when trying to detect protozoan parasites in water concentrates, and of particular importance was the finding of Bukhari and Smith (1995) that sucrose density centrifugation selectively concentrated viable, intact oocysts, with nonviable oocysts frequently being lost. This is presumably because oocysts with any fault in their wall will fill with sucrose and therefore be sedimented. The reason for the inefficiency of the density centrifugation step has not been elucidated, although it has been suggested that attachment to other particulates may change the buoyant density of oocysts and lead to their sedimentation. Indeed, Fricker (1995) demonstrated that the recovery of oocysts from water samples could be affected by the length of time for which they were in contact with the water concentrate, but that this was only the case when sucrose flotation was performed. Samples which were examined directly without density centrifugation gave similar recovery efficiencies, irrespective of whether they were examined immediately after seeding or after 48 hours contact with the concentrate. However, when sucrose flotation was used, the recovery of oocysts from raw waters concentrates fell from a mean of 55% to 18% after the same period of contact. This reduction in recovery efficiency was also seen with concentrates of reservoir water (67% to 23%) and fully treated water (80% to 52%). In addition to being extremely inefficient at recovering oocysts from water concentrates, the density centrifugation steps tend not to remove other biological material such as yeast or algal cells. In particular, autofluorescing algae can cause severe problems when examining slides for protozoa by epifluorescence microscopy, and thus more efficient methods for both recovery and separation of oocysts have been sought.

3.2.2. Immunomagnetic Separation

The principles behind this technology involve the attachment of specific antibodies to magnetizable particles, efficient mixing of the particles with the samples and then separation of the magnetizable particles from the remaining debris. If good mixing has occurred then magnetic particles will come into contact with any oocysts present and will bind through the attached antibodies. This will then allow the oocysts to be separated from 'non-target' material. The technique seems very simple, and indeed it is, but there are many issues that can cause it to fail. The first is the quality and specificity of the monoclonal antibodies that are available. Most of the commercially available monoclonal antibodies to *Cryptosporidium* are of the immunoglobulin M (IgM) isotype, are generally of low affinity, and present some problems with regard to conjugation and stability. When beads are being mixed with water concentrates the immunoglobulin–oocyst bonds are subjected to shear forces; therefore, the stronger the bond (or higher the antibody affinity) the more likely the bead is to remain in contact with the oocyst. The way in which the antibody is attached to the bead may also have an effect on recovery efficiency, since if the attachment between the bead and the antibody is weak, noncovalent, or otherwise unstable, the antibody may detach and the oocyst will not be recovered. The turbidity of the water concentrate appears to be the most critical factor associated with the recovery efficiency using IMS. Oocysts seeded into relatively clean suspensions are recovered efficiently, with recoveries of over 90% being reported. However, the real benefit of a good separation technique is with samples that have yielded a highly turbid concentrate, and it is in these samples that IMS does not appear to perform as efficiently. A study of immunomagnetic capture of *Giardia* cysts from water concentrates of various turbidities was conducted, and it was found that dramatic reductions in capture efficiency occurred above a turbidity of 600 ntu (Bifulco and Shaefer, 1993). A similar inverse relationship between recovery efficiency and concentrate turbidity is operative in the case of *Cryptosporidium* oocysts (J.H. Crabb, N.B. Turner, D.A. DuBourdieu, A. Jonas, and C.R. Fricker, unpublished observations). The use of antibodies of higher affinity may serve to improve the recovery efficiency of oocysts from high-turbidity samples, and recent experiments in our laboratories using a monoclonal IgG3 anti-*Cryptosporidium* oocyst antibody have demonstrated high recovery efficiencies with samples of greater than 5000 ntu (Crabb and colleagues, *ibid.*). In addition to the affinity of the antibody, the specificity is of course important, and this is particularly so because the surface of the oocyst is heavily glycosylated, and thus likely to possess some common antigens (Luft *et al.*, 1987). In fact, antibodies directed against the oocyst wall (and, in particular, the reagents approved for use in the EPA's ICR method) have been shown to react with other coccidian species as well as other species of *Cryptosporidium* (McDonald *et al.*, 1991; Nina *et al.*, 1992; Ortega-Mora *et al.*, 1992; Graczyk *et al.*, 1996a). Experiments have shown that commercially available monoclonal antibodies for the detection of *Cryptosporidium* cross-react with both yeast and algae (Rodgers *et al.*, 1995). The reactivity with yeast cells may not be due to a genuine cross-reactivity, but to the presence of 'antibody-binding proteins', which have been reported to be present in some species. These cross-reactions may reduce the effectiveness of IMS with certain types of sample, but they are more important when attempting to identify organisms. There have been monoclonal reagents described that are capable of distinguishing between *C. parvum* and *C. muris*, suggesting that development of reagents with enhanced specificity for *C. parvum* is a possibility (McDonald *et al.*, 1991).

3.3. Methods for Detection

3.3.1. *Immunofluorescence Microscopy*

Routine detection of *Cryptosporidium* oocysts relies on the use of epifluorescence microscopy. The approach differs somewhat in Europe and the USA, since the Europeans tend to examine material deposited on multi-well slides using a monoclonal antibody which has been directly labeled with FITC, whereas in the USA, water concentrates are generally examined on membrane filters after indirect staining using an anti-*Cryptosporidium* monoclonal antibody and an FITC-labeled anti-mouse antibody (ICR, or 'standard method'). There have been no definitive studies to compare the efficiencies of these procedures, but the tendency now is towards staining on glass slides with a directly labeled antibody. This tends to give less non-specific fluorescence, and less obscuring of fluorescent oocysts, due to embedment in the filter media. Several anti-*Cryptosporidium* antibodies are available commercially and, while most workers have their preference, there does not appear to be a single antibody which is preferred for all purposes. One specific failing of the commercially available antibodies is that they all apparently cross-react with all members of the genus, and therefore cannot be used to specifically identify *C. parvum*, as discussed above (Graczyk *et al.*, 1996a).

3.3.2. *Flow Cytometry*

Workers in the UK attempted to use flow cytometry with environmental samples in order to detect *Cryptosporidium* oocysts, but found that the sensitivity of the instruments was not great enough to distinguish oocysts from background noise (Vesey *et al.*, 1991). Further work, however, showed that the incorporation of a cell-sorting capability onto flow cytometers enabled oocysts to be sorted efficiently from background material (Vesey *et al.*, 1993b). Water concentrates are stained in suspension with an FITC-labeled anti-*Cryptosporidium* monoclonal antibody and passed through the fluorescent activated cell sorter (FACS). The sample stream is broken into a series of droplets, and droplets identified as containing particles of prespecified size, shape, and fluorescence characteristics are sorted away from those containing only background debris. In this way, the sorted material can be collected on a microscope slide and viewed by epifluorescence to confirm the presence of oocysts. The procedure is not specific enough to enable the count of sorted particles to provide a definitive number of oocysts present, since many other organisms of similar size may cross-react with the monoclonal antibody and have similar fluorescence characteristics. In addition, some water samples contain high numbers of autofluorescent algae, which may also mimic oocysts and therefore lead to incorrect conclusions if the instrument is used directly to produce counts of oocysts. However, epifluorescence microscopy can be used to confirm or negate the presence of oocysts. Microscopic examination of samples which have been sorted by FACS is much easier to perform than direct microscopy on nonsorted samples and, in particular, makes examination using differential interference contrast microscopy much simpler due to the removal of most of the background debris. An important consideration of the use of FACS is that the time taken to examine slides microscopically is considerably reduced and, since this is one of the most time-consuming parts of the analysis, samples can be analyzed more quickly. Furthermore, samples can be examined simultaneously for the presence of both *Cryptosporidium* oocysts and *Giardia* cysts (Vesey *et al.*, 1994) which makes the procedure more cost-effective. Despite the fact that some researchers from the USA have confirmed the benefits of FACS when examining water samples for the presence of oocysts (Danielson *et al.*, 1995; Hoffman *et al.*, in press), its use has not been widely adopted. However, in the UK, in particular, FACS is widely used, and it is also used in other parts of Europe and Australia. While it is clear that FACS has considerable benefits when examining water concentrates for oocysts, there are some considerations, in particular cost, which may restrict its widespread use. The instruments themselves may cost up to US $250 000 to purchase, have relatively high maintenance charges associated with them, and require skilled operators to use them effectively. Some less expensive instruments are being used, but it is not clear if these work as effectively as the more expensive instruments. One particular problem with some of the less expensive instrumentation is that the flow cell cannot easily be removed should it become blocked, or partially blocked, and this therefore necessitates the use of an 'in line filter' to prevent large particles from causing blockages. While this removes the problem of blockages, it is difficult to understand how a filter which is designed to remove large particles (~100 μm) does not begin to trap much smaller particles as it itself begins to become blocked during use. Preliminary work (C.R. Fricker, unpublished observations) suggests that these partial blockages of the filter can occur and that oocysts may be retained, thus preventing their detection. Clearly, a better understanding of the suitability of different FACS instruments for use in detecting oocysts in environmental samples is required. As mentioned earlier, one problem of using flow cytometry is that autofluorescent algae can give signals that may mimic those given by FITC-labeled oocysts. One possible way of overcoming this problem is to use antibodies labeled with two different fluorochromes, such that after excitation oocysts emit light at two different wavelengths. This approach would lead to a reduction in the number of false-positive signals being processed by the instrument, since it can be programmed to sort only those particles that fluoresce at two different, predefined wavelengths. Ideally, the approach would be to use two monoclonal antibodies raised against different and preferably spatially separated epitopes, although such an approach has been difficult to follow since the majority of commercially available antibodies appear to be directed against the same or very similar epitope. Thus, 'dual-labeling' experiments have been performed using the same antibody, labeled with two different fluorochromes. While this approach is still useful for reducing the number of false-positive events due to autofluorescing particles, other particles or organisms which cross-react or bind the antibody nonspecifically will continue to do so.

3.3.3. Polymerase Chain Reaction (PCR)

Perhaps one of the most widely applied procedures has been the use of the polymerase chain reaction (PCR) to detect specific sequences of nucleic acid which may be genus or species specific. Clearly, the ability to distinguish between *C. parvum* and other morphologically similar members of the genus has some benefit, and nucleic acid based techniques are the obvious choice for this problem. However, despite the exquisite specificity and sensitivity that PCR can offer, difficulties have been experienced with the application of PCR to water concentrates. This has largely been due to the degree of inhibition seen in a number of water types. PCR is sensitive to the concentration of many compounds within the reaction mixture, and those of particular concern to researchers working with water concentrates are divalent cations and humic and fulvic acids, compounds which are frequently found in water and which can cause a high degree of inhibition (Johnson *et al.*, 1995). Nonetheless, many workers have described protocols for the detection of *Cryptosporidium* oocysts by PCR, and a wide variety of primers have been described. These primers have been designed from various regions of the genome, and some which have apparent specificity include those from regions coding for the 18 S rRNA (Johnson *et al.*, 1995), coding for a protein containing repeating amino acid motifs (Ranucci *et al.*, 1993) and coding for the *Cryptosporidium* heat shock protein hsp70 (Stinear *et al.*, 1996). The use of PCR for the detection of oocysts in water concentrates does offer some advantages over that of direct microscopic examination, since the process can largely be automated and thus several samples can be processed simultaneously. Furthermore, the technique is, theoretically, sensitive down to a level of a single oocyst, and recent developments have suggested that it may be possible to distinguish viable from nonviable oocysts (Wagner-Wiening and Kimmig, 1995). Some workers claim to be able to detect a single oocyst in a water concentrate by using a procedure involving reverse transcription PCR (RT-oCR), where the target sequence codes for the *Cryptosporidium* heat shock protein Hsp 70 (Stinear *et al.*, 1996). This procedure uses the calcium carbonate flocculation procedure for concentrating samples, followed by induction of hsp, purification of the target sequence using immunomagnetic separation, where the beads are coated with a sequence complementary to the target m RNA sequence, followed by RT-PCR. The data presented showed that single viable oocysts could be detected by this procedure, even in the presence of PCR inhibitors. Such a method would be of considerable value to the water industry, facilitating rapid screening of samples, although as yet it is not quantitative and thus may be of limited value in some circumstances. Preliminary experiments in the authors' laboratory have not produced similar data, and comparative data from other laboratories are eagerly awaited. While the application of PCR to the difficult problem of detection of *Cryptosporidium* oocysts in water offers some potential advantages, several concerns have been voiced which may make the procedure less valuable. All the PCR procedures that have been described thus far require oocysts to be broken open, usually using a freeze–thaw lysis procedure, which prevents the possibility of examining the organisms microscopically. While attempts are being made to develop PCR procedures that are quantitative, the use of quantitative procedures in water concentrates where the levels being sought are extremely low seems unlikely to be available in the short term. Examination of water samples is likely to require that accurate counts are attainable due to the regulatory requirements of the tests. It has been suggested that DNA can survive in the environment for relatively long periods of time, and it has been demonstrated that, even in biologically active matrices, specific DNA sequences can be detected in material which dates back some 2000 years (Fricker *et al.*, 1997). Thus there is concern that false-positive results could be obtained either from nonviable oocysts or from free DNA. This objection can, of course, be overcome by the use of RT-PCR directed against mRNA, but the use of direct PCR against DNA may indeed generate false-positive results. Many parasitologists, and indeed water microbiologists who have by necessity become involved in testing for *Cryptosporidium* oocysts, favor a holistic approach, where the intact organism can be viewed directly in the first instance, with the possibility of using molecular techniques at a later stage. The likelihood is that a combined approach may be used whereby molecular techniques are used as a screening tool in the first instance on a portion of a water concentrate and, when positive results are generated, other approaches that involve microscopic examination for oocyst confirmation may then be used.

3.3.4. *Fluorescent* in situ *Hybridization (FISH)*

An alternative molecular approach which has been applied to the detection of oocysts is the use of nucleic acid probes that target the 18 S rRNA molecule. The procedure which uses fluorescent *in situ* hybridization (FISH) apparently was specific for *C. parvum* and detected only viable oocysts (Vesey *et al.*, 1995; in press). The benefits of targeting the 18 S rRNA molecule are that probes which are species or genus specific can be constructed: the 18 S molecule is present at a high copy number in viable cells and the short half-life of the molecule means that the copy will decline rapidly in nonviable oocysts. However, use of the currently available probes together with direct fluorescence labeling did not produce fluorescence bright enough to be used for primary detection, since autofluorescing particles, in particular algae, fluoresced more brightly. When combined with immunofluorescent antibody staining, the rRNA probes could be used to determine the viability of sporozoites. Clearly, the application of FISH to the detection of oocysts is in its infancy, but this type of molecular approach has much to offer, and improvements in labeling efficiency or the use of amplified signals may give the opportunity to label oocysts with two distinct fluorochromes that identify two completely different targets. This is a more holistic approach than the use of PCR, and offers the possibility of using flow cytometry or some other instrumental technique for detection and enumeration of oocysts, without the need for microscopic confirmation. However, for those workers who believe that microscopic examination is an essential part of the correct identification, the oocysts remain intact and can therefore be examined.

3.3.5. *Other Antibody-based Methods*

Several other techniques using anti-oocyst antibodies have also been described which include the use of enzyme-linked immunosorbent assay (ELISA), electrochemiluminescent assays, laser scanning, and cooled charge-couple devices (CCD) to detect oocysts labeled with chemiluminescent probes. The use of ELISA is not recommended for detecting oocysts in water samples, as the sensitivity attainable is not sufficient for this purpose, and its use is restricted to clinical applications. The other two techniques have shown promise in research environments (Campbell *et al.*, 1992; Reynolds and Fricker, in press), but are not yet routinely applicable. Furthermore, the performance of these techniques, as with immunofluorescence and IMS, is strictly dependent on the quality of the antibody employed in the method.

3.4. Methods for Determining Oocyst Viability

The significance of finding oocysts in treated and, to a lesser extent, raw waters is not always clear, since some of the organisms that are detected may be nonviable and thus pose no threat to public health, and therefore there has been considerable interest in developing *in vitro* methods which can determine oocyst viability. There have been many reports of the development of animal models for cryptosporidiosis, and clearly such models may be able to determine infectivity as well as viability, but such models are not suitable for determining the viability of oocysts in environmental samples. The most widely accepted *in vitro* procedure for determining oocyst viability, excystation, is also unsuitable for use with environmental samples. Therefore, other methods are being evaluated.

3.4.1. Vital Dye Methodology

Campbell *et al.* (1992) developed a procedure based on the inclusion/exclusion of the vital dyes propidium iodide (PI; membrane impermeant) and 4,6-diamidino-2-phenylindole (DAPI; membrane permeant), which gave extremely good correlation with *in vitro* excystation. Four classes of oocysts can be identified using the assay: those which are viable and include DAPI but exclude PI; those which are nonviable and include both DAPI and PI; and two classes which include neither DAPI or PI, those with internal contents (sporozoites) and those without, as determined by DIC microscopy. In environmental samples, 'empty' oocysts or shells are frequently seen, and clearly these are of no health significance, but if these shells are present in treated drinking water, then clearly they have passed through the treatment process and thus, if viable oocysts were present, these too may have passed through the treatment process. Since the recovery efficiency of the overall *Cryptosporidium* detection method is so poor, it is unwise to conclude that because the only oocysts seen in a treated water sample are empty or nonviable then there is no public health threat. At present, methods for determining viability are an aid to determining the risk of infection due to consumption of contaminated water, but if some form of disinfection is used in the future then the determination of the viability of oocysts will be essential. The DAPI/PI procedure is simple to perform and, while some workers have expressed some reservations over its applicability, its use is to be recommended for routine environmental work. The incorporation of DAPI into the nucleic acid acts as a further criterion for determining if a particle is an oocyst or not. As explained above, many other eukaryotic organisms of similar size may show the same fluorescence characteristics as oocysts, but these will have different nucleic acid distribution than oocysts in which four sporozoites can be seen. Other eukaryotes usually have a single large nucleus and thus these organisms can quickly be identified as not being *Cryptosporidium* oocysts.

3.4.2. Nucleic Acid Stains

An alternative to the DAPI/PI approach to determining viability has been suggested by Belosovic and Finch (1996), who have used new nucleic acid stains to differentiate between viable and nonviable oocysts. Two new stains have been identified: SYTO9, which stains nonviable oocysts green or bright yellow while viable oocysts have a green halo surrounding the cell while the interior remains unstained; and MPR71059, which stains nonviable oocysts red while viable oocysts remain unstained. These approaches have not been widely tested, although Belosevic and Finch (1996) have demonstrated that the results obtained with these dyes, correlate well with mouse infectivity using an outbred CD-1 neonatal mouse model. Since these assays are apparently simple and quick to perform, they may be suitable for incorporation into the methods for the detection of oocysts in water samples, but this has yet to be proven.

3.4.3. In Vitro *Excystation*

An important event in the life cycle of *Cryptosporidium*, and indeed any coccidian parasite, is excystation, which occurs within the lumen of the gut *in vivo*. This event can be performed *in vitro* by mimicking conditions found in the gut, by exposing oocysts to appropriate temperatures, levels of bile salts, and pancreatic enzymes, and several different procedures have been reported (Reducker and Speer, 1985; Woodmansee *et al.*, 1987; Blewett, 1989; Robertson *et al.*, 1993). The release of motile sporozoites can be observed microscopically and the ratio of sporozoites to totally or partially excysted oocysts can be calculated. This *in vitro* excystation can be useful in estimating the proportion of potentially viable or infectious oocysts within a given population, and has been used to determine the effect of various disinfectants and other environmental stresses (Blewett, 1989; Peeters *et al.*, 1989; Korich *et al.*, 1990; Robertson *et al.*, 1992; Finch *et al.*, 1993). While this technique may be useful when estimating the proportion of viable oocysts in a stock preparation, it has no place in examining the small numbers of oocysts found in environmental samples.

3.4.4. *Animal Infectivity*

The use of animal models to assess the infectivity of pathogenic microorganisms has been widely applied in medical microbiology and models exist for a wide range of organisms. A variety of animals has been assessed for use in determining the infectivity of *Cryptosporidium* oocysts, and mice have been chosen as the most suitable animal. However, several different strains of inbred and outbred mice have been used with widely differing responses (Sherwood *et al.*, 1982; Anderson, 1986; Blewett, 1989; Fayer *et al.*, 1991; Fayer, 1994). Many factors other than mouse strain may have been responsible for these differences, including variation in the administered doses and methods of determining that infection has occurred. There is at present no standard mouse infectivity model for *Cryptosporidium*, although the majority of workers now use neonatal mice, often of the CD-1 or BALB/c strains. These mouse models are expensive to perform, and unless very carefully controlled can lead to widely varying results. In addition, they measure infectivity, rather than viability, and may therefore give results which are quite different to those obtained using *in vitro* viability methods.

3.4.5. Cell Culture Methods

Recent refinements have been made to *in vitro* models of infectivity using tissue culture, improving the infectivity rate, and adapting the method for use with water concentrates (Upton *et al.*, 1994; Rochelle *et al.*, 1996; Slifko *et al.*, in press). For these assays, water samples are concentrated by normal procedures, and bacteria may be removed by exposure of the concentrate to concentrations of chlorine which are lethal to bacterial cells but which do not effect oocysts. The concentrates are then inoculated onto the tissue culture monolayer and left in contact for a period to allow potentially infectious oocysts to infect cells, before the remaining debris is washed away. The monolayer is then left for 24–48 hours before being examined for the presence of intracellular parasite antigen or nucleic acid. Immunofluorescent techniques have been used to identify cells which have become infected and this offers a way in which infection may potentially be quantified. However, it is not clear if the presence of a single infective oocyst will lead to one or more infected cells. In theory one might expect that an oocyst which excysts successfully would produce four infected tissue culture cells, but initial results have not demonstrated that this can be consistently achieved.

Other workers have adopted a somewhat different approach whereby they detect the presence of *Cryptosporidium* nucleic acids using the PCR (Rochelle *et al.*, 1996). While this method cannot be used to enumerate directly the oocysts present in any given sample, it can be applied in a 'most probable number' format to give an estimation of the number of oocysts present in a water concentrate. To achieve this, a number of aliquots of the water concentrate is applied to separate flasks of tissue culture cells, and the cells examined using the PCR after a suitable incubation period. Determination of the number of aliquots which contain infectious oocysts allows the number contained in the concentrate to be calculated.

3.4.6. Genetic Analysis

An alternative approach to the detection of *Cryptosporidium* oocysts and the determination of their viability is to attempt to induce the production of hsps and subsequently to detect the messenger RNA which codes for them. This approach has been used for the detection of *Giardia* in water concentrates and was deemed to be a sensitive and reliable test which used PCR to detect the specific mRNA species (Abbazsadegan *et al.*, 1997). A similar approach was described for *Cryptosporidium*, again using primers directed against the mRNA coding for hsp (Stinear *et al.*, 1996). Here, the authors reported that the procedure was relatively simple to perform and that interference from nontarget molecules could be removed without loss of sensitivity. Furthermore, it was reported that a single viable oocyst could be detected using this system and, while this has not been substantiated in peer-reviewed journals, such an approach holds much promise.

4. REGULATORY STATUS AND ENVIRONMENTAL LAWS

4.1. The USA

4.1.1. *SDWA, ESWTR, ICR*

The Safe Drinking Water Act (SDWA) was first enacted in 1974 to establish safety standards for water used for public consumption. The Environmental Protection Agency (EPA) was required to develop drinking water standards for over 80 different contaminants. The Act was amended in 1986 to require EPA to publish a 'maximum contaminant level' (MCL) goal for each contaminant which, in the judgment of the EPA, 'may have any adverse effect on the health of persons and which are known or anticipated to occur in public water systems'. This Act established the current philosophy of monitoring and treating water to remove such contaminants.

To meet the requirements of the SDWA, the EPA has implemented a series of rules regulating activities by the water companies. One such rule, the surface water treatment rule (SWTR), which became effective in 1989, required water treatment *in lieu* of water testing for contaminants which pose acute health risks to water consumers but are difficult to detect (USEPA, 1989). One somewhat ambiguous requirement of the rule was that the insufficient methods to detect a contaminant was not grounds for avoiding treatment. Hence, treatment was required without knowing if the contaminant to be treated for was present, and without knowing the effect of the treatment on the presence of such contaminant. Disinfection was required for surface water systems and for groundwater systems under the direct influence of surface water. Filtration was mandatory unless a utility could meet certain microbiological and turbidity requirements and had an effective watershed-protection program to control the potential for contamination by *Giardia* and viruses. In particular, the Rule specified a 3 log (inactivation rate) for *Giardia* and a 4 log (inactivation rate) for viruses (primarily enteric viruses) during the course of treatment (combined physical filtration and chemical inactivation). However, the Rule did not specify monitoring for the presence of such organisms, and the 'log reduction' requirement proved to be overly stringent in the case of very clean source waters, yet not stringent enough in the case of poor quality source water where microbiological contaminants exceeded safe levels even after compliance. The initial SWTR did not address the *Cryptosporidium* issue because of a lack of information on occurrence and treatment effectiveness.

As a result of continued outbreaks of giardiasis and cryptosporidiosis in the USA during development of the SWTR, the EPA proposed the enhanced surface water treatment rule (ESWTR) (USEPA, 1994). This legislation was designed to address some deficiencies in the SWTR, related primarily to protozoan control, and was intended for application to all surface-water systems serving populations greater than 10 000. The Proposed Rule focused primarily on 'treatment requirements for the water-borne pathogens *Giardia*, *Cryptosporidium*, and viruses'. In addition, the ESWTR sought to derive a balance between aggressive chemical disinfection and the risk of cancer posed by the disinfection byproducts. Unlike the SWTR (which specified treatment and inactivation rates), the Proposed ESWTR considered various treatment alternatives designed to choose a surrogate organism, whose containment would indicate equal or higher levels of containment of other dangerous microbes. Alternatives could be based on *Giardia*, *Cryptosporidium*, or viruses; within a group, MCLs required could be triggered by the actual number of organisms found in the source water. For example, <1 oocyst of *Cryptosporidium* per 100 l of raw water would require, a 3 log reduction; 1–9 oocysts per 100 l would trigger a 4 log requirement, 10–99 oocysts trigger a 5 log requirement, and >99 would trigger a 6 log reduction. There are obvious problems in both accurately determining the source levels and in affecting this level of reduction. More data were needed on occurrence, infectivity, and treatment effectiveness with regard to *Cryptosporidium*.

To generate the data needed to move forward with further rule-making on the ESWTR (from a proposed rule to an interim rule), an information collection rule (ICR) was enacted in 1996 after being delayed for 2 years due to the lack of an accurate and sensitive method for detecting *Cryptosporidium* in waters. This rule has the goal of establishing occurrence data on disinfection by-products and microbial pathogens (especially *Cryptosporidium*) in drinking water, and is required for all surface water utilities serving populations of >100 000. While originally targeted for utilities serving populations of >10 000, the EPA did not want to place undue burden on the smaller utilities, particularly given the costly ICR method for monitoring for *Cryptosporidium*, and the questionable results likely to be generated (due to the poor recoveries using the method). The EPA expects to use the occurrence and cost data from the ICR to select and justify the regulatory actions to be promulgated under an 'Interim' ESWTR in 1998 and ultimately a 'Final' ESWTR, originally estimated to be effective in the year 2000. The EPA stated that the interim rule may make use of 'less than ideal' methods, but that the final rule would make use of the best methods then available. These dates will likely be delayed due to the delay in implementation of the ICR.

The testing framework provided by the ICR is rigid, inflexible, and provides no room for new methods evaluation. The ICR method has been described above (Sections 2 and 3). Interestingly, the EPA acknowledges that this method has severe limitations, but that given the 'large number of public water systems which would be generating data . . . an 8% recovery rate [of *Cryptosporidium*] can provide a reliable adjustment factor from which to estimate national occurrence'. Obviously, the high public pressure to address the *Cryptosporidium* problem in drinking water played a major role in this rule being promulgated.

While the ICR method is fixed and cannot be changed, the proposed ESWTR provides a mechanism to evaluate new methods. Furthermore, several EPA sponsored studies are planned to address comparative evaluations of new methods (including a 'survey' study of over 50 smaller water utilities [10 000–100 000 population] in which new candidate method(s) can be used in place of the ICR method). The proposed ESWTR states: 'Following the full compilation and analysis of all data collected under the ICR rule *and from other research findings*, EPA would propose a long-term ESWTR with which systems serving fewer than 10 000 people would comply', (italics added).

4.2. The UK

The regulatory position with regard to *Cryptosporidium* and water supplies is somewhat unclear since the current European drinking water directive requires that 'water intended for human consumption should not contain pathogenic organisms' (Departments of Environment and Health, 1990). Clearly such a requirement is not enforceable, since it is not possible to determine if pathogenic organisms are absent in the total volume of water supplied. Therefore the proposed draft of the European directive makes it a general requirement that 'water intended for human consumption does not contain pathogenic micro-organisms and parasites in numbers which constitute a potential danger to health'. Such a standard is not particularly useful since the level of many of the organisms which may be present in water and which constitute a potential danger to human health is not adequately identified. For example, the infectious dose of *Cryptosporidium* is not clear for either the immunocompetent or immunocompromised population, and therefore it could be argued that only if an outbreak occurs has there been a breach of the regulations. However, until more information is available regarding the infectious dose of *Cryptosporidium* and the recovery efficiencies of the procedures used to detect the organism, it is unlikely that numerical standards will be set.

5. WATER TREATMENT OPTIONS

5.1. Current Municipal Water Treatment Practices

Elimination of oocysts during water processing and treatment occurs in three stages: (1) reduction of oocyst contamination of source water or treatment plant intakes; (2) physical removal; and (3) chemical or physical disinfection. Current practices for the treatment of water consist of a compilation of a number of unit processes designed to provide safe and esthetically pleasing drinking water and the exact combination of these processes which is required depends to a large extent on the quality of the raw water. Furthermore, other constraints such as land availability and cost have impacted the choice of treatment options. However, a multiple barrier approach is almost universally taken in an attempt to remove a wide variety of potential contaminants including, color, turbidity, pesticides and other organic molecules, algae, zooplankton, and pathogens, including bacteria, viruses, and protozoa. In addition, effective water treatment should remove unwanted tastes and odors. Typically the types of treatment process available might include reservoir storage, microstraining, flocculation, rapid filtration, slow sand filtration, and disinfection. All or some of these processes may be incorporated into the treatment process, depending on the raw water quality.

Ground waters derived from deep wells and boreholes are generally considered to be hygienically safe, since the water has to pass through soil and rock to get to the abstraction point. However, rocks which are fissured can allow the passage of small particles into the ground water and even in well maintained, deep boreholes, *Cryptosporidium* oocysts, bacteria, and viruses have been found. Contamination of ground waters is of particular concern, since this type of water usually receives minimal treatment, often merely chlorine disinfection, which offers no protection against the threat of *Cryptosporidium* infection. Without vast capital expenditure little can be done, in terms of treatment, to prevent the threat of contamination causing disease, and so attention to detail in the maintenance of good-quality ground water is essential. The quality of the raw water should be monitored, well linings inspected regularly, and any potential sources of contamination (e.g. grazing cattle, slurry storage) kept as far removed from the well-heads as possible.

Where space permits, reservoir storage of water is often used to improve water quality, largely due to the removal of particulate material by sedimentation and die-off of enteric bacteria due to the relatively harsh environment and/or predation by zooplankton. However, theoretical calculations suggest that sedimentation of *Cryptosporidium* oocysts is unlikely to have a significant effect on their removal from a body of water unless they are attached to other particulates (Anon., 1990). Nonetheless, reservoir storage can result in significant reductions in the concentration of oocysts, which may be a result of predation, death, sedimentation when in contact with other particulates, or a combination of factors. In addition, the use of reservoir storage has the beneficial effect of diluting any high concentrations of oocysts which may be present in surface waters from time to time, and therefore, with respect to the removal of oocysts, its use is preferable to direct river abstraction wherever possible, although other considerations may prevent this. Microstrainers are sometimes used to remove particulates, frequently algal cells, from raw water prior to further treatment. Typically, the pore size of these filters is 25–35 μm and thus they cannot be relied upon to remove *Cryptosporidium* oocysts. Nonetheless, as other particulates are trapped by the microstrainers, the pore size will decrease and oocysts may be removed, and thus, while this form of treatment cannot be considered reliable for removing oocysts, the washings from the filters may well contain oocysts and other pathogenic organisms and must be treated accordingly. Extrapolation from data on the removal of other particles suggests that microstrainers can effect removal of up to 60% of oocysts (Parker *et al.*, 1990). As described above, particles can be removed from water by storage in a reservoir and, while this is effective, the small size of many of the particles means that their sedimentation rate is so slow that they are not effectively removed. To increase the rate of sedimentation of particles in a body of water, flocculation followed by clarification is often used. This process involves the addition of chemicals to the water, causing a floc to form and helping to drag down particles. Typically the chemicals used are ferric or aluminum salts, and there appears to be no real difference in the effectiveness of aluminum sulfate, polyaluminum chloride, ferric sulfate, or ferric chloride in removing oocysts and similarly sized particles (Anon., 1990). Rapid sand filtration is a common treatment process used to remove particulates, and when operating efficiently is thought to be capable of 3 log removal of *Cryptosporidium* oocysts (Anon., 1990). Other investigations have given a range of removal rates, including 91% (Rose *et al.*, 1986), 65–90% (Gregory *et al.*, 1991), and >99.999% (Hall *et al.*, 1994), with the higher removal rates being achievable when coagulant dosing was applied to the water prior to filtration. Slow sand filtration relies on a much slower filtration of water (about 2% of that of rapid gravity filtration) using fine sand and in general no chemicals are used to condition particles prior to slow sand filtration as these would tend to clog the pores. No data are available for removal of oocysts in full-scale plants, but a number of pilot-scale studies have been completed where the removal efficiencies were generally good. Ives *et al.* (1993) demonstrated removal of 97.6–99.7% in laboratory-scale sand columns, while Hall *et al.* (1994) demonstrated removals of >99.98%. In another study using surface water to supply the population of London, England, heat-inactivated oocysts were added at a concentration of 4000/l and no oocysts were found in the filtrate. At the end of the study, intact oocysts were found only in the first 2.5 cm of the sand filter (Timms *et al.*, 1995).

5.2. Disinfection Options

Given the refractory nature of *Cryptosporidium* oocysts to standard water chlorination, significant effort has been expended to find new chemical agents, combinations, or conditions that could be used to effectively treat water. The necessary limitation to simply increasing C_t values (concentration × time required for a specific level of inactivation) of a given agent is that aggressive chemical disinfection must be balanced by the risk of harmful disinfection by-products. An illustration of the impracticality of simply increasing the C_t value of currently used chemicals is provided by the following sample calculation of C_t values. A recent evaluation of chlorine treatment demonstrated that purified oocysts remain infectious for neonatal mice after exposure to 5.25% sodium hypochlorite (commercial strength laundry bleach, about 5×10^4 ppm) for 2 hours at 21°C (Fayer, 1995). The mice in this study received 1.5×10^5 oocysts; given that reproducible infections can be elicited in neonatal mice by administration of as few as 100–500 oocysts (Finch *et al.*, 1993b), the hypochlorite treatment was capable of achieving a maximum of 3 log (inactivation) (99.9%). The C_t value for 99.9% inactivation in this case is 6×10^6 (mg/l × time in minutes). This value indicates that water chlorinated at 1 ppm (standard level for drinking water) would need a contact time of about 12 years to affect a 99.9% inactivation. This illustration is probably an overestimation, as treatment in this study was conducted without pH adjustment, which would have significantly improved the inactivation. Alternatively, this study was conducted in demand-free water, and organics material that would be expected to be present in raw water would exhaust the chlorine to a significant extent. Other reported C_t values for chlorine inactivation are 7.2×10^3 (Lisle and Rose, 1995), and 9.6×10^3 (Korich *et al.*, 1990) for 99% inactivation. While these values are about 2 log lower than the illustration above for equivalent inactivation, they confirm that chlorine treatment alone (as hypochlorite) cannot be expected to provide reasonable disinfection for *Cryptosporidium* oocysts.

Results obtained with chlorine in the form of chloramine were not significantly better than the results obtained with hypochlorite; results with chlorine dioxide were somewhat more encouraging, but still inadequate when this chemical was used alone (Peeters *et al.*, 1989; Korich *et al.*, 1990; Lisle and Rose, 1995). Ozone treatment has shown the most promise when considering single agents, with C_t values reported to be 1–50 mg min/l for inactivation rates of 1–4 log (Peeters *et al.*, 1989; Korich *et al.* 1990; Finch *et al.*, 1993a; Parker *et al.*, 1993). The effectiveness of ozone appears to be highly dependent on temperature and the presence of organic material, and some of the variable data were due to the use of different measures of viability or infectivity. Given that a common dosage for ozonation of water is to achieve a C_t of about 2.5 mg min/l residual ozone (Peeters *et al.*, 1989), this level would appear to be suboptimal for significant log reductions of *Cryptosporidium*, especially at lower temperatures.

Sequential treatments using the above-mentioned agents appear significantly more promising than single agents alone. Recent data have emerged indicating that initial treatment with ozone is capable of rendering the oocyst sensitive to subsequent inactivation with chloramine or chlorine dioxide at much lower concentrations, presumably because of damage to the oocyst wall, improving permeability to the second treatment (Finch, 1996). Ozone is commonly used in bottled water operations and smaller treatment facilities. However, these data must be validated in a water-treatment setting, investigating the effects of temperature and dissolved organic material on the efficacy of such treatments on *Cryptosporidium* oocysts using infectivity studies, before justifying the expense of installation of this capability in large treatment facilities.

Physical methods for inactivation of oocysts that have been found to be useful in certain circumstances include ultraviolet (UV) light and heat. While the former has received considerable attention with regard to water treatment, the cost of the latter would be prohibitive in a water-treatment setting, given the energy required to pasteurize drinking water (Fayer, 1994). UV light treatment of water appears to be a technique that could be implemented to water treatment, if only for smaller systems. Depending on the configuration of the system, UV light treatment could require up to several hours to achieve the desired level of inactivation (Lorenzo-Lorenzo *et al.*, 1993). In contrast, 2–3 log of inactivation was observed using higher intensity UV light systems designed specifically for inactivation of oocysts in drinking water (Campbell *et al.*, 1995).

5.3. Watershed Management

Management of agricultural waste contamination into metropolitan water supplies is being recognized as a very cost-effective mechanism of controlling the level of *Cryptosporidium* in source waters, particularly in cases where water is not filtered (e.g. New York City water supply), and where installation of filtration would cost several billion dollars. Present efforts are focusing on dairy farms operating in watershed areas. These farms could be purchased and shut down as one means of controlling waste runoff; alternatively, the utility or the government could provide assistance to farms to help contain and control runoff, particularly in winter and spring when runoff is very high and often out of control. Current thinking with regard to these alternatives can be observed in the situation faced by metropolitan New York City. Within the approximately 1600 square miles of watershed north of the city, there are over 500 farms that could potentially contribute *Cryptosporidium*-contaminated runoff to the reservoir system. Since the City effectively shut down many farms when the reservior system was constructed in the late 19th and early 20th centuries, there has been significant animosity towards the city by watershed farmers who feel their livelihood is in jeopardy. Public outcry has led to a major effort to preserve farming in the watershed. Several studies are underway to understand better the nature of runoff contamination and to help farmers improve the management of waste runoff (Powers, 1996). Such studies are interdisciplinary: (1) molecular epidemiological tools are beginning to be used to trace sources of oocysts; (2) survey studies are attempting to classify the shedding potential of various animals and to segregate them into specially contained areas (e.g. calf-rearing barns with heightened containment and diversion of wastes); (3) watershed hydrology studies hope to focus management efforts to the farms with the highest risk of reservoir contamination; and (4) improved methods for detecting oocysts (as discussed above) are required to measure the magnitude of the problem and the results of the improved management. These studies should all contribute to a more efficient and cost-effective 'partnership' with the farms to minimize contamination of the water. As population densities increase in watershed areas, such information will be critical to maintain the best source water possible.

5.4. Theoretical Considerations

It is clear that *Cryptosporidium* oocysts can pass through water-treatment plants and into the distribution system and thus that the risk of water-borne cryptosporidiosis is a real one which cannot be eliminated completely. However, it is the responsibility of the water providers to minimize the risk through the use of best practice. The measures which can be taken to reduce the risk of water-borne cryptosporidiosis are not particularly different from those which are required for other water-borne pathogens, although the lack of a suitable disinfectant has a considerable effect on the way in which water suppliers must act. As detailed above, efforts to maintain a high quality of raw water are essential, since the lower the concentration of oocysts in the raw water, the easier it becomes to reduce them to acceptable levels. Watershed management should include efforts to improve the microbiological quality of sewage as well as agricultural discharges to receiving waters and to ensure that the quality of raw water is monitored regularly with respect to its *Cryptosporidium* content. In terms of treatment, the multiple barrier approach is still to be recommended, since each barrier plays a part in reducing the overall risk. It is essential that the efficiency of treatment is monitored such that coagulation and filtration, for example, are operating optimally. The method of monitoring is still not clear but, in addition to specific *Cryptosporidium* monitoring, continuous monitoring of turbidity or particle counting will play an important role in maintaining optimum performance for some parts of the treatment process. There appears to be little hope that a disinfection system will become available in the near future which will be effective enough to be relied upon and thus physical treatment systems are still the most important at present. It is therefore important that such systems are not overloaded, and to this end care must be taken when dealing with recycled water from filter or microscreen backwashes. Whenever possible this water should not be recycled back to the treatment works, but if this essential, then it is imperative that the water is treated to remove the majority of oocysts. Treatments for this type of water might include lagooning or some form of filtration.

For treated drinking water, the ultimate barrier for *Cryptosporidium* oocysts is nanofiltration, where the pore size of the filter is in the nanometer range. These membranes are principally used for removal of unwanted molecular components which may effect the aesthetic characteristics of water, but obviously will also be able to prevent the passage of microorganisms if operated and maintained properly. However, the cost of applying such technology for the removal of *Cryptosporidium* on a large treatment plant is prohibitive, and for the present we must rely on operating treatment systems as effectively as possible. Other measures that are important are ensuring that water-storage facilities (post-treatment) and the water distribution system are secure, and that no breaches of integrity occur that may allow ingress of contaminated water. Finally, water providers must be able to give advice to their customers, particularly those which may be at a higher risk, about the potential dangers of *Cryptosporidium*. An integral part of providing suitable advice is for the water company to monitor the raw-water quality regularly (using reliable methods), monitor the treatment process, and have a documented plan of action to follow should anything unusual occur. It is essential when dealing with a possible outbreak of cryptosporidiosis that a clear plan of action is followed as quickly as possible in order to limit the number of potential infections.

REFERENCES

Abbaszadegan, M., Hubner, M.S., Gerba, C.P. and Pepper, I.L. (1997). Detection of viable *Giardia* cysts by amplification of heat shock-induced mRNA. *Applied and Environmental Microbiology* **63**, 324–328.

Anderson, B.C. (1986). Effect of drying on the infectivity of cryptosporidia-laden calf feces for 3 to 7 day old mice. *American Journal of Veterinary Research* **47**, 2272–2273.

Anon. (1990). *Cryptosporidium* in Water Supplies. HMSO. London.

Anon. (1995). Largest reported water-borne outbreak of 1995. *Cryptosporidium Capsule* **1**(1), 7.

Anon. (1996a). Over one hundred infected with *Cryptosporidium* in Ontario. *Cryptosporidium Capsule* **1**(6), 6.

Anon. (1996b). *Cryptosporidium* outbreaks between 1993 and 1994. *Cryptosporidium Capsule* **1**(6), 6–7.

Anon. (1996c). CDC release study linking water supply to last summer's outbreak in Worcester, MA. *Cryptosporidium Capsule* **1**(6), 7.
Anon. (1996d) Cryptosporidiosis on Vermont farm and distortion in the media. *Cryptosporidium Capsule* **1**(7), 1–2.
Anon. (1996e) Largest recreational water-borne outbreak. *Cryptosporidium Capsule* **1**(8), 10–11.
Anon. (1996f) *Cryptosporidium* outbreak in British Columbia, Canada. *Cryptosporidium Capsule* **1**(10), 1–3.
Anon. (1996g) Largest *Cryptosporidium* outbreak ever reported in British Columbia, Canada. *Cryptosporidium Capsule* **1**(11), 1–3.
Anon. (1996h) Cryptosporidiosis at a California water park. *Cryptosporidium Capsule* **1**(12), 5.
Anon. (1996i) Dual outbreak reported in Florida. *Cryptosporidium Capsule* **2**(1), 1–2.
Anon. (1996j) Apple cider outbreak in New York. *Cryptosporidium Capsule* **2**(1), 5.
Anon. (1996k) Potential outbreak in Idaho. *Cryptosporidium Capsule* **2**(2), 1.
Arrowood, M.J. and Sterling, C.R. (1987). Isolation of *Cryptosporidium* oocysts and sporozoites using discontinuous sucrose and isopycnic percoll gradients. *Journal of Parasitology* **73**, 314–319.
Belosevic, M. and Finch, G. (1996) Development of a vital stain methodology for *Giardia* and *Cryptosporidium*. *American Water Works Association Annual Conference*, 23 June, 1996, Toronto, Canada
Besser-Wiek, J.W., Forfang, J., Hedberg, C.W., Korlath, J.A., Osterholm, M.T., Sterling, C.R. and Garcia, L. (1996) Foodborne outbreak of diarrheal illness associated with *Cryptosporidium parvum* — Minnesota, 1995. *Morbidity and Mortality Weekly Report* **45**, 783–784.
Bifulco, J.M. and Shaefer, F.W. (1993). Antibody-magnetite method for selective concentration of *Giardia lamblia* cysts from water samples. *Applied and Environmental Microbiology* **59**, 772–776.
Blewett, D.A. (1989) Disinfection and oocysts. In: *Proceedings of the First International Workshop on Cryptosporidiosis, Edinburgh* (K.W. Angus and D.A. Blewett, eds), pp. 107–115. Edinburgh: Animal Diseases Research Association.
Bukhari, Z. and Smith, H.V. (1995). Effect of three concentration techniques on viability of *Cryptosporidium parvum* oocysts recovered from feces. *Journal of Clinical Microbiology* **33**, 2592–2595.
Campbell, A.T., Robertson, L.J. and Smith, H.V. (1992). Viability of *Cryptosporidium parvum* oocysts: correlation of *in vitro* excystation with inclusion or exclusion of fluorogenic vital dyes. *Applied and Environmental Microbiology* **58**, 3488–3493.
Campbell, A.T., Robertson, L.J., Smith, H.V. and Girdwood, R.W.A. (1994). Viability of *Cryptosporidium parvum* oocysts concentrated by calcium carbonate flocculation. *Journal of Applied Bacteriology*, **76**, 638–639.
Campbell, A.T., Robertson, L.J., Snowball, M.R. and Smith, H.V. (1995). Inactivation of oocysts of *Cryptosporidium parvum* by ultraviolet light. *Water Research* **29**, 2583–2586.
CDC (1982) Cryptosporidiosis: assessment of chemotherapy of males with acquired immune deficiency syndrome (AIDS). *Morbidity and Mortality Weekly Report* **31**, 589–602.
CDC (1996). Surveillance for water-borne disease outbreaks — United States, 1993–1994. *Morbidity and Mortality Weekly Report* **45**, (SS-1), 1–30.
Clancy, J.L., Gollnitz, W.D. and Tabib, Z. (1994) Commercial labs: how accurate are they? *Journal of the American Water Works Association* **86**, 89–97.
Current, W.L., Reese, N.C., Ernst, J.V., Bailey, W.S., Heyman, M.B. and Weinstein, W.M. (1983). Human cryptosporidiosis in immunocompetent and immunodeficient

persons. Studies of an outbreak and experimental transmission. *The New England Journal of Medicine* **308**, 1252.

Danielson, R.E., Cooper, R.C, and Riggs, J.L. (1995). *Giardia* and *Cryptosporidium* analysis: a comparison of microscopic and flow cytometric techniques. *Proceedings of the Water Quality Technology Conference*, 1673–1678.

Dawson, D.J., Maddocks, M., Roberts, J. and Vidler, J.S. (1993). Evaluation of recovery of *Cryptosporidium parvum* oocysts using membrane filtration. *Letters in Applied Microbiology* **17**, 276–279.

Departments of Environment and Health (1990). *Cryptosporidium* in water supplies. *Report of the Group of Experts*. London: HMSO.

DuPont, H.L., Chappell, C.L., Sterling, C.R., Okhuysen, P.C., Rose, J.B. and Jakubowski, W. (1995). The infectivity of *Cryptosporidium parvum* in healthy volunteers. *New England Journal of Medicine* **332**, 855–859.

Fayer, R. (1994). Effect of high temperature on infectivity of *Cryptosporidium parvum* oocysts in water. *Applied and Environmental Mocrobiology* **60**, 2732–2735.

Fayer, R. (1995). Effect of sodium hypochlorite exposure on inefectivity of *Cryptosporidium parvum* oocysts for neonatal BALB/c mice. *Applied Environmental Microbiology* **61**, 844–846.

Fayer, R., Nerad, T., Rall, W., Lindsay, D.S. and Blagburn, B.L. (1991). Studies on cryopreservation of *Cryptosporidium parvum*. *Journal of Parasitology* **77**, 357–361.

Finch, G.R (1996). Chemical disinfection of *Cryptosporidium* oocysts. *American Water Works Association Annual Conference*, 23 June, 1996, Toronto, Canada.

Finch, G., Black, K., Gyurek, L. and Belosevic, M. (1993a). Ozone inactivation of *Cryptosporidium parvum* in demand-free phosphate buffer determined by *in vitro* excystation and animal infectivity. *Applied and Environmental Microbiology* **59**, 4203.

Finch, G.R., Daniels, E.K., Black, E.K., Scgafer, F.W. and Belosevic, M. (1993b). Dose response of *Cryptosporidium parvum* in outbred neonatal CD-1 mice. *Applied and Environmental Microbiology* **59**, 3661–3665.

Fricker, C.R. (1995). Detection of *Cryptosporidium* and *Giardia* in water. In: *Protozoan parasites in water* (W.B. Betts, D. Casemore, C.R. Fricker, H.V. Smith and J. Watkins eds). London, The Royal Society of Chemistry. pp. 91–96

Fricker, C.R., Turner N.B., Rolchigo P.M., Margolin A.B. and Crabb J.H. (1995). Improved recovery of *Cryptosporidium* oocysts from drinking water using vortex-flow filtration combined with immunomagnetic separation. *The American Journal of Tropical Medicine and Hygiene.* **53** (suppl.), 285.

Fricker, E.J., Spigelman, M. and Fricker, C.R. (in press) The detection of *Escherichia coli* DNA in the ancient remains of Lindow Man using the polymerase chain reaction. *Letters in Applied Microbiology.*

Fries, L., Hillman, K., Crabb, J., Linberg, S., Hamer, D., Griffith, J., Keusch, G., Soave, R. and Peterson, C. (1994). Clinical and microbiologic effects of bovine anti-*Cryptosporidium* immunoglobulin (BACI) on cryptosporidial diarrhea in AIDS. *34th Interscience Conference on Antimicrobial Agents and Chemotherapy* **100**, 198.

Goldstein, S., Juranek, D., Ravenholt, O., Hightower, A., Martin, D., Mesnik, J., Griffiths, S., Bryant, A., Reich, R. and Herwaldt, B. (1996). Cryptosporidiosis: an outbreak associated with drinking water despite state-of-the-art water treatment. *Annals of Internal Medicine* **124**, 459–468.

Graczyk, T.K., Cranfield, M.R. and Fayer, R. (1996a). Evaluation of commercial enzyme immunoassay (EIA) and immunofluorescent antibody (IFA) test kits for detection of *Cryptosporidium* oocysts of other species other than *Cryptosporidium parvum*. *The American Journal of Tropical Medicine and Hygiene* **54**, 274–279.

Graczyk, T.K., Cranfield, M.R., Fayer, R. and Anderson, M.S. (1996b). Viability and infectivity of *Cryptosporidium parvum* oocysts are retained upon intestinal passage

through a refractory avian host. *Applied and Environmental Microbiology* **62**, 3234–3237.
Gregory, J., Ives, K.J., Scutt, J.E. and Makanjuola, D.B. (1991). *Removal of* Cryptosporidium *Oocysts by Water Treatment Methods at Laboratory Scale*. Final Report to the Department of the Environment, London.
Haas, C.N. and Rose, J.B. (1996). Distribution of *Cryptosporidium* oocysts in a water supply. *Water Research* **30**(10), 2251–2254
Hall, T., Pressdee, J. and Carrington (1994). *Removal of* Cryptosporidium *Oocysts by Water Treatment Processes*. Report No. FRO 0457. Foundation for Water Research.
Halladay, D. (1994) *US. Dairy Statistics*.
Hoffman, R.M., Standridge, J.H., Prieve, A.F., Cucunato, J.C. and Bernhardt, M. (in press). The use of flow cytometry for detection of *Cryptosporidium* and *Giardia* in water samples. *Journal of the American Water Works Association*.
Ives, K.J., Gregory, J.E. and Pugh, H. (1993). A microsphere in water: *Cryptosporidium parvum*. In: *Proceedings of the 6th World Filtration Congress, Nogoya, Japan*, pp. 224–231. Society of Chemical Engineers, Japan.
Jakubowski, W. and Erickson, T.H. (1979). Methods for detection of *Giardia* cysts in water supplies. In: *Water-borne Transmission of Giardiasis* (W. Jakubowski and J.C. Hoff, eds), pp. 193–210. Environmental Protection Agency Report 600/9-79-001. Washington, DC: EPA.
Johnson, D.W. Pieniazek, N.J., Griffin, D.W., Misener, L. and Rose, J.B. (1995). Development of a PCR protocol for sensitive detection of *Cryptosporidium* oocysts in water samples. *Applied and Environmental Microbiology* **61**, 3849–3855.
Jonas, A, Crabb, J.H., Turner, N.B. and Fricker, C.R. (in press a). Use of vortex flow filtration for improved recovery of *Cryptosporidium* oocysts. *Water Science and Technology*.
Jonas, A., Crabb, J., Murrin, K., Slade, J.S. and Fricker, C.R. (in press b). Simultaneous concentration and seperation of protozoan parasites and viruses using vortex flow filtration and immunomagnetic seperation. *Proceedings of the 1996 Water Quality Technology Conference*. American Water Works Association.
Korich, D.G., Mead, J.R., Madore, M.S., Sinclair, N.A. and Sterling, C.R. (1990). Effects of ozone, chlorine dioxide, chlorine, and monochloramine on *Cryptosporidium parvum* oocyst viability. *Applied and Environmental Microbiology* **56**, 1423–1428.
Kuhls, T., Mosier, D., Crawford, D. and Griffiths, J. (1994). Seroprevalence of cryptosporidial antibodies during infancy, childhood, and adolescence. *Clinical Infectious Diseases* **18**, 731–735.
LeChevallier, M.W., Lee, R.G. and Norton, D.N. (1991a) Occurence of *Giardia* and *Cryptosporidium* spp. in surface water supplies. *Applied Environmental Microbiology* **57**, 2610–2616.
LeChevallier, M.W., Lee, R.G., and Norton, D.N. (1991b) Occurence of *Giardia* and *Cryptosporidium* spp. in filtered drinking water supplies. *Applied Environmental Microbiology* **57**, 2617–2621.
Lisle, J.T. and Rose, J.B. (1995) *Cryptosporidium* contamination of water in the USA and UK: a mini-review. *Journal of Water SRT — Aqua* **44**, 103–177.
Lorenzo-Lorenzo, M.J., Ares-Mazas, M.E., Villiacorta-Martinez de Maturana and Duran-Oreiro, D. (1993). Effect of ultraviolet disinfection of drinking water on the viability of *Cryptosporidium parvum* oocysts. *Journal of Parasitology* **79**, 67–70.
Luft, B.J., Payne, D., Woodmansee, D. and Kim, C.W. (1987). Characterization of the antigens from sporulated oocysts of *Cryptosporidium parvum*. *Infection and Immunity* **55**, 2436–2441.
MacKenzie, W., Hoxie, N., Proctor, M, Gradus, M., Blair, K., Peterson, D., Kazmierczak, J., Addiss, D., Fox, K., Rose, J. and Davis, J. (1994a). A massive outbreak

in Milwaukee of *Cryptosporidium* infection transmitted through the public water supply. *The New England Journal of Medicine* **331**, 161–167.

MacKenzie, W., Schell, W., Blair, K., Addiss, D., Peterson, D., Hoxie, N., Kazmierczak, J. and Davis, J. (1994b). Massive outbreak of water-borne *Cryptosporidium* infection in Milwaukee, Wisconsin: recurrence of illness and risk of secondary transmission. *Clinical Infectious Diseases* **21**, 57–62.

MacKenzie, W.R., Kazmierczak, J.J. and Davis, J.P. (1995). An outbreak of cryptosporidiosis associated with a resort swimming pool. *Epidemiology Infectious* **115**, 545–553.

Madore, M.S., Rose, J.B., Gerba, C.P., Arrowood, M.J. and Sterling, C.R. (1987). Occurrence of *Cryptosporidium* oocysts in sewage effluents and select surface waters. *Journal of Parasitology* **73**, 702.

McDonald, V., Deer, R.M.A., Nina, J.M.S., Wright, S., Chiodini, P.L. and McAdam, K.P.W.J. (1991). Characteristics and specificity by hybridoma antibodies against oocyst antigens of *Cryptosporidium parvum* from man. *Parasite Immunology* **13**, 251–259.

McTigue, N., LeChevallier, M., Clancy, J. and Fredricksen, D. (1996). Assessment of particle removals by conventional filtration. *American Water Works Association Annual Conference*, 23 June 1996, Toronto, Canada.

Millard, P.S., Gensheimer, K.F., Addiss, D.G., Sosin, D.M., Beckett, G.A., Houck-Jankoski, A. and Hudson, A. (1995). An outbreak of cryptosporidiosis from freshpressed apple cider. *Journal of the American Medical Association* **8**, 776.

Morgan, D., Allaby, M., Crook, S., Casemore, D., Healing, T.D., Soltanpoor, N., Hill, S. and Hooper, W. (1995). Waterborne cryptosporidiosis associated with a borehole supply. *Community Disease Report (CDR) Review* **23**, 5.

Musial, C.E., Arrowood, M.J., Sterling, C.R. and Gerba, C.P. (1987). Detection of *Crytposporidium* in water using polypropylene cartridge filters. *Applied and Environmental Microbiology* **53**, 687–692.

Nieminski, E.C., Schaefer, F.W. and Ongerth, J.E. (1995). Comparison of two methods for detection of *Giardia* cysts and *Cryptosporidium* oocysts in water. *Applied and Environmental Microbiology* **61**, 1714–1719.

Nime, F.A., Burek, J.D., Page, D.L., Holscher, M.A. and Yardley, J.H. (1976). Acute enterocolitis in a human being infected with the protozoan *Cryptosporidium*. *Gastroenterology* **70**, 592–598.

Nina, J.M.S., McDonald, V., Dyson, D.A., Catchpole, J., Uni, S., Iseki, M., Chiodini, P.L. and McAdam, K.P.W.J. (1992). Analysis of oocyst wall and sporozoite antigens from three *Cryptosporidium* species. *Infection and Immunity* **60**, 1509–1513.

O'Donoghue, P. (1995). *Cryptosporidium* and Cryptosporidiosis in man and animals. *International Journal for Parasitology* **25**, 139–195.

Ongerth, J.E. and Stibbs, H.H. (1987). Identification of *Cryptosporidium* oocysts in river water. *Applied and Environmental Microbiology* **61**, 1714–1719.

Ortega-Mora, L.M., Troncoso, J.M., Rojo-Vazquez, F.A. and Gomez-Baustista, M. (1992). Cross-reactivity of polyclonal serum antibodies generated against *Cryptosporidium parvum* oocysts. *Infection and Immunity* **60**, 3442–3445.

Parker, J.F.W., Carrington, E.G. and Smith, H.V. (1990). *Progress Report on the Removal of* Cryptosporidium *Oocysts by Water Treatment Processes*. Report No. FRO 0155. Foundation for Water Reasearch.

Parker, J., Greaves, G. and Smith, H.V. (1993) The effect of ozone on the viability of *Cryptosporidium parvum* oocysts and a comparison of experimental methods. *Water Science and Technology* **27**, 93.

Peeters, J.E., Mazas, E.A., Masschelein, W.J., Martinez de Maturana, I.V. and Debacker, E. (1989). Effect of disinfection of drinking water with ozone or chlorine

dioxide on survival of *Cryptosporidium parvum* oocysts. *Applied and Environmental Microbiology* **55**, 1519–1522
Powers, M. (1996). College keeps city's taps running clean. *Cornell Focus* **5**(2), 4–11.
Ranucci, L., Muller, H.-M., La Rosa, G., Reckman, I., Gomez Morales, M.A., Spano, F., Pozio, E. and Crisanti, A. (1993) Characterization and immunolocalization of a *Cryptosporidium* protein containing repeated amino acid motifs. *Infection and Immunity* **61**, 2347–2356.
Reduker, D.W. and Speer, C.A. (1985). Factors influencing excystation in *Cryptosporidium* oocysts from cattle. *Journal of Parasitology* **71**, 332–339.
Reynolds, D. and Fricker, C.R. (in press). Use of laser scanning for the detection of waterborne microorganisms. In: *Proceedings of the Water Quality Technology Conference*.
Robertson, L.J., Campbell, A.T. and Smith, H.V. (1992). Survival of oocysts of *Cryptosporidium parvum* under various environmental pressures. *Applied and Environmental Microbiology* **58**, 3493–3500.
Robertson, L.J., Campbell, A.T. and Smith, H.V. (1993). In vitro excystation of *Cryptosporidium parvum*. *Parasitology* **106**, 13–19.
Rochelle, P.A., Ferguson, D.M., Handojo, T.J., De Leon, R., Stewart, M.H. and Wolfe, R.L. (1996). Development of a rapid detection procedure for *Cryptosporidium*, using *in vitro* cell culture combined with PCR. *Journal of Eukaryotic Microbiology* **43**, S72.
Rodgers, M.R., Flanigan, D.J. and Jakbowski, W. (1995). Identification of algae which interfere with the detection of *Giardia* cysts and *Cryptosporidium* oocysts and a method for alleviating this interference. *Applied and Enviromental Microbiology* **61**, 3759–3763.
Rolchigo, P.M. (1995). Vortex flow filtration. *Separation and Filtration Systems* **1**(3), 22–28
Rose, J.B., Kayed, D., Madore, M.S., Gerba, C.P., Arrowood, M.J. and Sterling, C.R. (1988). Methods for the recovery of *Giardia* and *Cryptosporidium* from environmental waters and their comparative occurrence. In: *Advances in Giardia Research* (W. Wallace and B. Hammond, eds). Calgary, Canada: University of Calgary Press.
Rose, J.B., Landeen, L.K., Riley, K.R. and Gerba, C.P. (1989). Evaluation of immunofluorescence techniques for detection of *Cryptosporidium* oocysts and *Giardia* cysts from environmental samples. *Applied and Environmental Microbiology* **55**, 3189–3196.
Rose, J.B., Gerba, C.P. and Jakubowski, W. (1991). Survey of potable water supplies for *Cryptosporidium* and *Giardia*. *Environmental Science and Technology* **25**, 1393.
Sheather, A.L. (1923). The detection of protozoan and mange parasites by a flotation technique. *Journal of Comparative Pathology* **36**, 266–267.
Shepherd, K.M. and Wyn-Jones, A.P. (1996). An evaluation of methods for the simultaneous detection of *Cryptosporidium* oocysts and *Giardia* cysts from water. *Applied and Environmental Microbiology* **62**, 1317–1322.
Sherwood,D., Angus, K.W., Snodgrass, D.R. and Tzipori, S. (1982). Experimental cryptosporidiosis in laboratory mice. *Infection and Immunity* **38**, 471–475.
Slifko, T.R., Friedman, D.E. and Rose, J.B. (in press). Unique cultural methods used to detect viable *Cryptosporidium parvum* oocysts in environmental samples. *Water Science and Technology*.
Solo-Gabriele, H. and Neumeister, S. (1996). US outbreaks of Cryptosporidiosis. *The Journal of the American Water Works Association* **88**(9), 76–86.
Sorvillo, F.J., Lieb, L.E., Kernat, P.R. and Ash, L.R. (1994). Epidemiology of cryptosporidiosis among persons with acquired immune deficiency syndrome in Los Angeles County. *American Journal of Tropical Medicine and Hygiene* **51**, 326–331.
Stinear, T., Matusan, A., Hines, K. and Sandery, M. (1996). Detection of a single

viable *Cryptosporidium parvum* oocyst in environmental concentrates by reverse transcription-PCR. *Applied and Environmental Microbiology* **62**, 3385–3390.

The National *Cryptosporidium* Survey Group (1992). A survey of *Cryptosporidium* oocysts in surface and groundwaters in the UK. *Journal of International Water and Engineering Management* **6**, 697.

Timms, S., Slade, J.S. and Frcker, C.R. (1995). Removal of *Cryptosporidium* Oocysts by slow sand filtration. *Water Science and Technology* **32**, 81–84

Tyzzer, E.E. (1907) A sporozoan found in the peptic glands of the common mouse. *Proceedings of the Society for Experimental Biology and Medicine* **5**, 12.

Upton, S.J., Tilley, M. and Brillhart, D.B. 1994 Comparative development of *Cryptosporidium parvum* (Apicomplexa) in eleven continuous host lines. *FEMS Microbiology Letters*, **118**, 233–236.

USDA (1993). Cryptosporidium *is Common in Dairy Calves*. Fort Collins, CO: National Dairy Helfer Evaluation Project, National Animal Health Monitoring System.

USEPA (1989). National primary drinking water regulations; filtration, and disinfection; turbidity; *Giardia* lamblia, viruses, *Legionella*, and heterotrophic bacteria. *Federal Register* **54**, 27 486–27 541.

USEPA (1994). Enhanced surface water treatment requirements. *Federal Register* **59**, 38 832–38 858.

Vesey, G. and Slade, J.S. (1991). Isolation and identification of *Cryptosporidium* from water. *Water Science and Technology* **24**, 425–429.

Vesey, G., Slade, J.S. and Fricker, C.R. (1991). Taking the eye strain out of environmental *Cryptosporidium* analysis. *Letters in Applied Microbiology* **13**, 62–65.

Vesey, G., Slade, J.S., Byrne, M., Shepherd, K. and Fricker, C.R. (1993a). A new method for the concentration of *Cryptosporidium* oocysts from water. *Journal of Applied Bacteriology* **75**, 82–86.

Vesey, G., Slade, J.S., Byrne, M., Shepherd, K., Dennis, P.J. and Fricker, C.R. (1993b). Routine monitoring of *Cryptosporidium* oocysts in water using flow cytometry. *Journal of Applied Bacteriology* **75**, 87–90.

Vesey, G., Hutton, P.E., Champion, A.C., Ashbolt, N.J., Williams, K.L., Warton, A. and Veal, D.A. (1994). Application of flow cytometric methods for the routine detection of *Cryptosporidium* and *Giardia* in water. *Cytometry* **16**, 1–6.

Vesey, G., Ashbolt, N., Wallner, G., Dorsch, M., Williams, K.L. and Veal, D.A. (1995). Assessing *Cryptosporidium parvum* oocyst viability with fluorescent *in-situ* hybridization using ribosomal RNA probes and flow cytometry. In: *Protozoan Parasites and Water* (W.B. Betts, D. Casemore, C. Fricker, H. Smith, and J. Watkins, eds), pp. 133–138. Cambridge: Royal Society of Chemistry.

Vesey, G., Ashbolt, N., Fricker, E.J., Deere, D., Williams, K.L., Veal, D. and Dorsch, M. (in press). The use of ribosomal rRNA targeted oligonucleotide probes for fluorescent labeling of viable *Cryptosporidium* oocysts. *Applied and Environmental Microbiology*.

Wagner-Wiening, C. and Kimmig, P. (1995). Detection of viable *Cryptosporidium parvum* oocysts by PCR. *Applied and Environmental Microbiology* **61**, 4514–4516

Whitmore, T.N. (1995). Rapid techniques for the recovery of *Cryptosporidium*. In: *Protozoan Parasites in Water* (W.B. Betts, D. Casemore, C.R. Fricker, H.V. Smith and J. Watkins, eds), pp. 139–142. Cambridge: The Royal Society of Chemistry.

Woodmansee, D.B., Powell, E.C., Pohlenz, J.F.L. and Moon, H.W. (1987). Factors affecting motility and morphology of *Cryptosporidium* sporozoites *in vitro*. *Journal of Protozoology* **34**, 295–297.

PART 2

Enterocytozoon bieneusi and other microsporidia

INTRODUCTION TO MICROSPORIDIA

The response to an emerging pathogen is a true test of medical science. Through knowledge of the fundamental biology of infectious agents, rational therapies to control the problem are possible. Classical scientists demonstrated their genius by using clinical problems to demonstrate the benefits of antisepsis, vaccination, etc., changing the practice of medicine as a result. The situation is somewhat different in the case of microsporidiosis, where interest converged from diverse disciplines. Several laboratories around the world have advanced the field both in parallel and in an integrated fashion. In the author's case, interest in microsporidiosis proceeded from clinical observations of diarrhea and wasting associated with acquired immune deficiency syndrome (AIDS). An initially casual encounter with an ultrastructural pathologist led to the realization that microsporidiosis was a common, not rare, complication of AIDS, and grew into a web of associations among other clinicans, gastroenterologists, infectious disease physicians, nutrition specialists, basic biologists, taxonomists, veterinary biologists, and others. The flow of information and cooperation has been multidirectional. The result has been rapid advances in knowledge, with consensus in most areas. The aim of this section is to synthesize the information on microsporidiosis and discuss the place of the organisms in the phylogenetic tree as well as their role as pathogens.

The first chapter in this section, by Drs Elizabeth Didier, Karen Snowden, and John Shadduck, reviews the basic biology of microsporidians that infect mammals, including their unique morphologic characteristics, life cycles, and taxonomy. Host–parasite relationships of mammalian microsporidia will be discussed, including host specificity, adaptations that promote parasite survival, and host immune response. Drs Donald Kotler and Jan Orenstein follow with a discussion of clinical aspects of microsporidiosis, based mainly upon observations made in patients with AIDS. Particular emphasis is given to intestinal disease, which is the most common problem associated with microsporidiosis. The epidemiology is discussed, including an evaluation of the evidence for and against microsporidia as enteric pathogens. The pathogenesis of intestinal and hepatobiliary disease is discussed, as are its clinical–pathogical correlations and natural course. The currently used diagnostic tests are presented, including light and electron microscopic examinations of clinical specimens. Finally, the treatment of patients with microsporidiosis is discussed, including studies of specific antiparasitic therapies.

The emergence of microsporidia as a clinical problem has coincided with advances in the field of molecular biology, the techniques of which promise greatly enhanced sensitivity and specificity. These techniques, which are equally valuable to basic scientists and cliniciansm, have forced re-examination of dogma in scientific disciplines. Molecular biology is the tool that will forever change the practice of medicine. The final chapter in this section, by Drs Louis Weiss and Charles Vossbrink, demonstrates the application of molecular biology to microsporidia. The authors demonstrate the abilities as well as the limitations of molecular detection and analysis in detail, and either corroborate or challenge traditional methods in taxonomy, epidemiology, clinical diagnosis, and pharmaceutical research.

D. KOTLER

Biology of Microsporidian Species Infecting Mammals

Elizabeth S. Didier,[1] Karen F. Snowden[2] and John A. Shadduck[3]

[1]*Department of Microbiology, Tulane Regional Primate Research Center, Covington, LA 70433, USA, and Departments of* [2]*Veterinary Parasitology and* [3]*Veterinary Pathobiology, Texas A&M University, College Station, TX 77843, USA*

1. Introduction 284
2. Morphology 285
   2.1. Polar filament/tubule. 285
   2.2. Anchoring disk and polaroplast 288
   2.3. Nucleus. 288
   2.4. Endoplasmic reticulum and ribosomes 288
   2.5. Posterior vacuole 289
   2.6. Spore wall. 289
3. Life Cycle 290
   3.1. Transmission 291
   3.2. Germination 292
   3.3. Merogony 293
   3.4. Sporogony 293
   3.5. Secondary cycles of infection 295
   3.6. Growth *in vitro* of microsporidia 295
4. Taxonomy 296
   4.1. *Enterocytozoon* 298
   4.2. *Encephalitozoon* 299
   4.3. *Pleistophora* 300
   4.4. *Trachipleistophora* 301
   4.5. *Nosema* 301
   4.6. *Vittaforma* 302
   4.7. *Thelohania* 302
   4.8. *Microsporidium*. 303
5. Mammalian Host–Parasite Relationships 303
   5.1. Parasite survival 304
   5.2. Mechanisms of host resistance 305

6. Summary and Conclusions 309
Addendum 310
References 310

Microsporidia (phylum Microspora) are obligate intracellular protozoan parasites that infect a wide range of vertebrate and invertebrate hosts. Over 1000 species have been classified into approximately 100 genera, and at least 13 species have been reported to infect mammals. Phylogenetically, the microsporidia are early eukaryotes because they have a true nucleus, possess prokaryote-like ribosomes, and lack mitochondria. The species that infect mammals are relatively small, measuring 2.0–7.0 µm long and 1.5–5.0 µm wide. The mature organism is the spore, which is enclosed by a chitinous coat, making it relatively resistant to the environment. Infections often occur by fecal–oral or urinary–oral transmission, although vertical transmission is quite common in the carnivores. Host cells become infected through a process of germination in which the spore propels its contents through the everting and unwinding polar filament into the host cell. The polar filament is unique to the microsporidia. With a few exceptions, microsporidiosis is typically chronic and subclinical in immunologically competent hosts. Young carnivores infected with microsporidia, however, develop severe and sometimes lethal renal disease, and immunodeficient laboratory animals (e.g. athymic and SCID mice) develop ascites and die from microsporidiosis. This review describes the morphology, life cycle, taxonomy, and host–parasite relationships of the species of microsporidia that infect mammals.

1. INTRODUCTION

Microsporidia are single-celled obligate intracellular protozoan parasites belonging to the phylum Microspora within which exist over 1000 species classified into approximately 100 genera (Canning and Lom, 1986; Sprague *et al.*, 1992). Microsporidia commonly infect arthropods and may infect members of all classes of vertebrates. To date, at least 13 species of microsporidia have been reported to infect mammals (Table 1).

Historically, microsporidia infections in silkworms, honey bees, and salmonid fish have been responsible for significant economic losses (see Canning and Lom, 1986; Canning and Hollister, 1992; Weber *et al.*, 1994). Microsporidiosis also has interfered with interpretation of animal experiments. This occurred in part because commonly used laboratory animals such as rabbits and mice with microsporidiosis often show few or no clinical signs of disease, and results attributed to, or associated with, the pathogen in question have been found later to be due to subclinical microsporidian infections (Shadduck and Pakes, 1971). More recently, microsporidiosis has been recognized as an important infectious disease in animals (Shadduck *et al.*, 1996). In human beings, it is an opportunistic infection in immunologically compromised persons including transplant patients and those with acquired immune deficiency syndrome (AIDS) (Canning and Hollister, 1987; Shadduck 1989; Shadduck and Greeley, 1989; Rabodonirina *et al.*, 1996; Shadduck *et al.*,

1996). The purpose of this chapter is to describe the basic biology of microsporidian species that have been reported to infect mammals.

2. MORPHOLOGY

The mature spores of microsporidia infecting mammals are oval to pyriform in shape and measure 2.0–7.0 μm long and 1.5–5.0 μm wide. Proliferative stages may be rounder and slightly larger. Spores appear refractile and green under the light microscope and, by histochemical staining, are Gram positive and acid fast.

2.1. Polar Filament/Tubule

The characteristic that, above all, defines an organism as being a microsporidian is the polar filament, a tube coiled within the mature spore (Canning and Lom, 1986; Sprague et al., 1992). The polar filament originates from the anchoring disc or polar sac at the anterior end of the spore and forms coils at the posterior region of the spore. During the infection process the polar filament everts and propels the contents of the spore into the host cell without destroying the host-cell membrane. Although used interchangeably, the term 'polar filament' is often used to describe the coiled structure and the term 'polar tube' generally refers to the everted or discharged structure. The discharged polar tube extends approximately 50–100 μm and is about 0.1–0.15 μm wide (Weidner, 1972; Weidner and Byrd, 1982). The polar filament is shorter when still coiled within the spore. The number of turns of the polar filament varies among species.

The polar filament appears to be generated from endoplasmic reticulum and Golgi-like vesicles (Vávra, 1968, 1972; Vernick et al., 1977; Desportes-Livage et al., 1996; Takvorian and Cali, 1996). It is surrounded by a membrane and filled with electron-dense, unassembled polar tube protein(s). As seen by transmission electron microscopy, the polar filament consists of several concentric layers around a central core (Weidner, 1982; Canning and Hollister, 1987; Silveira and Canning, 1995). The electron-dense polar tube proteins seem to assemble at the growing tip of the tube during discharge and the discharging polar filament resembles a cylinder within another cylinder, so that there appear to be several layers separated by a lucent ring from a multilayered inner tube (Canning and Hollister, 1987; Canning et al., 1992). During germination, the polar tube displays plasticity as the sporoplasm passes through (Weidner, 1976). The extruded polar tube appears hollow. Disulfide linkages are important in stabilizing the polar tube proteins since incubation with sodium dodecyl sulfate left polar filaments/

Table 1 Species of microsporidia reported to infect animals.

Species	Mammalian hosts	Sites of infection	Growth in vitro	References
Enterocytozoon bieneusi	Humans, pigs	Small intestine, gallbladder, liver	No[a]	Desportes et al. (1985), Cali and Owen (1990), Visvesvara et al. (1995), Deplazes et al. (1996)
Encephalitozoon intestinalis (previously *Septata intestinalis*)	Humans	Disseminated	Yes	Cali et al. (1993), Hartskeerl et al. (1995), Didier et al. (1996b)
Encephalitozoon hellem	Humans	Disseminated	Yes	Friedberg et al. (1990), Yee et al. (1990), Didier et al. (1991a), Hollister et al. (1993)
Encephalitozoon cuniculi		Disseminated	Yes	Levaditi et al. (1923), Shadduck (1969), Shadduck et al. (1978), Stewart et al. (1979), Canning and Lom (1986), Van Rensburg et al. (1991), De Groote et al. (1995), Didier et al. (1995), Hollister et al. (1995), Deplazes et al. (1996), Didier et al. (1996c), Mathis et al. (1996)
Strain I	Rabbits, mice, humans			
Strain II	Mice, blue foxes			
Strain III	Domestic dogs, humans			
	Rodents, goats, sheep, swine,			
	Horses, foxes, cats, human and nonhuman primates[b]			
Nosema connori	Humans	Disseminated	NA	Margileth et al. (1973), Sprague et al. (1974)
Nosema ocularum	Humans	Corneal stroma	NA	Cali et al. (1991a)
Nosema-like sp.	Humans	Corneal stroma, skeletal muscle	NA	Bryan et al. (1990), Cali et al. (1996)
Vittaforma corneae (previously *Nosema corneum*)	Humans	Corneal stroma	Yes	Davis et al. (1990), Shadduck et al. (1990), Silveira and Canning (1995)
Pleistophora sp.	Humans	Skeletal muscle	NA	Ledford et al. (1985), Chupp et al. (1993)
Trachipleistophora hominis	Humans	Skeletal muscle, nasal	Yes	Field et al. (1996), Hollister et al. (1996)

Table 1 Continued.

Species	Mammalian hosts	Sites of infection	Growth in vitro	References
Thelohania apodemi	Field mice	Skeletal muscle, brain	NA	Doby et al. (1963)
Thelohania-like sp.	Humans	Cardiac muscle, liver, brain	NA	Yachnis et al. (1996)
Microsporidium spp.[c]	Humans			Canning and Lom (1986), Weber et al. (1994)

NA, culture not attempted.
[a]Short-term growth of *Ent. bieneusi* has been reported (Visvesvara et al. 1995b)
[b]Hosts infected with *Enc. cuniculi* of unknown strain; see Canning and Lom (1986) for complete listing of mammalian hosts.
[c]'Catch-all' genus for microsporidia that could not be classified (Canning and Lom, 1986).

tubes intact whereas incubation with 2-mercaptoethanol or dithiothreitol led to dissociation of the polar filament/tube (Weidner, 1976; Keohane et al., 1996). In one study, Keohane et al. (1996) identified four polar tube proteins of 23, 27, 34, and 43 kDa.

2.2. Anchoring Disk and Polaroplast

The anchoring disk (also called the polar cap) is found in the foremost portion of the spore. The anchoring disc is the most carbohydrate–rich structure in the microsporidia (Desportes-Livage et al., 1996) and is responsible for the acid-fast staining. Just posterior to the anchoring disc is an array of lamellar membranes called the polaroplast. These membranes also are rich in carbohydrates and are continuous with the outer membrane of the polar filament (Vernick et al., 1977; Canning and Lom, 1986). The polaroplast becomes the outer membrane of the sporoplasm, which is the sac that becomes filled with the spore contents after polar filament extrusion (Weidner et al., 1984).

2.3. Nucleus

At the end of the polar filament is the nucleus which exists either singly (e.g. *Encephalitozoon, Enterocytozoon, Pleistophora, Trachipleistophora*) or as a diplokaryon (e.g. *Nosema, Thelohania, Vittaforma*). The diplokaryon consists of two closely aligned nuclei which function as a single unit (Sprague et al., 1992). Nuclei divide with mitotic spindles but lack centrioles (Canning and Lom, 1986; Shadduck and Greeley, 1989). Between 8 (e.g. *Vairimorpha* sp. and *Nosema costelytrae*, which infect various insects) and 16 (e.g. *Glugea atherinae*, which infects fish) chromosomal bands of deoxyribonucleic acid (DNA) have been detected in several species of microsporidia using pulsed-field gel electrophoresis (Munderloh et al., 1990; Malone and McIvor, 1993; Biderre et al., 1994, 1995). The only microsporidian species infecting mammals whose chromosomes have been characterized is *Enc. cuniculi*, which has 11 chromosomal DNA bands yet has the smallest nuclear genome (2.9 Mbp) among the microsporidia with a single nucleus (Biderre et al., 1995). For comparison, the *Enc. cuniculi* haploid genome size is smaller than that of *Escherichia coli* (4.7 Mbp) while the largest microsporidian genome size measured to date is 19.5 Mbp in *G. atherinae* (Biderre et al., 1994, 1995).

2.4. Endoplasmic Reticulum and Ribosomes

Microsporidia are eukaryotes but contain electron-dense ribosomes that resemble those of prokaryotes (Curgy et al., 1980). Vossbrinck and colleagues

(Vossbrinck and Woese, 1986; Vossbrinck et al., 1987) found that *Vairimorpha necatrix* (a microsporidian of butterflies) contains small 70S ribosomes which have 16S and 23S subunits, but lack a separate 5.8S subunit typically found in eukaryotes. The 23S subunit, however, contained sequences of the typical 5.8S subunit. The 16S subunit found in this microsporidian is smaller than the 18S subunit typically found in eukaryotes and there exists little homology between this microsporidian ribosomal ribonucleoprotein (rRNA) gene sequence and those of other eukaryotes.

Sporonts and sporoblasts are particularly rich in smooth and rough endoplasmic reticulum. In *Enterocytozoon bieneusi*, the endoplasmic reticulum arises at the same time as the polar filament discs and appears to aid in the formation and alignment of the polar filaments (Cali and Owen, 1990; Desportes-Livage et al., 1996; Takvorian and Cali, 1994, 1996).

2.5. Posterior Vacuole

The posterior vacuole is formed from Golgi vesicles that coalesce late in sporogony (spore maturation) after generation of the polar filament (Canning and Lom, 1986). The posterior vacuole provides a diagnostic characteristic for detecting microsporidia in clinical specimens using light microscopy methods (Weber et al., 1992a). The function of the posterior vacuole is unknown but it, along with the polaroplast, swells just before germination (Lom and Vávra, 1963; Undeen, 1990; Undeen and Frixione, 1990).

2.6. Spore Wall

The mature spore is surrounded by an outer electron-dense exospore composed primarily of (glyco)proteins and an inner electron-lucent endospore which contains chitin (Canning and Lom, 1986; Weber et al., 1994). The endospore is relatively thin at the anterior region near the anchoring disc. A plasma or unit membrane within the spore wall encloses the spore contents.

The spore wall affords the microsporidian a level of resistance to environmental influences. For example, some spores can remain viable in distilled water for up to 10 years. Shadduck and Polley (1978) tested a variety of conditions on the infectivity of *Encephalitozoon cuniculi* spores and found that they survived at least 9 days when incubated in phosphate-buffered saline at 37°C, more than 24 days when incubated at 4°C or 20°C, and longer than 6 months when stored at −70°C. Spores survived heating to 56°C for 60 minutes but not 120 minutes. In addition, the *Enc. cuniculi* spores survived treatment for 24 hours at pH 9 or 4, and also survived sonication of four 10-mA bursts of 30 seconds as well as five freeze–thaw cycles in liquid nitrogen. *Enc. cuniculi* spores were killed, however, if treated for 10 minutes with 2% (v/v)

Lysol, 10% (v/v) formalin, or 70% (v/v) ethanol. Storage in CsCl (1.35 g/ml) or 40% (w/v) sucrose for 24 hours resulted in loss of *Enc. cuniculi* infectivity *in vitro*.

The spore wall is somewhat plastic, being larger before germination and slightly smaller after germination. The spore wall also provides some measure of resistance against which sufficient pressure can be maintained within the spore to facilitate germination or extrusion of the polar filament (Undeen and Frixione, 1991).

3. LIFE CYCLE

The life cycle of microsporidia in mammalian hosts is relatively simple (Figure 1). After transmission of the microsporidian, infection occurs via

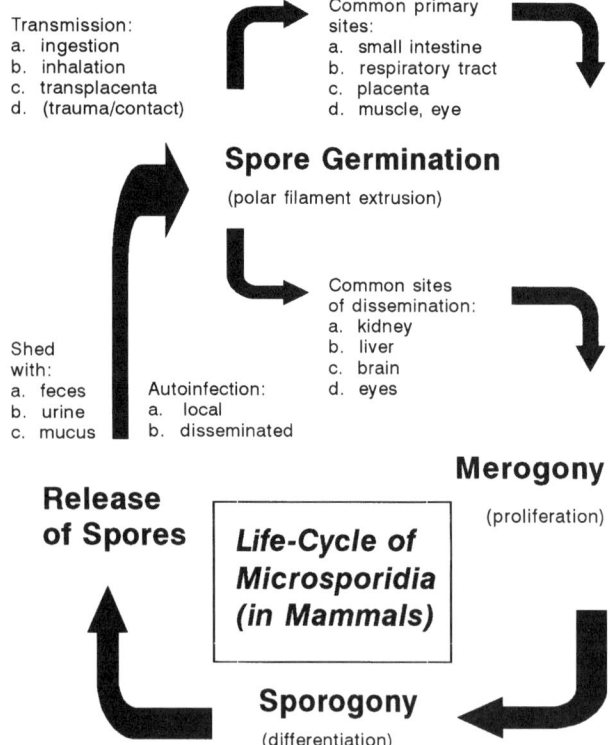

Figure 1 Life cycle of microsporidian species in mammals.

polar filament extrusion, a process called germination. Organisms may also be internalized by phagocytosis, since macrophages are easily infected with some microsporidia (e.g. *Encephalitozoon* spp.). After entering the host cell the microsporidia enter the proliferative phase, called merogony or schizogony, followed by differentiation into spores, which is termed sporogony. Released spores continue to replicate in the host (called autoinfection) either by infecting adjacent cells or disseminating to other tissue sites. Spores are also shed with urine, feces, or respiratory secretions for transmission to new hosts.

3.1. Transmission

The most common routes of infection in mammals are ingestion or inhalation of spores (i.e. horizontal transmission). The intestine is the primary site of many human microsporidian infections with *Enterocytozoon bieneusi* and *Encephalitozoon intestinalis* (see Orenstein *et al.*, 1990; Orenstein, 1991; Kotler and Orenstein, 1995). Inhalation probably occurs since organisms (e.g. *Encephalitozoon* spp. and *Ent. bieneusi*) are present in the sputum, bronchoalveolar lavage fluid, and bronchoalveolar epithelium of infected individuals (Orenstein *et al.,* 1992; Weber *et al.,* 1992b, 1993, 1994; Schwartz *et al.,* 1993). Experimentally, monkeys, mice, rats, rabbits, dogs, and cats have been infected with *Enc. cuniculi* by oral inoculation (Levaditi *et al.,* 1923; Nelson, 1967; Petri, 1966; McCully *et al.,* 1978; Cox *et al.,* 1979; Pang and Shadduck, 1985; Van Dellen *et al.,* 1989). Intranasal transmission of infection with *Enc. cuniculi* has been reported in mice (Perrin, 1943; Nelson, 1967) but, in both cases, infection may have occurred as a result of swallowing or inhaling the organisms (Canning and Lom, 1986). In addition, mice became infected after being caged with infected animals (Liu *et al.,* 1988, 1989), and a child seroconverted after coming into close contact with *Enc. cuniculi*-infected dogs that were shedding organisms in the urine (McInnes and Stewart, 1991).

Evidence for transplacental transmission of *Enc. cuniculi* has been reported in mice, rabbits, blue foxes, domestic dogs, horses, and nonhuman primates (Plowright, 1952; Plowright and Yeoman, 1952; Basson *et al.,* 1966; Anver *et al.,* 1972; Hunt *et al.,* 1972; Nordstoga, 1972; Mohn *et al.,* 1974; 1981, 1982; Shadduck *et al.,* 1978; Zeman and Baskin, 1985; Van Rensburg *et al.,* 1991). Canning and colleagues (Canning and Lom, 1986; Canning and Hollister, 1987), however, commented that transplacental transmission appears to be more common in carnivores than in rodents and, furthermore, stated that transplacental transmission probably depends 'on the time that the infection was acquired in relation to pregnancy and the degree of infection' (Canning and Hollister, 1987).

Other less common routes of infections also have been described. Experimentally, rabbits could be infected with *Enc. cuniculi* given intrarectally (Fuentealba *et al.,* 1992). It is likely that ocular infections may be transmitted by contact with contaminated hands, and hand washing has been recommended to reduce spread of infection (Bryan, 1995). It is unknown how cases of *Nosema* and *Pleistophora* were transmitted to the corneal stroma and skeletal muscles, respectively, in humans, but trauma may be a possible mechanism for transmission (Ledford *et al.*, 1985; Davis *et al.*, 1990; Chupp *et al.*, 1993).

3.2. Germination

Under appropriate conditions, the microsporidian spores will germinate and infect host cells. Because microsporidia, particularly *Encephalitozoon* spp., can replicate in macrophages, it is possible that organisms may be phagocytozed before germination. *In vitro*, discharge occurs in less than 1 second if the pH of the surrounding medium changes from acid/neutral to alkaline (Lom and Vávra, 1963; Weidner, 1972, 1982). Both the polaroplast (the membranous structure contiguous with the anchoring disc) and the posterior vacuole swell just before germination (Lom and Vávra, 1963; Undeen, 1990; Undeen and Frixione, 1990). Osmotic pressure builds to approximately 6060 kPa (60 atm) as water is drawn into the spore (Pleshinger and Weidner, 1985; Undeen and Frixione, 1991), although the mechanism by which osmotic pressure increases to cause polar filament extrusion is not completely understood. Monovalent ions move passively through the spore wall and do not appear directly to cause the swelling needed to create sufficient pressure for filament extrusion. More likely, trehalose, which exists in relatively high concentrations in the nongerminated spore, is converted to glucose which could then exert sufficient osmotic pressure to draw water into the spore (Undeen, 1990; Undeen and Frixione, 1990).

Ca^{2+} also contributes to polar filament discharge. Ca^{2+} ions easily traverse the spore wall through ion channels, and bind to receptors on the spore wall and spore plasma membrane (Pleshinger and Weidner, 1985). In addition, Weidner and Byrd (1982) observed that $^{45}CaCl_2$ became incorporated into the membranes of expanded, but not contracted, polaroplasts. Addition of the calcium ionophore A21837 induced spore discharge (Weidner and Byrd, 1982), whereas incubation with the ion chelator ethylene glycol-bis (β-aminoethyl ether) (EGTA) blocked it (Pleshinger and Weidner, 1985). Furthermore, Leitch *et al.* (1993a,b) suggested that Ca^{2+} and its interaction with cytoskeletal components influenced spore discharge because treatment with the microtubule inhibitor demecolcine, the microfilament disruptor cytochalasin D, or the Ca^{2+} channel blocker nifedipine, inhibited *Encephalitozoon hellem* polar filament discharge.

3.3. Merogony

The process of merogony, or schizogony, is generally described as the proliferative phase of development (Figure 2). Sprague et al. (1992) suggested that the term 'merogony' be applied to proliferation of diplokaryotic stages while 'schizogony' should be applied to stages with unpaired single nuclei. The term merogony is used more commonly to describe the proliferative phase in general and will be so used here. Merogony occurs following sporoplasm entry into a susceptible host cell. The early stages, referred to as meronts, may have irregular membrane surfaces and tend to be larger and more rounded than the mature spore, although dividing forms may appear elongated. In general, meronts contain poorly organized rough and smooth endoplasmic reticulum. Division occurs either by binary fission (e.g. *Encephalitozoon, Nosema, Thelohania, Vittaforma*), or karyokinesis may occur before delayed cytokinesis, resulting in multinucleated forms called merogonial plasmodia (e.g. *Enterocytozoon, Pleistophora, Trachipleistophora*). Merogony may occur in direct contact with the host cell cytoplasm (sometimes referred to as 'in the hyaloplasm'; e.g. *Nosema, Enterocytozoon*), within a host-cell-derived parasitophorous vacuole (e.g. *Encephalitozoon*), within an amorphous surface coat deposited by the parasites (e.g. *Pleistophora, Trachipleistophora, Thelohania*), or with the individual organisms intimately surrounded by endoplasmic reticulum (e.g. *Vittaforma*) (Weidner, 1975; Issi, 1986; Canning and Hollister, 1992; Sprague et al., 1992; Silveira and Canning, 1995; Field et al., 1996; Hollister *et al.*, 1996).

3.4. Sporogony

The process of sporogony begins as the outer surface of the developing organism starts to become more electron dense (Canning and Lom, 1986). These stages are referred to as sporonts and may divide a limited number of times by binary fission (e.g. *Encephalitozoon, Nosema, VIttaforma*), sometimes so rapidly that chains of four or eight connected organisms result. Multinucleated sporonts may develop within what are referred to as sporogonial plasmodia (e.g. *Enterocytozoon, Pleistophora*). *Trachipleistophora* has multinucleated merogonial plasmodia that undergo plasmotomy to form uninucleated late meronts, so that only uninucleated sporonts are seen in this genus. Of the microsporidia infecting mammals, only *Thelohania* spp. undergo meiosis during sporogony (see Sprague et al., 1992). When sporogony results in the deposition or formation of parasite-generated sporophorous vesicle, the microsporidia are characterized as pansporoblastic (e.g. *Pleistophora, Trachipleistophora, Thelohania*). The microsporidia which undergo

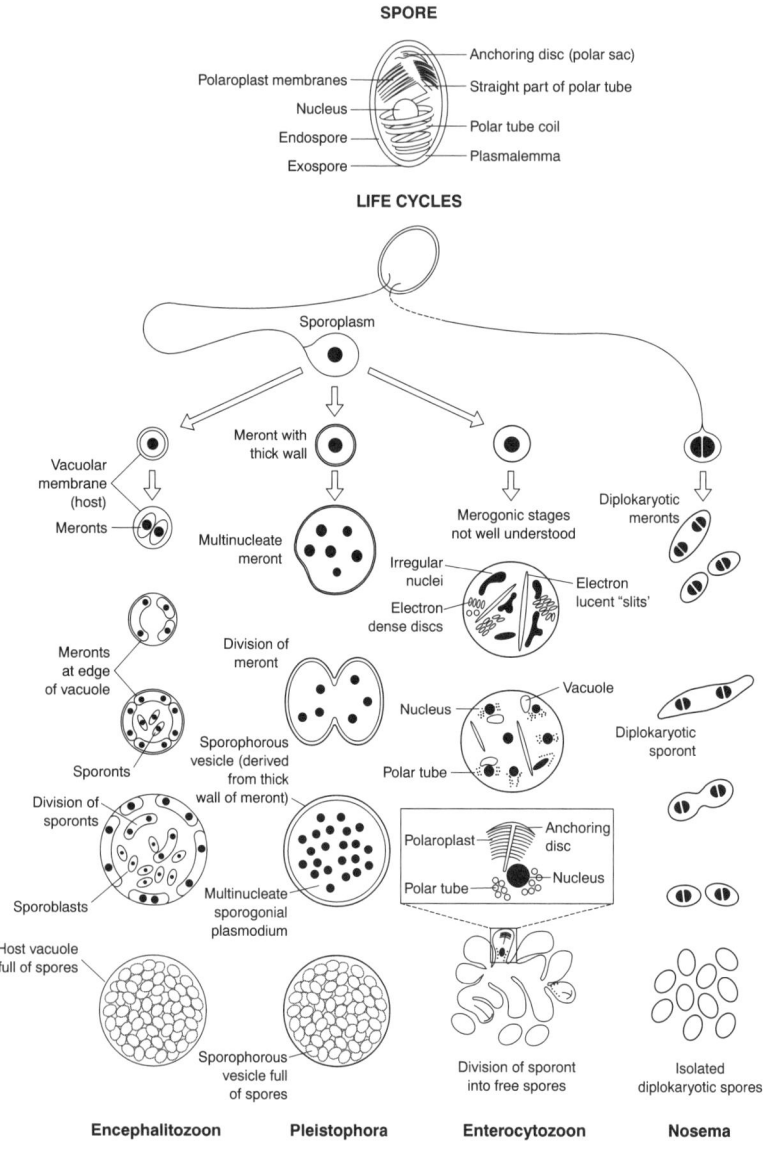

Figure 2 Diagram of microsporidian development. *Vittaforma corneae* replication is similar to that of *Nosema*, except that host cell endoplasmic reticulum surrounds each of the developing organisms and 4–8 sporoblasts are generated from each sporont. *Trachipleistophora hominis* replication is pansporoblastic, which is similar to that of *Pleistophora*, except that the sporophorous vesicles are not multilayered as they are in *Pleistophora*. In addition, sporonts of *Tr. hominis* are uninucleated. Reproduced by permission from E.U. Canning and W.S. Hollister (1992). Human infections with microsporidia. *Reviews in Medical Microbiology* **3**, 35–42.

sporogony in the absence of a parasite-derived outer coat are characterized as apansporoblastic (e.g. *Encephalitozoon, Enterocytozoon, Vittaforma, Nosema*). Stages in which organelles can be recognized are referred to as sporoblasts. Smooth and rough endoplasmic reticulum increase and produce Golgi-like vesicles which appear to generate the polar filament. Further development of the spore wall can be discerned as the electron-dense exospore becomes distinguishable from the electron-lucent endospore. Mature spores develop from sporoblasts as a result of bisporogony (binary fission), tetrasporogony (in which a second binary fission occurs before the first binary fission was completed), or polysporogony in which many spores separate from a multinucleated sporogonial plasmodium (Canning and Lom, 1986; Canning and Hollister, 1987; Van Gool *et al.*, 1994). Whereas meronts, sporonts, and sporoblasts display metabolic activities, little metabolic activity has been detected in mature spores.

3.5. Secondary Cycles of Infection

After the microsporidia replicate to a point when the host cell is no longer able to contain the parasites, the host cell ruptures to release spores and immature organisms. The mature spore is considered to be the infectious stage, although less mature forms may be infectious or may have developed sufficiently to complete differentiation outside the host cell and thus later become infectious. Organisms are either shed with the stool or with respiratory secretions or autoinfect the host (either locally or by disseminating). The microsporidia enter new cells by injecting the spore contents directly into the host cells, a process called germination, or the host cell may phagocytose the microsporidia, after which germination occurs. *Enterocytozoon, Pleistophora*, and *Nosema* remain relatively localized in the small intestine, skeletal muscle, and corneal stroma, respectively, although some exceptions have been reported (see Orenstein, 1991; Weber *et al.*, 1994). *Encephalitozoon* species generally disseminate (Orenstein, 1991; Weber *et al.*, 1994), probably by infecting macrophages (Canning and Lom, 1986). Secondary sites of infection are commonly the kidney, liver, spleen, brain, peritoneum, and nasal sinuses. After the kidneys become infected, microsporidia are shed with the urine, providing another means of transmitting the infection to new hosts. Essentially, all tissues appear to be vulnerable to infection with at least one genus of microsporidia (Orenstein, 1991; Kotler and Orenstein, 1995; Yachnis *et al.*, 1996).

3.6. Growth *In Vitro* of Microsporidia

Several species of microsporidia that infect mammals can be grown *in vitro* (see Table 1). *Enc. cuniculi* was the first mammalian microsporidian to be grown in culture (Shadduck, 1969) and several strains of *Enc. cuniculi* were subsequently grown in culture and identified (Shadduck *et al.*, 1978; Stewart *et al.*, 1979; Hollister *et al.*, 1995; Didier *et al.*, 1995). In addition, *Enc. hellem, Enc. intestinalis, Vittaforma corneae*, and *Trachipleistophora hominis* can be grown in culture (Shadduck *et al.*, 1990; Didier *et al.*, 1991a,b; 1996b; Silveira *et al.*, 1993; Hartskeerl *et al.*, 1995; Visvesvara *et al.*, 1995a; Hollister *et al.*, 1996), whereas attempts to culture *Ent. bieneusi* have met with only limited success (Visvesvara *et al.*, 1995b). The most commonly utilized host cells for growing microsporidia include primary and established kidney fibroblasts (eg. RK-13, MDCK) from rabbits and canines, but a wide range of epithelial cell lines will support growth of microsporidia (Montrey *et al.*, 1973; Canning and Lom, 1986; Hartskeerl *et al.*, 1995; Didier *et al.*, 1996b).

4. TAXONOMY

As discussed by Sprague *et al.* (1992) and Baker *et al.* (1995), the taxonomy of the microsporidia does not necessarily correlate with their evolution, primarily due to lack of sufficient information. What is known about the molecular phylogeny of the microsporidia is discussed by Weiss and Vossbrinck. Microsporidia are considered to be early or primitive eukaryotes because they have a nucleus, contain prokaryote-like ribosomes, and lack mitochondria and peroxisomes (Vossbrinck and Woese, 1986; Canning and Hollister, 1987; Vossbrinck *et al.*, 1987; Shadduck and Greeley, 1989).

The microsporidia were first placed into the phylum Microspora by Sprague (1969, 1977), and were further classified by the Committee on Systematics and Evolution of the Society of Protozoologists in 1980 (Levine *et al.*, 1980). The most recent complete review of the taxonomy of the microsporidia was published by Sprague *et al.* (1992), who divided them into two classes: the Dihaplophasea, which contains species with a diplokaryon during some stage of development, and the Monohaplophasea, which contains species with a single nucleus during all stages of development. As described by Sprague *et al.* (1992), these higher categories of microsporidia were grouped according to chromosome cycle. In earlier taxonomic classifications, however, the microsporidia were divided into the two orders Pansporoblastina and Apansporoblastina, based on the presence or absence, respectively, of the sporophorous vesicle (Sprague, 1982; Canning and Lom, 1986; Bryan *et al.*, 1990), and these categories are still commonly used when describing new species of microsporidia that infect mammals. In the most recent taxonomic description

Table 2 Taxonomic classification of microsporidian species that infect mammals.[a]

Phylum	Microspora						
Class	Dihaplophasea				Haplophasea		
Order	Meiodihaplophasida	Dissociodihaplophasida			Glugeida		
Superfamily	Thelohanioidea	Nosematoidea					
Family	Thelohanidae	Nosematidae	Not placed into higher taxa		Pleistophoridae	Encephalitozoonidae	Enterocytozoonidae
Genus	*Thelohania*	*Nosema*	*Vittaforma*	*Trachipleistophora*	*Pleistophora*	*Encephalitozoon*	*Enterocytozoon*
Species	*Th. apodemi* *Thelohania*-like sp.	*N. connori* *N. ocularum* *Nosema*-like sp.	*V. corneae*	*Tr. hominis*	*Pleistophora* sp.	*Enc. cuniculi* (I, II, III) *Enc. hellem* *Enc. intestinalis*	*Ent. bieneusi*

[a] This table is based on the taxonomy of Sprague *et al.* (1992). These microsporidia also are commonly classified in the Apansporoblastina (families Nosematidae, Encephalitozoonidae, and Enterocytozoonidae, and genus *Vittaforma*) and the Pansporoblastina (families Thelohanidae and Pleistophoridae, and genus *Trachipleistophora*).

by Sprague *et al.* (1992), new orders were established, and families were defined on the basis of host–parasite relationships, characteristics of sporogony, and spore morphology. The taxonomic classification of microsporidia known to infect mammals used by Sprague *et al.* (1992) is shown in Table 2. Assignments into genera, species, and strains (subspecies) are still being debated as new information is assimilated. The combined characters typically used to classify microsporidia into genera, however, include the type of host, mode of transmission, the host–parasite interface, microsporidian development (merogony and sporogony), and spore morphology (size, nuclear arrangement, number of turns of the polar filament); these characters are given below for each genus containing microsporidia that infect mammals.

4.1. Enterocytozoon

To date, *Ent. bieneusi* is the only species in the genus (*Enterocytozoon salmonis* was recently transferred to *Nucleospora salmonis* by Kent *et al.*, 1996). The type host for *Ent. bieneusi* is humans. *Ent. bieneusi* is probably the most commonly detected microsporidian in persons with AIDS (Orenstein *et al.*, 1990; Orenstein, 1991; Asmuth *et al.*, 1994; Weber *et al.*, 1994) and it has been recorded only in humans, with one exception: Deplazes *et al.* (1996) reported it in pigs. Transmission of *Ent. bieneusi* infections generally occurs via the oral–fecal route, as spores are shed with the feces of infected hosts. Inhalation may also occur (Weber *et al.*, 1992b), while vertical (transplacental) transmission has not been reported for this organism.

The primary site of infection is the enterocyte of the small intestine (jejunum and duodenum) in which the organism develops between the host cell nucleus and the intestinal lumen. The organisms replicate in direct contact with the host cell cytoplasm and develop as multinucleated meronts and then sporonts (also referred to as merogonial and sporogonial plasmodia, respectively). The plasmodia contain electron-lucent slits which are believed by Canning and Hollister (1992) to be expanded endoplasmic reticulum. At the host–parasite interface there are numerous host cell mitochondria, which appear to provide nutrients to the developing microsporidia (Kotler and Orenstein, 1995). Sporogony begins as polar filaments form from electron-dense discs and surround the randomly spaced nuclei (Desportes *et al.*, 1985; Cali and Owen, 1990; Desportes-Livage *et al.*, 1996). The nuclei remain single throughout the development of *Ent. bieneusi*. The sporogonial plasmodium eventually separates into individual organisms which develop exospores as they mature into spores.

Spores are typically shed with the feces after being released into the intestinal lumen, or they may infect adjacent cells of the small intestine. Occasionally, *Ent. bieneusi* infects the gallbladder, pancreas, or liver, but generally infections remain localized in the small intestine (see Orenstein, 1991; Weber et al., 1994). The mature *Ent. bieneusi* spore measures only 1.5 × 1.0 µm, making it the smallest known microsporidian. The polar filament coils approximately six times in two rows. The endospore and exospore are relatively thin compared with those of other microsporidia (Canning and Hollister, 1992; Sprague et al., 1992).

4.2. Encephalitozoon

The genus *Encephalitozoon* presently contains three species: *Enc. cuniculi*, the type species; *Enc. hellem*; and *Enc. intestinalis* (first named *Septata intestinalis* Cali, Kotler and Orenstein 1983). *Enc. cuniculi* infects a wide range of mammalian hosts, including rodents, rabbits, dogs, blue foxes, non-human primates, and man. *Enc. cuniculi* is further divided into three strains (Didier et al., 1995). Strain I was originally isolated and cultured from rabbits (Shadduck, 1969). Strain II came originally from mice but was later also identified in Norwegian blue foxes (*Alopes lagopus*) (Didier et al., 1995; Mathis et al., 1996), and strain III originated from domestic dogs (Shadduck et al., 1978). These strains of *Enc. cuniculi* are not particularly host specific, human infections of *Enc. cuniculi* in Switzerland are of strain I, while those in the USA are of strain III (Deplazes et al., 1996; Didier et al., 1996c). *Enc. hellem* had been reported only in humans until recently, when it was detected in budgerigars (*Melopsittacus undulatus*) in the USA (Black et al., 1997). To date, *Enc. intestinalis* has been reported in humans only (Didier et al., 1996b). Spores of the three species of *Encephalitozoon* usually cannot be distinguished morphologically, but are separable by immunological, biochemical, and molecular characteristics (Didier et al., 1991a,b, 1995, 1996a,b,c; Schwartz et al., 1992; Vossbrinck et al., 1993; Hartskeerl et al., 1995; Visvesvara et al., 1995a, b).

The most common modes for transmitting infections with *Encephalitozoon* spp. are ingestion and inhalation. Transplacental transmission has been reported for *Enc. cuniculi* and occurs more often in canines than in rodents (Canning and Lom, 1986; Canning and Hollister, 1987). It is possible, though not proven, that ocular *Enc. hellem* or *Enc. intestinalis* infections in humans are transmitted by contact with aerosols, or perhaps by contact with hands contaminated with urine, because the eye infections are usually associated with the presence of spores in urine (Schwartz et al., 1993; Bryan, 1995).

Members of *Encephalitozoon* replicate within a parasitophorous vacuole which is believed to originate from host cell endoplasmic reticulum (Issi, 1986) and which often becomes closely surrounded by host cell mitochondria (Shadduck and Pakes, 1971). In addition to infecting epithelial and endothelial cells, *Encephalitozoon* spp. can replicate within macrophages, suggesting that organisms may be phagocytosed and replicate within the phagocytic vacuole. Meronts divide repeatedly by binary fission in close association with the limiting membrane of the parasitophorous vacuole. Sporogony begins as the outer surface of the organism thickens and organelles begin to develop. Sporonts are often localized in the central region of the parasitophorous vacuole and divide to produce two sporoblasts (bisporogony) (Cali, 1991; Sprague *et al.*, 1992; Cali *et al.*, 1993; Van Gool *et al.*, 1994), although tetrasporogony occurs occasionally in all three species (Vávra *et al.*, 1972; Cali *et al.*, 1993; Van Gool *et al.*, 1994; Canning *et al.*, 1994; Didier *et al.*, 1991b). The parasitophorous vacuoles of *Enc. intestinalis* are filled with fibrillar material that surrounds each of the developing organisms in a honeycomb-type matrix (Cali *et al.*, 1993), while the parasitophorous vacuoles of *Enc. cuniculi* and *Enc. hellem* contain substantially less, or no, fibrillar material (Weber *et al.*, 1994); this material has not yet been characterized.

Eventually, the parasitophorous vacuole and host cell membranes rupture to release spores. In some reports describing development of *Enc. hellem* and *Enc. intestinalis*, the parasitophorous vacuole membrane ruptures before the host cell plasma membrane, so that organisms appear free in the host cell cytoplasm (Desser *et al.*, 1992; Canning *et al.*, 1992; Cali *et al.*, 1993). This finding has been used putatively to identify new *Encephalitozoon* spp, but in all these cases the host cell organelles appeared abnormal or degraded. This suggested to Van Gool *et al.* (1994) that care must be taken to characterize intracellular development of microsporidia in well-preserved tissues. Spores may be released with the feces early after infection (Orenstein, 1991; Cali *et al.*, 1993; Franzen *et al.*, 1996), but are usually seen in the urine after dissemination, probably via trafficking within monocytes/macrophages (Canning and Lom, 1986). Secondary sites of infection typically include the kidney, liver, spleen, and brain, but *Encephalitozoon* spp. may infect almost any site (Orenstein, 1991; Orenstein *et al.*, 1992; Weber *et al.*, 1994; Kotler and Orenstein, 1995).

Mature spores of *Encephalitozoon* spp. are slightly larger than those of *Ent. bieneusi*, and measure approximately 1.5–2.0 μm × 2.5–3.0 μm. The polar filament typically coils from three to six times in *Enc. cuniculi* and four to seven times in *Enc. hellem* and *Enc. intestinalis*.

4.3. Pleistophora

Pleistophora typicalis is the type species of this genus (Canning and Hazard, 1982). Species of *Pleistophora* often infect the skeletal muscles of fish, and a few infections have been reported in humans (Ledford *et al.*, 1985; Chupp *et al.*, 1993; Hollister *et al.*, 1996). Meronts secrete an amorphous surface coat that is permeated by channels and undergo nuclear division resulting in multinucleated meronts (Canning and Nicholas, 1980). The outer coat divides with the meronts to generate plasmodia with variable numbers of nuclei.

Sporogony begins as the outer coat develops three layers and it is then referred to as the sporophorous vesicle (Canning and Nicholas, 1980). Sequential divisions generating multinucleated sporogonial plasmodia continue (a process called plasmotomy) until each sporoblast contains one nucleus. *Pleistophora* spp. are considered pansporoblastic because of this outer coat surrounding the organisms. These microsporidia undergo polysporogony to generate many sporoblasts from a single sporogonial plasmodium (Canning and Hazard, 1982; Sprague *et al.*, 1992). The nuclei remain unpaired at all stages.

The mode of transmission is not well understood, since the primary site of infection in mammals is skeletal muscle (Ledford *et al.*, 1985; Sprague *et al.*, 1992; Chupp *et al.*, 1993). Infections within a host appear to spread to adjacent or nearby cells. Mature spores measure approximately 3.5–4.4 µm × 2.3–2.8 µm and contain approximately 10–12 turns of the polar filament (Canning and Hollister, 1992; Sprague *et al.*, 1992).

4.4. Trachipleistophora

Tr. hominis is the only species in this genus, which was named from the Greek '*trachis*', meaning 'rough', due to the surface coat which covers all stages of this microsporidian, and 'pleistophora' from its similarities with the genus *Pleistophora* (see Hollister *et al.*, 1996; Field *et al.*, 1996). *Tr. hominis* was isolated and cultured from a person with AIDS and no other host has been identified at present. Sites of infection include skeletal muscle, corneal epithelium, and nasopharyngeal tissues. The transmission mode is not known.

Meronts contain two or more nuclei within an electron-dense outer surface coat which bears branched processes (unlike *Pleistophora*). The meronts divide by binary fission or plasmotomy until each organism contains one nucleus. Sporogony continues as the surface coat develops into the sporophorous vesicle, which is not multilayered (as it is in *Pleistophora*) and expands to contain the sporonts, which divide by binary fission. Multinucleated sporogonial forms do not develop in *Trachipleistophora* (as opposed to *Pleistophora*).

The sporophorous vesicle separates to generate mature spores which measure 4.0 × 2.4 µm. The exospore is described by Hollister *et al.* (1996) as being slightly rugose and the endospore is relatively thick. The polar filament coils approximately 11 times, and measures 0.1 µm wide and at least 75 µm long when extruded. *Tr. hominis* has not yet been placed into any higher taxon.

4.5. Nosema

The type species of this genus is *Nosema bombycis* Naegeli, 1857. Members of the genus primarily infect insects, but two species have been identified in humans. *N. connori* caused a disseminated infection in an athymic child (Margileth *et al.*, 1973; Sprague *et al.*, 1974) and *N. ocularum* was described as infecting the corneal stroma of an individual who was seronegative for the human immunodeficiency virus (HIV) (Bryan *et al.*, 1990; Cali *et al.*, 1991a). More recently, a *Nosema*-like microsporidian was reported as causing myositis in an AIDS patient (Cali *et al.*, 1996). The mode of transmission in mammals is not known.

Nosema is characterized by having a diplokaryon in all stages of development. Meronts divide numerous times by binary fission while in direct contact with the host cell cytoplasm. During sporogony, the microsporidian plasma membrane thickens, and sporonts divide to produce two sporoblasts; the species is therefore considered disporoblastic. Replication continues until the host cell ruptures, releasing organisms that infect adjacent cells. Mature spores of *N. connori* and *N. ocularum* are 4.0–5.0 μm × 2.0–3.0 μm in size and contain 9–12 coils of the polar filament (Margileth *et al.*, 1973; Bryan *et al.*, 1990; Cali *et al.*, 1991b).

4.6. Vittaforma

This genus was established after the microsporidian *Nosema corneum* Shadduck, Meccoli, Davis and Font, 1990 was more completely characterized and found to differ sufficiently from the diagnostic characters assigned to *Nosema* (see Silveira and Canning, 1995). This organism, *Vittaforma corneae*, has been reported to infect the corneal stroma of an HIV-seronegative individual (Davis *et al.*, 1990). The mode of transmission is unknown.

Diplokaryotic meronts are scattered throughout the host cell cytoplasm and divide by binary fission, but, instead of being in direct contact with host cell cytoplasm (like *Nosema*), individual *Vittaforma* meronts and sporonts are surrounded or enveloped by host cell endoplasmic reticulum studded with ribosomes (unlike *N. bombycis*). Sporogony begins as the outer membrane thickens and is polysporoblastic, resulting in the generation of four to eight sporoblasts arranged linearly. The nuclei form diplokarya in all stages (Silveira and Canning, 1995). Mature spores measure 3.8 × 1.2 μm and display five to seven turns of the polar filament. *V. corneae* will not be placed into a family or higher taxon until additional information (e.g. the 16S RNA gene sequence) becomes available (Silveira and Canning, 1995).

4.7. Thelohania

The type species of this genus is *Thelohania giardi* Henneguy, 1982. Species of *Thelohania* primarily infect mosquitoes and marine animals (shrimp and blue crabs), and only rarely infect mammals (Vernick *et al.*, 1977; Sprague *et al.*, 1992). *Th. apodemi* Doby, Jeannes and Rault, 1963 was described in brains and skeletal muscles of field mice (Canning and Hollister, 1987), and *Thelohania*-like organisms were described in the brain, skeletal muscle, liver, and kidney of two individuals with AIDS (Yachnis *et al.*, 1996). The mode of transmission of *Thelohania* infections in mammals is unknown (Sprague *et al.*, 1992).

The life cycle of *Thelohania* is more complicated than that of the other microsporidian species infecting mammals, as both sexual and asexual stages develop. Meronts (diplokaryons) divide by binary division, and sporogony occurs within a pansporoblastic membrane or envelope. Sporonts undergo meiosis and sporogony is octosporoblastic. Sporonts are diplokaryotic although the zygote stage is unikaryotic. A complete taxonomic description of the genus is given by Sprague *et al.* (1992). Mature spores of *Th. apodemi* are pyriform, measure 5–6 µm long, have relatively thick endospores, and are often released from the host cell while still contained within the sac that originally served as the sporophorous vesicle (Sprague *et al.*, 1992; Yachnis *et al.*, 1996).

4.8. *Microsporidium*

Microsporidia which are incompletely characterized for classification into new or established genera are typically placed in the 'catch-all' genus, *Microsporidium*. Several such microsporidia have been reported in mammals (see Canning and Lom, 1986; Cali, 1991; Weber *et al.*, 1994).

5. MAMMALIAN HOST–PARASITE RELATIONSHIPS

Among animals, microsporidia most commonly infect rabbits, rodents, and young carnivores, but rarely infect ruminants (Shadduck and Orenstein, 1993). Humans are believed to be most at risk if immunologically compromised (e.g. suffering from AIDS). The most complete lists of mammalian hosts with microsporidiosis are given by Canning and Lom (1986) and Canning and Hollister (1987). Much of what is known about mammalian microsporidiosis is based upon clinical observations in human patients and experimental studies with *Enc. cuniculi*. *Enc. cuniculi* was the first microsporidian found to infect mammals (Wright and Craighead, 1922; Levaditi *et al.*, 1923); it probably has the widest host range among mammals (Canning and Lom, 1986; Canning and Hollister, 1987), and was the first mammalian microsporidian to be grown in culture (Shadduck, 1969). Three types of host–parasite relationships exist in mammals infected with microsporidia: (1) young hosts who develop acute, and often lethal disease; (2) immunologically competent adults who develop chronic, subclinical infections; and (3) immunologically deficient hosts who develop clinically significant infections that may be lethal.

Transplacentally transmitted *Enc. cuniculi* infections in domestic dogs, Norwegian blue foxes, and squirrel monkeys are often lethal. These animals develop neurological disorders including ataxia, paralysis, and convulsions and may die from wasting or cerebral hemorrhage, often within a few months (see Canning and Lom, 1986; Canning and Hollister, 1987). Initially, the parasites appear to replicate because the host's immune system is immature. In cases where the infections become protracted, parasite burdens appear to decrease as the immune system matures. Blue foxes and domestic dogs then develop hypergammaglobulinemia and hypersensitivity responses (type III), which appear to contribute to renal failure (Mohn and Nordstoga, 1975; Arneson and Nordstoga, 1977). Vertical transmission is rare in rabbits and rodents and, to date, has not been proven in humans.

Most cases of microsporidiosis occur as a result of horizontal transmission and, if the hosts are immunologically competent, mild acute infections develop that resolve clinically but persist as chronic clinically silent infections. Such infections typically occur in rabbits and rodents. Organisms may be shed with the urine and, at necropsy, the parasites may be seen either contained within host cells or, sometimes, scattered within randomly distributed granulomatous lesions. Chronically infected rabbits and rodents may also be identified by the presence of specific serum antibodies (Shadduck and Pakes, 1971; Shadduck, 1989; Shadduck and Greeley, 1989).

Microsporidial infections in immunodeficient animals and humans are usually lethal, as the microsporidia replicate without restraint. Athymic and SCID mice suffer lethal disease when inoculated with *Enc. cuniculi*, *Enc. hellem*, or *V. corneae*, and interferon-γ receptor 'knockout' mice develop clinical signs after infection with *Enc. intestinalis* (Gannon, 1980; Schmidt and Shadduck, 1983, 1984; Koudela *et al.*, 1993; Silveira *et al.*, 1993; Didier *et al.*, 1994; Achbarou *et al.*, 1996). Persons with HIV infections and CD4+ T cell levels below 100/mm^3 are particularly susceptible to microsporidiosis (Orenstein, 1991; Weber *et al.*, 1994; Kotler and Orenstein, 1995). Whether such infections in persons with AIDS are due to reactivation of previous chronic infections or to new exposures has yet to be determined.

5.1. Parasite Survival

In immunologically competent hosts such as rabbits and mice, microsporidia are still able to persist and cause chronic infections (Shadduck and Pakes, 1971; Cox *et al.*, 1979). The mechanisms by which microsporidia survive are not well characterized, but they do affect host cell functions. The various types of host–parasite interface provide a level of protection from host cell lysosomes and also must allow for the influx of metabolic products to sustain the microsporidia. Cells infected with microsporidia appear to have more ribosomes and endoplasmic reticulum, suggesting that the microsporidia induced an increase in host cell metabolism (Silveira and Canning, 1995). Furthermore, host cell mitochondria seem to increase in number and are found in juxtaposition to the replicating microsporidia (Shadduck and Pakes, 1971; Kotler and Orenstein, 1995). Envelopment of some microsporidian species with host cell endoplasmic reticulum may be protective due to interference with recognition signals (Silveira and Canning, 1995) and, in other cases, the microsporidia are able to block phagosome acidification by translocation of pH-sensitive receptors, thereby preventing phagosome–lysosome fusion (Weidner and Sibley, 1985). Host cells infected with *V. corneae in vitro* become multinucleated, suggesting that some microsporidia may be able to arrest host cell cytokinesis (Silveira and Canning, 1995). Furthermore, *Enc. cuniculi* can cause short-term suppression of immune responses in mice (Cox, 1977; Niederkorn *et al.*, 1981; Didier and Shadduck, 1988).

5.2. Mechanisms of Host Resistance

The immune response appears to be critical in controlling microsporidiosis in animals and humans. Immunologically competent animals (Shadduck, 1989; Shadduck and Greeley, 1989) and humans (Sandfort *et al.*, 1994; Sowerby *et al.*, 1995; Sobottka *et al.*, 1995) may show some signs of disease, but generally resolve infections with microsporidia. Infections with *Enc. cuniculi, Enc. hellem,* and *V. corneae* are lethal, however, in immunodeficient hosts such as athymic and SCID mice (Gannon, 1980; Schmidt and Shadduck, 1983, 1984; Koudela *et al.*, 1993; Silveira *et al.*, 1993). Persons with AIDS, particularly as immunodeficiency progresses, are also susceptible to microsporidiosis (Bryan *et al.*, 1990; Orenstein, 1991; Weber *et al.*, 1994; Kotler and Orenstein, 1995). Conversely, hypersensitivity responses to microsporidia may contribute to disease. Blue foxes and domestic dogs with protracted microsporidiosis display hypergammaglobulinemia, and the resulting immune complexes subsequently contribute to renal failure (Mohn and Norsdstoga, 1975; Arneson and Nordstoga, 1977; Mohn *et al.*, 1982; Stewart *et al.*, 1986). A well-regulated immune system therefore appears to be critical for maintaining a balanced host–parasite relationship in mammalian hosts of microsporidia.

5.2.1. *Humoral Immune Responses*

In immunologically competent laboratory animals (mice, rhesus monkeys), immunoglobulin M (IgM) responses develop within 2 weeks, while IgG levels reach a peak after about 5–6 weeks and generally persist for the life of the host (Shadduck and Pakes, 1971; Schmidt and Shadduck, 1983; Liu *et al.*, 1988, 1989; Didier *et al.*, 1994). The continued presence of high levels of specific antibodies has been useful for antemortem diagnosis of infected animals, which permits their removal from animal colonies (Cox *et al.*, 1977; Shadduck, 1989; Shadduck and Greeley, 1989). Injection of mice expressing high levels of serum antibodies to *Enc. cuniculi* with hydrocortisone resulted in re-expression of clinical microsporidiosis (Innes *et al.*, 1962), and inoculation of mice with heat-killed *Enc. cuniculi* resulted in low antibody levels which declined rapidly (Liu and Shadduck, 1988), providing evidence that antibody expression parallels parasite persistence in animals.

Immunologically competent humans have been reported to express antibodies to *Enc. cuniculi* (Bergquist *et al.*, 1984; Hollister *et al.*, 1991), but it was not possible to determine if the presence of antibodies correlated with true infection because microsporidia were not observed directly. Recently, microsporidia were detected in individuals who were not infected with HIV (Sandfort *et al.*, 1994; Sobottka *et al.*, 1995). No information is presently available, however, as to whether these individuals remained chronically (subclinically) infected and whether they continued to produce specific antibodies.

Immunodeficient animals such as SCID mice produced no detectable antibody after infection with *Enc. cuniculi* (see Hermanek *et al.*, 1993). In rhesus monkeys with simian immunodeficiency virus (SIV) infection and HIV-infected humans who are also infected with microsporidiosis, antibody levels were variable and of little diagnostic value (Didier *et al.*, 1991c, 1993, 1994). In addition, SIV-infected rhesus monkeys who were already immunodeficient (on the basis of CD4+/CD29+ T cell levels falling below 10%) at the time of *Enc. cuniculi* or *Enc. hellem* inoculation failed to produce specific antibodies above an enzyme-linked immunosorbent assay (ELISA) titer of 1 : 800 (a specific ELISA titer ≥1 : 800 was considered by Hollister *et al.* (1991) to be consistent with microsporidiosis in immunologically competent hosts). If these SIV-infected monkeys were inoculated with microsporidia before becoming immunodeficient, high levels of specific antibodies were generated, but these declined approximately 1 month before death (Didier *et al.*, 1994, and unpublished observations).

HIV-infected individuals with proven microsporidian infections, as well as HIV-seronegative individuals with no history of microsporidiosis, produced variable antibody responses to microsporidia (Didier *et al.*, 1991a, b, 1993). This variability is believed to depend in part upon the immune status of the host when the microsporidian infection first occurred. Some individuals expressed high levels of specific antibodies (ELISA titer ≥1 : 800) 2 weeks before death, while others developed only low antibody levels (Didier *et al.*, 1991a, b, and unpublished observations). Conversely, casual exposure to microsporidia through ingestion of food or by insect stings (McDougall *et al.*,

1993), or the presence of subclinical infection, may have accounted for the relatively high expression of specific antibodies in immunologically competent persons with no known history of microsporidiosis.

Two protective functions have been associated with specific antibodies. Antibodies generated against microsporidia may function as opsonins to facilitate macrophage-mediated phagocytosis. *Enc. cuniculi* organisms treated with normal rabbit serum were taken up by lagomorph peritoneal macrophages and the resulting phagocytic vacuoles failed to fuse with the lysosomes. However, phagocytic vacuoles containing microsporidia treated with immune serum did fuse with the lysosomes, resulting in parasite destruction (Niederkorn and Shadduck, 1980). Specific murine antibodies raised against *Enc. cuniculi* also fixed complement and killed organisms *in vitro* (Schmidt and Shadduck, 1984).

Specific antibodies to microsporidia probably play a role in resistance but, alone, do not protect the host from disease. For example, transfer of hyperimmune serum to athymic mice infected with *Enc. cuniculi* failed to prolong their survival (Schmidt and Shadduck, 1984), indicating that other protective mechanisms must exist. Furthermore, some antibody responses probably contribute to disease, as observed in blue foxes and domestic dogs infected with *Enc. cuniculi*, which developed immune complex disease (Mohn and Norsdstoga, 1975; Arneson and Nordstoga, 1977; Mohn *et al.*, 1982; Stewart *et al.*, 1986).

5.2.2. Cell-mediated Immune Responses

Cell-mediated immune responses are critical in preventing lethal disease in mice, as shown by the fact that adoptive transfer of sensitized syngeneic T-cell-enriched spleen cells protected *Enc. cuniculi*-infected athymic BALB/c or SCID mice (Schmidt and Shadduck, 1984; Hermanek *et al.*, 1993). Adoptive transfer of naive syngeneic splenic lymphocytes failed to protect athymic mice if given at the same time as the *Enc. cuniculi* infection (Schmidt and Shadduck, 1984), but did protect SCID mice if transferred 2 weeks before *Enc. cuniculi* infection (Hermanek *et al.*, 1993). Cytotoxic T cells specific for *Enc. cuniculi*-infected cells could not be detected, but lymphocytes were found to release cytokines or other factors that could activate macrophages to kill the *Enc. cuniculi* (Schmidt and Shadduck, 1984). Murine macrophages that were activated by incubation with lipopolysaccharide (10 ng/ml) and interferon-γ (100 units/ml) or tumor necrosis factor α (1000 units/ml) were able to kill, or control the growth of, *Enc. cuniculi in vitro*, and this macrophage-mediated killing was dependent on the generation of nitrogen intermediates (Didier and Shadduck, 1994; Didier *et al.*, 1994; Didier, 1995). Desportes-Livage and colleagues found that interferon-γ receptor 'knockout' BALB/c mice were significantly more susceptible to infection with *Enc. intestinalis* than the wild-type BALB/c mice (Achbarou *et al.*, 1996), further indicating an important role for interferon-γ in resistance to microsporidiosis *in vivo*. The authors cautioned, however, that, because these mice were not killed by *Enc. intestinalis*, other factors, which have yet to be characterized, must be important for resistance to microsporidiosis (e.g. host specificity factors or local IgA).

Little is known about the role that T cells play in humans infected with microsporidiosis. There appears to be some correlation between CD4+ T cell levels and clinical signs associated with microsporidiosis (Orenstein, 1991; Kotler and Orenstein, 1995). Cytokine mRNA expression in intestinal biopsies has been evaluated by Snijders *et al.* (1995), but no significant difference was seen between levels of tumor necrosis factor or interleukins 1β, 6, 8, or 10 in six HIV-infected individuals with microsporidiosis and diarrhea and versus those in seven HIV-infected individuals with diarrhea of unknown etiology and in HIV-seronegative individuals. The authors suggested that immunotherapy would therefore not be likely to relieve the microsporidiosis, but no comparison was made with HIV-seronegative individuals infected with microsporidia.

The role of natural killer (NK) cells as mediators of resistance/susceptibility has also been examined. Niederkorn *et al.* (1983) found that BALB/c mice inoculated with *Enc. cuniculi* displayed a greater increase in NK activity against YAC-1 tumor cells than did non-infected BALB/c mice, when compared with infected C57BL/6 mice. The authors, however, stated that the role of NK cells in resistance was questionable because uninfected C57BL/6 mice expressed a higher baseline NK activity yet were more susceptible to *Enc. cuniculi* infection. In addition, NK-defective beige mice did not die of *Enc. cuniculi* infection and athymic mice, which died after infection with *Enc. cuniculi*, expressed higher NK activities than their euthymic counterparts, indicating further that specific T lymphocytes play the major role in resistance to microsporidiosis.

5.2.3. *Host Specificity*

Host specificity appears to play some part in resistance or susceptibility to infection with some species of microsporidia that infect mammals. Among the more than 1000 species of microsporidia, relatively few have been found to infection mammals. *Enc. hellem* had been reported only in humans until it was recently described in parakeets (Black *et al.*, 1997). To date, *Enc. intestinalis* has been described in humans only (Didier *et al.*, 1996b). *Ent. bieneusi* infects and causes disease in human beings, and has only once been reported in a nonhuman host (pig) (Deplazes *et al.*, 1996). *Enc. cuniculi*, however, has a wide host range among mammals (Canning and Lom, 1986). As research continues and better diagnostic methods become available, many of these microsporidia may be recognized in nonhuman hosts. However, some degree of host specificity must exist because many unsuccessful attempts have been made to inject *Ent. bieneusi* into a variety of immunodeficient strains of mice (e.g. athymic, SCID, beige) (unpublished observations).

Host genetics may also play some role in the susceptibility (or, conversely, resistance) to microsporidiosis. Niederkorn *et al.* (1981) found that C57BL/6 mice were significantly more susceptible to *Enc. cuniculi* than BALB/c mice, and that this disparity resulted from genes located outside the major histocompatibility (H-2) complex. C57BL/6 mice infected with *Enc. cuniculi* expressed lower splenic lymphocyte blast responses to mitogens (concanavalin A and phytohemagglutinin) than did lymphocytes from infected BALB/c mice. Furthermore, the antibody response to a second antigen (sheep erythrocytes) was lower in infected C57BL/6 mice than in infected BALB/c mice. In subsequent studies, Liu *et al.* (1989) found that the susceptibility of C57BL/6 mice was due to a delay in production of cytokines that induced macrophage-mediated killing, and this delay or immune suppression was, in turn, due to prostaglandin production by the accessory cells.

The host-specific mechanisms responsible for the hypersensitivity responses in blue foxes and domestic dogs that lead to disease are not well understood. To address this issue, Liu and Shadduck (1988) infected the autoimmune-prone MRL/MPJ-LPR mice with *Enc. cuniculi* in an attempt to develop a hypersensitivity model for microsporidiosis. MRL/MPJ-LPR mice bear the *lpr* gene which results in an age-related development of hypergammaglobulinemia, antinuclear antibodies, and circulating immune complexes that cause autoimmune disease (Murphy and Roths, 1978). They found, however, that the *lpr* gene did not appear to play a role in progression of microsporidiosis-associated disease in these mice.

6. SUMMARY AND CONCLUSIONS

Microsporidiosis is an example of a relatively recent human disease that has commonly been recognized in other animals (Shadduck *et al.* 1996). In immunologically competent hosts, microsporidiosis develops into a balanced host–parasite relationship such that the host survives with few or no clinical signs of disease and the parasite persists. In the absence of a competent immune system or, conversely, if hypersensitivity immune responses develop, the host–parasite relationship becomes unbalanced, resulting in clinical signs of disease. While there is some debate about associating the presence of microsporidia with the development of clinically significant disease in humans (Rabeneck *et al.*, 1993, 1995), much can be learned from natural infections in animals. Results from serological studies suggested that humans were naturally infected with microsporidia (Bergquist *et al.*, 1984; Hollister *et al.*, 1991), as occurs in laboratory animals (Shadduck and Pakes, 1971; Canning and Lom, 1986; Shadduck, 1989; Shadduck and Greeley, 1989;

Shadduck and Orenstein, 1993). With the recent improvements in diagnostic methods for detecting microsporidia, reports are being published about immunocompetent humans who have become infected with microsporidia and subsequently resolved their infections (Sandfort et al., 1994; Sobottka et al., 1995; Sowerby et al., 1995). As more is learned about the epidemiology, immunology, and pathology of microsporidiosis, advances will hopefully be made in the prevention and control of microsporidiosis in susceptible mammalian hosts.

ADDENDUM

Since the submission of this review, natural (i.e. not experimental) *Ent. bieneusi* infections of the hepatobiliary system were reported in several species of macaque monkeys (*Macaca mulatta, M. nemestrina, M. cyclopis*) by Mansfield et al. (1997). Furthermore, persistent *Ent. bieneusi* infections were established experimentally in SIV-infected *M. mulatta* (Tzipori et al., 1997) and in immunosuppressed gnotobiotic piglets (S. Tzipori, personal communication). The significance of these findings is that the animal models for *Ent. bieneusi* will permit studies to improve diagnostic and therapeutic strategies and to reveal more about the biology of these infections.

A second important recent development is the observation that TNF-α may play a role in the pathogenesis of intestinal microsporidiosis. Sharpstone et al. (1997) found that fecal TNF-α levels were significantly higher in HIV-infected individuals with microsporidiosis than in HIV-infected individuals without microsporidiosis. Treatment with thalidomide, a TNF-α inhibitor, resulted in relief of symptoms (e.g. reduced numbers of stools) and the generation of abnormally-shaped microsporidia. These results are significant for a better understanding of the beneficial versus harmful effects of the immune system in hosts with microsporidiosis and offer some insight for developing immunotherapeutic strategies for treatment of microsporidiosis.

REFERENCES

Achbarou, A., Ombrouck, C., Gneragbe, T., Charlotte, F., Renia, L., Desportes-Livage, I. and Mazier, D. (1996). Experimental model for human intestinal microsporidiosis in interferon gamma receptor knockout mice infected by *Encephalitozoon intestinalis*. *Parasite Immunology* **18**, 387–392.
Anver, M.R., King, N.W. and Hunt, R.D. (1972). Congenital encephalitozoonosis in a squirrel monkey (*Saimiri sciureus*). *Veterinary Pathology* **9**, 475–480.

Arnesen, K. and Nordstoga, K. (1977). Ocular encephalitozoonosis (nosematosis) in blue foxes, polyarteritis nodosa and cataract. *Acta Ophthalmologica* **55**, 641–651.
Asmuth, D.M., DeGirolami, P.C., Federman, M., Ezrathy, C.R., Pleskow, D.K., Desai, G. and Wanke, D.A. (1994). Clinical features of microsporidiosis in patients with AIDS. *Clinical Infectious Diseases* **18**, 819–825.
Baker, M.D., Vossbrinck, C.R., Didier, E.S., Maddox, J.V. and Shadduck, J.S. (1995). Small subunit ribosomal DNA phylogeny of various microsporidia with emphasis on AIDS related forms. *Journal of Eukaryotic Microbiology* **42**, 564–570.
Basson, P.A., McCully, R.M. and Warnes, W.E.J. (1966). Nosematosis: report of a canine case in the Republic of South Africa. *Journal of the South African Veterinary Medical Association* **37**, 3–9.
Bergquist, R., Morfeldt-Masson, L., Pehrson, P.O., Petrini, B. and Wasserman, J. (1984). Antibody against *Encephalitozoon cuniculi* in Swedish homosexual men. *Scandinavian Journal of Infectious Diseases* **16**, 389–391.
Biderre, C., Pagès, M., Metenier, G., David, D., Bata, J., Prensier, G. and Vivares, C.P. (1994). On small genomes in eukaryotic organisms: molecular karyotypes of two microsporidian species (Protozoa) parasites of vertebrates. *Comptes Rendus de l'Académie des Sciences* **317**, 399–404.
Biderre, C., Pagès, M., Metenier, G., Canning, E.U. and Vivares, C.P. (1995). Evidence of the smallest nuclear genome (2.9 Mb) in the microsporidium *Encephalitozoon cuniculi*. *Molecular and Biochemical Parasitology* **74**, 229–231.
Black, S.S., Steinohrt, L.A., Bertucci, D.C., Rogers, L.B. and Didier, E.S. (1997). *Encephalitozoon hellem* in budgerigars (*Melopsittacus undulatus*). *Veterinary Pathology* **34**, 189–198.
Bryan, R.T. (1995). Microsporidiosis as an AIDS-related opportunistic infection. *Clinical Infectious Diseases* **21**, supplement 1, S62–S65.
Bryan, R.T., Cali, A., Owen, R.L. and Spencer, H.C. (1990). Microsporidia: opportunistic pathogens in patients with AIDS. In: *Progress in Clinical Parasitology* (T. Sun, ed.). Vol. 2, pp. 1–26. Philadelphia: Field and Wood.
Cali, A. (1991). General microsporidian features and recent findings on AIDS isolates. *Journal of Protozoology* **38**, 625–630.
Cali, A. and Owen, R.L. (1990). Intracellular development of *Enterocytozoon*, a unique microsporidian found in the intestine of AIDS patients. *Journal of Protozoology* **37**, 145–155.
Cali, A., Meisler, D., Lowder, C.Y., Lembach, R., Ayers, L., Takvorian, P.M., Rutherford, I., Longworth, D.L., McMahon, J.T. and Bryan, R.T. (1991a). Corneal microsporidioses: characterization and identification. *Journal of Protozoology* **38**, S215–S217.
Cali, A., Meisler, D.M., Rutherford, I., Lowder, C.Y., McMahon, J.T., Longworth, D.L. and Bryan, R.T. (1991b). Corneal microsporidiosis in a patient with AIDS. *American Journal of Tropical Medicine and Hygiene* **44**, 463–468.
Cali, A., Kotler, D.P. and Orenstein, J.M. (1993). *Septata intestinalis* n.g., n.sp., an intestinal microsporidian associated with chronic diarrhea and dissemination in AIDS patients. *Journal of Protozoology* **40**, 101–112.
Cali, A., Takvorian, P.M., Lewin, S., Rendel, M., Sian, C., Wittner, M. and Weiss, L.M. (1996). Identification of a new *Nosema*-like microsporidian associated with myositis in an AIDS patient. *Journal of Eukaryotic Microbiology* **43**, 108S.
Canning, E.U. and Hazard, E.I. (1982). Genus *Pleistophora* Gurley, 1893: an assemblage of at least three genera. *Journal of Protozoology* **29**, 39–49.
Canning, E.U. and Hollister, W.S. (1987). Microsporidia of mammals—widespread pathogens or opportunistic curiosities? *Parasitology Today* **3**, 267–273.

Canning, E.U. and Hollister, W.S. (1992). Human infections with microsporidia. *Reviews in Medical Microbiology* **3**, 35–42.

Canning, E.U. and Lom, J. (1986). *The Microsporidia of Vertebrates.* New York: Academic Press.

Canning, E.U. and Nicholas, J.P. (1980). Genus *Pleistophora* (Phylum Microspora): redescription of the type species, *Pleistophora typicalis* Gurley, 1893 and ultrastructural characterization of the genus. *Journal of Fish Diseases* **3**, 317–338.

Canning, E.U., Curry, A., Lacey, C.J. and Fenwick, D. (1992). Ultrastructure of *Encephalitozoon* sp. infecting the conjunctival, corneal, and nasal epithelia of a patient with AIDS. *European Journal of Protistology* **28**, 226–237.

Canning, E.U., Field, A.S., Hing, M.C. and Marriott, D.J. (1994). Further observations on the ultrastructure of *Septata intestinalis. European Journal of Protistology* **40**, 414–422.

Chupp, G.L., Alroy, J., Adelman, L.S., Breen, J.C. and Skolnik, P.R. (1993). Myositis due to *Pleistophora* (microsporidia) in a patient with AIDS. *Clinical Infectious Diseases* **16**, 15–21.

Cox, J.C. (1977). Altered immune responsiveness associated with *Encephalitozoon cuniculi* infection in rabbits. *Infection and Immunity* **15**, 392–395.

Cox, J.C., Gallichio, H.A., Pye, D. and Walden, N.B. (1977). Application of immunofluorescence to the establishment of an *Encephalitozoon cuniculi*-free rabbit colony. *Laboratory Animal Science* **27**, 204–209.

Cox, J.C., Hamilton, R.C. and Attwood, H.D. (1979). An investigation of the route and progression of *Encephalitozoon cuniculi* infection in adult rabbits. *Journal of Protozoology* **26**, 260–265.

Curgy, J.J., Vávra, J. and Vivares, C. (1980). Presence of ribosomal RNAs with prokaryotic properties in microsporidia, eukaryotic organisms. *Biology of the Cell* **38**, 49–52.

Davis, R.M., Font, R.L., Keisler, M.S. and Shadduck, J.A. (1990). Corneal microsporidiosis. A case report including ultrastructural observations. *Ophthalmology* **97**, 953–957.

DeGroote M.A., Visvesvara, G.S., Wilson M.L., Pieniazek, N.J., Slemenda, S.B., Da Silva, A.J., Leitch, G.J., Bryan, R.T. and Reves, R. (1995). Polymerase chain reaction and culture confirmation of disseminated *Encephalitozoon cuniculi* in a patient with AIDS: successful therapy with albendazole. *Journal of Infectious Diseases* **171**, 1375–1378.

Deplazes, P., Mathis, A., Mueller, C. and Weber, R. (1996). Molecular epidemiology of *Encephalitozoon cuniculi* and first detection of *Enterocytozoon bieneusi* in faecal samples in pigs. *Journal of Eukaryotic Microbiology* **43**, 93S.

Desportes, I., Le Charpentier, Y., Galian, A., Bernard, F., Cochand-Priollet, B., Lavergne, A., Ravisse, P. and Modigliani, R. (1985). Occurrence of a new microsporidian, *Enterocytozoon bieneusi* n.g., n.sp., in the enterocytes of a human patient with AIDS. *Journal of Protozoology* **32**, 250–254.

Desportes-Livage, I., Chilmonczyk, S., Hedrick, R., Ombrouck, C., Monge, D., Maiga, I. and Gentilini, M. (1996). Comparative development of two microsporidian species: *Enterocytozoon bieneusi* and *Enterocytozoon salmonis*, reported in AIDS patients and salmonid fish, respectively. *Journal of Eukaryotic Microbiology* **43**, 49–60.

Desser, S.S., Hong, H. and Yang, Y.J. (1992). Ultrastructure of the development of a species of *Encephalitozoon* cultured from the eye of an AIDS patient. *Parasitology Research* **78**, 677–683.

Didier, E.S. (1995). Nitrogen intermediates implicated in the inhibition of *Encepha-*

litozoon cuniculi (phylum Microspora) replication in murine peritoneal macrophages. *Parasite Immunology* **17**, 405–412.

Didier, E.S. and Shadduck, J.A. (1988). Modulated immune responsiveness associated with experimental *Encephalitozoon cuniculi* infection in BALB/c mice. *Laboratory Animal Science*. **38**, 680–684.

Didier, E.S. and Shadduck, J.A. (1994). IFN-γ and LPS induce murine macrophages to kill *Encephalitozoon cuniculi in vitro*. *Journal of Eukaryotic Microbiology* **41**, 34S.

Didier, E.S., Didier, P.J., Friedberg, D.N., Stenson, S.M., Orenstein, J.M., Yee, R.W., Tio, F.O., Davis, R.M., Vossbrinck, C., Millichamp, N. and Shadduck, J.A. (1991a). Isolation and characterization of a new human microsporidian, *Encephalitozoon hellem* (n.sp.), from three AIDS patients with keratoconjunctivitis. *Journal of Infectious Diseases* **163**, 617–621.

Didier, E.S., Shadduck, J.A., Didier, P.J., Millichamp, N.J. and Vossbrinck, C.R. (1991b). Studies on ocular Microsporidia. *Journal of Protozoology* **38**, 635–638.

Didier, P.J., Didier, E.S., Orenstein, J.M. and Shadduck, J.A. (1991c). Fine structure of a new human microsporidian, *Encephalitozoon hellem*, in culture. *Journal of Protozoology* **38**, 502–507.

Didier, E.S., Kotler, D.P., Dieterich, D.T., Orenstein, J.M., Aldras, A.M., Davis, R., Friedberg, D.N., Gourley, W.K., Lembach, R., Lowder, C.Y., Meisler, D.M., Rutherford, I., Yee, R.W. and Shadduck, J.A. (1993). Serological studies in human microsporidia infections. *AIDS* **7**, S8–S11.

Didier, E.S., Varner, P.W., Didier, P.J., Aldras, A.M., Millichamp, N.J., Murphey-Corb, M., Bohm, R. and Shadduck, J.A. (1994). Experimental microsporidiosis in immunocompetent and immunodeficient mice and monkeys. *Folia Parasitologica* **41**, 1–11.

Didier, E.S., Vossbrinck, C.R., Baker, M.D., Rogers, L.B., Bertucci, D.C. and Shadduck, J.A. (1995). Identification and characterization of three *Encephalitozoon cuniculi* strains. *Parasitology* **111**, 411–422.

Didier, E.S., Rogers, L.B., Brush, A.D., Wong, S. and Bertucci, D.C. (1996a). Diagnosis of disseminated microsporidian *Encephalitozoon hellem* infection by PCR–Southern analysis and successful treatment with albendazole and fumagillin. *Journal of Clinical Microbiology* **34**, 947–952.

Didier, E.S., Rogers, L.B., Orenstein, J.M., Baker, M.D., Vossbrinck, C.R., Van Gool, T., Hartskeerl, R., Soave, R. and Beaudet, L.M. (1996b). Characterization of *Encephalitozoon (Septata) intestinalis* isolates cultured from nasal mucosa and bronchoalveolar lavage fluids of two AIDS patients. *Journal of Eukaryotic Microbiology* **43**, 34–43.

Didier, E.S., Visvesvara, G.S., Baker, M.D., Rogers, L.B., Bertucci, D.C., DeGroote, M.A. and Vossbrinck, C.R. (1996c). A microsporidian isolated from an AIDS patient corresponds to the *Encephalitozoon cuniculi* strain III originally isolated from domestic dogs. *Journal of Clinical Microbiology* **34**, 2835–2837.

Doby, J.-M., Jeannes, A. and Rault, B. (1963). *Thelohania apodemi* n.sp., première microsporidie du genre *Thelohania* observée chez un Mammifère. *Comptes Rendus de l'Academie des Sciences* **257**, 248–251.

Field, A.S., Marriott, D.J., Milliken, S.T., Brew, B.J., Canning, E.U., Kench, J.G., Darveniza, P. and Harkness, J.L. (1996). Myositis associated with a newly described microsporidian, *Trachipleistophora hominis*, in a patient with AIDS. *Journal of Clinical Microbiology* **34**, 2803–2811.

Franzen, C., Muller, A., Hartmann, P., Kochaneck, M., Diehl, V. and Fatkenheuer, G. (1996). Disseminated *Encephalitozoon (Septata) intestinalis* infection in a patient with AIDS. *New England Journal of Medicine* **335**, 1610–1611.

Friedberg, D.N., Stenson, S.M., Orenstein, J.M., Tierno, P.M. and Charles, N.C.

(1990). Microsporidial keratoconjunctivitis in acquired immunodeficiency syndrome. *Archives of Ophthalmology* **108**, 504–508.
Fuentealba, I.C., Mahoney, N.T., Shadduck, J.A., Harvill, J., Wicher, V. and Wicher, K. (1992). Hepatic lesions in rabbits with *Encephalitozoon cuniculi* administered per rectum. *Veterinary Pathology* **29**, 536–540.
Gannon, J. (1980). The course of infection of *Encephalitozoon cuniculi* in immunodeficient and immunocompetent mice. *Laboratory Animals* **14**, 189–192.
Hartskeerl, R.A., Van Gool, T., Schuitema, A.R.J., Didier, E.S. and Terpstra, W.J. (1995). Genetic and immunological characterization of the microsporidian *Septata intestinlis* Cali, Kotler, and Orenstein, 1993: reclassification to *Encephalitozoon intestinalis*. *Parasitology* **110**, 277–285.
Hermanek, J., Koudela, B., Kucerova, A., Ditrich, O. and Travnicek, J. (1993). Prophylactic and therapeutic immune reconstitution of SCID mice infected with *Encephalitozoon cuniculi*. *Folia Parasitologica* **40**, 287–291.
Hollister, W.S., Canning, E.U. and Wilcox, A. (1991). Evidence for widespread occurrence of antibodies to *Encephalitozoon cuniculi* (Microspora) in man provided by ELISA and other serological tests. *Parasitology* **102**, 33–43.
Hollister, W.S., Canning, E.U., Colbourn, N.I. and Lacey, C.J.M. (1993). Characterization of *Encephalitozoon hellem* (Microspora) isolated from the nasal mucosa of a patient with AIDS. *Parasitology* **107**, 351–358.
Hollister, W.S., Canning, E.U., Colbourn, N.I. and Aarons, E.J. (1995). *Encephalitozoon cuniculi* isolated from the urine of an AIDS patient, which differs from a canine isolate. *Journal of Eukaryotic Microbiology* **42**, 367–372.
Hollister, W.S., Canning, E.U., Weidner, E., Field, A.S., Kench, J. and Marriott, D.J. (1996). Development and ultrastructure of *Trachipleistophora hominis* n.g., n.sp. after *in vitro* isolation from an AIDS patient and inoculation into athymic mice. *Parasitology* **112**, 143–154.
Hunt, R.D., King, N.W. and Foster, H.L. (1972). Encephalitozoonosis: evidence for vertical transmission. *Journal of Infectious Diseases* **126**, 212–214.
Innes, J.R., Zeman, W., Frenkel, J.K. and Borner, G. (1962). Occult endemic encephalitozoonosis of central nervous system in mice. *Journal of Neuropathology and Experimental Neurology* **21**, 519–533.
Issi, I.V. (1986). Microsporidia as a phylum of parasitic protozoa. *Academy of Sciences of the USSR: Protozoology (Leningrad)* **10**, 6–136.
Kent, M.L., Hervo, D.M.L., Docker, M.F. and Devlin, R.H. (1996). Taxonomy studies and diagnostic tests for myxosporean and microsporidian pathology of salmonid fishes utilizing ribosomal DNA sequence. *Journal of Eukaryotic Microbiology* **43**, 98S–99S.
Keohane, E.M., Takvorian, P.M., Cali, A., Tanowitz, H.B., Wittner, M. and Weiss, L.M. (1996). Identification of a microsporidian polar tube protein reactive monoclonal antibody. *Journal of Eukaryotic Microbiology* **43**, 26–31.
Kotler, D.P. and Orenstein, J.M. (1995). Microsporidia. In: *Infections of the Gastrointestinal Tract* (M.J. Blaser, P.D. Smith, J.I. Ravdin, H.B. Greenberg and R.L. Guerrant, eds), pp. 1129–1140. New York: Raven Press.
Koudela, B., Vitovec, J., Kucerova, S., Ditrich, O. and Travnicek, J. (1993). The severe combined immunodeficient mouse as a model for *Encephalitozoon cuniculi* microsporidiosis. *Folia Parasitologica* **40**, 279–286.
Ledford, D.K., Overman, M.D., Gonzalo, A., Cali, A., Mester, W. and Lockey, R.F. (1985). Microsporidiosis myositis in a patient with acquired immunodeficiency syndrome. *Annals of Internal Medicine* **102**, 628–630.
Leitch, G.J., He, Q., Wallace, S. and Visvesvara, G.S. (1993a). Inhibition of the spore

polar filament extrusion of the microsporidian, *Encephalitozoon hellem*, isolated from an AIDS patient. *Journal of Eukaryotic Microbiology* **40**, 711–717.
Leitch, G.J., Visvesvara, G.S. and He, Q. (1993b). Inhibition of microsporidian spore germination. *Parasitology Today* **9**, 422–424.
Levaditi, C., Nicolau, S. and Schoen, R. (1923). L'agent étiologique de l'encephalite épizootique du lapin (*Encephalitozoon cuniculi*). *Comptes Rendus de l'Académie des Sciences* **89**, 984–986.
Levine, N.D., Corliss, J.O., Cox, F.E.G., Deroux, G., Grain, J., Honigberg, B.M., Leedale, G.F., Loeblich, A.R., III, Lom, J., Lynn, D., Merinfeld, E.G., Page, F.C., Poljansky, G., Sprague, V., Vavra, J. and Wallace, F.G. (1980). A newly revised classification of the protozoa. *Journal of Protozoology* **27**, 37–58.
Liu, J. and Shadduck, J.A. (1988). *Encephalitozoon cuniculi* infection in MRL/MPJ-LPR (lymphoproliferation) mice. *Laboratory Animal Science* **38**, 685–688.
Liu, J.J., Greeley, E.H. and Shadduck, J.A. (1988). Murine encephalitozoonosis: the effect of age and mode of transmission on occurrence of infection. *Laboratory Animal Science* **38**, 675–679.
Liu, J.J., Greeley, E.H. and Shadduck, J.A. (1989). Mechanisms of resistance/susceptibility to murine microsporidiosis. *Parasite Immunology* **11**, 241–256.
Lom, J. and Vávra, J. (1963). The mode of sporoplasm extrusion in microsporidian spores. *Acta Protozoologica* **1**, 81–89.
Malone, A.L. and McIvor, C.A. (1993). Pulsed-field electrophoresis of DNA from four microsporidian isolates. *Journal of Invertebrate Pathology* **61**, 203–205.
Mansfield, K.G., Carville, A., Shvetz, D., Mackay, J., Tzipori, S. and Lackner, A.A. (1997). Identification of an *Enterocytozoon bieneusi*-like microsporidian parasite in simian-immunodeficiency-virus-inoculated macaques with hepatobiliary disease. *American Journal of Pathology* **150**, 1395–1405.
Margileth, A.M., Strano, A.J., Chandra, R., Neafie, R., Blum, M. and McCully, R.M. (1973). Disseminated nosematosis in an immunologically compromised infant. *Archives in Pathology* **95**, 341–343.
Mathis, A., Akerstedt, J., Tharaldsen, J., Odegaard, O. and Deplazes, P. (1996). Isolates of *Encephalitozoon cuniculi* from farmed blue foxes (*Alopex lagopus*) from Norway differ from isolates from Swiss domestic rabbits (*Oryctolagus cuniculus*). *Parasitology Research* **82**, 727–730.
McCully, R.M., Van Dellen, A.F., Basson, P.A. and Lawrence, J. (1978). Observations on the pathology of canine microsporidiosis. *Onderstepoort Journal of Veterinary Research* **45**, 75–91.
McDougall, R.J., Tandy, M.W., Boreham, R.E., Stenzel, D.J. and O'Donoghue, P.J. (1993). Incidental finding of a microsporidian parasite from an AIDS patient. *Journal of Clinical Microbiology* **31**, 436–439.
McInnes, E.F. and Stewart, C.G. (1991). The pathology of subclinical infection of *Encephalitozoon cuniculi* in canine dams producing pups with overt encephalitozoonosis. *Journal of the South African Veterinary Association* **62**, 51–54.
Mohn, S.F. and Nordstoga, K. (1975). Electrophoretic patterns of serum proteins in blue foxes with special reference to changes associated with nosematosis. *Acta Veterinaria Scandinavica* **16**, 297–306.
Mohn, S.F., Nordstoga, K., Krogsrun, J. and Helgebostad (1974). Transplacental transmission of *Nosema cuniculi* in the blue fox (*Alopex lagopus*). *Acta Pathologica Scandinavica* **82**, 299–300.
Mohn, S.F., Landsverk, T. and Nordstoga, K. (1981). Encephalitozoonosis in the blue fox – morphological identification of the parasite. *Acta Pathologica Microbiologica Scandinavica* **89**, 117–122.

Mohn, S.F., Nordstoga, K. and Dishington, I.W. (1982). Experimental encephalitozoonosis in the blue fox. Clinical, serological and pathological examination of vixens after oral and intrauterine inoculation. *Acta Veterinaria Scandinavica* **23**, 490–502.

Montrey, R.D., Shadduck, J.A. and Pakes, S.P. (1973). In vitro study of host range of three isolates of *Encephalitozoon* (*Nosema*). *Journal of Infectious Diseases* **127**, 450–454.

Munderloh, U.G., Durti, T.J. and Ross, S.E. (1990). Electrophoretic characterization of chromosomal DNA from two microsporidia. *Journal of Invertebrate Pathology* **56**, 243–248.

Murphy, E.D. and Roths, J.B. (1978). Autoimmunity and lymphoproliferation: induction by mutant gene *lpr* and acceleration by a male-associated factor in strain BXSB mice. In: *Genetic Control of Autoimmune Disease* (M.R. Rose, P.E. Bigassi and N.L. Warner, eds), pp. 207–220. Amsterdam: North Holland.

Naegeli, C. (1857). Uber die neue Krankheit der Seidenraupe und vervandte Organismen. *Botanische Zeitung* **15**, 760–761.

Nelson, J.B. (1967). Experimental transmission of a murine microsporidian in Swiss mice. *Journal of Bacteriology* **94**, 1340–1345.

Niederkorn, J.Y. and Shadduck, J.A. (1980). Role of antibody and complement in the control of *Encephalitozoon cuniculi* infection by rabbit macrophages. *Infection and Immunity* **17**, 995–1002.

Niederkorn, J.Y., Shadduck, J.A. and Schmidt, E.C. (1981). Susceptibility of selected inbred strains of mice to *Encephalitozoon cuniculi*. *Journal of Infectious Diseases* **144**, 249–253.

Niederkorn, J.Y., Brieland, J.K. and Mayhew, E. (1983). Enhanced natural killer activity in experimental murine encephalitozoonosis. *Infection and Immunity* **41**, 302–307.

Nordstoga, K. (1972). Nosematosis in the blue fox. *Veterinary Medicine* **24**, 21–24.

Orenstein, J.M. (1991). Microsporidiosis in the acquired immunodeficiency syndrome. *Journal of Parasitology* **77**, 843–864.

Orenstein, J.M., Chiang, J., Steinberg, W., Smith, P.D., Rotterdam, H. and Kotler, D.P. (1990). Intestinal microsporidiosis as a cause of diarrhea in human immunodeficiency virus-infected patients: a report of 20 cases. *Human Pathology* **21**, 475–481.

Orenstein, J.M., Tenner, M., Cali, A. and Kotler, D.P. (1992). A microsporidian previously undescribed in humans, infecting enterocytes and macrophages, and associated with diarrhea in an acquired immunodeficiency syndrome patient. *Human Pathology* **23**, 722–728.

Pang, V.F. and Shadduck, J.A. (1985). Susceptibility of cats, sheep, and swine to a rabbit isolate of *Encephalitozoon cuniculi*. *American Journal of Veterinary Research* **46**, 1071–1077.

Perrin, T.L. (1943). Spontaneous and experimental *Encephalitozoon* infection in laboratory animals. *Archives of Pathology* **36**, 559–567.

Petri, M. (1966). The occurrence of *Nosema cuniculi* (*Encephalitozoon cuniculi*) in the cells of transplantable, malignant ascites tumours and its effect upon tumour and host. *Acta Pathologica Microbiologica Scandinavica* **66**, 13–30.

Pleshinger, J. and Weidner, E. (1985). The microsporidian spore invasion tube. IV. Discharge activation begins with pH-triggered Ca^{2+} influx. *Journal of Cell Biology* **100**, 1834–1838.

Plowright, W. (1952). An encephalitis–nephritis syndrome in the dog, probably due to congenital *Encephalitozoon* infection. *Journal of Comparative Pathology* **62**, 83–92.

Plowright, W. and Yeoman, G. (1952). Probable *Encephalitozoon* infection of the dog. *Veterinary Record* **64**, 381–388.
Rabeneck, L., Gyorkey, F., Genta, R.M., Gyorkey, P., Foote, L.W. and Risser, J.M. (1993). The role of microsporidia in the pathogenesis of HIV-related chronic diarrhea. *Annals of Internal Medicine* **119**, 895–899.
Rabeneck, L., Genta, R.M., Gyorkey, F., Clarridge, J.E., Gyorkey, P. and Foote, L.W. (1995). Observations on the pathological spectrum and clinical course of microsporidiosis in men infected with the human immunodeficiency virus: follow-up study. *Clinical Infectious Diseases* **20**, 1229–1235.
Rabodonirina, M., Bertocchi, M., Desportes-Livage, I., Cotte, L., Levrey, H., Piens, M.A., Monneret, M., Celard, G., Mornex, J.F. and Mojon, M. (1996). *Enterocytozoon bieneusi* as a cause of chronic diarrhea in a heart–lung transplant recipient who was seronegative for human immunodeficiency virus. *Clinical Infectious Diseases* **23**, 114–117.
Sandfort, J., Hannemann, A., Gelderblom, H., Stark, K., Owen, R.L. and Ruf, B. (1994). *Enterocytozoon bieneusi* infection in an immunocompetent patient who had acute diarrhea and who was not infected with the human immunodeficiency virus. *Clinical Infectious Diseases* **19**, 514–516.
Schmidt, E.C. and Shadduck, J.A. (1983). Murine encephalitozoonosis model for studying the host–parasite relationship of a chronic infection. *Infection and Immunity* **40**, 936–942.
Schmidt, E.C. and Shadduck, J.A. (1984). Mechanisms of resistance to the intracellular protozoan *Encephalitozoon cuniculi* in mice. *Journal of Immunology* **133**, 2712–2719.
Schwartz, D.A., Bryan, R.T., Hewan-Lowe, K.O., Visvesvara, G.S., Weber, R., Cali, A. and Angritt, P. (1992). Disseminated microsporidiosis (*Encephalitozoon hellem*) and acquired immunodeficiency syndrome. *Archives of Pathology and Laboratory Medicine* **116**, 660–668.
Schwartz, D.A., Visvesvara, G.S., Leitch, G.J., Tashjian, L., Pollack, M. Holden, J. and Bryan, R.T. (1993). Pathology of symptomatic microsporidial (*Encephalitozoon hellem*) bronchiolitis in AIDS: a new respiratory pathogen diagnosed from lung biopsy, bronchoalveolar lavage, sputum, and tissue culture. *Human Pathology* **24**, 937–943.
Shadduck, J.A. (1969). *Nosema cuniculi*: in vitro isolation. *Science* **166**, 516–517.
Shadduck, J.A. (1989). Human microsporidiosis and AIDS. *Reviews of Infectious Diseases* **11**, 203–207.
Shadduck, J.A. and Greeley, E. (1989). Microsporidia and human infections. *Clinical Microbiology Review* **2**, 158–165.
Shadduck, J.A. and Orenstein, J.M. (1993). Comparative pathology of microsporidiosis. *Archives of Pathology and Laboratory Medicine* **117**, 1215–1219.
Shadduck, J.A. and Pakes, S.P. (1971). Spontaneous diseases of laboratory animals which interfere with biomedical research: encephalitozoonosis (nosematosis) and toxoplasmosis. *American Journal of Pathology* **64**, 657–671.
Shadduck, J.A. and Polley, M.B. (1978). Some factors influencing the *in vitro* infectivity and replication of *Encephalitozoon cuniculi*. *Journal of Protozoology* **25**, 491–496.
Shadduck, J.A., Bendele, R. and Robinson, G.T. (1978). Isolation of the causative organism of canine encephalitozoonosis. *Veterinary Pathology* **15**, 449–460.
Shadduck, J.A., Meccoli, R.A., Davis, R. and Font, R.L. (1990). First isolation of a microsporidian from a human patient. *Journal of Infectious Diseases* **162**, 773–776.
Shadduck, J.A., Storts, R. and Adams, L.G. (1996). Selected examples of emerging

and reemerging infectious diseases in animals. *American Society for Microbiology News* **62**, 586–588.

Sharpstone, D., Rowbottom, A., Francis, N., Tovey, G., Ellis, D., Barrett, M., Gazzard, B. (1997). Thalidomide: a novel therapy for microsporidiosis. *Gastroenterology* **112**, 1823–1829.

Silveira, H. and Canning, E.U. (1995). *Vittaforma corneae* n.comb. for the human microsporidium *Nosema corneum* Shadduck, Meccoli, Davis & Font, 1990, based on its ultrastructure in the liver of experimentally infected athymic mice. *Journal of Eukaryotic Microbiology* **42**, 158–165.

Silveira, H., Canning, E.U. and Shadduck, J.A. (1993). Experimental infection of athymic mice with the human microsporidian *Nosema corneum*. *Parasitology* **107**, 489–496.

Snijders, F., Van Deventer, S.J., Bartelsman, J.F., Den Otter, P., Jansen, J., Mevissen, M.L., Van Gool, T., Danner, S.A. and Reiss, P. (1995). Diarrhoea in HIV-infected patients: no evidence of cytokine-mediated inflammation in jejunal mucosa. *AIDS* **9**, 367–373.

Sobottka, I., Albrecht, H., Schottelius, J., Visvesvara, G.S., Laufs, R. and Schwartz, D.A. (1995). Disseminated *Encephalitozoon* (*Septata*) *intestinalis* infection in a patient with AIDS: novel diagnostic approaches and autopsy-confirmed parasitological cure following treatment with albendazole. *Journal of Clinical Microbiology* **33**, 2948–2952.

Sowerby, T.M., Conteas, C.N., Berlin, O.G.W. and Donovan, J. (1995). Microsporidiosis in patients with relatively preserved CD4 counts. *AIDS* **9**, 975–984.

Sprague, V. (1969). Need for drastic revision of the classification of subphylum Amoebagena. *Progress in Protozoology. Proceedings of the 3rd International Congress on Protozoology, Leningrad*, p. 372.

Sprague, V. (1974). *Nosema connori* n.sp., microsporidian parasite of man. *Transactions of the American Microscopical Society* **93**, 400–403.

Sprague, V. (1977). Annotated list of species of microsporidia. 2. Systematics of the Microsporidia. In: *Comparative Pathology* (L.A. Bulla and T.C. Cheng, eds), pp. 31–46. New York: Plenum Press.

Sprague, V. (1982). Microspora. In: *Synopsis and Classification of Living Organisms* (S.B. Parker, ed.), pp. 589–594. London: McGraw-Hill.

Sprague, V., Becnel, J.J. and Hazard, E.I. (1992). Taxonomy of phylum Microspora. *Critical Reviews in Microbiology* **18**, 285–395.

Stewart, C.G., Van Dellen, A.G.F. and Botha, W.S. (1979). Canine encephalitozoonosis in kennels and the isolation of *Encephalitozoon* in tissue culture. *Journal of the South African Veterinary Association* **50**, 165–168.

Stewart, C.G., Reyers, F. and Snyman, H. (1986). The relationship in dogs between primary renal disease and antibodies to *Encephalitozoon cuniculi*. *Journal of the South African Veterinary Association* **59**, 19–21.

Takvorian, P.M. and Cali, A. (1994). Enzyme histochemical identification of the Golgi apparatus in the microsporidian, *Glugea stephani*. *Journal of Eukaryotic Microbiology* **41**, 63S–64S.

Takvorian, P.M. and Cali, A. (1996). Polar tube formation and nucleoside diphosphatase activity in the microsporidian, *Glugea stephani*. *Journal of Eukaryotic Microbiology* **43**, 102S–103S.

Tzipori, S., Carville, A., Widmer, G., Kotler, D., Mansfield, K. and Lackner, A. (1997). Transmission and establishment of a persistent infection of *Enterocytozoon bieneusi*, derived from a human with AIDS, in simian immunodeficiency virus-infected rhesus monkeys. *Journal of Infectious Diseases* **175**, 1016–1020.

Undeen, A.H. (1990). A proposed mechanism for the germination of microsporidian (Protozoa: Microspora) spores. *Journal of Theoretical Biology* **142**, 223–235.
Undeen, A.H. and Frixione, E. (1990). The role of osmotic pressure in the germination of *Nosema algerae* spores. *Journal of Protozoology* **37**, 561–567.
Undeen, A.H. and Frixione, E. (1991). Structural alteration of the plasma membrane in spores of the microsporidium *Nosema algerae* on germination. *Journal of Protozoology* **38**, 511–518.
Van Dellen, A.F., Stewart, C.G. and Botha, W.S. (1989). Studies of encephalitozoonosis in vervet monkeys (*Cercopithecus pygerythrus*) orally inoculated with spores of *Encephalitozoon cuniculi* isolated from dogs (*Canis familiaris*). *Onderstepoort Journal of Veterinary Research* **56**, 1–22.
Van Gool, T., Canning, E.U., Gilis, H., Van Den Bergh Weermen, M.A., Eeftinck Schattenkerk, J.K.M. and Dankert, J. (1994). *Septata intestinalis* frequently isolated from stool of patients with a new cultivation method. *Parasitology* **109**, 281–289.
Van Rensburg, I.B., Volkmann, D.H., Soley, J.T. and Stewart, C.G. (1991). *Encephalitozoon* infection in a still-born foal. *Journal of the South African Veterinary Association* **62**, 130–132.
Vávra, J. (1968). Ultrastructural features of *Caudospora simulii* Weiser (Protozoa, Microsporidia). *Folia Parasitologica* **15**, 1–9.
Vávra, J. (1972). Detection of polysaccharides in microsporidian spores by means of the periodic acid–thiosemicarbazide–silver proteinate test. *Journal of Microscopy* **14**, 357–360.
Vávra, J., Bedrnik, P. and Cinatl, J. (1972). Isolation and *in vitro* cultivation of the mammalian microsporidian *Encephalitozoon cuniculi*. *Folia Parasitologica* **19**, 349–354.
Vernick, S.H., Sprague, V. and Krause, D. (1977). Some ultrastructural and functional aspects of the Golgi apparatus of *Thelohania* sp. (Microsporida) in the shrimp *Pandalus jordani* Rathbun. *Journal of Protozoology* **24**, 94–99.
Visvesvara, G.S., Da Silva, A.J., Croppo, G.P., Pieniazek, N.J., Leitch, G.J., Ferguson, D., De Moura, H., Wallace, S., Slemenda, S.B., Tyrrell, I., Moore, D.F. and Meador, J. (1995a). *In vitro* culture and serologic and molecular identification of *Septata intestinalis* isolated from urine of a patient with AIDS. *Journal of Clinical Microbiology* **33**, 930–936.
Visvesvara, G.S., Leitch, G.J., Pieniazek, N.J., Da Silva, A.J., Wallace, S., Slemenda, S.B., Weber, R., Schwartz, D.A., Gorelkin, L., Wilcox, C.M., and Bryan, R.T. (1995b). Short-term *in vitro* culture and molecular analysis of the microsporidian, *Enterocytozoon bieneusi*. *Journal of Eukaryotic Microbiology* **42**, 564–570.
Vossbrinck, C.R. and Woese, C.R. (1986). Eukaryotic ribosomes that lack a 5.8S RNA. *Nature* **320**, 287–288.
Vossbrinck, C.R., Maddox, J.V., Friedman, S., Debrunner-Vossbrinck, B.A. and Woese, C.R. (1987). Ribosomal RNA sequence suggests microsporidia are extremely ancient eukaryotes. *Nature* **326**, 411–414.
Vossbrinck, C.R., Baker, M.D., Didier, E.S., Debrunner-Vossbrinck, B.A. and Shadduck, J.A. (1993). Ribosomal DNA sequences of *Encephalitozoon hellem* and *Encephalitozoon cuniculi*. Species identification and phylogenetic construction. *Journal of Eukaryotic Microbiology* **40**, 354–362.
Weber, R., Bryan, R.T., Owen, R.L., Wilcox, C.M., Gorelkin, L. and Visvesvara, G.S. (1992a). Improved light-microscopical detection of microsporidia spores in stool and duodenal aspirates. *New England Journal of Medicine* **326**, 161–166.
Weber, R., Kuster, H., Keller, R., Bachi, T., Spycher, M.A., Briner, J., Russi, E. and Luthy, R. (1992b). Pulmonary and intestinal microsporidiosis in a patient with the

acquired immunodeficiency syndrome. *American Review of Respiratory Diseases* **146**, 1603–1605.

Weber, R., Kuster, H., Visvesvara, G.S., Bryan, R.T., Schwartz, D.A. and Luthy, R. (1993). Disseminated microsporidiosis due to *Encephalitozoon hellem*: pulmonary colonization, microhematuria, and mild conjunctivitis in a patient with AIDS. *Clinical Infectious Diseases* **17**, 415–419.

Weber, R., Bryan, R.T., Schwartz, D.A. and Owen, R.L. (1994). Human microsporidial infections. *Clinical Microbiology Reviews* **7**, 426–461.

Weidner, E. (1972). Ultrastructural study of microsporidian invasion into cells. *Zeitschrift für Parasitenkunde* **40**, 227–242.

Weidner, E. (1975). Interactions between *Encephalitozoon cuniculi* and macrophages: parasitophorous vacuole growth and the absence of lysosomal fusion. *Zeitschrift für Parasitenkunde* **47**, 1–9.

Weidner, E. (1976). The microsporidian spore invasion tube. The ultrastructure, isolation, and characterization of the protein comprising the tube. *Journal of Cell Biology* **71**, 23–34.

Weidner, E. (1982). The microsporidian spore invasion tube, III. Tube extrusion and assembly. *Journal of Cell Biology* **93**, 976–979.

Weidner, E. and Byrd, W. (1982). The microsporidian spore invasion tube. II. Role of calcium in the activation of invasion tube discharge. *Journal of Cell Biology* **93**, 970–975.

Weidner, E. and Sibley, D. (1985). Phagocytized intracellular microsporidian blocks phagosome acidification and phagosome–lysosome fusion. *Journal of Protozoology* **32**, 311–317.

Weidner, E., Byrd, W., Scarborough, A., Pleshinger, J. and Sibley, D. (1984). Microsporidian spore discharge and the transfer of polaroplast organelle membrane into plasma membrane. *Journal of Protozoology* **31**, 195–198.

Weiss, L.M. and Vossbrinck, C.R. (1998). Microsporidiosis: molecular and diagnostic aspects. *Advances in Parasitology* **40**, 351–395.

Wright, J.H. and Craighead, E.M. (1922). Infectious motor paralysis in young rabbits. *Journal of Experimental Medicine* **36**, 135–140.

Yachnis, A.T., Berg, J., Martinez-Salazar, A., Bender, B.S., Diaz, L., Rojiani, A.M., Eskin, T.A. and Orenstein, J.M. (1996). Disseminated microsporidiosis especially infecting the brain, heart, and kidneys. *Clinical Microbiology and Infectious Disease* **106**, 535–543.

Yee, R.W., Tio, F.O., Martinez, A., Held, K.S., Shadduck, J.A. and Didier, E.S. (1990). Resolution of microsporidial epithelial keratopathy in a patient with AIDS. *Ophthalmology* **98**, 196–201.

Zeman, D.H. and Baskin, G.B. (1985). Encephalitozoonosis in squirrel monkeys (*Saimiri sciureus*). *Veterinary Pathology*. **22**, 24–31.

Clinical Syndromes Associated with Microsporidiosis

Donald P. Kotler[1] and Jan M. Orenstein[2]

[1]*Gastrointestinal Division, Department of Medicine, St Luke's – Roosevelt Hospital Center, College of Physicians and Surgeons, New York, NY 10032, USA, and* [2]*Department of Pathology, George Washington University School of Medicine, Washington, DC, 20037, USA*

1. Introduction..322
2. Microsporidial Species Causing Intestinal Disease in AIDS..................323
3. Epidemiology...323
   3.1. Prevalence 323
   3.2. Microsporidia as enteric pathogens 324
   3.3. Mode of transmission...327
4. Pathogenesis of Intestinal Injury.......................................327
   4.1. Primary infection...327
   4.2. Small intestinal injury...328
   4.3 Hepatobiliary disease 329
5. Immune Response to Microsporidia 330
6. Clinical Illness 330
   6.1. Small intestinal disease 330
   6.2. Hepatobiliary disease 331
   6.3. Other sites of infection 332
7. Diagnosis..333
   7.1. Electron microscopy..333
   7.2. Light microscopy 335
   7.3. Special preparations and procedures 339
   7.4. Stool examination..340
   7.5. Molecular techniques 340
8. Treatment 341
   8.1. Drug therapy..341
   8.2. Fluid, electrolyte, and nutritional therapies........................341
9. Conclusion...342
   References 343

Microsporidia are ubiquitous in nature. Several clinical syndromes have been associated with microsporidiosis, especially in HIV-infected individuals, and include enteropathy, keratoconjunctivitis, sinusitis, tracheobronchitis, encephalitis, interstitial nephritis, hepatitis, cholecystitis, osteomyelitis, and myositis. Diarrhea and malabsorption are the most common clinical problems. Enterocytozoon bieneusi is the most common microsporidial cause of intestinal disease. A second species, Encephalitozoon intestinalis *(originally named* Septata intestinalis*) is associated with disseminated as well as intestinal disease. Microsporidiosis has been seen worldwide, and is recognized as a frequent enteric infection in patients with AIDS. The pathogenesis of intestinal disease is related to excess death of enterocytes as a result of cellular infection. Clinically, microsporidiosis most often presents with diarrhea and weight loss as a result of small intestinal injury and malabsorption. However, microsporidia have been detected in virtually all organs, and may provoke symptoms related to their specific localization. The diagnosis of microsporidiosis is made histologically, either from tissue biopsies or secretions. While transmission electron microscopy was required for diagnosis in the past, special stains and light microscopy, as well as immunohistochemical and molecular techniques are capable of providing a firm diagnosis. Therapeutic options are limited.* Enc. intestinalis *responds well to albendazole, while no antiparasitic therapy has documented efficacy in* Ent. bieneusi *infections.*

1. INTRODUCTION

Microsporidia are obligate, intracellular, protozoal parasites (Vossbrink *et al.*, 1987) that are ubiquitous in nature, infecting species in all five classes of vertebrates as well as invertebrates. Microsporidial forms were first recognized in tissue sections by light microscopy about 70 years ago (Levaditi *et al.*, 1923), and linked to tissue injury and human disease almost 40 years ago (Matsubayashi *et al.*, 1959). *Encephalitozoon cuniculi* was the first microsporidian shown to cause infection in humans, and has a relatively broad host range in mammals (Bergquist *et al.*, 1983). Reports of clinical disease related to any microsporidial infections in man were rare before 1985. Six genera have been identified as pathogens in man, three of which are related to species causing disease in animals (see the chapter by Didier and colleagues in this volume, 1998).

The emergence of the acquired immunodeficiency syndrome (AIDS) has been accompanied by a heightened appreciation of the spectrum of potential pathogens facing man. Starting in 1985, reports of diarrheal syndromes associated with microsporidiosis were published (Desportes *et al.*, 1985; Dobbins and Weinstein, 1985). Other clinical syndromes associated with microsporidiosis have been reported in patients with AIDS, including keratoconjunctivitis, sinusitis, tracheobronchitis, encephalitis, interstitial nephritis, hepatitis, cholecystitis, osteomyelitis, and myositis. However, diarrhea and malabsorption are still the most common clinical problems in this population. The aim of this chapter is to describe the clinical and pathogenic features of microsporidiosis in patients with AIDS and to discuss the diagnostic modalities and treatment of such patients.

2. MICROSPORIDIAL SPECIES CAUSING INTESTINAL DISEASE IN AIDS

Enterocytozoon bieneusi were observed within villus epithelial cells in small intestinal biopsies from AIDS patients in Texas and in Washington, DC, in 1982 (Orenstein, 1991). The first literature reports appeared in 1985 from the USA and France (Desportes *et al.*, 1985; Dobbins and Weinstein, 1985). Several case reports and small case series followed (Modigliani *et al.*, 1985; Curry *et al.*, 1988; Orenstein *et al.*, 1990; Bernard *et al.*, 1991; Michiels *et al.*, 1991; Simon *et al.*, 1991b; Ullrich *et al.*, 1991). By late 1991, almost 100 cases were known, and over 500 cases were recognized by the middle of 1993. Cases of *Ent. bieneusi* have been reported from all continents with prevalence rates comparable to other AIDS-associated opportunistic enteric infections, such as *Cryptosporidium*. *Ent. bieneusi* has been identified in patients with other immune deficiencies (Sax *et al.*, 1995), as well as in an immunocompetent individual with a self-limited diarrheal illness (Sandfort *et al.*, 1994). These observations suggest that this microsporidial infection is more widespread than previously suspected.

In 1988, we identified a microsporidia by transmission electron microscopy that differed ultrastructurally from *Ent. bieneusi* (Orenstein *et al.*, 1992a). Two years later, three additional cases were identified. Since then, the second species, *Encephalitozoon intestinalis* (originally named *Septata intestinalis*) (Cali *et al.*, 1993), has been identified in the USA, Europe, and Australia. In contrast to infection with *Ent. bieneusi*, *Enc. intestinalis* infection is not limited to small intestinal enterocytes, but infects macrophages and disseminates widely (Orenstein *et al.*, 1992c). The biology of these organisms is discussed in greater detail in the accompanying chapters.

3. EPIDEMIOLOGY

3.1. Prevalence

Intestinal microsporidiosis in AIDS has been seen worldwide. Although initially regarded as an uncommon pathogen, it is recognized now as a frequent enteric infection in patients with AIDS. Case series have been published from North America, Europe, Africa, and Australia. Prevalence rates among AIDS patients have varied between 2% and 50% (Canning and

Hollister, 1990; Greenson *et al.*, 1991; Simon *et al.*, 1991b; Swenson *et al.*, 1993; Cotte *et al.*, 1993; Rabenek *et al.*, 1993; Molina *et al.*, 1993; Kotler and Orenstein, 1994b; Drobniewski *et al.*, 1995), depending on the study group and methods of diagnosis. Our own prospective, sequential, series of gastrointestinal evaluations in AIDS patients revealed an incidence of 33% in AIDS patients with chronic diarrhea (Kotler and Orenstein, 1994). The incidence of infection with *Ent. bieneusi* was about 10 times higher than that of *Enc. intestinalis* in early biopsy studies but more recent stool studies suggest a more equal incidence (Franzen *et al.*, 1996). The epidemiology of the two species will be considered together, though there are no studies which demonstrate that it is the same for both organisms.

3.2. Microsporidia as Enteric Pathogens

3.2.1. Determining a Cause–Effect Relationship

Defining the relationship between a candidate pathogen and the presence of clinical disease may be difficult, especially in HIV infection, since a large number of novel agents have been identified in such patients. Organisms such as cytomegalovirus and *Mycobacterium avium* complex originally were suspected as being commensal, rather than true disease-causing agents. Establishing a cause–effect relationship may be difficult in situations where animal models for a specific disease do not exist. The current situation is reminiscent of the middle and late 19th century, when a large number of potential microbial pathogens were first identified. At the time, several investigators proposed criteria by which a candidate pathogen could be said to cause a specific disease. The best known of these criteria were advanced by Robert Koch (Fredericks and Relman, 1996), on the bais of his clinical observations in patients with tuberculosis and anthrax. Three criteria were stated: (1) that the microbe can be isolated in all cases of the disease and under circumstances that can account for the pathologic changes and clinical course; (2) that the microbe does not occur in another situation as a nonpathogenic agent; and (3) that introduction of the microbe to a new organism, after its isolation and propagation in pure culture, reproduces the disease. These criteria were most helpful for candidate bacterial pathogens that were culturable at that time. They are less helpful for infections in which culture techniques or suitable animal models are not available, or for noninfectious causes of disease.

Other investigators have proposed modifications, most notably Bradford-Hill (Bradford-Hill, 1965), whose criteria for establishing cause and effect go beyond the realm of infectious diseases and can apply to many conditions. The criteria include strength, consistency and specificity of associaton, temporality, biological gradient, plausability, coherence, reversibility, and analogy. None of these epidemiologic criteria is absolute.

3.2.2. Evidence for and against Microsporidia as Enteric Pathogens

Microsporidia originally were thought to cause intestinal disease based upon their identification in histologically abnormal small intestinal mucosa during the evaluation of chronic diarrhea in severely immunosuppressed AIDS patients. Strong clinical–pathologic correlations were reported by several investigators (Orenstein *et al.*, 1990; Greenson *et al.*, 1991; Eeftinck Schattenkerk *et al.*, 1991; Kotler *et al.*, 1993). However, this association does not show specificity, since patients without intestinal symptoms usually are not evaluated by small intestinal biopsy. Recent studies have provided disparate information about a pathogenic role for *Ent. bieneusi*.

A prospective study of HIV-infected subjects with and without chronic diarrhea, defined as two or more loose stools per day for more than 30 days, showed *Ent. bieneusi* prevalence rates of over 20% for each group, irrespective of peripheral blood CD4+ lymphocyte counts, co-infections or other clinical parameters (Rabenek *et al.*, 1993). On follow-up studies, diarrhea did not invariably develop in patients with documented microsporidiosis, nor did they invariably have histological abnormalities (Rabenek *et al.*, 1995). The authors concluded that *Ent. bieneusi* infection is not associated with clinical disease in all cases, and suggested that microsporidia may not be true enteric pathogens. However, they did not distinguish between the presence of microsporidiosis as a commensal organism or the possibility that microsporidia may exist in a clinically latent state as well as being capable of producing tissue damage, as is seen in many animal models. However, close examination of their data suggests a biological gradient between parasite burden and the risk of having clinical symptoms, suggesting that other factors may influence the development of clinically overt disease.

In order to define the pathologic features of intestinal microsporidiosis, we studied a series of controls and AIDS patients, the latter selected on the basis of having microsporidiosis or cryptosporidiosis or neither organism on transmission electron microscopy of small intestinal biopsies (Kotler *et al.*, 1993). Partial villus atrophy and crypt hyperplasia were seen in the patients with microsporidiosis (and cryptosporidiosis) but not in the controls or in AIDS patients without enteric pathogens. These results suggest that the development of small intestinal injury in an AIDS patient is related to the presence of disease complications rather than to the immune deficient state *per se*. The mucosal histologic changes were similar in the patients with cryptosporidiosis and microsporidiosis. Alterations in the specific activities of selected mucosal digestive enzymes (sucrase, lactase, maltase) mirrored the histologic changes. These results show both an association of small intestinal damage and enteric infections, as well as a pattern of intestinal injury consistent with the mechanism of cellular injury, thus satisfying the plausability criterion of Bradford-Hill.

A more recent study compared the abilities of the polymerase chain reaction (PCR) and transmission electron microscopy to detect microsporidia in intestinal biopsies. Comparisons were made in a series of AIDS patients with and without diarrhea, the former group with and without microsporidiosis (Coyle et al., 1996). All patients with microsporidiosis, except one, had chronic diarrhea, and the final patient had objective evidence of small intestinal injury on biopsy and by D-xylose absorption testing. There were near-perfect correlations between the electron microscopic and PCR findings. Both *Ent. bieneusi* and *Enc. intestinalis* could be distinguished. These results demonstrate both strength and specificity of association. Other studies have documented the reversal of renal failure and dysfunction during therapy of *Enc. intestinalis* (see below) with albendazole (Orenstein et al., 1993; Aarons et al., 1994), thus fulfilling the criterion for reversibility, at least for *Enc. intestinalis*.

Other factors may influence the risk of developing clinical disease. Most studies found that microsporidiosis typically is diagnosed in patients with severe depletion of CD4+ lymphocytes (15–30 cells/mm^3) (Field et al., 1990; Eeftinck Schattenkerk et al., 1991; Molina et al., 1993; Kotler and Orenstein, 1994). The one study finding similar prevalence rates in patients with and without diarrhea showed much higher CD4+ lymphocyte counts (113 and 192 cells/mm^3 in patients with and without diarrhea, respectively) (Rabenek et al., 1993). These results suggest that the level of immune function may influence the probability of developing clinically overt disease.

Further information about the pathogenicity of microsporidia has been obtained from an animal model. Tzipori and colleagues (Tzipori et al., in press) were able to establish a persistent infection in rhesus monkeys infected with the simian immunodeficiency virus (SIV) using microsporidial spores isolated from symptomatic human AIDS patients by intestinal lavage. Infection was documented by shedding of spores in the stool and by *in situ* hybridization studies of intestinal mucosa. They were unable to transmit the infection to an SIV-negative, immunocompetent monkey. Of note, the infected monkeys showed no clinical signs of disease and had relatively preserved CD4+ lymphocyte counts, which are higher than seen in symptomatic AIDS patients (see below), and more in line with Rabenek's study (Rabenek et al., 1993), which also demonstrated asymptomatic infections in patients with relatively high CD4+ lymphocyte counts.

The same group reviewed autopsy pathology of monkeys dying of simian AIDS and noted the frequent presence of hepatic histopathology. Examination of their material disclosed a 30% prevalence of hepatic and small intestinal microsporidiosis in these animals, and not in autopsied animals without hepatic histopathology (K. Mansfield, A. Carville, D. Shvetz, J. MacKay, S. Tzipori, and A. Lackner, unpublished observations). The organism found very strongly resembled *Ent. bieneusi*, by both ultrastructual and molecular analysis. The localization of the infection generally was the same as in AIDS patients, though the hepatobiliary findings were predominant in the monkeys, while small intestinal disease usually is more significant in humans. The authors concluded that infection with an *Ent. bieneusi*-like organism occurs in immune deficient monkeys as a natural infection and can lead to tissue damage. In other studies, *Encephalitozoon hellem* and *Enc. cuniculi* could also be transmitted to immune athymic mice and SIV-infected macaques (Didier *et al.*, 1994).

Based upon the published studies, one can conclude, with fair certainty, that microsporidia are a cause of intestinal and hepatobiliary disease in patients with AIDS. Symptomatic infection also may occur, as a self-limited disease in the absence of immune deficiencies, and in immune deficiencies other than AIDS. Since infection is not always associated with symptoms, other factors besides infection alone determine the clinical course. It is possible that chronic latent infections occur, with clinical activation related to progressive immune deficiency or other factors (Franzen *et al.*, 1996).

3.3. Mode of Transmission

Neither the mode of transmission nor the natural reservoirs of *Ent. bieneusi* has been defined. Transmission of the infection to an immune deficient monkey was accomplished by the oral route. Most investigators suspect the parasite is acquired by ingestion of contaminated food and water. Although the majority of cases have been diagnosed in homosexual males, diagnoses also have been made in heterosexual women and children.

4. PATHOGENESIS OF INTESTINAL INJURY

4.1. Primary Infection

Little is known about primary infection. An animal study, discussed above, indicates that the infection is transmitted by the oral route. Factors that may influence infectivity, such as resistance to drying, acid, etc., have not been studied for *Ent. bieneusi* or *Enc. intestinalis*. The intraluminal factors affecting excystation are uncertain. After passage of ingested spores into the intestinal lumen, the polar filament is extruded, possibly in response to the higher pH in the small intestinal lumen and other local factors (Canning and Hollister, 1987). On occasion, spores with extruded polar filaments are seen in intestinal aspirates (D.P. Kotler, unpublished observations). The polar filament probably facilitates infection of epithelial cell by direct penetration across the brush border membrane and injection of sporoplasm. The life cycles of other microsporidia have been measured *in vitro* and are complete within a few hours (Canning and Lom, 1986). Such studies have not been done in human infections. Extrusion of spores into the lumen or sloughing of infected cells returns the spores to the intestinal lumen (Figure 1). Spores may travel from cell to cell underneath the mucus layer, infecting contiguous cells, or may enter the bulk luminal fluid and potentially infect cells at distant sites or be excreted in feces. It is unclear which mechanism is more important or if alternative mechanisms of cell-to-cell transmission occur.

4.2. Small Intestinal Injury

Current understanding of disease pathogenesis is based upon clinicopathologic correlations and the pattern of intestinal injury. Small intestinal injury is characterized by villus atrophy, presumed due to chronic excess loss of epithelial cells. The decrease in surface area is reflected in subnormal serum concentrations of D-xylose following an oral test dose (Kotler et al., 1990a, 1993). Intestinal dysfunction in Ent. bieneusi or Enc. intestinalis infections

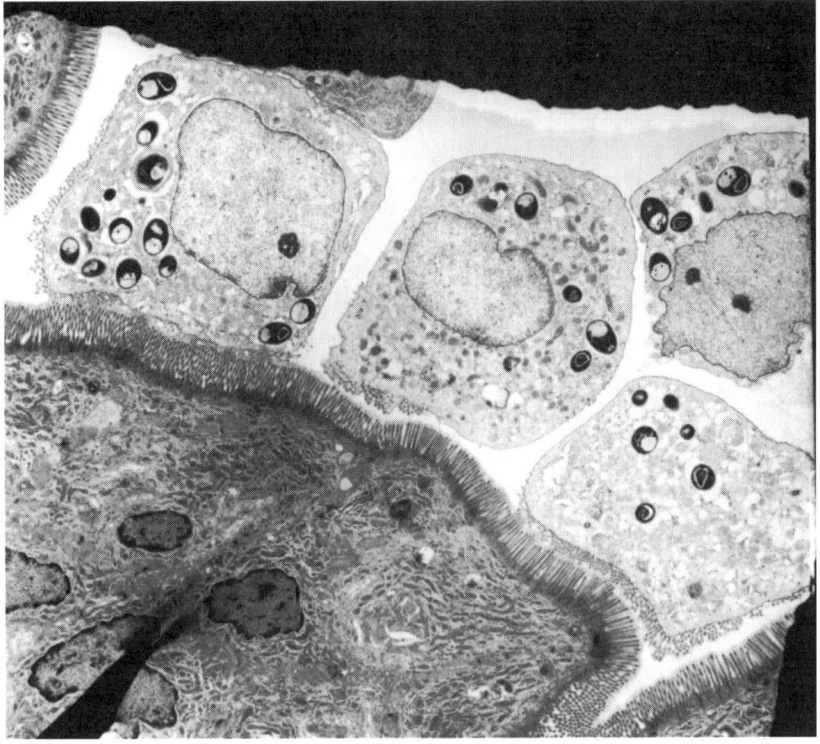

Figure 1 Transmission electron microscopy of jejunal biopsy performed after intestinal lavage. Several shed epithelial cells contain microsporidial spores. ×2250.

resembles that of patients with tropical sprue and celiac disease, two diseases also characterized by excess losses of villus enterocytes (Brunner et al., 1970). This relationship, together with the clinical–pathological correlations, represents the criterion of analogy, as described by Bradford-Hill (1965).

Under normal circumstances, the rates of epithelial cell production and loss on the villus are coordinated in order to maintain homeostasis, and are regulated by both discrete and interdependent influences (Johnson, 1988). The average turnover time for an epithelial cell is approximately 72 hours, during which time cell maturation and senescense occurs. The time required for maturation of specific enterocyte enzymes varies. For example, while the specific activities of the brush border disaccharidases maltase and sucrase are roughly equivalent along the length of the villus, the specific activities of lactase and enzymes of lipid metabolism are expressed more slowly, and are absent in the lower villus and highest in the upper villus (Shiau et al., 1979; Boyle et al., 1980).

Cell proliferation and loss are affected by many physiologic and pathologic stimuli. Under conditions of increased cell loss, such as those due to microsporidial infection, compensatory crypt hyperplasia occurs and returns villus architecture towards normal, resulting in partial villus atrophy and elongated crypts. The migration rate of newly formed enterocytes is increased, leaving insufficient time for complete functional maturation. Clinically, greater deficits in lactose and fat absorption occur than in starch and sucrose absorption in diseases producing villus atrophy and crypt hyperplasia.

4.3. Hepatobiliary Disease

Opportunistic infection of the bile ducts in patients with AIDS was first described in 1983. Several etiologic agents have been identified, including Cytomegalovirus, *Cryptosporidium parvum*, mycobacteria, *Isospora belli*, as well as malignancies, though about one half of all cases were unexplained in most series (Cello et al., 1989; McWhinney et al., 1991; Beaugerie et al., 1992; Orenstein et al., 1992a; Pol et al., 1993). Both *Ent. bieneusi* and *Enc. intestinalis*, are less commonly reported causes of AIDS-related cholangitis and acalculous cholecystitis, occurring in less than 10% in most series. Pol et al. (1993) reported on a series of patients with AIDS-related cholangitis in whom no opportunistic infection had been identified but who were subsequently found to have microsporidiosis. Microsporidiosis is associated with AIDS-related cholangitis in about one-third of cases in the author's experience (D.P. Kotler, personal observations). Given the difficulties in establishing the diagnosis of microsporidiosis at many centers and the finding of frequent hepatobiliary microsporidiosis in simian AIDS, it is likely that a significant proportion of patients with AIDS-related cholangitis have microsporidiosis as the cause.

The pathogenesis of AIDS-related cholangitis is not known. Many authors believe that sclerosing cholangitis in human immunodeficiency virus (HIV) infection and other clinical situations represents a stereotypic pattern of injury in the bile duct. Suggested mechanisms include injury related to desquamation of the epithelium and exposure of the lamina propria to bile salts and other compounds in the bile, which produces an inflammatory reaction and concentric fibrosis.

5. IMMUNE RESPONSE TO MICROSPORIDIA

Little is known about immunity to *Ent. bieneusi* and *Enc. intestinalis*. Antibodies to microsporidia have been determined in serologic studies, but the specificity of the antibody response is uncertain (Didier *et al.*, 1993; Aldras *et al.*, 1994). There is no published information on the role of cell mediated immunity, though clinical specimens typically show increased numbers of intraepithelial lymphocytes in areas containing *Ent. bieneusi*. Symptomatic illness in AIDS patients usually is associated with severe depletion of CD4+ lymphocytes, suggesting a crucial role for cell-mediated immunity. The topic is covered in greater detail in the chapter by Didier and colleagues (this volume, 1998).

6. CLINICAL ILLNESS

6.1. Small Intestinal Disease

The severity of clinical illness in patients with microsporidiosis is quite variable. Clinically, intestinal infection with *Enc. intestinalis* is similar to *Ent. bieneusi* infection, and they produce typical symptoms of malabsorption (Kotler, 1991). A small percentage of patients do not complain of diarrhea, though objective evidence of enteropathy and malabsorption may be present in the absence of diarrhea. Patients typically complain of 3–10, nonbloody bowel movements of variable volume and consistency, occurring at irregular intervals during the day. Bowel movements tend to cluster during one portion of the day, usually late evening or early morning, and nocturnal diarrhea is uncommon. Some of the bowel movements are watery and of large volume, while formed stools are rarely passed. Excessive flatus and an alteration in the odor of feces and flatus are often noted. Patients with mild disease may describe intolerance to lactose and fat, whereas those with more severe symptoms are affected by almost all food intake. When severe, there is associated dehydration and electrolyte abnormalities, predominantly hypokalemia, hypomagnesemia, and decreased serum bicarbonate concentrations. There is no fever. Appetite may be preserved, but calorie counts often reveal inadequate intake. Clinically, a prolonged satiety phase is observed, similar to that of other clinical and experimental malabsorption syndromes (Sclafani et al., 1978). This alteration in appetitive behavior may be related to the presence of unabsorbed nutrients in the lower intestine, changes in gastric and pancreatic secretion, or disturbances in gastric emptying and small intestinal transit (Burn-Murdoch et al., 1978; Owyang et al., 1983; Spiller et al., 1984). Malabsorption of sugars and fats can be detected (Kotler et al., 1990a, 1993; Asmuth et al., 1994). Weight loss is slow and progressive. Rapid changes in weight are usually related to alterations in hydration status, usually associated with a change in the intensity of diarrhea or rehydration therapy (Babameto et al., 1994).

6.2. Hepatobiliary Disease

The major clinical findings are abnormal liver function tests and abdominal pain suggestive of cholecystitis or biliary colic. The diagnosis of AIDS-related cholangitis is based on typical findings on cholangiography. Cello (1989) described four distinct patterns of AIDS-related cholangitis: papillary stenosis (15–20%); focal strictures and dilatation of the intra- and extrahepatic ducts (sclerosing cholangitis) (20%); combined papillary stenosis and sclerosing cholangitis (50%); and long extrahepatic bile duct strictures (15%). Liver biopsy findings in AIDS-related cholangitis are nonspecific and may be normal in some patients. Portal fibrosis and nonspecific inflammatory cell infiltrates have been described.

There is no curative therapy for AIDS-associated cholangitis. The major therapeutic possibility is the treatment of pain. Substantial relief pain is associated with sphincterotomy in a majority of patients who demonstrate papilary stenosis on cholangiogram. In other patients with findings consistent with cholecystitis, cholecystectomy by the open or laparoscopic routes may bring relief of symptoms. However, the underlying process of sclerosing cholangitis is progressive.

The prognosis in patients with AIDS-related cholangitis of any etiology is poor, as it is a manifestation of late-stage AIDS. Short survivals have been reported, with the specific cause of death unrelated to the cholangitis. However, prolonged survival of up to 32 months has been seen in a minority of patients. In these patients, death was directly related to the cholangitis, with liver failure in several patients and refractory bacterial cholangitis in another (D.P. Kotler, unpublished observations).

The authors have seen a case of subacute pancreatitis associated with chronic pain in a pediatric patient with *Ent. bieneusi* infection, although histologic examination of the pancreas or pancreatic ducts was not performed. As noted above, immunodeficient monkeys with microsporidiosis may have prominent injury of both intrahepatic and extrahepatic biliary epithelium, as well as small intestinal enterocytes, gallbladder epithelium, and pancreatic ductal epithelium.

6.3. Other Sites of Infection

Patients with *Enc. intestinalis* infections are prone to develop disseminated disease (Orenstein *et al.*, 1992c; Cali *et al.*, 1993; Aarons *et al.*, 1994; Molina *et al.*, 1995; Dore *et al.*, 1995; Gunnarson *et al.*, 1995). The first case seen developed fulminant multiorgan failure, resembling a hyperinfection syndrome such as seen with *Strongyloides stercoralis*. The most prominent site of injury is the kidney. Although patients may be asymptomatic, some have flank pain and symptoms of urethritis. Renal failure has been ascribed to microsporidiosis (Aarons *et al.*, 1994). Renal involvement has been confirmed by the identification of spores, both intracellular and free, in urinary sediment (Color Plate 1). Ultrastructural analysis of cell preparations revealed infection of both tubule cells and transitional cells. Autopsy material from a patient showed interstitial nephritis, with *Enc. intestinalis* in the proximal and distal convoluted tubule. Spores of *Enc. intestinalis* also have been identified in most other organs, including nasal and sinus mucosae, brain (pituitary), liver, spleen, as well as in a rectal ulcer (see below). Several case reports and case series have described pulmonary, nasal, corneal, and conjunctival infections with other microsporidia, including *Enc. cuniculi* and the morphologically identical *Enc. hellum* (Cali *et al.*, 1991; Schwartz *et al.*, 1992; Weber *et al.*, 1992a; Schwartz *et al.*, 1993; Weber *et al.*, 1993; Remadi *et al.*, 1995). Isolated case reports of peritonitis and myositis also have been reported (Ledford *et al.*, 1985; Zender *et al.*, 1989). These infections are notable for the relatively minor inflammatory reaction and their chronicity. Disseminated disease associated with serious multisystem failure was observed in the first patient with *Enc. intestinalis* to be recognized. Disseminated infections with serious multisystem involvement, including the brain, were reported, with etiologic agents being *Enc. cuniculi* in one case (Mertens *et al.*, 1997), and a *Pleistophora*-like organism in another (Yachnis *et al.*, 1996).

7. DIAGNOSIS

Diagnostic capabilities have undergone dramatic changes over the past 10 years. Initially, the diagnosis was based exclusively on transmission electron microscopy, thus limiting the ability to make diagnoses to few specialized centers. Using electron microscopy as a reference, light microscopic changes suggestive of microsporidiosis became apparent (Lucas *et al.*, 1989; Peacock *et al.*, 1991; Simon *et al.*, 1991a). The development of stains capable of detecting spores or earlier developmental forms (Rijpstra *et al.*, 1988; Giang *et al.*, 1993; Bryan and Weber, 1993; Field *et al.*, 1993; Kotler *et al.*, 1994; Franzen *et al.*, 1995) coincided with a marked increase in the number of cases being diagnosed. Diagnostic techniques have been applied to fecal specimens, thus broadening the pool of patients in whom a search for microsporidiosis is possible (van Gool *et al.*, 1990; Orenstein, 1991; Verre *et al.*, 1992; Weber *et al.*, 1992b, 1994; van Gool *et al.*, 1993; DeGirolami *et al.*, 1995; Claridge *et al.*, 1996). The development of molecular techniques carries the promise of greatly increased diagnostic sensitivity and specificity, as well as providing a tool for use in epidemiologic studies (Zhu *et al.*, 1993; Weiss *et al.*, 1994; Franzen *et al.*, 1995a; Fedorko *et al.*, 1995). Ongoing studies to determine the sensitivities, specificities and predictive values for the various diagnostic modalities will allow their standardization.

7.1. Electron Microscopy

The different stages in the life cycle of *E. bieneusi* and *Enc. intestinalis* have been characterized by transmission electron microscopy in several laboratories (Orenstein *et al.*, 1990; Cali and Owen, 1990; Orenstein *et al.*, 1991). Since *Ent. bieneusi* cannot be propagated *in vitro*, the various developmental stages have been characterized in clinical specimens and their sequence inferred from studies of other microsporidia.

Ent. bieneusi infection is limited to epithelial cells in the small intestine and the hepatobiliary tree, sinuses, tracheobronchial tree and pancreas. Spores of *Ent. bieneusi* have been detected in the lamina propria macrophages (Schwartz *et al.*, 1995), probably as a result of cell lysis. There is no evidence for systemic dissemination of *Ent. bieneusi*. The earliest stages of *Ent. bieneusi* to be identified are meronts which are small, oval, membrane-bound inclusions containing free ribosomes, usually seen in the apical cytoplasm in association with mitochondria of host cell origin. A nucleus is the next structure to be recognized. Subsequently, nuclear division, emergence of rough endoplasmic reticulum, and development of electron-lucent inclusions are seen. Later in this stage, electron-dense, disk-like structures begin to develop from the clefts. Development of the polar tube and its association with an anchoring plate, polaroplast membrane, nucleus, and posterior vacuole occurs during sporogony (Figure 2). Through a complicated process of membrane invaginations, the sporogonial plasmodium divides into multiple sporonts. With development of the endospore and ectospore, first the

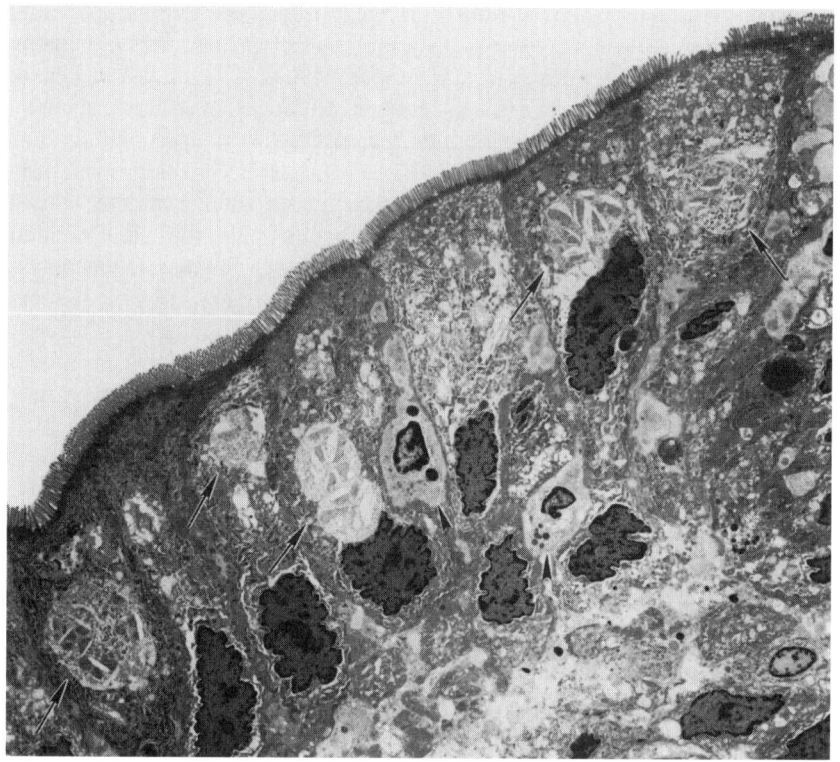

Figure 2(a) Low magnification transmission electron micrograph of villus enterocytes infected with the plasmodial stages of *Ent. bieneusi*. Note the clear clefts in the plasmodia, the vesiculated cytoplasm, the mostly intact brush border, and the intraepithelial lymphocytes (arrowheads). × 2178.

sporoblast and then the mature electron-dense, egg-shaped spore is formed (Figure 3).

The electron-dense disks, which appear at the beginning of sporogony develop into a coiled tubule. The coiled tube contains six turns in *Ent. bieneusi* and is organized into two tiers of three turns each. The two tiers are consistently out of register by about 45° (see Figure 3).

Mature spores may erupt through the enterocyte brush border membrane. However, some cells retain viability for a period of time after sloughing (see Figure 1). For this reason, it is possible that spore maturation might still proceed after the enterocyte has been extruded into the intestinal lumen.

Distinctive morphologic features of *Enc. intestinalis* include a unique development of individual cells within separate chambers of a parasitophorous vacuole (Figure 4). The developing organisms are separated by fibrillar septa of parasite origin. Other distinguishing features include the lack of

CLINICAL SYNDROMES ASSOCIATED WITH MICROSPORIDIOSIS 335

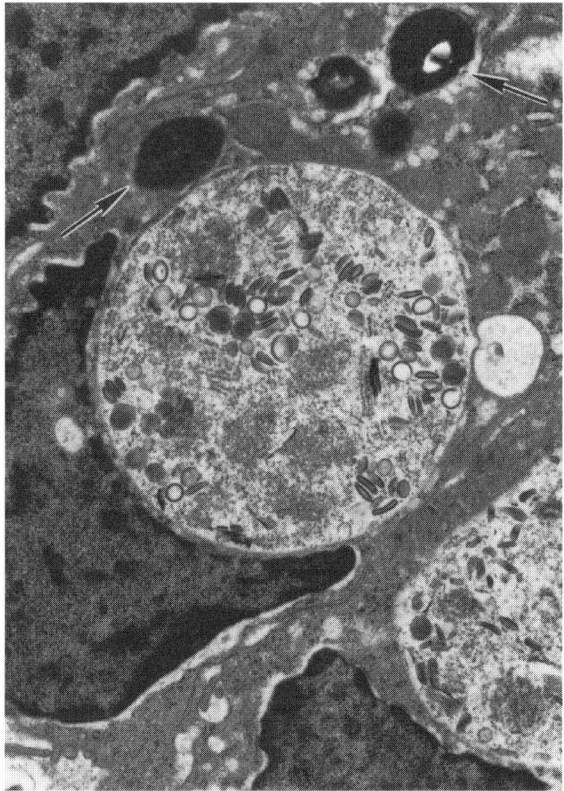

Figure 2(b) Transmission electron micrograph of a sporogonial plasmodium of *Ent. bieneusi* molding the apical pole of an enterocyte nucleus plus three electron-dense spores (arrows). The plasmodium contains dark disk-like structures that will form into spore polar tubes. × 10 800.

electron-lucent inclusions or electron-dense disks, the polar tubule with a single tier of 5–6 turns, and spores of 2.2 × 1.2 µm.

7.2. Light Microscopy

The routine histologic diagnosis of *Ent. bieneusi* infection is difficult by light microscopy, since the developing forms do not stain well with standard hematoxylin and eosin preparations. In addition, there is a gradient in the intensity of infection along the small intestine with the distal duodenum and proximal jejunum having higher parasite burdens than the proximal duodenum (Orenstein *et al.*, 1992b). Diagnosis by ileal biopsy has been made in several laboratories (Weber *et al.*, 1992c).

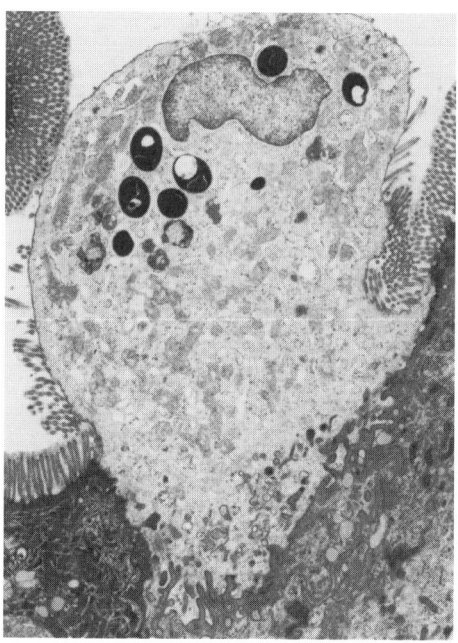

Figure 3(a) Transmission electron micrograph of a shedding enterocyte containing several oval spores of *Ent. bieneusi*, some with visible posterior vacuoles. The microvilli are mostly absent. × 3750.

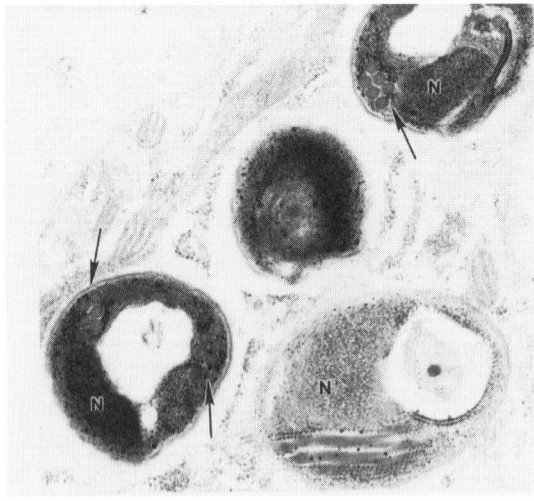

Figure 3(b) Transmission electron micrograph of four spores showing vacuoles, two rows of three turns of the polar tubes (arrows), nuclei (N), and a straight portion of a polar tube (arrowhead). × 30 360.

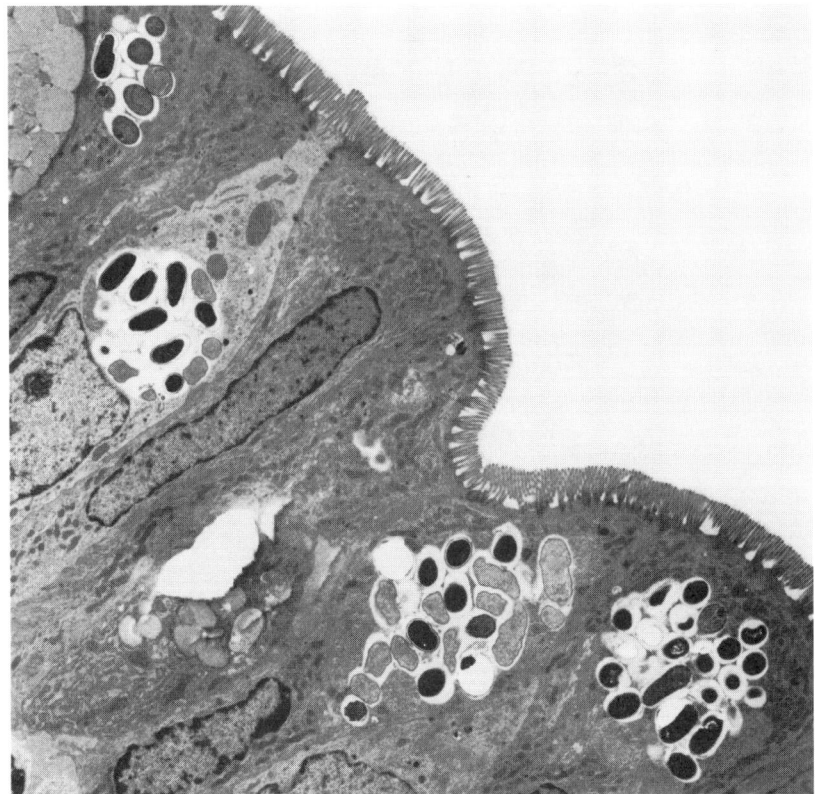

Figure 4(a) Transmission electron micrograph shows four villus enterocytes containing honeycombed parasitophorous vacuoles. The parasites (*Enc. intestinalis*) become more electron dense as they mature from meronts to spores. Note the intact brush border and the molded nucleus. × 3225.

A characteristic pattern of tissue injury and cytopathology has been found in AIDS patients evaluated for chronic diarrhea. *Ent. bieneusi* infection is usually associated with villus atrophy and crypt hyperplasia. In some cases, villus height is nearly normal and marked crypt hyperplasia is present. In other cases, significant villus atrophy is associated with less marked crypt hyperplasia. The reason for this variation is unknown, but may reflect differences in parasite burden or endogenous factors, such as nutritional status. Biopsies containing long, slender villi and short crypts are very unlikely to harbor *Ent. bieneusi*. The degree of injury is usually similar in different biopsies obtained from the same area.

Ent. bieneusi-associated cellular injury appears concentrated in the upper third of the villus, and is not seen in the crypt. There is no evidence of acute

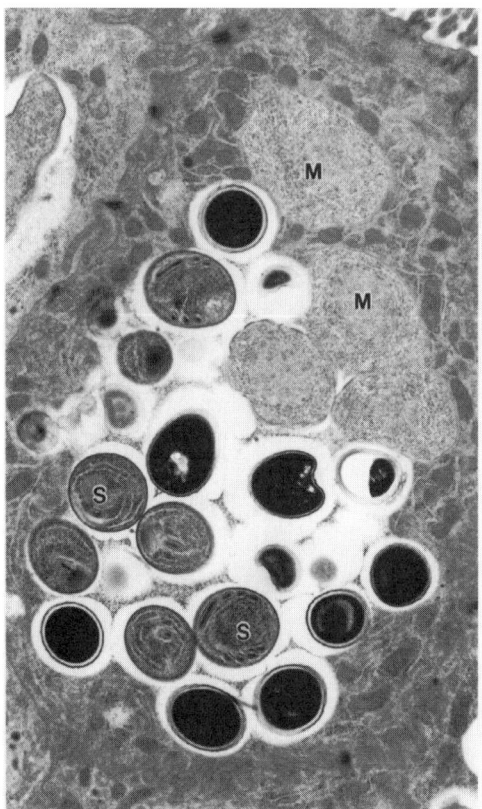

Figure 4(b) Higher magnification transmission electron micrograph shows scalloped, septated vacuoles with peripherally located meronts (M), sporoblasts (S), and electron-dense mature spores of *Enc. intestinalis.* × 8550.

inflammation, although the numbers of intraepithelial lymphocytes are increased in areas of greatest activity. There is epithelial disarray and excess nuclear debris. Individual sloughing cells appear as teardrop-shaped cells, invariably containing refractile spores (see Figure 3). This finding is characteristic of *Ent. bieneusi* and is its most readily recognized diagnostic feature. Villus enterocytes demonstrate cytopathic changes, including pleomorphism, hyperchromatic nuclei, and loss of the basal orientation of nuclei. There may also be increased numbers of lysosomes, vesiculation, vacuolization, and occasional lipid accumulation in the cytoplasm. The lamina propria may contain increased numbers of plasma cells and macrophages, but not neutrophils.

Since the organisms develop in the supranuclear and apical cytoplams, they may affect nuclear shape, producing a flattening or cupping of the apical pole

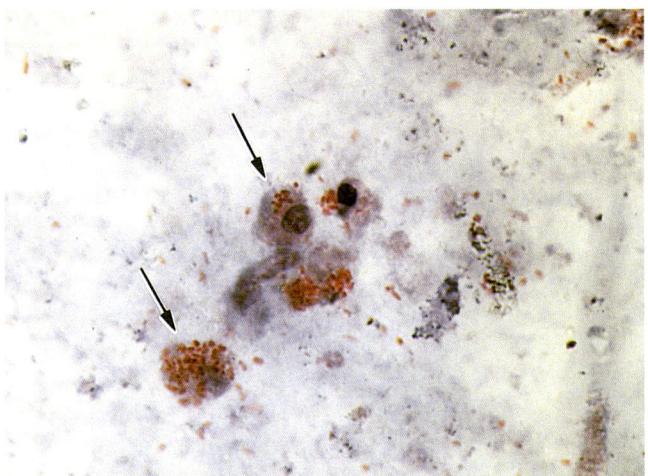

Color Plate 1 Spores of *Encephalitozoon intestinalis* stain red with the modified trichrome stain. In this urine sediment, there are spores within shed tubular epithelium (arrows) as well as free. × 640.

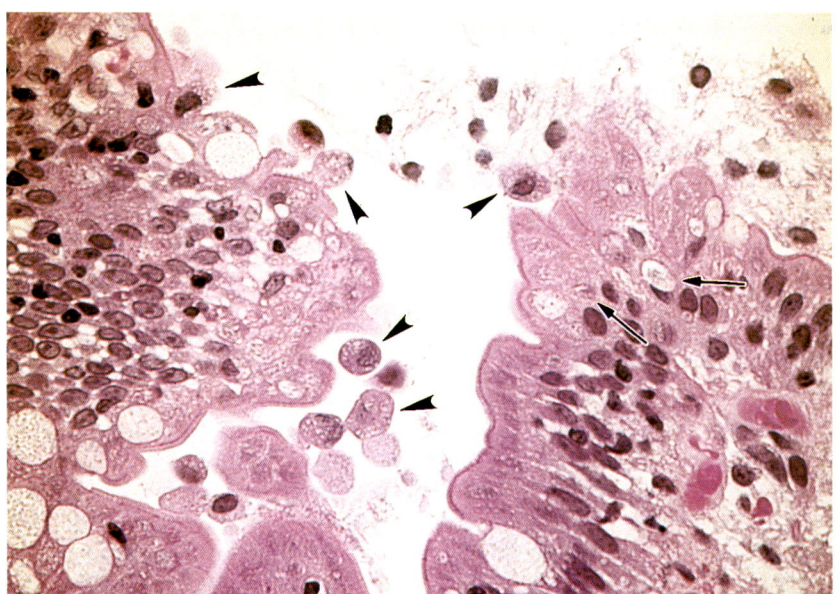

Color Plate 2 Tips of two villi heavily infected with *Enterocytozoon bieneusi*. Plasmodia (arrows) appear as bluish supranuclear structures often containing clear clefts. Mature spores are best identified in sections stained with hematoxylin and eosin as small refractile bodies within shedding enterocytes (arrowheads). × 576.

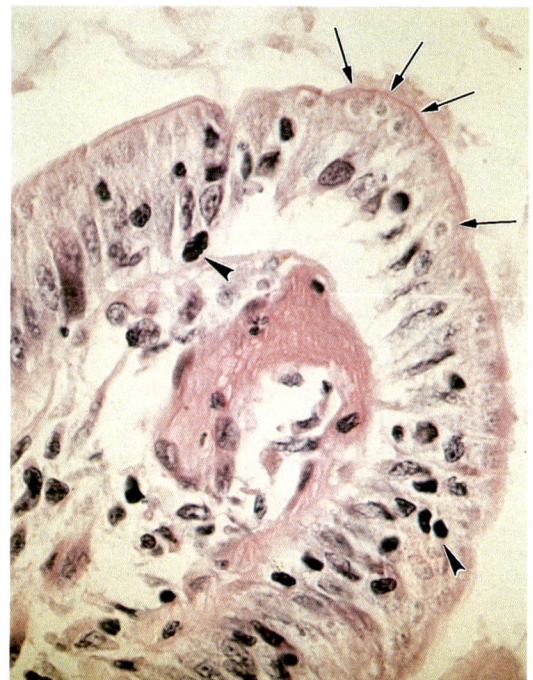

Color Plate 3 A strip of *Ent. bieneusi*-infected enterocytes is shedding from the tip of a villus. The plasmodia appear as light blue-staining supranuclear bodies (arrows). Some plasmodia appear to be within vacuoles, an artifact of cell degeneration. Note the intraepithelial lymphocytes (arrow-heads). Hematoxylin and eosin, × 512.

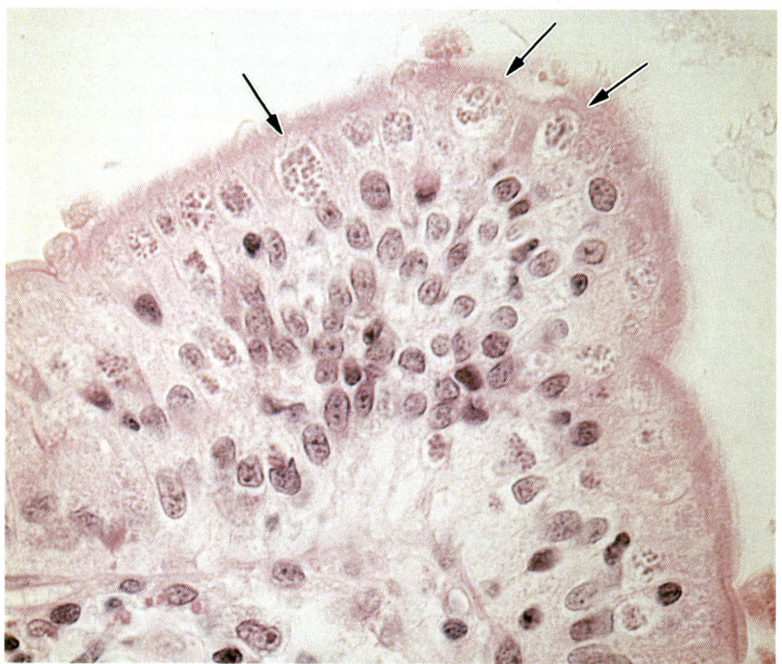

Color Plate 4(a) In this hematoxylin and eosin stained tangential section of the tip of a villus, both the parasitophorous vacuole and its developing parasites (*Enc. intestinalis*) are visible (e.g. arrows). × 640.

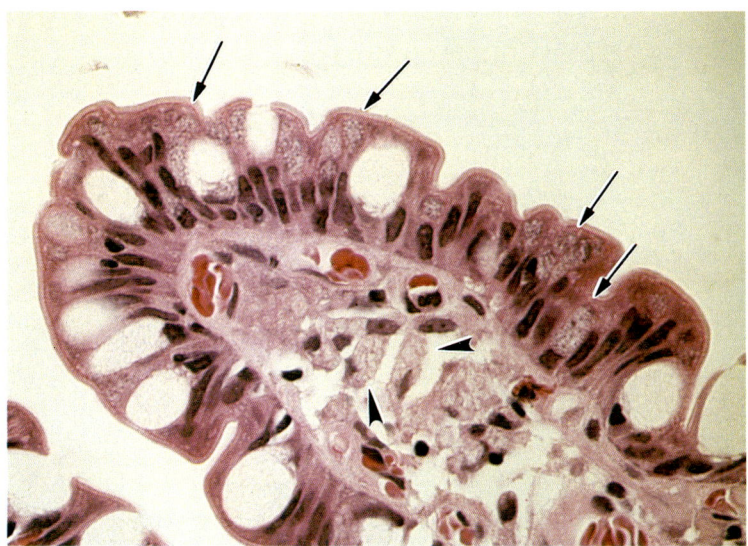

Color Plate 4(b) In this hematoxylin and eosin stained section, the organisms appear refractile (glassy). Note their supranuclear location and the intact brush border. The vacuoles in the lamina propria macrophages (arrows) presumably correspond to lysing spores. × 595.

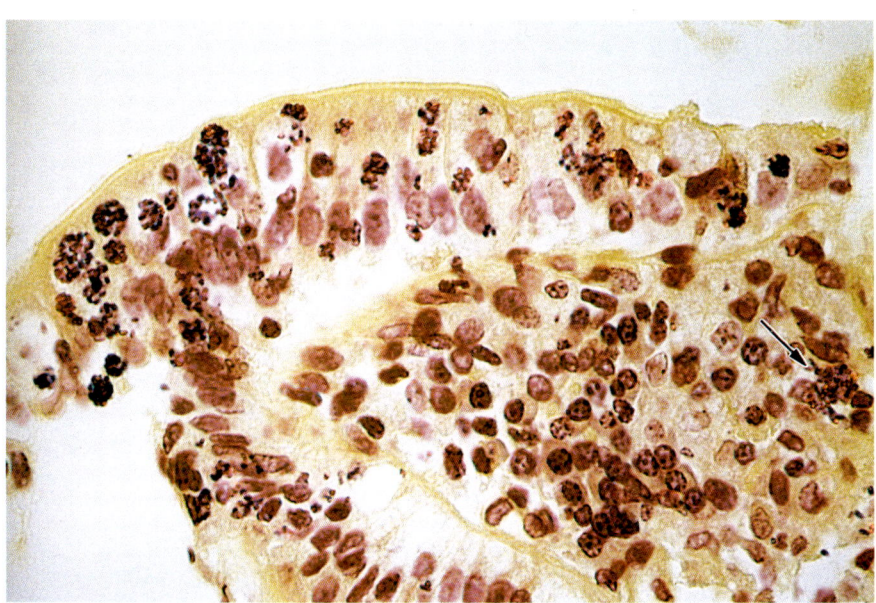

Color Plate 5 In this partially polarized Brown Benn (Gram) stained section supplied by AFIP, maturing parasites of *Enc. intestinalis* stain progressively darker red and are more birefringement (yellow). Note the polarizing spores in the lamina propria (arrows). Cell nuclei are stained in this overstained section. × 576.

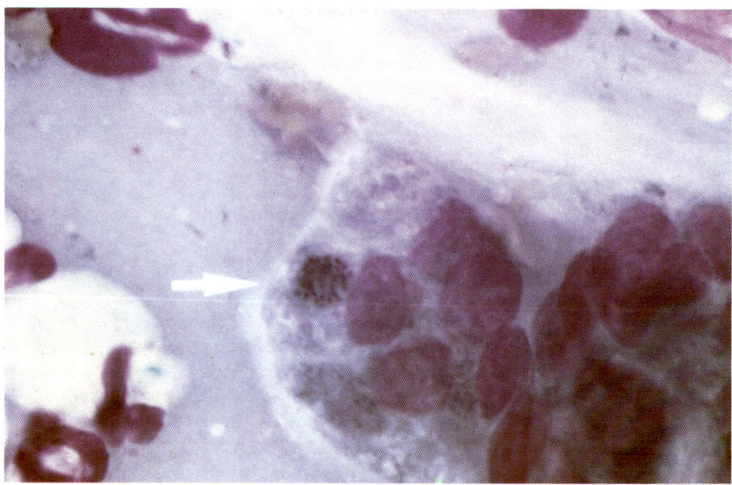

Color Plate 6 Giemsa stained touch preparation of an intestinal biopsy from an HIV-infected patient with diarrhea demonstrating *Ent. bieneusi* (arrow). Reproduced courtesy of Dr Murray Wittner, Albert Einstein College of Medicine.

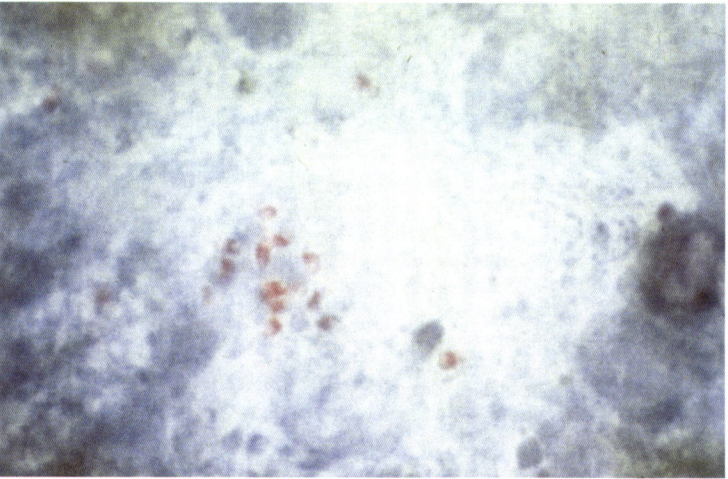

Color Plate 7 Chromotrope 2R stained preparation of a stool specimen, demonstrating microsporidia. Microsporidia stain light pink with a belt-like stripe girding them diagonally and equatorially. Reproduced courtesy of Dr Elizabeth Didier, Tulane Primate Center.

(see Figure 2b). The electron-lucent clefts also may be visible by light microscopy (see Color Plate 2), especially when flanked by material that picks up the hematoxylin stain. The result is a cat's-eye appearing structure in the apical portion of the cell. Although it is an artifact, plasmodia can appear to be within a vacuole (see Color Plate 3). Rarely, spores may be detected as clusters of negatively staining or refractile granular material in the cytoplasm of individual, but especially sloughing, cells.

Enc. intestinalis is diagnosed by the same techniques as *Ent. bieneusi*. The organism is usually easier to detect, due to its larger size, greater refractility, birefringence, enhanced staining, and greater parasite burden (Color Plates 4a and 4b). Macrophages (even endothelial cells and fibroblasts) containing spores can be seen in the lamina propria. Free spores and cells containing spores also can be found in urinary sediment (see Color Plate 1).

7.3. Special Preparations and Procedures

Special stains allow confirmation in a suspicious case, especially when transmission electron microscopy is not readily available. Useful stains include Giemsa, modified Warthin–Starry, Brown Benn, Gram, a modified tissue trichrome Chromotrope 2R stain (as described for examination of fecal specimens in paraffin-embedded tissues), and Giemsa-stained touch preparations of fresh mucosal biopsies (Simon *et al.*, 1991a; Giang *et al.*, 1993; Field *et al.*, 1993; Kotler *et al.*, 1994; Franzen *et al.*, 1995a). Examination under polarized light is another effective means of detecting *Ent. bieneusi*, as the spores are birefringent, particularly in Gram and Warthin–Starry stained sections. The spores of *Enc. intestinalis* are brightly birefringent, and stain to a much greater degree than *Ent. bieneusi* spores (Color Plates 5 and 6).

We evaluated the sensitivities, specificities, and predictive values for several special stains as compared to transmission electron microscopy (Kotler and Orenstein, 1994). The sensitivities and negative predictive values of the different techniques ranged from about 60–90%, while the specificities and positive predictive values were in the range 95–100%. Methods for quantitating parasite burden were devised. The parasite burden was lower in cases with false-negative results, compared to transmission electron microscopy, than in the true positives, suggesting that a false-negative diagnosis is related to a low parasite burden. Gram and Giemsa stains showed spores and some visualization of other intracellular forms, whereas the chromotrope stain, acid-fast stains, and touch preparations demonstrated only spores. A surprising finding was the excellent sensitivity of diagnosing *Ent. bieneusi* in hematoxylin and eosin stained sections when mixed with electron microscopy negative biopsies from AIDS patients, albeit by observers with large clinical experience. The ability to correctly diagnose microsporidiosis by hematoxylin and eosin stained sections has been noted by several investigators (Rijpstra *et al.*, 1988; Lucas *et al.*, 1989; Peacock *et al.*, 1991; Simon *et al.*, 1991a, b; Weber *et al.*, 1992).

Samples of intraluminal fluid and cytobrush preparations are also effective in detecting microsporidia, using Giemsa, Gram, Diff-Quik, and modified trichrome stains. The chromotrope 2R stain allows differentiation between microsporidial spores and bacteria, since bacteria (light-green counterstain) do not take up the chromotrope stain.

7.4. Stool Examination

The need for intestinal biopsy to diagnose microsporidiosis has limited its ready application in clinical situations and severely confounded most attempts to study the organism itself, due to its inherently invasive nature. The ability to accurately diagnose microsporidiosis would greatly facilitate work in the field. Several techniques have been applied successfully. Giemsa staining of a fecal preparation that was homogenized, sieved, centrifuged, and extensively washed, revealed *Ent. bieneusi* spores (van Gool *et al.*, 1990). Transmission electron microscopy of the stool also detected spores (Orenstein *et al.*, 1991). The chromotrope 2R modified trichrome stain also has been used to detect spores in stool (Weber *et al.*, 1992b). Occasional yeast forms or bacteria take up the chromotrope stain, but they differ from *Ent. bieneusi* in size and shape. The advantage of this stain is that specimens without special preparation or formalin-fixed stool may be used. However, attention to the staining conditions, especially decolorization, is required. A further modification of this method, staining at 56°C, decreases the time required for the parasite to take up the stain (Bryan *et al.*, 1991). An alternative method using the fluorochrome, Uvitex 2B, Calcifuor, Fungifluor, or Fungignal, is effective in staining chitin, which is a component of the wall of the microsporidial spore (van Gool *et al.*, 1993; Franzen *et al.*, 1995a). Although prior formalin fixation can reduce the staining, it is still within acceptable diagnostic levels.

7.5. Molecular Techniques

Several laboratories have successfully identified microsporidia in stool and clinical specimens using a variety of techniques, such as *in situ* hybridization and PCR (Zhu *et al.*, 1993; Weiss *et al.*, 1994; Franzen *et al.*, 1995b). These are described in detail in the chapter by Weiss and colleagues (this volume). Few studies have examined the relative sensitivities, specificities or predictive values of these methods.

8. TREATMENT

8.1. Drug Therapy

Few studies of drug therapy of *Ent. bieneusi* infection have been published. Symptomatic response to antiparasitic agents, such as metronidazole, has been reported, although histologic evidence of infection persists (Field *et al.*, 1992). There are anecdotal reports of other antimicrobials, such as humatin and trimethoprim–sulfamethoxazole, among others, being used (Dieterich *et al.*, 1993; Dionisio *et al.*, 1995).

A group from London reported positive results from preliminary treatment studies of albendazole, an anti-tubulin drug (Blanshard *et al.*, 1992). A prospective, published study of 29 patients and observations in a total of 66 patients with *Ent. bieneusi* confirmed that albendazole therapy leads to some symptomatic improvement and weight stabilization (Dieterich *et al.*, 1994). However, follow-up biopsies continued to show evidence of infection, and follow-up D-xylose absorption tests failed to show evidence of improvement. Confirmation of the beneficial effect of albendazole awaits the completion of a recently completed double-blind, placebo-controlled trial.

Enc. intestinalis differs from *Ent. bieneusi* in its uniformly excellent response to albendazole therapy (Orenstein *et al.*, 1993; Aarons *et al.*, 1994; Weber *et al.*, 1994; Molina *et al.*, 1995; Dore *et al.*, 1995). Although relatively few cases have been reported, all have responded clinically with the disappearance of diarrhea and a clearing of spores from the urine and stool. Follow-up biopsies have shown disappearance of spores and only ghosts of spores within macrophage lysosomes (Figure 5). Improvement in D-xylose absorption has been noted (Orenstein *et al.*, 1993). Whether or not viable sporoplasm and spores are being sequestered in some location is unknown.

8.2. Fluid, Electrolyte, and Nutritional Therapies

Patients with *Ent. bieneusi* infection are chronically dehydrated and may be depleted of both macronutrients and micronutrients. Electrolyte and mineral deficits, particulary K^+, Ca^{2+}, and Mg^{2+}, may be severe. Diet modification may be helpful in patients with mild-to-moderate disease. A recent study demonstrated stabilization of weight and body cell mass in patients receiving an oral semi-elemental diet. Parenteral nutritional therapy resulted in nutritional repletion in some patients (Kotler *et al.*, 1990b; Melchior *et al.*, 1996). Opiates such as diphenoxylate, paregoric, or tincture of opium may be effective, although the dose required sometimes causes excessive sedation.

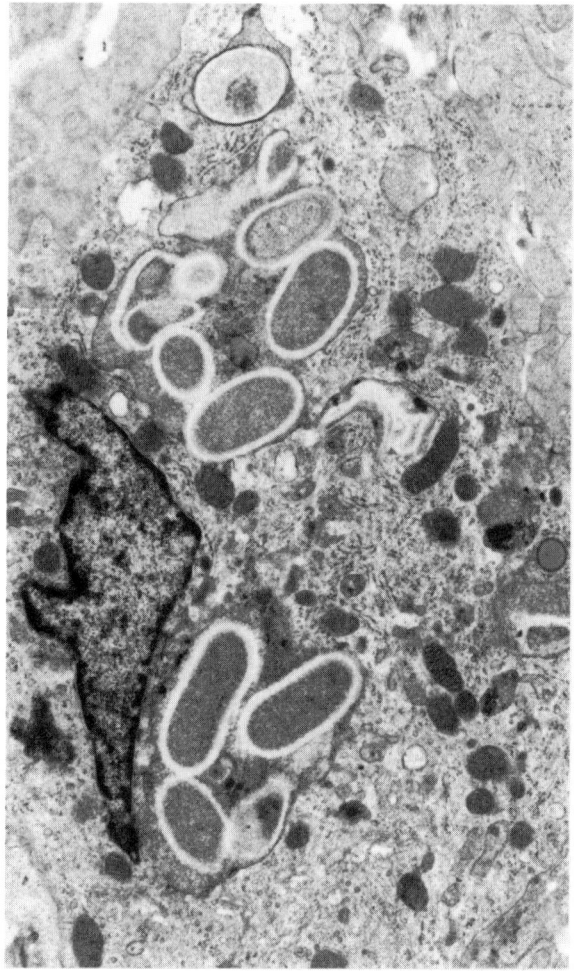

Figure 5 Transmission electron micrograph of a lamina propria macrophage that contains ghost-like remnants of spores of *Enc. intestinalis* within secondary lysosomes in a small bowel biopsy of a patient being treated with albendazole. × 8370.

9. CONCLUSION

Despite recognition that microsporidiosis is a potentially impoortant cause of intestinal disease in HIV-infected individuals, major gaps remain in our understanding of the tramsmission, life cycle, and immunobiology of microsporidia. Molecular techniques and the ability to propagate the organism *in vitro* should be helpful in addressing these issues and, ultimately, in developing effective therapies for microsporidiosis.

REFERENCES

Aarons, E.J., Woodrow, D., Hollister, W.S., Canning, E.U., Francis, N. and Gazzard, B.G. (1994). Reversible renal failure caused by a microsporidial infection. *AIDS* **8**, 1119–1121.
Aldras, A.M., Orenstein, J.M., Kotler, D.P., Shadduck, J.A. and Didier, E. (1994). Detection of Microsporidia by indirect imunofluorescence antibody test using polyclonal and monoclonal antibodies. *Journal of Clinical Microbiology* **32**, 608–612.
Asmuth, D.M., DeGirolami, P.C., Federman, M., Ezratty, C.R., Pleskow, D.K., Desai, G. and Wanke, C.A. (1994). Clinical features of microsporidiosis in patients with AIDS. *Clinical Infectious Diseases* **18**, 819–825.
Babameto, G., Kotler, D.P., Burastero, S., Wang, J. and Pierson, R.N. (1994). Alterations in hydration in HIV-infected individuals. *Clinical Research* **42**, 279A.
Beaugerie, L., Teilhac, M.-F., Deluol, A.-M., Fritsch, J., Girard, P.-M., Rozenbaum, W., Le Quintrec, Y. and Chatelet, F.-P. (1992). Cholangiopathy associated with *Microsporidia* infection of the common bile duct mucosa in a patient with HIV infection. *Annals of Internal Medicine* **117**, 401–402.
Bergquist, N.R., Waller, T., Mravak, S. and Meyer, U. (1983). Report of two recent cases of human microsporidiosis. In: *Abstracts. Annual Meeting, American Society of Tropical Medicine and Hygiene*, San Antonio, Texas.
Bernard, E., Michiels, J.F., Durant, J., Hoffman, P., Desalvador, F., Loubiere, R., Le Fichoua, Y. and Dellamonica, P. (1991). Intestinal microsporidiosis due to *Enterocytozoon bieneusi*: a new case report in an AIDS patient. *AIDS* **5**, 606–607.
Blanshard, C., Ellis, D.S., Tovey, D.G., Dowell, S., Gazzard, B.G. (1992). Treatment of intestinal microsporidiosis with albendazole in patients with AIDS. *AIDS* **6**, 311–313.
Boyle, J.T., Celano, P. and Koldovsky, O. (1980). Demonstration of a difference in expression of maximal lactase and sucrase activity along the villus in the adult rat jejunum. *Gastroenterology* **79**, 503–507.
Bradford-Hill, A. (1965). The environment and disease: association or causation? *Proceedings of the Royal Society of Medicine* **58**, 295–300.
Brunner, O., Edelman, S. and Klipstein, F.A. (1970). Intestinal morphology of rural Haitains: a comparison between overt tropical sprue and asymptomatic subjects. *Gastroenterology* **58**, 655–672.
Bryan, R.T. and Weber, R. (1993). Microsporidia. Emerging pathogens in immunodeficient persons. *Archives of Pathology and Laboratory Medicine* **117**, 1243–1245.
Bryan, R.T., Weber, R., Stewart, J.M., Angritt, P. and Visvesvara, G.S. (1991). New manifestations and simplified diagnosis of human microsporidiosis. *American Journal of Tropical Medicine and Hygiene* **45**, 133–134.
Burn-Murdoch, R.A., Fischer, M. and Hunt, J.N. (1978). The slowing of gastric emptying by proteins in test meals. *Journal of Physiology* **274**, 477–485.
Cali, A. and Owen, R.I. (1990). Intracellular development of *Enterocytozoon*, a unique microsporidian found in the intestine of AIDS patients. *Journal of Protozoology* **37**, 145–155.
Cali, A., Meisler, D.M., Rutherford, I., Lowder, C.Y., McMahon, J.T., Longworth, D.L. and Bryan, R.T. (1991). Corneal microsporidiosis in a patient with AIDS. *American Journal of Tropical Medicine and Hygiene* **44**, 463–468.
Cali, A., Kotler, D.P. and Orenstein, J.M. (1993). *Septata intestinalis*, n.g., n.sp., an intestinal microsporidian associated with chronic diarrhea and dissemination in AIDS patients. *Journal of Eukaryotic Microbiology* **40**, 101–112.

Canning, E.U. and Hollister, W.S. (1987). Microsporidia of mammals – widespread pathogens or opportunistic curiosities? *Parasitology Today* **3**, 267–273.
Canning, E.U. and Hollister, W.S. (1990). *Enterocytozoon bieneusi* (Microspora): prevalence and pathogenicity in AIDS patients. *Transactions of the Royal Society of Tropical Medicine* **84**, 181–186.
Canning, E.U. and Lom, J. (1986). *The Microsporidia of Vertebrates*. Orlando, FL: Academic Press.
Cello, J. (1989). Acquired immunodeficiency syndrome cholangiopathy: spectrum of disease. *American Journal of Medicine* **86**, 539–546.
Clarridge, J.E., Karkhanis, S., Rabeneck, L., Marino, B. and Foote, L.W. (1996). Quantitative light microscopic detection of *Enterocytozoon bieneusi* in stool specimens: a longitudinal study of human immunodeficiency virus-infected microsporidiosis patients. *Journal of Clinical Microbiology* **34**, 520–523.
Cotte, L., Rabodonirina, M., Piens, A., Perreard, M., Mofon, M. and Trepo, C. (1993). Prevalence of intestinal protozoons in French patients infected with HIV. *Journal of the Acquired Immunodeficiency Syndrome* **6**, 1024–1029.
Coyle, C.M., Orenstein, J.M., Wittner, M., Kotler, D.P., Noyer, C., Tanowitz, H. and Weiss, L.M. (1996). Prevalence of microsporidiosis in AIDS related diarrhea as determined by polymerase chain reaction to microspiridian ribosomal RNA. *Clinical Infectious Diseases* **23**, 1002–1006.
Curry, A., McWilliam, L.J., Haboubi, N.Y. and Mandal, B.K. (1988). Microsporidiosis in a British patient with AIDS. *British Medical Journal* **41**, 477–478.
DeGirolami, P.C., Ezratty, C.R., Desal, G., McCullough, A., Asmuth, D., Wanke, C. and Federman, M. (1995). Diagnosis of intestinal microsporidiosis by examination of stool and duodenal aspirate with Weber's modified trichrome and Unitex 2B stains. *Journal of Clinical Microbiology* **33**, 805–810.
Desportes, I., Le Charpentier, Y., Galian, A., Bernard, F., Cochand-Priollet, B., Lavergne, A., Ravisse, P. and Modigliani, R. (1985). Occurrence of a new microsporidan: *Enterocytozoon bieneusi* n.g., n.sp., in the enterocytes of a human patient with AIDS. *Journal of Protozoology* **32**, 250–254.
Didier, E.S., Kotler, D.P., Dieterich, D.T., Orenstein, J.M., Aldras, A.M., Davis, R., Friedberg, D.N., Gourley, W.K., Lembach, R., Lowder, C.Y., Meisler, D.M., Rutherford, I., Yee, R.W. and Shadduck, J.A. (1993). Serologic studies in human microsporidiosis. *AIDS* **7**, supplement 3, S8–S11.
Didier, E.S., Varner, P.W., Didier, P.J., Aldras, A.M., Millichamp, N.J., Murphy-Korb, M., Bohm, R. and Shadduck, J.A. (1994). Experimental microsporidiosis in immunocompetent and immunodeficient mice and monkeys. *Folia Parasitologia* **41**, 1–11.
Didier, E.S., Snowden, K.F. and Shadduck, J.A. (1998). Biology of microsporidian species infecting mammals. *Advances in Parasitology* **40**, 279–316.
Dieterich, D.T., Lew, E., Kotler, D.P., Poles, M. and Orenstein, J.M. (1993). Divergence between clinical and histologic responses during treatment of *Enterocytozoon bieneusi* infection with albendazole: prospective study and review of the literature. *AIDS* **7**, supplement 3, S43–S44.
Dieterich, D.T., Lew, E., Kotler, D.P., Poles, M. and Orenstein, J.M. (1994). Treatment with albendazole for intestinal disease due to *Enterocytozoon bieneusi* in patients with AIDS. *Journal of Infectious Disease* **169**, 173–183.
Dionisio, D., Sterrantino, G., Meli, M., Trotta, M., Milo, D. and Leoncini, F. (1995). Use of furazolidone for the treatment of microsporidiosis due to *Enterocytozoon bieneusi* in patients with AIDS. *Recenti Progresso Medicini* **86**, 394–397.
Dobbins, W. and Weinstein, W.M. (1985). Electron microscopy of the intestine and rectum in acquired immunodeficiency syndrome. *Gastroenterology*, 738–749.

Dore, G.J., Marriott, D.J., Hing, M.C., Harkness, J.L. and Field, A.S. (1995). Disseminated microsporidiosis due to *Septata intestinalis* in nine patients infected with the human immunodeficiency virus: response to therapy with albendazole. *Clinical Infectious Diseases* **21**, 70–76.

Drobniewski, F., Kelly, P., Carew, A., Ngwenya, B., Luo, N., Pankhurst, C. and Farthing, M. (1995). Human microsporidiosis in African AIDS patients with chronic diarrhea. *Journal of Infectious Disease* **171**, 515–516.

Eeftinck Schattenkerk, J.K.M., van Gool, T., van Ketel, R.J., Bartelsman, J.F.W.M., Kuiken, C., Terpstra, W.J. and Reiss, P. (1991). Clinical significance of small-intestinal microsporidiosis in HIV-1-infected individuals. *Lancet* **337**, 895–898.

Fedorko, D.P., Nelson, N.A. and Cartwright, C.P. (1995). Identification of microsporidia in stool specimens by using PCR and restriction endonucleases. *Journal of Clinical Microbiology* **33**, 1739–1741.

Field, A.S., Hing, M., Milliken, S.T. and Marriott, D.J. (1990). Microsporidia in the small intestine of HIV-infected patients. *Medical Journal of Australia*, **158**, 390–394.

Field, A.S., Harkness, A. and Marriott, D. (1992). Enteric microsporidiosis: incidence and response to albendazole or metronidazole. In: *VII International Conference on AIDS, Amsterdam*, abstract PoB 3344.

Field, A.S., Marriott, D.J. and Hing, M.C. (1993). The Warthin–Starry stain in the diagnosis of small intestinal microsporidiosis in HIV-infected patients. *Folia Parasitologia* **40**, 261–266.

Franzen, C., Muller, A., Salzberger, B., Fatkenheuer, G., Eidt, S., Mahrle, G., Diehl, V. and Schrappe, M. (1995a). Tissue diagnosis of intestinal microsporidiosis using a fluorescent stain with Uvitex 2B. *Journal of Clinical Pathology* **48**, 1009–1010.

Franzen, C., Muller, A., Hegener, P., Salzberger, B., Hartmann, P., Fatkenheuer, G., Diehl, V. and Schrappe, M. (1995b). Detection of microsporidia (*Enterocytozoon bieneusi*) in intestinal biopsy specimens from human immunodeficiency virus-infected patients by PCR. *Journal of Clinical Microbiology* **33**, 2294–2296.

Franzen, C., Kuppers, R., Muller, A., Salzberger, B., Fatkenheuer, G., Vetten, B., Diehl, V. and Schrappe, M. (1996). Genetic evidence for latent *Septata intestinalis* infection in human immunodeficiency virus-infected patients with intestinal microsporidiosis. *Journal of Infectious Disease* **173**, 1038–1040.

Fredericks, D.N. and Relman, D.A. (1996). Sequence-based identification of microbial pathogens: a reconsideration of Koch's postulates. *Clinical Microbiological Reviews* **9**, 18–31.

Giang, T., Kotler, D.P., Garro, M.L. and Orenstein, J.M. (1993). Tissue diagnosis of intestinal microsporidiosis using the chromotrope-2R trichrome stain. *Journal of Clinical Pathology* **117**, 1249–1253.

Greenson, J., Belitsos, P., Yardley, J. and Bartlett, J. (1991). AIDS enteropathy: occult enteric infections and duodenal mucosal alterations in chronic diarrhea. *Annals of Internal Medicine* **114**, 366–372.

Gunnarsson, G., Hurlbut, D., DiGirolami, P.C., Federman, M. and Wanke, C. (1995). Multiorgan microsporidiosis: report of five cases and review. *Clinical Infectious Diseases* **21**, 37–44.

Johnson, L.R. (1988). Regulation of gastrointestinal mucosal growth. *Physiologic Reviews* **68**, 456–469.

Kotler, D.P. (1991). Gastrointestinal complications of the acquired immunodeficiency syndrome. In: *Textbook of Gastroenterology* (T. Yamada, ed.), pp. 86–103. Philadelphia: Lippincott.

Kotler, D.P. and Orenstein, J.M. (1994). Prevalence of enteric pathogens in HIV-infected individuals referred for gastrointestinal evaluation. *American Journal of Gastroenterology* **89**, 1998–2002.

Kotler, D.P., Francisco, A., Clayton, F., Scholes, J. and Orenstein, J.M. (1990a). Small intestinal injury and parasitic disease in AIDS. *Annals of Internal Medicine* **113**, 444–449.

Kotler, D.P., Tierney, A.R., Wang, J. and Pierson, R.N. jr (1990b). Effect of home total parenteral nutrition upon body composition in AIDS. *Journal of Parenteral and Enteral Nutrition* **14**, 454–458.

Kotler, D.P., Reka, S., Chow, K. and Orenstein, J.M. (1993). Effects of enteric parasitoses and HIV infection upon small intestinal structure and function in patients with AIDS. *Journal of Clinical Gastroenterology* **16**, 10–15.

Kotler, D.P., Giang, T.T., Garro, M.L. and Orenstein, J.M. (1994). Light microscopic diagnosis of microsporidiosis in patients with AIDS. *American Journal of Gastroenterology* **89**, 540–504.

Ledford, D.M., Overman, M.D., Gonzalvo, A., Cali, A., Mester, S.W. and Lockey, R.F. (1985). Microsporidiosis myositis in a patient with the acquired immunodeficiency syndrome. *Annals of Internal Medicine* **102**, 628–629.

Levaditi, C., Nicolau, S. and Schoen, R. (1923). L'etiologie de l'encephalite. *Comptes Rendus de l'Academie des Sciences* **177**, 985–988.

Lucas, S.B., Papadaki, L., Conlon, C., Sewankambo, N., Goodgame, R. and Serwadda, D. (1989). Diagnosis of intestinal microsporidiosis in patients with AIDS. *Journal of Clinical Pathology* **42**, 885–887.

Matsubayashi, H., Koike, T., Mikata, I., Takei, H. and Hagiwara, S. (1959). A case of *Encephalitozoon*-like infection in man. *Archives of Pathology* **67**, 181–187.

McWhinney, P.H.M., Nathwani, D., Green, S.T., Boyd, J.F. and Forrest, J.A.H. (1991). Microsporidiosis detected in association with AIDS-related sclerosing cholangitis. *AIDS* **5**, 1394–1395.

Melchior, J.C., Chastang, C., Gelas, P., Carbonnel, F., Zazzo, J.F., Boulier, A., Cosnes, J., Bouletreau, B. and Messing, B. (1996). Efficacy of 2-month total parenteral nutrition in AIDS patients: a controlled randomized prospective trial. *AIDS* **10**, 379–384.

Mertens, R.B., Didier, E.S., Fishbein, M.C., Bertucci, D.C., Rogers, L.B. and Orenstein, J.M. (1997). Disseminated *Encephalitozoon cuniculi* microsporidiosis: infection of the brain, heart, kidneys, trachea, adrenal glands, urinary bladder, spleen, and lymph nodes in a patient with AIDS. *Modern Pathology* **10**, 68–77.

Michiels, J.F., Hofman, P., Saint Paul, M.C., Giorsetti, V., Bernard, E., Vinti, H. and Loubiere, R. (1991). Microsporidiose intestinale: 3 cas chez des sujets seropositifs pour le VIH. *Annals of Pathology* **11**, 169–175.

Modigliani, R., Bories, C., Le Charpentier, Y., Salmeron, M., Messing, B., Galian, A., Rambaud, J.C., Lavergne, A., Cochand-Priollet, B. and Desportes, I. (1985). Diarrhoea and malabsorption in acquired immune deficiency syndrome: a study of four cases with special emphasis on opportunistic protozoan infestations. *Gut* **26**, 179–187.

Molina, J.M., Sarfati, C., Beauvais, B., Lemann, M., Lesourd, A., Ferchal, F., Casin, I., Lagrange, P., Modigliani, R., Derouin, F. and Modai, J. (1993). Intestinal microsporidiosis in human immunodeficiency virus-infected patients with chronic unexplained diarrhea: prevalence and clinical and biologic features. *Journal of Infectious Disease* **167**, 217–221.

Molina, J.M., Oksenhendler, E., Beauvais, B., Sarfati, C., Jaccard, A., Derpouin, F. and Modai, J. (1995). Disseminated microsporidiosis due to *Septata intestinalis* in

patients with AIDS: clinical features and response to albendazole therapy. *Journal of Infectious Disease* **171**, 245-249.
Orenstein, J.M. (1991). Microsporidiosis in the acquired immunodeficiency syndrome. *Journal of Parasitology* **77**, 843-864.
Orenstein, J.M., Chiang, J., Steinberg, W., Smith, P.D., Rotterdam, H. and Kotler, D.P. (1990). Intestinal microsporidiosis as a cause of diarrhea in human immunodeficiency virus-infected patients: A report of 20 cases. *Human Pathology* **21**, 475-481.
Orenstein, J.M., Zierdt, W., Zierdt, C. and Kotler, D.P. (1991). Identification of spores of the Microspora, *Enterocytozoon bieneusi* in stool and duodenal fluid from AIDS patients with diarrhea. *Lancet*, **336**, 1127-1128.
Orenstein, J.M., Tenner, M., Cali, A. and Kotler, D.P. (1992a). A microsporidian previously undescribed in humans, infecting enterocytes and macrophages and associated with diarrhea in an AIDS patient. *Human Pathology* **23**, 722-728.
Orenstein, J.M., Tenner, M. and Kotler, D.P. (1992b). Localization of infection by the microsporidian *Enterocytozoon bieneusi* in the gastrointestinal tract of AIDS patients with diarrhea. *AIDS* **6**, 195-197.
Orenstein, J.M., Dieterich, D.T. and Kotler, D.P. (1992c). Systemic dissemination by a newly recognized microsporidia species in AIDS. *AIDS* **6**, 1143-1150.
Orenstein, J.M., Dieterich, D.T., Lew, E.A. and Kotler, D.P. (1993). Albendazole as a treatment for disseminated microsporidiosis due to *Septata intestinalis* in AIDS patients. *AIDS* **7**, supplement 3, S40-S42.
Owyang, C., Green, L. and Rader, D. (1983). Colonic inhibition of pancreatic and biliary secretion. *Gastroenterology* **84**, 470-475.
Peacock, C.S., Blanshard, C., Tovey, D.G., Ellis, D.S. and Gazzard, B.G. (1991). Histological diagnosis of intestinal microsporidiosis in patients with AIDS. *Journal of Clinical Pathology* **44**, 558-563.
Pol, S., Romana, C.A., Richard, S., Amouyal, P., Desportes-Livage, I., Carnot, F., Pays, J. and Berthelot, P. (1993). Microsporidia infection in patients with the human immunodeficiency virus and unexplained cholangitis. *New England Journal of Medicine* **328**, 95-99.
Rabeneck, L., Gyorkey, F., Genta, R., Gyorkey, P., Foote, L. and Risser, J.M.H. (1993). The role of Microsporidia in the pathogenesis of HIV-related chronic diarrhea. *Annals of Internal Medicine* **119**, 895-899.
Rabenek, L., Genta, R.M., Gyorkey, F., Clarridge, J.E., Gyorkey, P. and Foote, L.W. (1995). Observations on the pathological spectrum and clinical course of microsporidiosis in men infected with the human immunodeficiency virus: follow-up study. *Clinical Infectious Diseases* **20**, 1229-1235.
Remadi, S., Dumais, J., Wafa, K. and MacGee, W. (1995). Pulmonary microsporidiosis in a patient with the acquired immunodeficiency syndrome. A case report. *Acta Cytologia* **39**, 1112-1116.
Rijpstra, A.C., Canning, E.U., Van Ketel, R.J., Eeftinck Shattenkerk, J.K.M. and Laarman, J.J. (1988). Use of light microscopy to diagnose small-intestinal microsporidiosis in patients with AIDS. *Journal of Infectious Disease* **157**, 827-831.
Sandfort, J., Hannamen, A., Gelderblom, H., Stark, D., Owen, R.L. and Ruf, B. (1994). *Enterocytozoon bieneusi* in an immunocompetent patient who had acute diarrhea and was not infected with the human immunodeficiency virus. *Clinical Infectious Disease* **19**, 514-516.
Sax, P.E., Rich, J.D., Pieciak, W.S. and Trinka, Y.M. (1995). Intestinal microsporidiosis occuring in a liver transplant recipient. *Transplantation* **60**, 617-618.
Schwartz, D.A., Bryan, R.T., Hewan-Lowe, K.O., Visvesvara, G.S., Weber, R., Cali, A. and Angritt, P. (1992). Disseminated microsporidiosis (*Encephalitozoon hellem*)

and acquired immunodeficiency syndrome. Autopsy evidence for respiratory acquisition. *Archives of Pathology and Laboratory Medicine* **116**, 660–668.
Schwartz, D.A., Visvesvara, G.S., Diesenhouse, M.C., Weber, R., Font, R.L., Wilson, L.A., Corrent, G., Serdarevic, O.N., Rosberger, D.F. and Keenen, P.C. (1993). Pathologic features and immunofluorescent antibody demonstration of ocular microsporidiosis (*Encephalitozoon hellem*) in seven patients with acquired immunodeficiency syndrome. *American Journal of Ophthalmology* **115**, 285–292.
Schwartz, D.A., Abou-Elella, A., Wilcox, C.M., Gorelkin, L., Visvesvara, G.S., Thompson, S.E., Weber, R. and Bryan, R.T. (1995). The presence of *Enterocytozoon bieneusi* spores in the lamina propria of small bowel biopsies with no evidence of disseminated microsporidiosis. Enteric Opportunistic Infections Working Group. *Archives of Pathology and Laboratory Medicine* **119**, 424–428.
Sclafani, A., Koopmans, H.S., Vasselli, J. and Reichman, M. (1978). Effects of intestinal bypass surgery on appetite, food intake, and body weight in obese and lean rate. *American Journal of Physiology* **234**, E389–E398.
Shiau, Y.F., Kotler, D.P. and Levine, G.M. (1979). Can normal small bowel morphology be equated with normal function? *Gastroenterology* **76**, 1246.
Simon, D., Weiss, L., Tanowitz, H., Cali, A., Jones, J. and Wittner, M. (1991a). Light microscope diagnosis of human microsporidiosis and variable response to octreotide. *Gastroenterology* **100**, 271–273.
Simon, D., Weiss, L., Wittner, M., Cello, J., Basuk, P., Rood, R. and Cali, A. (1991b). Prevalence of microsporidia in AIDS patients with refractory diarrhea. *American Journal of Gastroenterology* **86**, 1348.
Spiller, R.C., Trotman, I.F., Higgins, B.E., Ghatei, M.A., Rimble, G.K., Lee, Y.C., Bloom, S.R., Misiewicz, J.J. and Silk, D.B.A. (1984). The ileal brake—inhibition of jejunal motility after ileal fat perfusion in man. *Gut* **25**, 365–374.
Swenson, J., MacLean, J.D., Kokoskin-Nelson, E., Szabo, J., Lough, J. and Gill, M.J. (1993). Microsporidiosis in AIDS patients. *Canadian Communicable Disease Reports* **19**, 13–15.
Tzipori, S., Carville, A., Widmer, G., Kotler, D., Mansfield, K. and Lackner A. (in press). Transmission and establishment of a persistent infection of *Enterocytozoon bieneusi* derived from a human with AIDS in SIV-infected rhesus monkeys. *Journal of Infectious Disease.*
Ullrich, R., Zeitz, M., Bergs, C., Janitschke, K. and Riecken, E.O. (1991). Intestinal microsporidiosis in a German patient with AIDS. *Klinische Wochenschrift* **69**, 443–445.
van Gool, T., Hollister, W.S., Schattenkerk, J.E., van den Bergh Weerman, M.A., Terpstra, W.J., van Ketel, R.J., Reiss, P. and Canning, E.U. (1990). Diagnosis of *Enterocytozoon bieneusi* microsporidiosis in AIDS patients by recovery of spores from faeces. *Lancet* **ii**, 697–698.
van Gool, T., Snijders, F., Reiss, P., Eeftinck Schattenkerk, J.K.M., van den Bergh Beerman, M.A., Bartlesman, J.F.W.M., Bruins, J.J.M., Canning, E.U. and Dankert, J. (1993). Diagnosis of intestinal and disseminated microsporidial infections in patients with HIV by a new rapid fluorescence technique. *Journal of Clinical Pathology* **46**, 694–699.
Verre, J., Marriott, D., Hing, M., Field, A. and Harkness, J. (1992). Evaluation of light microscopic detection of microsporidial spores in faeces from HIV infected patients. In: *Program and Abstracts from the Workshop in Intestinal Microsporidia in HIV Infection*, 15–16 December 1992, Paris, France.
Vossbrinck, C.R., Maddox, J.V., Friedman, S., Debrunner-Vossbrink, B.A. and Woerse, C.R. (1987). Ribosomal RNA sequence suggests microsporidia are extremely ancient eukaryotes. *Nature* **326**, 411–414.

Weber, R., Kuster, H., Keller, R., Bachi, T., Spycher, M.A., Briner, J., Russi, E. and Luthy, R. (1992a). Pulmonary and intestinal microsporidiosis in patients with the acquired immunodeficiency syndrome. *American Review of Respiratory Diseases* **146**, 1603–1605.
Weber, R., Bryan, R.T., Owen, R.L., Wilcox, C.M., Gorelkin, L. and Visvesvara, G.S. (1992b). Improved light-microscopical detection of microsporidia spores in stool and duodenal aspirates. *New England Journal of Medicine* **326**, 161–166.
Weber, R., Muller, A., Spycher, M.A., Opravil, M., Ammann, R. and Briner, J. (1992c). Intestinal *Enterocytozoon bieneusi* microsporidiosis in an HIV-infection patient: diagnosis by ileo-colonoscopic biopsies and long-term follow-up. *Clinical Investigation* **70**, 1019–1023.
Weber, R., Kuster, H., Visvesvara, G.S., Bryan, R.T., Schwartz, D.A. and Luthy, R. (1993). Disseminated microsporidiosis due to *Encephalitozoon hellem*: pulmonary colonization, microhematuria, and mild conjunctivitis in a patient with AIDS. *Clinical Infectious Diseases* **17**, 415–419.
Weber, R., Sauer, B., Spycher, M.A., Deplazes, P., Keller, R., Ammann, R., Briner, J. and Luthy, R. (1994). Detection of *Septata intestinalis* in stool specimens and coprodiagnostic monitoring of successful treatment with albendazole. *Clinical Infectious Diseases* **19**, 342–345.
Weiss, L.M., Zhu, X., Cali, A., Tanowitz, H.B. and Wittner, M. (1994). Utility of microsporidian rRNA in diagnosis and phylogeny: a review. *Folia Parasitologia* **41**, 81–90.
Yachnis, A.T., Berg, J., Martinez-Salazar, A., Bender, B.S., Diaz, L., Rojiani, A.M., Eskin, T.A. and Orenstein, J.M. (1996). Disseminated microsporidiosis especially infecting the brain, heart, and kidneys. *American Journal of Clinical Pathology* **106**, 535–543.
Zender, H.O., Arrigoni, E., Eckert, J. and Kapanci, Y. (1989). A case of *Encephalitozoon cuniculi* peritonitis in a patient with AIDS. *American Journal of Clinical Pathology* **92**, 352–356.
Zhu, X., Wittner, M., Tanowitz, H., Kotler, D.P., Cali, A. and Weiss, L.M. (1993). Small subunit rRNA sequence of *Enterocytozoon bieneusi* and its potential diagnostic role with use of the polymerase chain reaction. *Journal of Infectious Diseases* **168**, 1570–1575.

Microsporidiosis: Molecular and Diagnostic Aspects

Louis M. Weiss[1] and Charles R. Vossbrinck[2]

[1]*Departments of Pathology and Medicine, Albert Einstein College of Medicine, Bronx, NY 10461, USA, and* [2]*The Connecticut Agricultural Experiment Station, New Haven, CT 06504, USA*

1. Introduction 352
2. Molecular Biology 353
3. Phylogeny 358
   3.1. Placement of microsporidia among the Eukaryotes 358
   3.2. Molecular phylogeny of the phylum Microspora. 363
   3.3. Use of molecular versus morphological and ecological
       ('traditional') characters 368
4. Diagnosis. 369
   4.1. Cultivation. 369
   4.2. Serology 370
   4.3. Transmission electron microscopy 371
   4.4. Histology. 373
   4.5. Immunodetection 375
   4.6. Special stains 376
   4.7. Polymerase chain reaction 379
5. Summary. 385
   References 385

The term 'microsporidia' is a nontaxonomic designation which is used to refer to a group of intracellular parasites belonging to the phylum Microspora. These eukaryotic obligate intracellular protozoans have been described infecting every major animal group, especially insects, fish and mammals. They are important agricultural parasites in commercially important insects, fish, laboratory rodents, rabbits, fur-bearing animals, and primates. There is now an increasing recognition of microsporidia as important opportunistic pathogens in persons infected with the human immunodeficiency virus (HIV). Microsporidia possess ribosomes with features resembling prokaryotes. Phylogenetic analysis of the rRNA sequence from several of the microsporidia suggests that these organisms were early branches in the eukaryotic evolutionary line. The data on these molecular phylogenetic relationships are reviewed in this paper. Inroads have recently been made into the molecular biology of these organisms and these data are also presented. Diagnosis of microsporidia infection from stool examination is possible and has replaced biopsy as the initial diagnostic procedure in many laboratories. These staining techniques can be difficult, however, due to the small size of the spores. The specific identification of microsporidian species has classically depended on ultrastructural examination. With the cloning of the rRNA genes from the human pathogenic microsporidia it has been possible to apply polymerase chain reaction (PCR) techniques for the diagnosis of microsporidial infection at the species level. Both staining and PCR techniques for the diagnosis of microsporidia are reviewed.

1. INTRODUCTION

The term 'microsporidia' refers to a little studied but ubiquitous group of eukaryotic obligate intracellular protozoan parasites, infecting every major animal group. These organisms were first described by Nageli in 1857 as parasites of silkworms which he named *Nosema bombycis*. Microsporidia most often infect the digestive tract, but reproductive, respiratory, muscle, excretory, and nervous system infections are well documented (Sprague and Varva, 1977; Canning, 1990; Canning and Lom, 1986; Wittner *et al.*, 1993; Weber *et al.*, 1994a). In fact, microsporidia have been reported from every tissue and organ and their spores are common in environmental sources such as ditch water. Of the over 80 genera in the phylum Microspora, several have been demonstrated in human disease (Wittner *et al.*, 1993; Weber *et al.*, 1994a): *Nosema* (*N. corneum* has been recently renamed *Vittaforma corneae* (Silveira and Canning, 1995)), generally found in insects; *Pleistophora*, a pathogen of fish and insects; *Encephalitozoon*, found in many mammals; *Enterocytozoon* reported from AIDS patients (Desportes *et al.*, 1985) and several genera of fish (Chilmonczyk *et al.*, 1991); *Septata* (reclassified to *Encephalitozoon* (Baker *et al.*, 1995, Hartskeerl *et al.*, 1995)) reported from AIDS patients (Cali *et al.*, 1993); and *Trachipleistophora* reported from AIDs patients (Field *et al.*, 1996). The genus, *Microsporidium* has been used to designate microsporidia of uncertain taxonomic status. *Pleistophora*, *Nosema*, and *Trachiplesitophora* have been associated with myositis (Wittner *et al.*, 1993; Weber *et al.*, 1994a). The *Encephalitozoonidae* (*Enc. hellem*, *Enc. cuniculi* and *Enc. (Septata) intestinalis*) have been associated with disseminated disease as well as keratoconjunctivitis (Rastrelli *et al.*, 1994), sinusitis, respiratory disease, prostatic abscesses, and intestinal infection (Wittner *et al.*, 1993; Weber *et al.*, 1994a). *Nosema*, *Vittaforma*, and *Microsporidium* have been associated with stromal keratitis after trauma in immunocompetent hosts (Rastrelli *et al.*, 1994). *Enterocytozoon bieneusi*, first described in 1985

(Desportes et al., 1985), is associated with malabsorption and diarrhea, and has been described only in humans.

2. MOLECULAR BIOLOGY

Microsporidia have a number of unusual characteristics. They lack mitochondria and centrioles and possess prokaryotic size ribosomes (70S: consisting of a large (23S) and small (16S) subunit) (Ishihara and Hayashi, 1968; Curgy et al., 1990). In addition, the microsporidian *Vairimorpha necatrix* was found to lack a 5.8S ribosome RNA, but had sequences homologous to the 5.8S region in the 23S subunit (Vossbrinck and Woese, 1986). The absence of a 5.8S subunit has not been reported from any other eukaryotes. The small subunit (16S) rRNA of several microsporidia has been sequenced (Table 1) and found to be significantly shorter than both eurkaryotic and prokaryotic small subunit rRNA (Vossbrinck et al., 1987; Weiss et al., 1994). Microsporidian 16S rRNA diverges greatly from the small subunit rRNA sequences of other eukaryotes. Sequence data of rRNA (see Table 1) from *Enc. cuniculi*, *Enc. hellem*, *Ent. bieneusi*, and *Enc. intestinalis* (Hartskreel et al., 1993a,b, 1995; Vossbrinck et al., 1993; Zhu et al., 1993a,b,c, 1994; Vivesvara et al., 1994a; Weiss et al., 1994; Baker et al., 1995, Didier et al., 1995a, 1996a,b; Katiyar et al., 1995) have been used in developing diagnostic PCR primers (see Section 4.7) and in studying phylogenetic relationships (see Section 3.2). In *Enc. cuniculi* a set of tetranucleotide repeats (5'GTTT3') in the intergenic spacer (ITS) region has been found to vary between isolates from different hosts resulting in the definition of three isotypes (types I, II, and III) for this parasite (Didier et al., 1995b, 1996c). This ITS region heterogeneity may be useful in examining the epidemiology of human infection with *Enc. cuniculi*. One copy of the rDNA gene has been found on each chromosome of *Enc. cuniculi* (Vivares et al., 1996). Based on the absence of a binding site for paromomycin in the sequence of microsporidian rRNA, it has been suggested that microsporidia are not sensitive to this drug (Katiyar et al., 1995).

The karyotype of a few members of the phylum Microspora has been determined by pulsed field electrophoresis (Munderloh et al., 1990; Vivares et al., 1996). *Nosema pyrausta* and *Nosema furnaclais* have 13 chromosome bands ranging in size from 130 to 440 kb (Munderloh et al., 1990), *Glugea atherinae* has 16 bands of 2700 to 420 kb and *Spraguea lophii* has 12 bands of 980 to 230 kb (Biderre et al., 1994). The genome size of these microsporidia vary from 19.5 to 6.2 Mb. Recently, the genomic size of *Enc. cuniculi* has been demonstrated to be 2.9 Mb (Biderre et al., 1995), which suggests that it is the smallest eukaryotic nuclear genome so far indentified.

No introns have been found in microsporidian genes sequenced to date. It should be noted, however, that a unique U2 RNA homolog (DiMaria et al.,

Table 1 Microsporidian genes in GenBank.

Gene	Species	Reference
Small subunit (16S rRNA) rRNA genes		
U68474	Amblyospora sp.	Baker et al. (upublished)
U68473	Amblyospora californica	Baker et al. (unpublished)
L15741	Ameson michaelis	Zhu et al. (1994)
X98469	Encephalitozoon cuniculi (Stewart strain)	Hollister et al. (1996)
X98470	Encephalitozoon cuniculi (Donovan strain)	Hollister et al. (1996)
X98467	Encephalitozoon cuniculi (D. Owen strain)	Hollister et al. (1996)
L39107	Encephalitozoon cuniculi	Baker et al. (1995)
L17072	Encephalitozoon cuniculi	Visvesvara et al. (1994b)
L13295	Encephalitozoon cuniculi partial cds	Vossbrinck et al. (1993)
Z19563	Encephalitozoon cuniculi	Zhu et al. (1993b)
L07255	Encephalitozoon cuniculi	Hartskeerl et al. (1993)
L13393	Encephalitozoon hellem	Vossbrinck et al. (1993)
L39108	Encephalitozoon hellem	Baker et al. (1995)
L19070	Encephalitozoon hellem	Visvesvara et al. (1994b)
U39297	Encephalitozoon intestinalis partial cds	Franzen et al. (1996)
U09929	Encephalitozoon intestinalis	Visvesvara et al. (1995a)
L39113	Encephalitozoon intestinalis	Baker et al. (1995)
L19567	Encephalitozoon intestinalis	Zhu et al. (1993a)
L16867	Encephalitozoon sp.	R.A. Hartskeerl (unpublished)
L16866	Encephalitozoon sp.	Hartskeerl et al. (1993a)
L39109	Endoreticulatus shubergi	Baker et al. (1995)
L16868	Enterocytozoon bieneusi	Hartskeerl et al. (1993b)
L07123	Enterocytozoon bieneusi	Zhu et al. (1993c)
U10883	Enterocytozoon (Nucleospora) salmonis	Barlough et al. (1995)
U15987	Glugea atherinae	N.J. Pieniazek (unpublished)
L13293	Icthyosporidium giganteum partial cds	Vossbrinck et al. (1993)
L39110	Icthyosporidium sp.	Baker et al. (1995)
U26534	Nosema apis	N. Pieniazek (unpublished)
X73894	Nosema apis	L.A. Malone (unpublished)
U26158	Nosema bombycis partial cds	L.A. Malone (unpublished)
D85504	Nosema bombycis spore	T. Inoue (unpublished)
D14632	Nosema bombycis partial cds	Kawakami et al. (1994)
D85503	Nosema bombycis spore	T. Inoue (unpublished)
U26157	Nosema bombycis partial cds	L.A. Malone (unpublished)
L39111	Nosema bombycis	Baker et al. (1995)
U26533	Nosema ceranae	N. Pieniazek (unpublished)
U26532	Nosema furnacalis	N. Pieniazek (unpublished)
U11051	Nosema necatrix	Fries et al. (1996)
U27359	Nosema oulemate	N. Pieniazek (unpublished)
U09282	Nosema trichoplusiae	Pieniazek et al. (1996)
U09283	Nosema trichoplusiae	N. Pieniazek (unpublished)

Table 1 Continued.

Gene	Species	Reference
U11047	Nosema vespula	N. Pieniazek (unpublished)
L31842	Nosema vespula	J.A. Ninham (unpublished)
D85501	Nosema sp. spore	T. Inoue (unpublished)
U10883	Nucleospora (Enterocytozoon) salmonis	Barlough et al. (1995)
U78176	Nucleospora (Enterocytozoon) salmonis	Kent et al. (1996)
D85500	Pleistophora sp. spore	T. Inoue (unpublished)
U47052	Pleistophora anguillarum	W.H. Huang (unpublished)
U10342	Pleistophora sp. ATCC 500400	N. Pieniazek (unpublished)
X74112	Vairimorpha oncoperae	L.A. Malone (unpublished)
L13294	Vairimorpha lymantriae partial cds	
M24612	Vairimorpha necatrix	Vossbrinck et al. (1987)
Y00266	Vairimorpha necatrix	Vossbrinck et al. (1987)
L39114	Vairimorpha sp.	Baker et al. (1995)
L28977	Vairimorpha sp.	Baker et al. (1994)
L28976	Vairimorpha sp.	Baker et al. (1994)
D85502	Vairimorpha sp. spore	T. Inoue (unpublished)
X74112	Vavrai oncopertae	L.A. Malone (unpublished)
L39112	Vittaforma corneum	Baker et al. (1995)
U11046	Vittaforma corneum	N. Pieniazek (unpublished)

3'-end small subunit ITS and 5'-end large subunit (LSU) rRNA genes

L28960	Amblyospora sp.	Baker et al. (1994)
L20293	Ameson michaelis	Zhu et al. (1994)
X98466	Encephalitozoon cuniculi (Donovan strain)	Hollister et al. (1996)
X98468	Encephalitozoon cuniculi (D. Owen strain)	Hollister et al. (1996)
L29560	Encephalitozoon cuniculi	Katiyar et al. (1995)
L13332	Encephalitozoon cuniculi	Vossbrinck et al. (1993)
L13331	Encephalitozoon hellem	Vossbrinck et al. (1993)
L29557	Encephalitozoon hellem	Katiyar et al. (1995)
L20292	Encephalitozoon intestinalis	Zhu et al. (1994)
Y11611	Encephalitozoon intestinalis	Schnittger et al. (unpublished)
L20290	Enterocytozoon bieneusi	Zhu et al. (1994)
U61180	Enterocytozoon bieneusi ITS region	Deplazes et al. (unpublished)
L13430	Icthyosporidium giganteum	Vossbrinck et al. (1993)
U78815	Loma embiotocia	R.H. Devlin (unpublished)
U78736	Loma salmonae	R.H. Devlin (unpublished)
L28961	Nosema algerae	Baker et al. (1994)
L28962	Nosema bombycis	Baker et al. (1994)
D14631	Nosema bombycis partial cds	Kawakami et al. (1992)
L28963	Nosema distriae	Baker et al. (1994)
L28964	Nosema epilachnae	Baker et al. (1994)
L28965	Nosema heliothidis	Baker et al. (1994)
L28967	Nosema locustae	Baker et al. (1994)

Table 1 Continued.

Gene	Species	Reference
L28966	*Nosema kingi*	Baker *et al.* (1994)
L28968	*Nosema pyrausta*	Baker *et al.* (1994)
U78176	*Nucleospora* (*Enterocytozoon*) *salmonis*	Kent *et al.* (1996)
L28969	*Parathelohania anophelis*	Baker *et al.* (1994)
L28972	*Vairimorpha ephistiae*	Baker *et al.* (1994)
L28973	*Vairimorpha heterosporum*	Baker *et al.* (1994)
L13330	*Vairimorpha lymantriae*	Vossbrinck *et al.* (1993)
L28975	*Vairimorpha necatrix*	Baker *et al.* (1994)
L28971	*Vairimorpha* sp.	Baker *et al.* (1994)
L28970	*Vairimorpha* sp.	Baker *et al.* (1994)
Tubulin genes		
L47272	*Nosema locusta* α-tubulin partial cds	Li *et al.* (1996)
L47273	*Nosema locusta* β-tubulin partial cds	Li *et al.* (1996)
L31807	*Encephalitozoon cuniculi* β-tubulin partial cds	Edlind *et al.* (1994)
L47271	*Encephalitozoon hellem* β-tubulin full cds	Li *et al.* (1996)
L31808	*Encephalitozoon hellem* β-tubulin partial cds	Edlind *et al.* (1994)
L47274	*Encephalitozoon intestinalis* β-tubulin partial cds	Li *et al.* (1996)
Other gene sequences		
U28045	*Nosema bombycis* DNA fragment (unknown gene)	Malone and McIvor (1995)
U28046	*Nosema costelyatae* DNA fragment (unknown gene)	Malone and McIvor (1995)
L37097	*Nosema locustae* isoleucyl tRNA synthetase partial cds	Brown and Doolittle (1995)
D32139	*Glugea plecoglossi* elongation factor 1α	Kamaishi *et al.* (1996a)
D84253	*Glugea plecoglossi* peptide elongation factor 1α	Kamaishi *et al.* (1996a)
D79220	*Glugea plecoglossi* peptide elongation factor 2	Kamaishi *et al.* (1996b)
AF005490	*Nosema locustae* glutamyl-tRNA synthetase gene	Brown and Doolittle (1997)
AF005489	*Nosema locustae* glutamyl-tRNA synthetase gene	Brown and Doolittle (1997)
U97520	*Nosema locustae* mitochondrial-type HSP70 gene	Germot *et al.* (1997)
AF008215	*Varimorpha necatrix* mitochondrial HSP70 homolog	Hirt *et al.* (1997)

cds, coding sequence; ITS, internal transcribed spacer.

1996) is present, suggesting that introns also may be present. Most significantly, the study revealed that the cap structures of the U2 small nuclear RNAs and the mRNAs are neither 2,2,7-trimethylguanosine nor 7-methylguanosine; an addition to the list of unusual characters that set microsporidia apart from other eukaryotes. Identification of this unusual cap structure may well prove of importance, from both a basic molecular biological and antimicrosporidial drug development viewpoint. Polyadenylation occurs on mRNA in microsporidia as is true for all other eukaryotes studied to date (C.R. Vossbrinck, unpublished data).

The small microsporidian genomic size indicates that they may have developed strategies for packing genetic information tightly into the genome (e.g. virus-like genomic organization), or that they may have lost genetic information for metabolic pathways and depend on host-cell sources for these compounds. A study of chromosome 1 of *Enc. cuniculi* has demonstrated that in a 6 kbp region four protein coding genes were identified with open-reading frames (ORFs) in the same orientation which were only 195 bp apart (Vivares *et al.*, 1996). Partial gene clones for *cdc2* and aminopeptidase have been localized to chromosome 8 (272 kbp), and aminopeptidase, thymidylate synthase, serine hydroxymethyltransferase, and dihydrofolate reductase were localized to chromosome 1 (217 kbp) of *Enc. cuniculi* (Vivares *et al.*, 1996). These genes had a strong bias (75%) for a C or G residue in the third codon position (Vivares *et al.*, 1996).

The activity of several of the benzimidazoles (particularly albendazole) as antimicrosporidial agents has led to an interest in the β-tubulin genes of the microsporidia. β-tubulin is believed to be the target of these compounds. The tubulin genes available for several microsporidia which have been shown to be sensitive to benzimidazoles display the predicted amino acids (Cys165, Phe167, Glu198, Phe200, Arg241, His6) associated with this phenotype in other eukaryotes (Edlind *et al.*, 1994, 1996; Katiyar *et al.*, 1994; Li *et al.*, 1996). The β-tubulin gene of *Enc. cuniculi* appears to be present in two copies localized to chromosomes 2 (235 kbp) and 3 (241 kbp) (Vivares *et al.*, 1996). The β-tubulin gene of *Ent. bieneusi* has not been cloned but would be of particular interest given the poor efficacy of albendazole for the treatment of this microsporidian infection in humans. In addition to rRNA genes and several β-tubulin genes (Edlind *et al.*, 1994, 1996; L.M. Weiss, unpublished data), an α-tubulin gene (Li *et al.*, 1996), isoleucyl tRNA synthetase (Brown and Doolittle, 1995), elongation factor 1α (EF1α) (Kamiashi *et al.*, 1996a,b) and elongation factor 2 (EF2) (Hashimoto and Hasegawa, 1996) have been cloned from microsporidia.

3. PHYLOGENY

Sprague's classification system, proposed in 1977, and updated in 1982 and 1992 is commonly used for these organisms (Sprague and Varva, 1977; Levine *et al.*, 1980; Sprague *et al.*, 1992). However, as molecular analysis of rRNA genes becomes available, this classification system will likely be altered (Ragan, 1988; Baker *et al.*, 1995). Currently, taxonomy and species classification is based on ultrastructural features, including the size and morphology of the spores, the number of coils of the polar tube, the developmental life cycle, and the host–parasite relationship.

3.1. Placement of Microsporidia among the Eukaryotes

Based upon comparative rRNA analysis, it has been proposed (Vossbrinck *et al.*, 1987) that microsporidia, a group of eukaryotes lacking mitochondria, diverged from other eukaryotes prior to the introduction of oxygen by blue-green algae into the Earth's atmosphere and before eukaryotes developed a symbiotic relationship with the prokaryotic mitochondrion ancestors. Geologic evidence for the initial presence of oxygen in the Earth's atmosphere is based upon a layer of iron oxide estimated to have formed 2.5–2.8 billion years ago (Walker, 1983). The rRNA analysis demonstrated a remarkable degree of divergence of *Vair. necatrix* from all other eukaryotes for which rRNA sequences had been obtained (Vossbrinck *et al.*, 1987). Previous studies indicated that these organisms have a 'primitive' form of meiosis (Haig, 1993), lack mitochondria (Vavra, 1976), and lack a 5.8S rRNA (Vossbrinck and Woese, 1986). Based on this analysis, it was also proposed that other amitochondriate protozoa may also show a high degree of divergence and may therefore be primitively lacking mitochondria (Cavalier-Smith, 1987). Analysis of rRNA from the parasitic diplomonad *Giardia lamblia* (Sogin *et al.*, 1989) is consistent with Cavalier-Smith's (1987) proposal. *Vair. necatrix* branches first from the eukaryotic lineage according to maximum parsimony analysis of the rRNA sequences. However, because of its long branch length, it is believed that *Vair. necatrix* is changing at an evolutionary rate faster than the other organisms in the analysis, and therefore that maximum parsimony analysis is not as reliable as distance methods in these circumstances (Sogin *et al.*, 1989). Based on distance analysis results Sogin *et al.* (1989) conclude that *Giar. lamblia* branches first from the eukaryotic lineage followed by *Vair. necatrix*. A more recent analyses of small subunit rRNA (Leipe *et al.*, 1993), which includes additional protozoan groups, also supports the early divergence of the amitochondriate protozoa (Figure 1). However, Liepe *et al.* (1993) speculate that various factors may have an effect on altering the true tree topology. These factors include the parasitic nature of the amitochondriate protozoa, the high G + C content of *Giar. lamblia* (75% G + C), the low G + C content of *Vair. necatrix* (35% G + C) and the relative shortness of some of the sequences examined (Galtier *et al.*, 1995). Leipe *et al.* (1993) concluded that the high G + C content of *Giardia* does not bias its phylogenetic position, and that parasitism alone cannot explain the pattern of early diver-

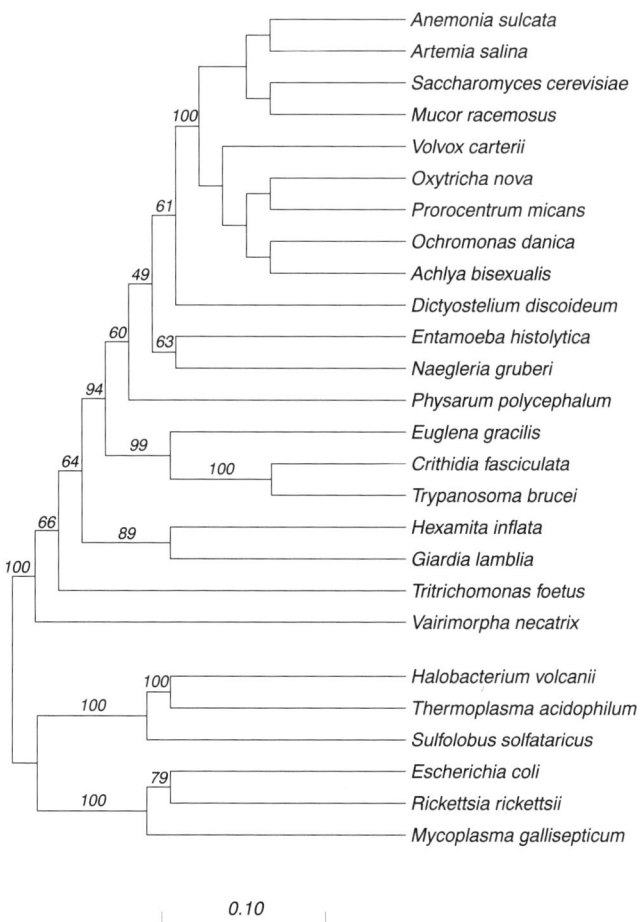

Figure 1 Multikingdom phylogeny inferred from 16S rRNA seqeunces. The phylogenetic tree was inferred using unambiguously aligned sequences using the neighbor-joining method in the PHYLIP program. Bootstrap support for topological elements in the tree are based upon 200 resamplings. Horizontal distances between nodes of the tree represent relative evolutionary distances. The scale bar corresponds to 10 changes per 100 positions. Reproduced, with permission, from Leipe *et al.* (1993).

gence; the diplomonad *Hexamita inflata*, which has a G + C composition of 51% and is free-living, diverges from the other eukaryotes with *Giardia* when prokaryotes with a normal range of G + C content are used as outgroups. They speculate that microsporidia may have diverged first from the eukaryotic line, but conclued that, because of its low G + C content, the position of the *Vair. necatrix* was still ambiguous.

A recent maximum likelihood analysis of EF1α (Figure 2) (Hashimoto and Hasegawa, 1996; Kamaishi *et al.*, 1996a, b) confirms the early branching pattern of both microsporidia and *Giardia* from the other eukaryotes. These authors provide a rigorous analysis of the amino acid sequence data and tested various models of amino acid substitution. They addressed the concern that the analysis of microsporidia could be biased by the low G + C content of *Vair. necatrix*, and instead used the microsporidian parasite of Ayu fish *Glugea plecoglossi* which has a G + C content of about 50%. They note previous concerns of the G + C bias and propose that phylogenies based on the analysis of amino acid sequences of proteins are not biased by genomic G + C content, giving a 'more robust estimation of the early divergence of eukaryotes' (Hashimoto and Hasegawa, 1996; Kamaishi *et al.*, 1996a,b). This early divergence of microsporidia was recently confirmed by additional analysis of the amino acid sequences of EF1α and EF2 (T. Hashimoto, personal communication). The major difference between these studies and previous studies using rRNA (Leipe *et al.*, 1993; Vossbrinck *et al.*, 1987) is the position of the flagellates, *Euglena* and *Trypanosoma*, as the first mitochondrion-

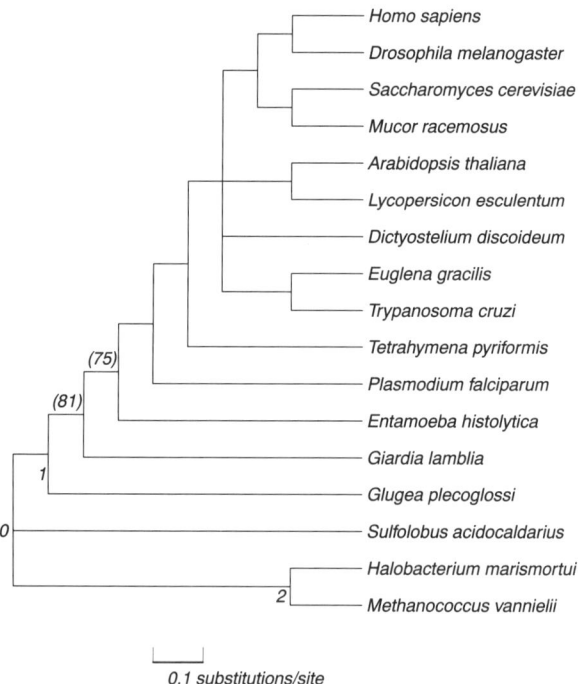

Figure 2 EF1α tree of eukaryotes with archaebacteria as an outgroup. The phylogenetic tree was constructed by the maximum-likelihood method based on the JTT model. The horizontal length of each branch is proportional to the estimated number of substitutions. Bootstrap probabilities are shown in parentheses. Reproduced, with permission, from Kamaishi *et al.* (1996a).

containing organisms to diverge from the eukaryotic line (rRNA analysis) or as a branch after the ciliates, *Tetrahymena* (EF1α analysis). In the rRNA phylogenies *Entamoeba* and *Naegleria* (which have mitochondria) are sister taxa, implying that these two sarcodines are related and that the mitochondria would have been secondarily lost from *Entamoeba*. In contrast, the EF1α tree places *Entamoeba* much lower, suggesting that these organisms never had mitochondria.

Maximum parsimony analysis and distance analysis of the β-tubulin amino acid sequence data (Edlind *et al.*, 1994, 1996; Katiyar *et al.*, 1994; Li *et al.*, 1996) places *Entamoeba histolytica* as the most divergent of the eukaryotes and relegates the microsporidia (in this case *Enc. hellem*) to the fungi. Both parsimony and distance analysis (PRODIST and NEIGHBOR) gave the same results. The unusual rRNA and the absence of typical fungal structures (cell walls, mitochondria, ergosterol-containing membranes) are believed to reflect secondary loss during evolution rather than primitive characteristics (Edlind *et al.*, 1996). Additional evidence for this view is the recent identification in *Enc. cuniculi* of separate genes for thymidylate synthase and dihydrofolate reductase, although these are on the same chromosome (Vivares *et al.*, 1996). It should be noted that *Candida albicans* was omitted from the analysis because the inclusion of its β-tubulin sequence yielded a very different phylogeny. Edlind *et al.* (1996) explain differences between the rRNA and β-tubulin phylogenies with respect to the microsporidia by postulating deletions in the microsporidial rRNA causing changes in adjacent, and possibly in distant, rRNA sequences, making it less reliable for phylogenetic analysis. With regard to differences between β-tubulin and rRNA phylogenies for *Entam. histolytica*, Edlind *et al.* (1996) state: 'It is more difficult to rationalize the dramatic difference in the placement of *E. histolytica* within the rRNA and tubulin (both α and β) phylogenies'. A number of factors point to the possibility that similarities among taxa in the β-tubulin phylogeny represent a case of convergent evolution due to selection pressure from naturally occurring antimicrobial compounds. The fact that some microsporidia and helminths are sensitive to albendazole while other microsporidia are not, and that some fungi are sensitive to the related compound nocodazole while others are not, implies that benzimidazole sensitivity is either plesiomorphic or polyphyletic in origin. Edlind *et al.* (1996) report the identification of a nocodazole resistant *Saccharomyces cerevisiae* isolate. The resistance correlates by comparative anlaysis with *Enc. hellem* and *Aspergillus nidulans* to a mutation from Val to Gly at amino acid residue 268, confirming that resistance to benzimidazole can be achieved by a single mutation. These facts lend credence to the idea that β-tubulin may be a target for antimicrobial activity and that, over the evolutionary time scale, organisms that were exposed to similar naturally occurring antimicrobial compounds developed similar mutations.

New information indicates that the microsporidia are likely to have lost mitochondria secondarily as is true for other amitochondric protozoa such as *Trichomonas vaginalis*. The recent identification in *Vair. necatrix* and *N. locustae* of prokaryotic homologues of heat shock protein 70, so called mitochondrial hsp70, indicates at one point the presence of mitochondria in the evolution of these organisms (Hirt *et al.*, 1997, in press; Brown and Doolittle, 1997). Maximum likelihood analysis provides relatively strong evidence that this prokaryotic hsp70 belongs to the mitochondrial group and demonstrated weak support for the fungal clade. This provides evidence for the hypothesis that microsporidia are derived from (i.e. related to) the fungi. In addition to the mitochondrial hsp70, tRNA synthetase genes consistent with the presence of mitochondria have been found in the microsporidia (T. Hashimoto, unpublished; Brown *et al.*, 1997). These data suggest that the microsporidia did not diverge from other eukaryotes before the advent of mitochondrial symbiosis.

Prior to molecular analysis of the eukaryotes, very little to nothing was stated with certainty about the relationships among the most distantly related taxa based on morphological and ecological characters. Molecular phylogenies do not by any means resolve all eukaryotic relationships, but do make an important addition to our knowledge. As more molecules are analyzed a consensus will develop concerning the best-guess relationship.

At present it is hard to understand how two distinct molecules, the rRNA and the EF1α proteins, could yield similar incorrect trees. It is important to present data on as many molecules as have been sequenced for a common set of organisms. The β-tubulin phylogeny as well as the affinity between microsporidian and fungal mitochondrial hsp70 not only stir up the field of eukaryotic molecular phylogeny but also substantiate findings where these molecular phylogenies agree. Currently, it is probably in poor judgement to use 'traditional' morphological and ecological characters to support or reject molecular phylogenies. It is more productive (and allows independent evaluation) to document on the molecular phylogeny the 'traditional' characteristics (Baker *et al.*, 1994, 1995).

There is little mention in the protozoan literature about the possibility that discerning the correct phylogeny may not be possible. Figure 3 shows a divergence of taxon B from taxon AC. From a cladistic view, the only characters that link taxa A and C and separate them from B are those which changed after the divergence of B from AC and before A and C split (period D). If period D is short then only the most variable sites will change. There is a good chance that these characters will change again during the period from the split of A and C to the present. Phylogenetic analysis has a limited resolution, even if all nucleotides (the entire genome) are sequenced for all organisms being analyzed. The fact that a phylogeny can be obtained for a group of organisms does not guarantee that it reflects evolutionary history. It has been shown for parsimony analysis and compatibility methods that, in some cases, when an incorrect tree is obtained, including additional data

Figure 3 Hypothetical phylogenetic tree.

verifies the incorrect phylogeny (Felsenstein, 1978). Portions of the phylogenies of the insect orders, the animal phyla, and the protozoa all include rapid radiation a relatively long time ago. Whether microsporidia are highly derived organisms or represent ancient eukaryotes is not resolved at this point. Analysis of two different molecules suggests microsporidia have a long evolutionary branch length, however, this could be due to rapid evolution based in part on the assumption of the correct outgroup. One must remember that there is no outgroup to the tree of life and that all three primary kingdoms (archaebacteria, bacteria and eukaryotes) are all equally likely to contain characters present in the ur-organism and it is equally possible that the organism had a nucleus and that, for the purposes of rapid reproduction, the nucleus as well as introns were lost in bacterial forms.

3.2. Molecular Phylogeny of the Phylum Microspora

In developing a classification of organisms it is first necessary to obtain a list of characters for all organisms to be classified. It is necessary to use the same set of characters for all the organisms. The inclusion of fossil data can be the key in determining minimum age of a particular characteristic and for concluding whether character states are primitive or derived. Unfortunately, there are no fossil data for the microsporidia and the morphological, life cycle and ecological characters change too rapidly to allow determination of phylogenetic relationships at higher taxonomic levels (above the family level). In addition, type specimens are often not available, identification of the material varies in reliability, and studies of various taxa are by necessity incomplete.

Taxonomic studies of the microsporidia depend on separating taxa based upon available ultrastructural and ecological characters. Traditional descriptions, obtained using the light microscope, are based primarily on spore characteristics such as size, shape, and the number of nuclei per spore, and on the host. Weiser (1977) lists 70 characters for the identification of microsporidia, which can be distinguished by light microscopy. The use of electron microscopy for taxonomic purposes, beginning in the mid-1970s, provides an additional, more discriminating set of characters. These include the structure of the polaroplast and the pansporoblastic membrane, the shape and number of coils in the polar filament and details of the exospore structure (for a review see Larsson, 1986). In addition to morphological studies of spores from what Larsson (1986) refers to as 'spontaneously infected hosts', life-cycle studies yield important information about vegetative growth and development (merogony and sporogeny) and growth in alternate hosts (Andreadis, 1983).

Modern classification schemes of the higher level taxa in the phylum Microspora are based upon structural features. Weiser (1977) separates the 'higher' microsporidia into two orders based on the presence of one (Pleistophoridida) or two (Nosematidida) nuclei in each spore. Issi (1986) defines taxa by a combination of characters. She divides the subclass Nosematidea, the bulk of the 'higher' microsporidia in her classification, into three orders based on a combination of features including spore shape, presence of the pansporoblastic membrane, structure of the polaroplast, number of nuclei/spore, and the formation of a rosette pattern during sporogeny. Among the researchers who use 'traditional' structural character states for classification of the microsporidia, Larsson (1986) is the only one to present phylogenetic trees. He considers the sporophorous vesicle and the diplokaryon of limited phylogenetic importance. Sprague and Vavra (1977) separate the 'higher' microsporidia into the Pansporoblastina and Apansporoblastina based on the presence or absence of a pansporoblastic membrane, and Sprague *et al.* (1992) emphasize chromosome cycle and separate the microsporidia into the Diphaplophasea, which have a diplokaryon in some phase of the life cycle, and the Haplophasea, which have unpaired nuclei in all stages of the life cycle. In most cases the authors indicate that their classifications are uncertain. Nonetheless, these publications present a foundation for microsporidiologists who are studying the classification, identification, and ecology of microsporidia.

Cloning of many microsporidian rRNA genes has been accomplished (see Table 1 for a list of genes and Table 2 for cloning and sequencing primers). The first use of comparative analysis of rRNA sequence data was presented as an unrooted tree of five microsporidial species (Vossbrinck *et al.*, 1993) using partial sequence data from the small and large subunits. *Enc. hellem* (Didier *et al.*, 1991a) had previously been characterized by Western bolt and sodium dodecyl sulfate–polyacrylamide gel electrophoresis (SDS-PAGE) analysis from three patients with the acquired immune deficiency syndrome (AIDS) showing differences from *Enc. cuniculi*. Ultrastructurally, *Enc. hellem* and *Enc. cuniculi* are almost identical (Didier *et al.*, 1991b). At that time it was not clear if *Enc. hellem* was simply an isolate of *Enc. cuniculi* or, at the other extreme, belonged in a separate genus. Analysis of *Enc. hellem*, *Enc. cuniculi*, *Vair. necatrix*, *Vair. lymantriae*, and *Ichthyosporidium giganteum* revealed that *Enc. hellem* and *Enc. cuniculi* were most likely distinct yet closely related species. *Vair. necatrix* and *Vair. lymantriae* were chosen because it was known that they were distinct, yet closely related, parasites of *Lepidoptera*. *I. giganteum* was chosen as a very different (outgroup) species. Such analysis also

Table 2 Primers for the identification and sequencing of microsporidian rDNA[a]

Primer[b]	Sequence
ss18f[c]	CACCAGGTTGATTCTGCC
ss18sf	GTTGATTCTGCCTGACGT
ss350f	CCAAGGA(T/C)GGCAGCAGGCGCGAAA
ss350r	TTTCGCGCCTGCTGCC(G/A)TCCTTG
ss530f	GTGCCAGC(C/A)GCCGCGG
ss530r	CCGCGG(T/G)GCTGGCAC
ss1047r	AACGGCCATGCACCAC
ss1061f	GGTGGTGCATGGCCG
ss1492r	GGTTACCTTGTTACGACTT (universal primer)
ss1537	TTATGATCCTGCTAATGGTTC
ls212rl	GTT(G/A)GTTTCTTTTCCTC
ls212r2	AATCC(G/A/T/C)(G/A)GTT(G/A)GTTTCTTTTCCTC
ls580r	GGTCCGTGTTTCAAGACGG

[a] Primers 18f and 1492r amplify most of the small subunit rRNA of the microsporidia. Primers 530f and 212r1 or 212r2 are used to amplify the small subunit rRNA and the ITS region. The remaining primers are used to sequence, with overlap, the forward and reverse strands of the entire small subunit rRNA and ITS region. ls580r amplifies a variable region of the 5'-end of the large subunit rRNA gene of many microspridia (e.g. *Nosema* and *Vairimorpha*), but it does not work on all microsporidia. ss1537 allows sequencing closer to the 3'-end of the small subunit rRNA of many, but not all, microsporidia. ss350f and ss350r may not be needed for sequencing reactions if 18f and 530r provide sufficient overlap to obtain clear sequence data.
[b] ss, primers in the small subunit rRNA gene; ls, primers in the large subunit rRNA gene; f, forward primer (positive strand); r, reverse primer (negative strand).
[c] Similar to V1 primer (Zhu *et al.*, 1993c; Weiss *et al.*, 1994).

demonstrated that *Enc. intestinalis* was a distinct organism (Vossbrinck *et al.*, 1993; Zhu *et al.*, 1993a).

By rRNA phylogeny *N. bombysis* the type for the genus *Nosema* and a parasite of the silk moth *Bombyx mori*, is closely related to the *Vairimorpha* species, which are also parasites of Lepidoptera (Baker *et al.*, 1994). In addition to highlighting the importance of host as an important taxonomic character, this study showed that an organism believed to be diplokaryotic throughout its life cycle is closely related to an organism that produces both isolated diplokaryotic spores and uninucleate octospores in packets surrounded by a pansporoblastic membrane. Several of the classification schemes mentioned above based on 'traditional' characters have divided *Vairimorpha* and *Nosema* into distantly related taxa because of these differences. At the same time this analysis demonstrated that *N. bombysis* is much more distantly related to other '*Nosema*' species such as *N. kingi*, *N. locustae*, and *N algerae*. Thus, neither the distinction of being diplokaryotic throughout its life cycle nor the presence of a pansporoblastic can be used alone to group microsporidia at higher levels. The genus *Nosema* defined as being diplokaryotic throughout its life cycle is, in all likelihood, a polyphyletic group of unrelated taxa (Sprague *et al.*, 1992).

Analysis of a variety of microsporidia, including five species isolated from AIDS patients (Figure 4) (Baker *et al.*, 1995) highlight the polyphyletic nature of AIDS-related microsporidia and brings into further doubt the use of any single character for developing higher taxonomic groupings. Figure 4 demonstrates that *Enc. intestinalis*, *Enc. cuniculi*, and *Enc. hellem* are three distinct, yet closely related, species. *Enc. hellem* and *Enc. cuniculi* are almost indistinguishable at the ultrastructural level, while *Enc. intestinalis* can be distinguised by the presence of an extracellular matrix surrounding the sporoblasts and spores (Cali *et al.*, 1993). Based on comparative rDNA analysis *Enc. intestinalis* and *Enc. cuniculi* are more similar to each other than to *Enc. hellem* (Baker *et al.*, 1995; Hartskeerl *et al.*, 1995).

Perhaps the most prevalent human microsporidial parasite is *Ent. bieneusi*. Comparative rDNA analysis (Baker *et al.*, 1995) demonstrates a close relationship between *Ent. bieneusi* and *Nucleospora* (*Enterocytozoon*) *salmonis* (Docker *et al.*, 1997), a parasite of salmonid fish. Ultrastructural similarities include precocious development of the polar tube before the division of the sporogonial plasmodium into sporoblasts and the lack of a pansporoblastic membrane (the growth of all stages of the parasite in direct contact with the host). The primary distinguishing feature is the growth of *Nucleospora salmonis* in the nucleus of the host cell rather than in the cytoplasm as seen in *Ent. bieneusi*. Only one case of infection with *Vitaforma corneae* has been reported from the corneal stroma of a nonimmunocompromised individual (Shadduck *et al.*, 1990). Ribosomal DNA analysis reveals a relatively close relationship to *Endoreticulatus schubergi* and a '*Pleistophora*' species, both parasites of insects (Baker *et al.*, 1995).

These molecular relationships may be useful in suggesting the environmental reservoir for the microsporidia found in humans. *Encephalitozoonidae* appear to be found in many animals, and infection may stem from contact with other humans, pets, or food sources. *Ent. bieneusi* may be a water- or food-borne pathogen related to fish pathogens. *Vitaforma corneae* infection may represent a random opportunistic event initiated by direct contact with spores in the environment (which normally would infect insects).

Encephalitozoon cuniculi, first isolated from rabbits, has since been isolated from a number of mammals (Canning and Lom, 1986; Didier *et al.*, 1995b; Deplazes *et al.*, 1996). Examination of the ITS region of several *Enc. cuniculi* isolates (Didier *et al.*, 1995b) reveals differences in the number of GTTT repeats between mouse (two repeats), rabbit (three repeats), and dog (four repeats) isolates (Figure 5). ITS region sequences of additional *Enc. cuniculi* isolates (two from mouse, two from rabbit, one from dog) show correspondence between host and number of GTTT repeats. The sequence of the ITS region of *Enc. cuniculi* from AIDS patients from two independent studies (Didier *et al.*,

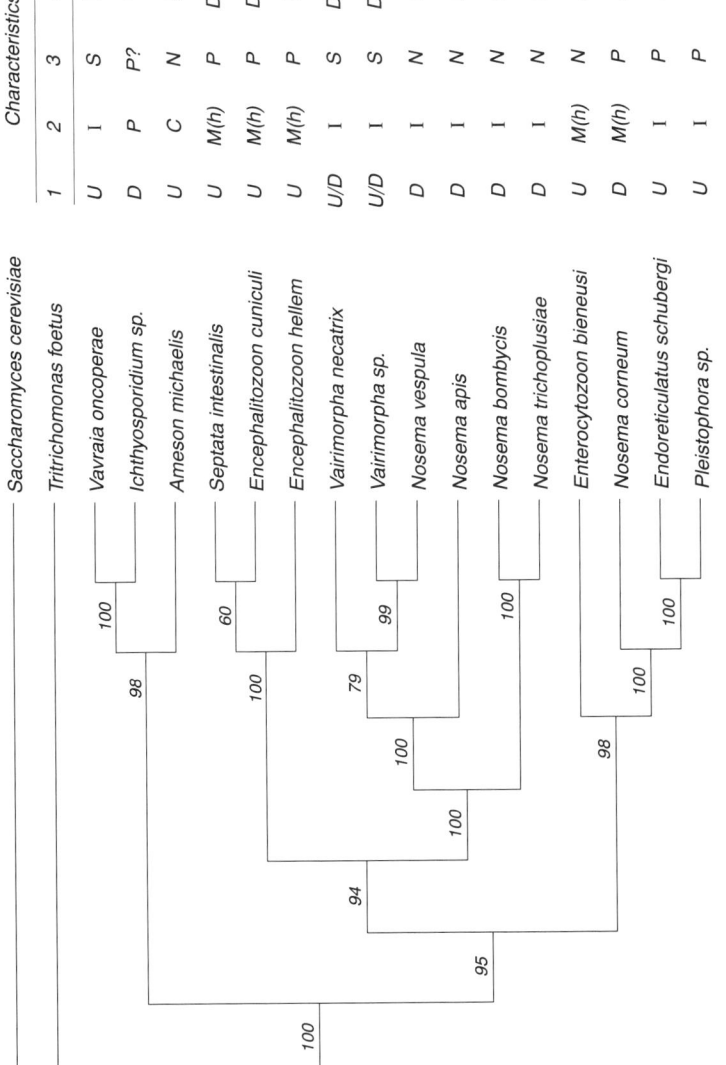

Figure 4 Microsporidian phylogeny-inferred rRNA analysis. Bootstrap analysis (400 replicates) of the most parsimonious tree. Numbers represent the percentage of bootstrap replicates. (1) Nuclear condition of the spore: D, diplokaryotic; U, uninucleate. (2) Host: C, crustacea; I, insecta; M, mammalia (h, human); P, pisces. (3) Membrane surrounding spores: N, none; P, parasitophorous vacuole; S, sporophorous vesicle. (4) Sporogony: D, disporous; O, octosporous; P, polysporous; T, tetrasporous. (5) Chromosome cycle: D1, dihaplophasic/haplophasic with meiosis; D2, dihaplophasic/haplophasic with nuclear dissociation; H, haplophasic only. *Nosema* sp. characteristics are based on the genomic placement. (Note: *Septata intestinalis* has been renamed *Enc. intestinalis*.) Reproduced, with permission, from Baker *et al*. (1995).

Mouse 5' TGTTGTTGTGTTTTGATGGATGTTTGTTT-------- GTGG 3'
Rabbit 5' TGTTGTTGTGTTTTGATGGATGTTTGTTTGTTT---- GTGG 3'
Dog   5' TGTTGTTGTGTTTTGATGGATGTTTGTTTGTTTGTTTGTGG 3'

Figure 5 Internal transcribed spacer sequence of *Enc. cuniculi* isolates demonstrating the variation in the number of GTTT repeats.

1996c; Hollister *et al.*, 1996) demonstrated the four GTTT repeats, as seen in the dog strains. Eleven isolates of *Enc. cuniculi* from rabbits in Switzerland (Deplazes *et al.*, 1996) all had three GTTT repeats, adding more evidence to the correlation between the number of repeats and host. The same study reported six isolates from humans which had three GTTT repeats, as found in rabbits, rather than four, as found in dogs. This may imply different environmental sources of *Enc. cuniculi* in AIDS patients.

3.3. Use of Molecular versus Morphological and Ecological ('Traditional') Characters

With the advent of molecular characters, and the large discrepancies found between the molecular and the traditional characters, much interest has developed regarding the resolution of these discrepancies. The resolution is simple when we look at the extremes of the problem. The molecular data provide an excellent means for the identification of a species. Should a species description include a complete small subunit rRNA sequence, and especially if the ITS region is included, the assurance of reisolating the same species in the future would be much greater than if a comparison was made with electron micrographs from a previous study. More significantly, the sequence data provide a large number of unambiguous characters, each with four clear character states, and provide the sorely needed base for phylogenetic construction. On the other hand, a phylogeny with an assortment of names at the terminal branches is meaningless without the morphological and ecological data that characterize the organism. For this reason, presenting a molecular phylogeny with the 'traditional' morphological and ecological characters might be the clearest and most unbiased method of presenting data (Baker *et al.*, 1995). Future debate over the content of a species description will be fruitless unless it centers around the most efficient means of collecting microsporidial data. Clearly, the phylogenetic analysis of the sequence data will play a key role in determining the taxonomic and evolutionary significance of morphological and ecological characters and in the reordering of microsporidial systematics in the future.

In the meantime the comparative phylogenetic analysis or rDNA sequence data are facilitating the determination of the origin of microsporidia both directly, by showing us which of the sequenced species are most closely related, and indirectly, by showing us the significance of the traditional morphological and ecologial characters so that the vast information compiled by previous microsporidial studies can better be used to track down the source of these human infections.

4. DIAGNOSIS

The diagnosis of microsporidial infections has been steadily improving over the past few years with the development of noninvasive diagnostic techniques. Currently, diagnosis of microsporidiosis is dependent on the demonstration of the organisms by light or electron microscopy. Serology has not proven useful for the diagnosis of microsporidiosis.

4.1. Cultivation

Many microsporidial species can be maintained in the laboratory in their natural hosts. Axenic culture has not been described for these obligate intracellular parasites. However, *in vitro* cultivation is possible for many microsporidian species. Cultivation from laboratory specimens is possible, but it can take 3–6 weeks before microsporidia are clearly identified. The following microsporidia, pathogenic in humans, have been cultivated in tissue culture: *N. corneum* (*Vit. corneae*) (Shadduck *et al.*, 1990), *Enc. cuniculi* (Shadduck, 1969), *Enc. hellem* (Didier *et al.*, 1991a,b; Visvesvara *et al.*, 1994b), *Enc. intestinalis* (van Gool *et al.*, 1994; Visvesvara *et al.*, 1994b, 1995a; Doultree *et al.*, 1995), and *Trachipleistophora hominis* (Field *et al.*, 1996). Short-term cultivation of *Ent. bieneusi* has been reported (Visvesvara *et al.*, 1995b). Adenovirus can mimic the cytopathologic effect of microsporidia and have been misidentified as *Ent. bieneusi* (Visvesvara *et al.*, 1996). The related microsporidia *Nucleospora* (*Enterocytozoon*) *salmonis* (Chilmonczyk *et al.*, 1991) can be grown *in vitro* in salmon lymphoid cells. A cultivation method for the isolation of microsporidia from stool using RK13 cells grown on Transwells allowed the identification of *Enc. intestinalis* infection in the presence of suspected *Ent. bieneusi* infection (van Gool *et al.*, 1994), suggesting that co-infection with different microsporidia can occur.

4.2. Serology

Tests for detecting immunoglobulin G (IgG) and IgM antibodies to *Enc. cuniculi* include the carbon immunoassay (CIA), the indirect fluorescent antibody (IFA) method, and the enzyme-linked immunosorbent assay

(ELISA). These assays are commercially used for the testing of rabbits to assure *Encephalitozoon*-free animals. Antibodies to *Enc. cuniculi* have been found in humans, but it is uncertain whether these represent infection, cross-reactivity to other microsporidian species, or nonspecific reactions (Niederkorn *et al*; 1980; Singh *et al*., 1982; Waller and Bergquist, 1982; Bergquist *et al*., 1984a,b; Hollister and Canning, 1987; Langley *et al*., 1987). Serologic cross-reactivity among microsporidia has been demonstrated by both immunofluorescence (Langley *et al*., 1987; Aldras *et al*., 1994) and Western blotting (Weiss *et al*., 1992; Ombrouck *et al*., 1995). Nonetheless, the human serologic data are intriguing. Singh *et al*. (1982) found positive titers in 6 of 69 healthy adults in England, 38 of 89 Nigerians with tuberculosis, 13 of 70 Malaysians with filariasis, and 33 of 92 Ghanians with malaria. In Swedish studies, 14 of 115 travelers returning from the tropics, but none of 48 nontravelers, were seropositive (WHO Parasitic Diseases Surveillance, 1983). In another Swedish study, 10 of 30 HIV-positive males were seropositive; all of them had traveled to the tropics (Bergquist *et al*., 1984b). These data suggest that microsporidia are common in man and, like other gastrointestinal pathogens, are associated with travel or residence in the tropics or developing nations. Diagnosis of microsporidiosis by the detection of IgM antibody was reported in a case of disseminated *Enc. cuniculi* infection with encephalitis in an immunocompetent 2-year-old male (Bergquist *et al*., 1984b). In a study of 12 AIDS patients with *Ent. bieneusi*, two AIDS patients with *Enc. intestinalis*, and two immunocompetent patients with *Vit. corneae*, ELISA titers to *Enc. hellem*, *Enc. cuniculi*, or *Vit. corneae* were not useful for diagnosis (Didier *et al*., 1993). False-negative titers were present in seven of the patients with microsporodiosis, and half of the control patients (without clinical microsporidiosis) had positive serology to microsporidia (Didier *et al*., 1993). This is consistent with other AIDS-associated infections in which serology has not proven useful. Western-blot analysis of the serologic response of animals to cultured microsporidia has been useful in speciation of microsporidia (Didier *et al*., 1991a, 1995b, 1996b).

4.3. Transmission Electron Microscopy

Curently, taxonomy and species classification is based on ultrastructural features, including the size and morphology of the spores, the number of coils of the polar tube, the developmental life cycle, and the host–parasite relationships. The life cycle of microsporidia has three phases: a proliferative (merogony) phase, the spore production (sporogony) phase, and the transmission (spore) phase (Cali and Owen, 1988). The spore is characteristic for the phylum, in that it is unicellular, with a resistant spore wall, a uninucleate or binucleate sporoplasm, and an extrusion apparatus consisting of a single polar tube with an anterior attachment complex (Levine et al., 1980). In general, spores range in size from 1 to 12 μm (Cali and Owen, 1988). The spore coat consists of an electron-dense, proteinaceous exospore, an electron-lucent endospore composed of chitin and protein (Kudo, 1921; Dissanaike and Canning, 1957; Vavra, 1976), and an inner membrane or plasmalemma (Canning and Lom, 1986; Cali and Owen, 1988). The extrusion apparatus consists of a coiled polar tube which is attached to the inside of the anterior end of the spore by an anchoring disk, and, depending on the species, forms from 4 to approximately 30 coils around the sporoplasm in the spore. While inside the spore, the core of the polar tube contains fine particulate, electron-dense material, and consequently the polar tube is sometimes referred to as a polar filament prior to discharge (Lom, 1972; Weidner, 1972; Canning and Lom, 1986). At the anterior end of the spore, below the anchoring disk, the polar tube is surrounded by membranous structures called the lamellar polaroplast and tubular polaroplast. The sporoplasm may have a single nucleus or may have two abutted nuclei (diplokaryon). A posterior vacuole and numerous ribosomes, arranged in helical coils or sheets, also are present in the sporoplasm (Cali and Owen, 1988).

Within host cells microsporidia display four types of host–parasite relationship, which are used in classification of these organisms. The first, demonstrated by *Nosema* and *Enterocytozoon*, is direct contact of the parasite plasmalemma and host cell cytoplasm throughout replication until sporogony, when the plasmalemma thickens due to secretion of spore coat by the parasite. The second is parasite-induced indirect contact demonstrated by *Pleistophora*, in which the parasites initiate replication in direct contact with the host cell cytoplasm, but a parasite-induced vesicle develops that isolates the parasite from the host cell cytoplasm at some point during development. Such microsporidia are termed pansporoblastic. The third type of relationship is demonstrated by Encephalitozoonidae, in which the parasites are enclosed in a host-produced phagosome during all stages of development. The fourth category is known as host- and parasite-induced indirect contact; at present no microsporidia that infect humans have displayed this relationship (Cali and Owen, 1988).

For transmission electron microscopy (TEM), specimens can be fixed using standard techniques such as 2.5% glutaraldehyde (Schwartz et al., 1994). Encephalitozoonidae are characterized by: proliferative cells containing one to several nuclei, with each sporont producing two sporoblasts, and each sporoblast producing one spore. The spore containing a single nucleus with five or six polar tube coils (Figure 6) measures 1.5–2.0 μm long (Cali and Owen, 1990). All stages are found in the cytoplasm of the host cell inside a phagosome or a host-formed parasitophorous vacuole. In contrast, *Nosema* and *Nosema*-like organisms are characterized by: spores with 11–13 coils of the polar filament, paired abutted nuclei, and developing organisms in direct contact with the host cell cytoplasm (Cali, 1991b; Schwartz et al., 1994;

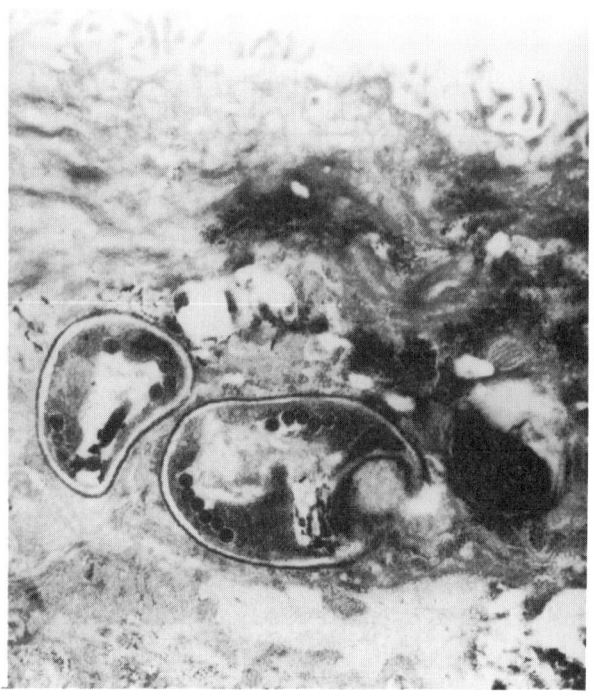

Figure 6 Transmission electron micrograph of a conjunctival biopsy from an HIV-infected patient with punctate keratoconjuctivitis. An *Enc. hellem* spore, with five to six polar tube coils in cross-section, is present. Reproduced by courtesy of Dr Ann Cali and Dr Peter Takvorian, Rutgers University.

Deportes *et al.*, 1995). *Pleistophora* and *Trachipleistophora* have a single nucleus and are pansporoblastic. Enterocytozooonidae are characterized by the formation of two types of plasmodial stage (proliferative and sporogonic), electron-lucent inclusions, and electron-dense disks. The 1–2 μm spore contains a single nucleus and 5–7 coils of polar tubes arranged in two rows (Desportes *et al.*, 1985; Cali, 1991). TEM has been utilized to identify *Ent. bieneusi* spores in stool specimens (van Gool *et al.*, 1990; Weber *et al.*, 1994a).

4.4. Histology

In gastrointestinal microsporidiosis and disseminated disease, microsporidia spores are discernible to experienced observers in touch preparations (Color Plate 6) and paraffin-embedded or semi-thin plastic sectons stained with hematoxylin and eosin, Chromotrope 2R (Figure 7), Warthin–Starry silver

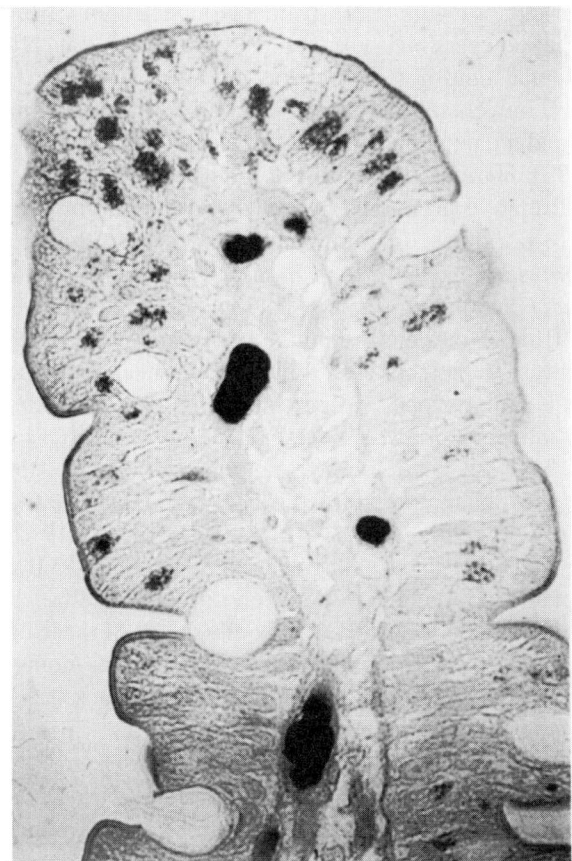

Figure 7 Chromotrope 2R stained preparation of an intestinal biopsy from an HIV-infected patient with diarrhea demonstrating *Enc. intestinalis*. Reproduced courtesy of Dr Donald P. Kotler, St. Lukes Roosevelt Hospital Center, College of Physicians and Surgeons, Columbia University and Dr Jan M. Orenstein, George Washington University.

stain, Giemsa stain, or Brown–Brenn or Brown–Hopps tissue Gram stains (Rijpstra *et al.*, 1988; Lucas *et al.*, 1989; Weir and Sullivan, 1989; Orenstein *et al.*, 1990b; Weber *et al.*, 1992, 1994a; Field *et al.*, 1993a,b; Wittner *et al.*, 1993; Franzen *et al.*, 1995a). In an examination of 34 TEM positive patients it was demonstrated that Giemsa staining had a sensitivity of 57% and specificity of 100%, that the chromotrope 2R stain had a sensitivity of 83% and specificity of 100%, and that false-negative specimens were from patients with lower parasite burdens on TEM (Kotler *et al.*, 1994). Specific identification of microsporidian species ultimately depends on ultrastructural examination (i.e. TEM examination of biopsy material). It is also possible, however, to use *in situ* hybridization with rRNA primers to identify microsporidia in tissue sections (S. Tizpori, personal communication). Spores can often be demonstrated in duodenal fluid obtained at the time of endoscopy. Histologic sections from small bowel tend to reveal a mild inflammatory infiltrate (primarily lymphocytic) and a patchy distribution of infected enterocytes (Shadduck and Orenstein, 1993). Various developmental stages (spores, sporoblasts, sporogonic and proliferative plasmodia) have been identified in human tissues.

Ocular microsporidiosis in immunocompetent hosts has been associated with: *Microsporidium ceylonesis* (Ashton and Wirasinha, 1973), *Microsporidium africanus* (Pinnolis *et al.*, 1981), *Vit. corneae* (Shadduck *et al.*, 1990; Cali *et al.*, 1991a; Silveira and Canning, 1995), and *N. ocularum* (Cali *et al.*, 1991b). In contrast to patients with AIDS and corneal microsporidiosis, non-AIDS patients have developed deeper, ulcerative lesions with microsporidia invading deep into the corneal stroma causing an interstitial keratitis. Histologic sections have revealed findings ranging from intact stromal lamellae with a paucity of inflammatory cells to frank necrotizing keratitis (Ashton and Wirasinha, 1973; Pinnolis *et al.*, 1981; Davis *et al.*, 1990; Cali *et al.*, 1991a,b). Highly refractile organisms of 3.5–4.5 × 1.5–3.0 μm are found within macrophages or lying free between corneal lamellae, but not within superficial epithelial cells as in AIDS patients. On TEM these organisms were large, binucleate, and lacked parasitophorous vacuoles compared to those seen in AIDS-related cases (Davis *et al.*, 1990; Shadduck *et al.*, 1990).

In HIV-1 infected patients, ocular microsporidial infection has been restricted to the superficial epithelium of the cornea and conjuctiva, i.e. superficial keratoconjunctivitis. This keratitis rarely progresses to corneal ulceration. Slit-lamp examination reveals fine to coarse punctate epithelial opacities, granular epithelial cells with irregular fluoroescein uptake, marked conjunctival injection that may be confined to the inferior fornix, and superficial corneal infiltrates. The anterior chamber is uninflamed or has minimal cells and flare (Friedberg *et al.*, 1990; Orenstein *et al.*, 1990a; Lowder *et al* 1990; Cali *et al.*, 1991a,b; Didier *et al.*, 1991a; Yee *et al.*, 1991; Lacy *et al.*, 1992; Canning and Hollister, 1992; Metcalfe *et al.*, 1992; Schwartz *et al.*, 1993; Weber *et al.*, 1993). Diagnosis of corneal infections in patients with AIDS has been technically less difficult than diagnosis of intestinal microsporidiosis. Surrounding inflammatory reactions have been notably absent and invasion beyond the epithelial layer is rare. In conjunctival tissue, on the other hand, inflammatory cells are evident and organisms are present within conjunctival macrophages. Biopsy specimens are not needed for diagnosis in the majority of cases but, if performed, can be viewed by using a touch preparation technique with Gram stain, Giemsa stain, or chromotrope 2R (Rastrelli *et al.*, 1994; Weber *et al.*, 1994a; Shah *et al.*, 1996). Fresh tissue may also be examined by phase-contrast microscopy, due to their thick wall; unstained spores are refractile, appearing green, and may be birefringent (Tiner, 1988).

4.5. Immunodetection

Diagnosis of microsporidia infection by stool examination is difficult due to the small size of spores (van Gool et al., 1990; van Gool et al., 1993; Wittner et al., 1993; Weber et al., 1994a). Immunologic detection of spores is feasible and a polyclonal serum to *Enc. hellem* has been used to demonstrate this organism in tissue samples (Schwartz et al., 1993). Polyclonal serum prepared to *Enc. cuniculi* or *Enc. hellem* also demonstrate cross-reaction with *Ent. bieneusi* with immunofluorescent techniques (Weiss et al., 1992; Zierdt et al., 1993). Monoclonal antibodies to *Enc. intestinalis* and *Enc. hellem* have been prepared and have demonstrated good reactivity with these organisms in clinical specimens (Aldras et al., 1994; Visvesvara et al., 1994b; Beckers et al., 1996). Monoclonal antibodies to *Enc. hellem* (IgM isotype, C12/E9/E11) had some cross-reaction with *Enc. cuniculi* and *Vit. cornea* by immunofluorescence, but were not useful for *Ent. bieneusi* (Aldras et al., 1994). Monoclonal antibodies to *Ent. bieneusi* have not been described. Polyclonal murine antiserum to Encephalitozoonidae had a much lower sensitivity and specificity when compared to chromotrope 2R or chemofluorescent brighteners in examining stool containing *Ent. bieneusi* (Garcia et al., 1994; Didier et al., 1995a). High background fluorescence, a problem in immunofluorescent staining, can be reduced by absorbing the antisera with fecal debris, but background fluorescence is always present (Didier et al., 1995a). The sensitivity of these techniques is likely to be much higher when using a species-specific antibody, i.e. when diagnosing *Enc. intestinalis* infection with an *Enc. intestinalis* antiserum. In fact, monoclonal antibodies to *Enc. intestinalis* (Si93/Si13) have been produced that display no cross reactivity to other *Encephalitozoonidae* (Beckers et al., 1996). Such monospecific antibodies may enable species identification without the need for ultrastructural studies (Schwartz et al., 1993; Visvesvara et al., 1994b). A monoclonal antibody, 3B6, reactive to Encephalitozoonidae exospore has also been described (Enriquez et al., 1997). This antibody reacts with *Enc. hellem, Enc. cuniculi, Enc. intestinalis, Nosema* sp., and *Vairimorpha* sp., but not with *Ent. bieneusi* (Enriquez et al., 1997). This antibody may prove useful for the diagnosis of suspected *Encephalitozoonidae* infections, especially in the eye, sinus, and urinary system.

4.6. Special Stains

Effective morphologic demonstration of microsporodia by light microscopy can be accomplished by staining methods that produce differential contrast between the spores of the microsporidia and the cells and debris in clinical samples, such as stool, in which they are found. In addition, adequate magnification (i.e. ×1000) given the small size of the spores (1–5 µm) is required for visualization. Chromotrope 2R (Weber et al., 1992), calcofluor white (fluorescent brightener 28) (Vavra et al., 1993a,b), and Uvitex 2B (Weir and Sullivan, 1989; van Gool, 1993; Vavra et al., 1993a) have all been reported to be useful as selective stains for microsporidia in stool specimens as well as other body fluids. Microsporidia in body fluids other than stool have also been visualized with Giemsa, Brown-Hopps Gram stain, acid-fast staining, or Warthin–Starry silver staining. Although microsporidia in stool specimens can be visualized with Giemsa stain, they are difficult to distinguish from bacteria (Beauvais et al., 1993).

The procedures routinely used to concentrate parasites in stool specimens have not concentrated microsporidial spores, and in fact some of them have decreased spore numbers resulting in false-negative results. However, in one study (Carter et al., 1996), the addition of KOH as a mucolytic prior to stool centrifugation increased the sensitivity of chromotrope 2R staining.

The chromotrope stain developed by Weber et al. (1994a) has been used to identify successfully microsporidia in a variety of clinical samples, including duodenal aspirates, urine, bronchoalveolar lavage fluid, sputum, conjunctival smears, fresh stool, and 10% formalin fixed stool (Color Plate 7). This stain is a modified trichrome stain utilizing a 10-fold higher concentration of chromotrope 2R with a prolonged staining time (Weber et al., 1992). Specimens are fixed with methanol for 5 minutes and stained with chromotrope 2R solution (prepared by adding 6 g chromotrope 2R, 0.15 g fast green, and 0.7 g phosphotungstic acid to 3 ml glacial acetic acid, leaving for 30 minutes, and then adding 100 ml distilled water) for 90 minutes followed by rinsing for 10 seconds in acid alcohol (4.5 ml acetic acid per 1000 ml 90% ethanol), 5 minutes in 95% ethanol, 5 minutes in 100% alcohol, and 10 minutes in xylene (or a xylene substitute; Hemo-De, Fisher Scientific, Pittsburgh, PA). Ryan and colleagues (1993) have reported that a further modification of this staining procedure, in which aniline blue (0.5 g/100 ml) is substituted for fast green and the level of phosphotungstic acid is reduced (0.25 gm/100 ml), gives improved results. It has been suggested that the staining characteristics of microsporidia with this stain can be improved by staining at 50°C for 10 minutes (Kokoskin et al., 1994). Using this stain spores appear light pink against a green (Weber et al., 1992) or blue (Ryan et al., 1993) background, with a belt-like stripe girding them diagonally and equatorially (Weber et al., 1992, 1994a,b; Rastrelli et al., 1994). Other fecal elements may stain reddish with these stains, but can be distinguised from microsporidia spores by their size, shape, and staining pattern. Moura et al. (1996) have described a Gram-chromotrope stain that can also be used to detect microsporidia and can be completed in 15–20 minutes. This technique is performed by heat fixing a thin smear of the material to be examined and then staining the slide with gentian violet for 1 minute, Gram's iodine for 1 minute, and then using a decolorizer solution until all excess stain is removed. The Gram-stained material is then stained with chromotrope stain (Weber et al., 1992) for 5 minutes, rinsed in acid alcohol, and then washed for 1 minute in 95% and then 100% ethanol. Mircrosporidia spores stained with this method have a violet appearance and retain the characteristic belt-like stripe. An acid-fast trichrome stain that allows the detection of *Cryptosporidium parvum* and *Isospora belli* as well as microsporidia has also been described (Ignatius *et al.*, 1997). For this technique methanol fixed slides are stained for 10 minutes in carbol–fuchsin solution (25 g phemol, 500 ml distilled water, 25 ml saturated alcholic fuchsin solution (consisting of 2.0 g basic fuchsin in 25 ml 96% ethanol)), washed with water, decolorized with 0.5% HCl/alcohol, washed with water and stained for 30 minutes with chromotrope 2R solution (prepared as described above) (Weber *et al.*, 1992a), rinsed with acid alcohol (4.5 ml acetic acid in 995.5 ml 90% ethanol), and then washed with 95% ethanol. With this stain microsporidia appear pink with a pale diagonal or horizontal strip, and the oocysts of *C. parvum* or *Iso. belli* appear violet or bright pink (Ignatius *et al.*, 1997).

Alternatively, spores may be visualized by ultraviolet (UV) microscopy using chemofluorescent optical brightening agents such as calcofluor white M2R (fluorescent brightener 28, Fungi-Fluor), and Uvitex 2B (Fungiqual A, Dieter Reinehr and Manfred Rembold, Spezialchemikalien fur die Medizinische Diagnostik, Kandern, Germany), which stain chitin in the spore wall (endospore layer) (Weir and Sullivan, 1989; van Gool *et al.*, 1993; Vavra *et al.*, 1993a,b; Weber *et al.*, 1994b; DeGirolami *et al.*, 1995; Franzen *et al.*, 1995b; Luna *et al.*, 1995). Blue fluorescence filters (e.g. excitation 390–420 nm, chromatic beam splitter 450 nm, barrier filter 470 nm) are required to achieve the necessary specificity with these stains (Figure 8). Fungi and other fecal elements may also stain with chemofluorescent optical brightening agents, but can be distinguished from microsporidian spores by their size and morphology. Slides can be examined immediately after adding these agents as they bind rapidly to the microsporidian spore wall. Vavra *et al.* (1993a) studied the effects of fixation, stain concentration, and pH on these chemofluorescent stains, and found that 1% dye concentration (in 0.1 M phosphate buffer, pH 7 or 8.5) was needed for optimal fluorescence. Evans's blue as a counterstain diminished fluorescence and that fixation did not affect stain performance. Staining of spores, even those stored for a prolonged period of time, was enhanced by applying chemofluorescent dyes dissolved in 1 N NaOH (instead of 0.1 M phosphate buffer) at room temperature.

Using 50 stool specimens positive for microsporidia by TEM, both the chromotrope 2R and chemofluorescent brightening stains identified 100% of specimens if at least 50 high power (i.e. × 1000) fields were examined (Didier *et al.*, 1995a) The intensity of infection correlates with the sensitivity of these methods. In a study employing Uvitex 2B (van Gool *et al.*, 1993), all of the 186 stool samples examined from 19 patients with biopsy-proven *Ent. bieneusi* infection were positive, and none of the 55 stool samples from 16 biopsy-negative patients were positive. In another study, Uvitex 2B staining

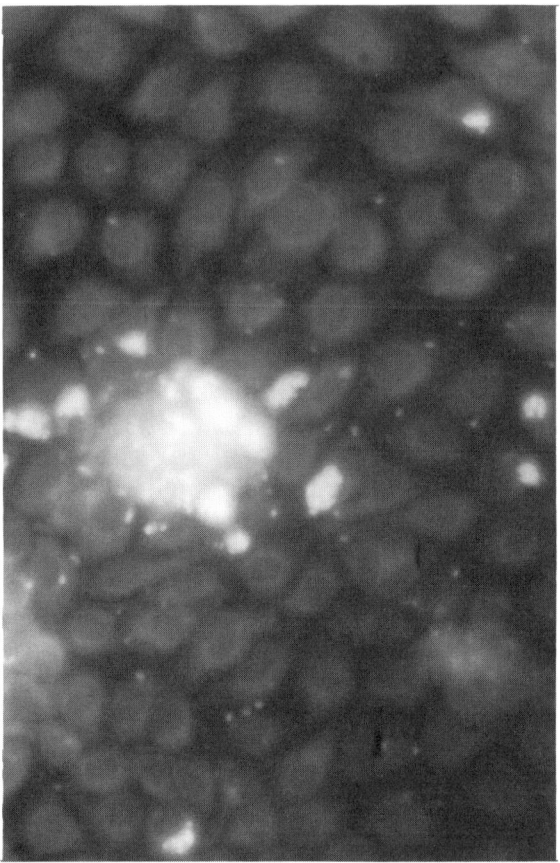

Figure 8 Calcofluor white stained preparation of *Enc. cuniculi* in tissue culture.

detected microsporidia in all chromotrope 2R-positive samples, and also identified seven additional samples (from three patients) as positive (DeGirolami *et al.*, 1995). On re-examination, stool samples from these patients were also positive with the chromotrope 2R stain. All patients with positive duodenal biopsies were positive on stool examination by chromotrope or chemofluorescent methods (DeGirolami *et al.*, 1995). Of the six patients with negative duodenal biopsies who were identified as positive by chromotrope 2R or chemofluorescent staining, four had confirmation of microsporidia in the stool by TEM (DeGirolami *et al.*, 1995). In infections with low numbers of spores it is likely that these noninvasive methods have a higher sensitivity than biopsy techniques. The limit of detecting microsporidia by these techniques appears to be 5×10^4 organisms/ml (500 organisms/10 µl) (Didier *et al.*, 1995a). Overall, the sensitivity of the chemofluorescent brightener based stains is slightly higher than chromotrope based stains (especially when low numbers of spores are present in a sample); however, the specificity of the chemofluorescent stains is lower—90% vs 100% in the study by Didier *et al.* (1995a). Based on an analysis of the performance characteristics of these smears, it has been suggested that chemofluorescent stains should be utilized for screening stool specimens, and that all positive specimens be confirmed by a chromotrope 2R stain before being considered a true positive (Didier *et al.*, 1995a). Neither the chromotrope nor chemofluorescent stains provide information on the species of microsporidia being identified. A recent report illustrates that these stains may give false-positive results due to insect microsporidia contaminating ingested food (McDougall *et al.*, 1993).

4.7. Polymerase Chain Reaction

As discussed in Section 3, a number of microsporidian small subunit rRNA sequences have been determined. Homology PCR cloning of the rRNA genes of several mircrosporidia has been accomplished using primers complementary with conserved sequences in *Vair. necatrix* (Vossbrinck *et al.*, 1987) and other phylogenically conserved primers (see Table 2). The small subunit (SSU) rRNA gene sequences for the human pathogens *Enc. cuniculi, Enc. hellem, Enc. intestinalis, Ent. bieneusi,* and *Vit. corneae* (Vossbrinck *et al.*, 1993; Hartskeerl *et al.*, 1993a,b, 1995; Zhu *et al.*, 1993a,b,c 1994; Vivesvara *et al.*, 1994a; Weiss *et al.*, 1994; Baker *et al.*, 1995; Katiyar *et al.*, 1995; Didier *et al.*, 1995a, 1996a, b) have been determined and are available in the GenBank database (see Table 1). It has been possible to design PCR primers to these small subunit rRNA genes to identify microsporidia in clinical samples (Table 3). It is also possible by this method to identify microsporidia at the species level without the use of ultrastructural examination. In the amplification of microsporidia from clinical specimens, it is clear that routine extraction methods are sufficient for tissue biopsies, corneal scrapings, cultured organisms, and urine specimens. For the amplification of microsporidia from stool, mechanical disruption with 500-µm glass beads (Mini-Bead Beater, Biospec Products Inc., Bartlesville, OK) and/or harsh extraction conditions have generally been required to obtain reproducible amplification and to eliminate nonspecific inhibitors of Taq polymerase. Successful methods of obtaining amplification of different pathogens from stool have included the use of sodium hypochlorite (Fedorko et al., 1995), guanidine thiocyanate (Boom *et al.*, 1990; Kock *et al.*, 1997), DTT (Katzwinkel-Wladarsch *et al.*, 1996), and hexadecyltrimethylammonium bromide (Carville *et al.*, in press). Two main approaches have been employed in the construction of PCR primers for microsporidia: the design of universal panmicrosporidia primers, and the design of species-specific primer pairs.

Table 3 Diagnostic PCR primers for microsporidia pathogenic in humans.

Microsporidia amplified	Primers	Name[a]	ΔT [b]	Amplicon[c]	References
Encephalitozoonidae and Ent. bieneusi	5'CACCAGGTTGATTCTGCCTGAC3' 5'CCTCTCCGGAACCAAACCTG3'	PMP1 (VI) PMP2	60	Eb 250 Ec 268 Ei 270 Eh 279	Fedorko et al. (1995)
Encephalitozoonidae and Ent. bieneusi	5'TGAATG(G/T)GTCCTGT3' 5'TCACTCGCGCTACT3' 5'GTTCATCGCACTACT3'	MSP1 MSP2A MSP2B	58	Eb 508 Ec 289 Ei 305	Katzwinkel-Wladarsch et al. (1996)
	5'GGAATTCACACCGCCCGTC(A/G)(C/T)TAT3' 5'CCAAGCTTATGCTTAAGT(C/T)(A/C)AA(A/G)GGGT3' 5'CCAAGCTTATGCTTAAGT CCAGGGAG3'	MSP3 MSP4A MSP4B			
Encephalitozoonidae and Ent. bieneusi	5'CCAGGUTGATUCTGCCUGACG3' 5'TUACCGGCGCUGCUGGCAC3' 5'AAGGAGCCTGAGAGATGGCT3' 5'CAATTGCTTCACCCTAAGGTC3' 5'GACCCCTTTGCACTCGCACAC3' 5'TGCCCTCCAGTAAATCACAAC3' 5'CCTCCAATCAATCTCGACTC3'	Mic3U Mic421U Mic266 Eb379 Ec378 Eh410 Ei395	65/62	Eb 132 Ec 113 Eh 134 Ei 128	Kock et al. (1997)
Encephalitozoonidae	5'TGCAGTTAAAATGTCCGTAGT3' 5'TTTCACTCGCCGCTACTCAG3'	int530f int580r	40	1000	Didier et al. (1996a)
Enc. intestinalis	5'CACCAGGTTGATTCTGCCTGAC3' 5'CTCGCTCCTTTACACTCGAA3'	V1 Si500	58	375	Weiss et al. (1994)
Enc. intestinalis	5'GGGGGTAGGAGTGTTTTG3' 5'CAGCAGGCTCCCTGCCATC3'	3 3	65	930	Schuitema et al. (1996)

Table 3 Continued.

Microsporidia amplified	Primers	Name[a]	ΔT [b]	Amplicon[c]	References
Enc. cuniculi	5'ATGAGAAGTGATGTGTGTGCG3' 5'TGCCATGCACTCACAGGCATC3'		55	549	Visvesvara et al. (1994b), De Groote et al. (1995)
Enc. hellem	5'TGAGAAGTAAGATGTTTAGCA3' 5'GTAAAAAGACTCTCACACTCA3'		55	547	Visvesvara et al. (1994b)
Ent. bieneusi	5'GAAACTTGTCCACTCCTTACG3' 5'CCATGCACCACTCCTGCCATT3'	EBIEF1 EBIER1	55	607	DaSilva et al. (1996)
Ent. bieneusi	5'CACCAGGTTGATTCTGCCTGAC3' 5'ACTCAGGTGTTATACTCACGTC3'	V1 EB450	48	353	Zhu et al. (1993c), Coyle et al. (1996)
Ent. bieneusi	5'CACCAGGTTGATTCTGCCTGAC3' 5'CAGCATCCACCATAGACAC3'	V1 Mic3	54	446	Mansfield et al. (1997), Carville et al. (1997)
Ent. bieneusi	5'TCAGTTTTGGGTGTGGTATCGG3' 5'GCTACCCATACACACATCATTC3'	Eb.gc Eb.gt	49	210	Velasquez et al. (1996)
Ent. bieneusi	5'GCCTGACGTAGATGCTAGTC3' 5'ATGGTTCTCCAACTGAAACC3'	2 2	55	1265	Schuitema et al. (1996)

[a] The designation of the primers given in the reference.
[b] Annealing temperature in PCR.
[c] Size of amplified fragment in base pairs. Eb, Ent. bieneusi; Ec, Enc. cuniculi; Ei, Enc. intestinalis; Eh, Enc. hellem.

A primer set (V1::1492r) has been designed that amplifies the SSU rRNA gene sequence of many microsporidia to produce a product of 1200–1500 bp for most microsporidia (Weiss *et al.*, 1994). For phylogenetic studies a second set of primers (530f::580r) has been used to amplify part of the SSU rRNA gene, the ITS region, and part of the large subunit rRNA gene of many microsporidia (Vossbrinck *et al.*, 1993; Zhu *et al.*, 1993c; Baker *et al.*, 1994, 1995; Weiss *et al.*, 1994). These 'universal' primer pairs are extremely useful for obtaining rRNA sequence data of unknown, new, or novel microsporidia, and thereby facilitate their classification by molecular means (see Table 2). Unfortunately, these primers are not useful for diagnosis as the large size of the product (i.e. the amplicon) limits their sensitivity. This is particularly evident in formalin-fixed tissue where reliable amplification is best obtained when primers are separated by 100–400 bp. In fresh tissue, however, primers separated by 700–1000 bp can still provide adequate sensitivity.

A recent publication describes a nested PCR across the SSU rRNA gene, ITS region, and large subunit rRNA gene region for the detection of microsporidia from stool specimens that can amplify *Ent. bieneusi*, *Enc. cuniculi*, and *Enc. intestinalis* using the primers MSP1::MSP2A and MSP2B for the first PCR and MSP3::MSP4A and MSP4B for the second (i.e. nested) PCR (Katzwinkel-Wladarsch *et al.*, 1996). Using these primers *Ent. bieneusi* yields a 508-bp amplicon, and Encephalitozoonidae a 300-bp amplicon. MnlI restriction digestion of the amplicon can be used to identify the microsporidian species present (*Enc. cuniculi*, 289 bp; *Enc. intestinalis*, 305 bp; no data provided for *Enc. hellem*). This nested PCR was able to detect 3–100 *Enc. cuniculi* spores in 0.1 g of stool (Katzwinkel-Wladarsch *et al.*, 1996). A pan-microsporidian primer set PMP1::PMP2 (Fedorko *et al.*, 1995) has been desiged for the Encephalitozoonidae (*Enc. cuniculi*, *Enc. hellem*, *Enc. intestinalis*), *Ent. bieneusi*, and *Vit. corneae* (Fedorko and Hijazi, 1996) which yields a 250–279 bp product depending on the species of microsporidia. Unique restriction products of the amplicons are produced when they are digested by PstI or HaeIII, thus allowing species identification; however, these restriction fragments differ by 2–11 bp. Due to the small difference in restriction fragment sizes, known microsporidian controls should be used and acrylamide electrophoresis is useful. This primer set has been used successfully on formalin-fixed stool specimens treated with sodium hypochlorite. Another nested PCR has recently been described to identify human pathogenic microsporidia (Kock *et al.* 1997). In this procedure an initial PCR (annealing temperature 65°C) with primers Mic3U::Mic 421U is followed by four species specific PCRs (annealing temperature 62°C) using Mic266 with Eb379, Ec378, Eh410 or Ei395 which allows the identification of the species of microsporidia in the sample. This PCR is reported to detect as few as 200 spores in a gram of feces.

A pan-Encephalitozoonidae primer set (int530f::int580r) has been described that amplifies a 1000-bp amplicon only from microsporidia belonging to the family Encephalitozoonidae (Vossbrinck et al., 1993; Didier et al., 1996a). Species-specific diagnosis of the amplicon can be obtained by using either: a set of species-specific oligonucleotides for Southern blotting (Enc. cuniculi, 5'TAGCGGCTGACGAAGCTGC3'; Enc.hellem, 5'TGAGTGT-GAGAGTGTTTTTACAT3'; Enc. intestinalis, 5'CGGGCAAGGAGAAC-GAGGACGG3'), digestion to create species-specific restriction patterns (Vossbrinck et al., 1993) or by heteroduplex mobility shift analysis of the amplicon (Didier et al., 1996a). This primer set has been used successfully to amplify microsporidia from urine, conjunctiva, and tissue culture.

Primer sets specific to Enc. hellem (Visvesvara et al., 1994), Enc. cuniculi (De Groote et al., 1995), and Enc. intestinalis (Schuitema et al., 1993; Weiss et al., 1994; Schuitema et al., 1996) have been described. The Enc. hellem primer pair (5'TGAGAAGTAAGATGTTTAGCA3'::5'GTAAAAACACT-CTCACACTCA3') has been used to amplify a 547-bp product from Enc. hellem (CDC:0291:V213) spores obtained from tissue culture. The Enc. cuniculi primer pair (5'ATGAGAAGTGATGTGTGTGCG3'::5'TGCCATG-CACTCACAGGCATC3') was used to amplify a 549-bp product from Enc. cuniculi spores obtained from tissue culture (Visvesvara et al., 1994; De Groote et al., 1995). Both these primer pairs are species specific. The Enc. intestinalis primer set (V1::SI500) has been used to amplify a 375-bp amplicon from intestinal biopsies, duodenal fluid, and formalin-fixed stool from patients with Enc. intestinalis (Weiss et al., 1994; Franzen et al., 1996; Ombrouck et al., 1996, 1997; L.M. Weiss, unpublished data). The Enc. intestinalis primer set '3', which amplifies a 930-bp region, has also been used in tissue biopsies, duodenal fluid, and stool specimens (Schuitema et al., 1993, 1996). Very little data are available on the sensitivity and specificity of these primers. In a recent study using the Enc. intestinalis primer set (V1::SI500), amplification occurred for all five TEM confirmed cases (Coyle et al., 1996), and it has been reported that this primer set can detect a single spore in a 'spiked' stool specimen (Ombrouck et al., 1996, 1997).

Several specific primer sets have been developed for the diagnosis of Ent. bieneusi infections. Primer set V1::EB450 has been used to amplify a 353-bp amplicon from Ent. bieneusi infected tissue, duodenal fluid, and stool (Zhu et al., 1993c; Franzen et al., 1996; Ombrouck et al., 1996, 1997). This primer is specific for Ent. bieneusi and does not amplify Enc. hellem (Coyle et al., 1996), Enc. cuniculi, or Enc. intestinalis (Zhu et al., 1993c; Coyle et al., 1996). In a recent study, all 25 TEM-confirmed cases of Ent. bieneusi were identified using this primer pair. The sensitivity of PCR is probably higher than TEM, as patients initially identified only by PCR (i.e. negative by TEM) were subsequently demonstrated on follow-up pathology to have this parasite (thus the positive PCR was not a false-positive reaction) (Coyle et al., 1996). This primer pair has been reported to detect 10–100 spores in 'spiked' stool specimens (Ombrouck et al., 1997). There is a difference of about 1% between the sequence for Ent. bieneusi on which these primers were based and the other *Ent. bieneusi* rRNA sequence in GenBank. This may represent either true variation or sequencing error. Based on this observation, a modified EB450 primer with a C instead of a G in position 9 has been suggested (DaSilva *et al.*, 1996). A second primer set, EB1EF1::EB1ER1, amplified a 607-bp fragment from duodenal fluid and bile fluid from a patient with *Ent. bieneusi* and culture supernatant from a short-term *Ent. bieneusi* in vitro culture (Da Silva *et al.*, 1996). Primer set EB1EF1::EB1ER1 may have a higher sensitivity than V1::EB450 (Da Silva *et al.*, 1996). An *Ent. bieneusi*-like organism has been isolated from macaques infected with the Simian immunodeficiency virus (SIV) using EB1EF1::EB1ER1. In addition, V1::Mic3 amplifies a 446-bp product from *Ent. bieneusi* (Mansfield *et al.*, 1997), and this amplicon has been used to demonstrate *Ent. bieneusi* by *in situ* hybridization. Primer set V1::Mic3 has also amplifed *Ent. bieneusi* from the the stool of SIV-infected macaques (S. Tzipori, personal communication) and in formalin-fixed stool (Carville *et al.*, 1997). The primer set '2' amplifies a 1265-bp DNA from *Ent. bieneusi* and has worked on clinical specimens (tissue and stool) (Schutiema *et al.*, 1996). The primer set Eb.gc::Eb.gt amplifies a 210-bp fragment in the intergenic spacer of *Ent. bieneusi* (Velasquez *et al.*, 1996). This primer set has identified two *Ent. bieneusi* stool specimens and did not amplify four *Enc. intestinalis* stool specimens. Primers specific for *Nucleospora (Enterocytozoon) salmonis*, a microsporidian of the family Enterocytozoonidae which is a pathogen of salmonid fish, have also recently been described (Barlough *et al.*, 1995; Docker *et al.*, 1997).

The use of PCR for the diagnosis of microsporidiosis shows great promise. It is likely that PCR will be adapted large-scale epidemiologic studies on these pathogens once better techniques have been developed to eliminate PCR inhibitors and to reduce the amount of manipulation currently required to process stool specimens. These techniques have already been applied to investigations on animal reservoirs of *Ent. bieneusi*, and recently both SIV-infected macaques (Mansfield *et al.*, 1997) and pigs (Deplazes *et al.*, 1996) have been identified as being infected with *Ent. bieneusi*. Sequencing of the rRNA genes of the *Ent. bieneusi*-like organisms identified in these studies will confirm if these are identical to the human pathogen or a closely related organism. It is possible that, as more isolates of *Ent. bieneusi* are identified, strain differences will be demonstrable in the intergenic spacer region of this organism, as has been demonstrated for *Enc. cuniculi* (Didier *et al.*, 1996c).

5. SUMMARY

Knowledge about the phylum Microspora and its relationship to humans is in a period of rapid growth. In the last few years noninvasive diagnostic techniques for microsporidia have been developed and are being applied to both clinical and epidemiologic studies. As techniques have improved new microsporidian pathogens are being identified and the reservoir hosts of the microsporidia infecting humans are being sought. Studies on the molecular phylogeny of the microsporidia have demonstrated that they display a large evolutionary distance from most eukaryotes and as more molecular phylogenies are developed their place in the 'tree of life' may be clarified. Recently, work has begun on the molecular biology of these organisms, and this work is beginning to suggest that the genomic organization of these organisms may be different from that of other eukaryotes.

REFERENCES

Aldras, A.M., Orenstein, J.M., Kotler, D.P., Shadduck, J.A. and Didier, E.S. (1994). Detection of microsporidia by indirect immunofluorescence antibody test using polyclonal and monoclonal antibodies. *Journal of Clinical Microbiology* **32**, 608–612.
Andreadis, T.G. (1983). Life cycle and epizootiology of *Amblyospora* sp. (Microspora, Amblyosporidae) in the mosquito *Aedes cantator*. *Journal of Protozoology* **30**, 509–518.
Ashton, N. and Wirasinha, P.A. (1973). Encephalitozoonosis of the cornea. *British Journal of Ophthalmology* **57**, 669–674.
Baker, M.D., Vossbrinck, C.R., Maddox, J.V. and Undeen, A.H. (1994). Phylogenetic relationships among *Vairimorpha* and *Nosema* specis (Microspora) based on ribosomal RNA sequence data. *Journal of Invertebrate Pathology* **64**, 100–106.
Baker, M.D., Vossbrinck, C.R., Didier, E.S., Maddox, J.V. and Shadduck, J.A. (1995). Small subunit ribosomal DNA phylogeny of various microsporidia with emphasis on AIDS related forms. *Journal of Eukaryotic Microbiology* **42**, 564–570.
Barlough, J.E., McDowell, T.S., Milani, A., Bigornia, L., Slemenda, S.B., Pieniazek, N.J. and Hedrick, R.P. (1995). Nested polymerase chain reaction for detection of *Enterocytozoon salmonis* genomic DNA in chinook salmon *Oncorynchus tshawytscha*. *Diseases of Aquatic Organisms* **23**, 17–23.
Beauvais, B., Sarfati, C., Molina, J.M., Lesourd, A., Lariviere, M. and Derouin, F. (1993). Comparative evaluation of five diagnostic methods for demonstrating microsporidia in stool and intestinal biopsy specimens. *Annals of Tropical Medicine and Parasitology* **87**, 99–102.
Beckers, P.J.A., Derks, G.J.M.M., Van Gool, T., Rietveld, F.J.R. and Sauerwein, R.W. (1996). *Encephalitozoon intestinalis* specific monoclonal antibodies for laboratory diagnosis of microsporidiosis. *Journal of Clinical Microbiology* **34**, 282–285.
Bergquist, R., Morfeldt-Mansson, L., Pehrson, P.O., Petrini, B. and Wasserman, J. (1984a). Antibody against *Encephalitozoon cuniculi* in Swedish homosexual men. *Scandinavian Journal of Infectious Diseases* **16**, 389–591.
Bergquist, N.R., Stintzing, G., Smedman, L., Waller, T. and Andersson, T. (1984b). Diagnosis of encephalitozoonsis in man by serological tests. *British Medical Journal* **288**, 902.
Biderre, C., Pages, M., Metenier, G., David, D., Bata, J., Prensier, G. and Vivares, C.P. (1994). On small genomes in eukaryotic organisms: molecular karyotypes of two microsporidian species (Protozoa) parasites of vertebrates. *Comptes Rendus des Seances de l'Academie des Sciences, Serie III, Sciences de la Vie* **317**, 399–404.

Biderre, C., Pages, M., Metenier, G., Canning, E.U. and Vivares, C.P. (1995). Evidence for the smallest nuclear genome (2.9 Mb) in the microsporidium *Encephalitozoon cuniculi*. *Molecular and Biochemical Parasitology* **74**, 229–231.

Boom, R., Sol, C.J.A., Salimans, M.M.M., Jansen, C.L., Wertheim-van Dillen, P.M.E. and Van Der Noordaa, J. (1990). Rapid and simple method for purification of nucleic acids. *Journal of Clinical Microbiology* **28**, 495–503.

Brown, J.R. and Doolittle, W.F. (1995). Root of the universal tree of life based on ancient aminoacyl-tRNA synthetase gene duplications. *Proceedings of the National Academy of Sciences USA* **92**, 2441–2445.

Brown, J.R. and Doolittle, W.F. (1997). Gene descent, duplication, and horizontal transfer in the evolution of glutamyl-tRNA and glutamyl-tRNA synthetases. *J. Mol. Evol.* in press.

Cali, A. (1991). General microsporidian features and recent findings on AIDS isolates. *Journal of Protozoology* **38**, 625–630.

Cali, A. and Owen, R.L. (1988). Microsporidiosis In: *Laboratory Diagnosis of Infectious Diseases: Principles and Practice*, Vol. 1. (A. Ballows, W.J. Hausler, jr, M. Ohashi and H. Turano, eds), pp. 929–947. New York: Springer-Verlag.

Cali, A. and Owen, R. (1990). Intracellular development of *Enterocytozoon bieneusi*. A unique microsporidian found in the intestine of AIDS patients. *Journal of Protozoology* **37**, 145–155.

Cali, A., Meisler, D.M., Lowder, C.Y., Lembach, R., Ayers, L., Takvorian, P.M., Rutherford, I., Longworth, D.L., McMahon, J. and Bryan, R.T. (1991a). Corneal microsporidiosis: Characterization and identification. *Journal of Protozoology* **39**, 215.

Cali, A., Meisler, D.M., Rutherford, I., Lowder, C.Y., McMahon, J.T., Longwirth, D.L. and Bryan, R.T. (1991b). Corneal microsporidiosis in a patient with AIDS. *American Journal of Tropical Medicine and Hygiene* **44**, 463–468.

Cali, A., Kotler, D.P. and Orenstein, J.M. (1993). *Septata intestinalis* n.g., n.sp., an intestinal microsporidian associated with chronic diarrhea and dissemination in AIDS patients. *Journal of Eukayotic Microbiology* **40**, 101–112.

Canning, E.U. (1990). Phylum Microspora. In: *Handbook of Proctoctista* (L. Margulis, J.O. Corliss, M. Melkonian, D.J. Chapman and H.I. McKhann, eds), pp. 53–72. Boston: Jones and Bartlett.

Canning, E.U. and Hollister, W.S. (1992). Human infections with microsporidia. *Reviews in Medical Microbiology* **3**, 35–42.

Canning, E.U. and Lom, J. (1986). *The Microsporidia of Vertebrates*. New York: Academic Press.

Carter, P.L., MacPherson, D.W. and McKenzie, R.A. (1996). Modified technique to recover microsporidian spores in sodium acetate–acetic acid–formalin-fixed fecal samples by light microscopy and correlation with transmission electron microscopy. *Journal of Clinical Microbiology* **34**, 2670–2673.

Carville, A., Mansfield, K., Widner, G., Lackner, A., Kotler, D., Wiest, P., Gumbo, T., Sarbah, S. and Tzipori, S. (1997). Development and application of genetic probes for the detection of *Enterocytozoon bieneusi* in formalin-fixed stools and in intestinal biopsies of infected patients. *Clinical and Diagnostic Laboratory Immunology*, **4**, 405–408.

Cavalier-Smith, T. (1987). Eukaryotes with no mitochondria. *Nature* **6111**, 332–333.

Challier, S., Brown, S., Ombrouck, C., Desportes-Livage, I., De Nay, D. and Gentilini, M. (1994). Flow cytometry as a possible method of isolation of spores of the microsporidian *Enterocytozoon bieneusi*. *Journal of Eukayotic Microbiology* **41**, 27S.

Chilmonczyk, S.W., Cox, T. and Hedrick, R.P. (1991). *Enterocytozoon salmonis* n.sp.

an intracellular microsporidium from salmonid fish. *Journal of Protozoology* **31**, 264–269.

Coyle, C.M., Wittner, M., Dotler, D., Noyer, C., Orenstein, J.M., Tanowitz, H.B. and Weiss, L.M. (1996). Prevalence of microsporidiosis due to *Enterocytozoon bieneusi* and *Encephalitozoon (Septata) intestinalis* among patients with AIDS-related diarrhea: determination by polymerase chain reaction to the microsporidian small subunit-rRNA gene. *Clinical Infectious Diseases* **23**, 1002–1006.

Curgy, J.J., Vavra, J. and Vivares, C. (1990). Presence of ribosomal RNAs with prokaryotic properties in microsporidia, eukaryotic organisms. *Biologie Cellulaire*, **38**, 49–52.

DaSilva, A.J., Schwartz, D.A., Visvesvara, G.S., deMoura, H. Slemenda, S.B. and Pieniazek, N.J. (1996). Senitive PCR diagnosis of infections by *Enterocytozoon bieneusi* (microsporidia) using primers based on the region coding from small-subunit rRNA. *Journal of Clinical Microbiology* **34**, 986–987.

Davis, R.M., Font, R.L., Keisler, M.S. and Shadduck, J.A. (1990). Corneal microsporidiosis: a case report including ultrastructural observations. *Ophthalmology* **97**, 953–957.

DeGirolami, P.C., Ezratty, C.R., Desai, G., McCullough, A., Asmuth, D., Wanke, C. and Federman, M. (1995). Diagnosis of intestinal microsporidiosis by examination of stool and duodenal aspirate with Weber's modified trichrome and Uvitex 2B strains. *Journal of Clinical Microbiology* **33**, 805–810.

De Groote, M.A., Visvesvara, G., Wilson, M.L., Pieniazek, N.J., Slemenda, S.B., daSilva, A.J., Leitch, G.J., Bryan, R.T. and Reves, R. (1995). Polymerase chain reaction and culture confirmation of disseminated *Encephalitozoon cuniculi* in a patient with AIDS: successful therapy with albendazole. *Journal of Infectious Diseases* **171**, 1375–1378.

Deplazes, P., Mathis, A., Muller, C. and Weber, R. (1996). Molecular epidemiology of *Encephalitozoon cuniculi* and first detection of *Enterocytozoon bieneusi* in fecal samples of pigs. *Journal of Eukaryotic Microbiology* **43**, 93S.

Desportes, I., Le Charpentier, Y., Calian, A., Bernard, F., Cochand-Priollet, B., Lavergne, A., Ravisse, P. and Modigliani, R. (1985). Occurrence of a new microsporidian: *Enterocytozoon bieneusi* n.g., n.sp., in the enterocytes of a human patient with AIDS. *Journal of Protozoology* **26**, 179–187.

Didier, E.S., Didier, P.J., Friedberg, D.N. Stenson, S.M., Orenstein, J.M., Yee, R.W., Tio, F.W., Davis, R.M., Vossbrinck, C., Millichamp, N. and Shadduck, J.A. (1991a). Isolation and characterization of a new human microsporidian, *Encephalitozoon hellum* (n.sp.) from three AIDS patients with keratoconjunctivitis. *Journal of Infectious Diseases* **163**, 617–621.

Didier, P.J., Didier, E.S., Orenstein, J.M. and Shadduck, J.A. (1991b). Fine structure of a new human microsporidian *Encephalitozoon hellem* in culture. *Journal of Protozoology* **38**, 502–507.

Didier, E.S., Kotler, D.P., Dietrich, D.T., Orenstein, J.M., Aldras, A.M., Davis, R., Friedberg, D.N., Gourley, W.K., Lembach, R., Lowder, C.Y., Meisler, D.M., Rutherford, I., Yee. R.W. and Shadduck, J.A. (1993). Serologic studies in human microsporidiosis. *AIDS* **7**, S8–S11.

Didier, E.S., Orenstein, J.M., Aldras, A., Bertucci, D., Rogers, L.B. and Janney, F.A. (1995a). Comparison of three staining methods for detecting microsporidia in fluids. *Journal of Clinical Microbiology* **33**, 3138–3145.

Didier, E.S., Vossbrinck, C.R., Baker, M.D., Rogers, L.B., Bertucci, D.C. and Shadduck, J.A. (1995b). Identification and characterization of three *Encephalitozoon cuniculi* strains. *Parasitology* **111**, 411–421.

Didier, E.S., Rogers, L.B., Brush, A.D., Wong, S., Traina-Dorge, V. and Bertucci, D.

(1996a). Diagnosis of disseminated microsporidian *Encephalitozoon hellem* infection by PCR–Southern analysis and successful treatment with albendazole and fumagillin. *Journal of Clinical Microbiology* **34**, 947–952.

Didier, E.S., Rogers, L.B., Orenstein, J.M., Baker, M.D., Vossbrinck, C.R., Van Gool, T., Hartskeerl, R., Soave, R. and Beaudet, L.M. (1996b). Characterization of *Encephalitozoon (Septata) intestinalis* isolates cultured from nasal mucosa and bronchoalveolar lavage fluids of two AIDS patients. *Journal of Eukaryotic Microbiology* **43**, 34–43.

Didier, E.S., Visvesvara, G.S., Baker, M.D., Rogers, L.B., Bertucci, D.C., DeGroote, M.A. and Vossbrinck, C.R. (1996c). A microsporidian isolated from an AIDS patient corresponds to *Encephalitozoon cuniculi* III, originally isolated from domestic dogs. *Journal of Clinical Microbiology* **34**, 2835–2837.

DiMaria, P., Palic, B., Debrunner-Vossbrinck, B.A., Lapp, J. and Vossbrinck, C.R. (1996). Characterization of the highly divergent U2 RNA homolog in the microsporidian *Vairimorpha necatrix*. *Nucleic Acids Research* **24**, 515–522.

Dissanaike, A.S. and Canning, E.U. (1957). The mode of emergence of the sporoplasm in microsporidia and its relationship to the structure of the spore. *Parasitology* **47**, 92–99.

Docker, M.F., Kent, M.L., Hervio, D.M.L., Khattra, J.S., Leiss, L.M., Cali, A. and Devlin, R.H. (1997). Ribosomal DNA sequence of *Nucleospora salmonis* Hedrick, Groff and Baxa, 1991 (Microsporea: Enterocytozoonidae): implications for phylogeny and nomenclature. *Journal of Eukaryotic Microbiology* **44**, 55–60.

Doultree, J.C., Maerz, A.L., Ryan, N.J., Baird, R.W., Wright, E., Crowe, S.M. and Marshall, J.A. (1995). *In vitro* growth of the microsporidian *Septata intestinalis* from an AIDS patient with disseminated illness. *Journal of Clinical Microbiology* **33**, 463–470.

Edlind, T., Visvesvara, G., Li, J. and Katiyar, S. (1994). Cryptosporidium and microsporidial β-tubulin sequences: predictions of benzimidazole sensitivity and phylogeny. *Journal of Eukaryotic Microbiology* **41**, 38S.

Edlind, T.D., Li, J., Visvesvara, G.S., Vodkin, M.H., McLaughlin, G.L. and Katiyar, S.K. (1996). Phylogenetic analysis of β-tubulin sequences from amitochondrial protozoa. *Molecular Phylogenetics and Evolution* **5**, 359–367.

Enriquez, F.J., Ditrich, O., Palting, J.D. and Smith, K. (1997). Simple diagnosis of *Encephalitzoon* sp. microsporidial infections by using a panspecific antiexospore monoclonal antibody. *Journal of Clinical Microbiology* **35**, 724–729.

Fedorko, D.P. and Hijazi, Y.M. (1996). Application of molecular techniques to the diagnosis of microsporidial infection. *Emerging Infectious Diseases* **2**, 183–191.

Fedorko, D.P., Nelson, N.A. and Cartwright, C.P. (1995). Identification of microsporidia in stool specimens by using PCR and restriction endonucleases. *Journal of Clinical Microbiology* **33**, 1739–1741.

Felsenstein, J. (1978). Cases in which parsimony or compatibility methods will be positively misleading. *Systematic Zoology* **27**, 401–410.

Field, A.S., Hing, M.C., Milliken, S.T. and Marriott, D.J. (1993a). Microsporidia in the small intestine of HIV-infected patients. A new diagnostic technique and a new species. *Medical Journal of Australia* **158**, 390–394.

Field, A.S., Marriott, D.J. and Hing, M.C. (1993b). The Warthin–Starry stain in the diagnosis of small intestinal microsporidiosis in HIV-infected patients. *Folia Parasitolica (Praha)* **40**, 261–266.

Field, A.D., Marriott, D.J., Milliken, S.T., Brew, B.J., Canning, E.U., Kench, J.G., Darveniza, P. and Harkness, J.L. (1996). Myositis associated with a newly described microsporidian, *Trachipleistophora hominis*, in a patient with AIDS. *Journal of Clinical Microbiology* **34**, 2803–2811.

Franzen, C., Muller, A., Hegener, P., Salzberger, B., Hartmann, P., Fatkenheuer, G., Diehl, V. and Schrappe, M. (1995a). Detection of microsporidia (*Enterocytozoon bieneusi*) in intestinal biopsy specimens from human immunodeficiency virus-infected patients by PCR. *Journal of Clinical Microbiology* **33**, 2294–2296.
Franzen, C., Muller, A., Salzberger, B., Fatkenheuer, G., Eidt, S., Mahrle, G., Diehl, V. and Schrappe, M. (1995b). Tissue diagnosis of intestinal microsporidiosis using a fluorescent stain with Uvitex 2B. *Journal of Clinical Pathology* **48**, 1009–1010.
Franzen, C., Kuppers, R., Muller, A., Salzberger, B., Fatkenheuer, G., Vettern, B., Diehl, V. and Schrappe, M. (1996). Genetic evidence for latent *Septata intestinalis* infection in human immunodeficiency virus-infected patients with intestinal microsporidiosis. *Journal of Infectious Diseases* **173**, 1038–1040.
Friedberg, D.N., Stenson, S.M., Orenstein, J.M., Tierno, P.M. and Charles, N.C. (1990). Microsporial keratoconjunctivitis in acquired immunodefiency syndrome. *Archives of Ophthalmology* **108**, 504–508.
Fries, I.M., Feng, F., daSilva, A.J., Slemenda, S.B. and Pieniazek, H.J. (1996). *Nosema ceranae* n.sp. (Microsporidia, Nosematidae), morphological and molecular characterization of a microsporidian parasite of the Asian honey bee *Apis cerana* (Hymenoptera, Apidae) *Eur. J. Protistol.* **32**, 356–365.
Galtier, N. and Gouy, M. (1995). Inferring phylogenies from DNA sequences of unequal base compositions. *Proceedings of the National Academy of Sciences*, **92**, 11317–11321.
Germot, A. Philippe, H. and LeGuyader, H. (1997). Evidence for the loss of mitochondria from a mitochondrial-type HSP70 in *Nosema locustae*. *Mol. Biochem. Parasitol.* **87**, 159–168.
Haig, D. (1993). Alternatives to meiosis: the unusual genetics of red algae, microsporidia, and others. *Journal of Theoretical Biology* **163**, 15–31.
Hartskeerl, R.A., Schuitema, A.R. and deWachter, R. (1993a). Secondary structure of the small subunit ribosomal RNA sequence of the microsporidium *Encephalitozoon cuniculi*. *Nucleic Acids Research* **21**, 1489.
Hartskeerl, R.A., Schuitema, A.R., Van Gool, T. and Terpstra, W.J. (1993b). Genetic evidence for the occurrence of extra-intestinal *Enterocytozoon bieneusi* infections. *Nucleic Acids Research* **21**, 4150.
Hartskeerl, R.A., Van Gool, T., Schuitema, A.R., Didier, E.S. and Terpstra, W.J. (1995). Genetic and immunological characterization of the microsporidian *Septata intestinalis* Cali, Kotler and Orenstein, 1993: reclassification to *Encephalitozoon intestinalis*. *Parasitology* **110**, 277–285.
Hashimoto, T. and Hasegawa, M. (1996). Origin and early evolution of eukaryotes inferred from the amino acid sequences of translation elongation factors 1α/Tu and 2/G. *Advances in Biophysics* **32**, 73–120.
Hirt, R.P., Healy, D., Vossbrinck, C.R., Canning, E.U. and Embley, T.M. (1997) Identification of a mitochondrial Hsp70 orthologue in *Vairimorpha necatrix*: molecular evidence that microsporidia once contained mitochondria. *Current Biology* **7**, in press.
Hollister, W.S. and Canning, E.U. (1987). An enzyme-linked immunosorbent assay (ELISA) for detection of antibodies to *Encephalitozoon cuniculi* and its use in determination of infections in man. *Parasitology* **94**, 209–219.
Hollister, W.S., Canning, E.U. and Anderson, C.L. (1996). Identification of microsporidia causing human disease. *Journal of Eukaryotic Microbiology* **43**, 104S–105S.
Ignatius, F., Lehmann, M., Minksits, K., Regnath, T., Arvand, M., Engelmann, E., Futh, U., Hahn, H. and Wagner, J. (1997). A new acid-fast trichrome stain for

simultaneous detection of *Cryptosporidium parvum* and microsporidial species in stool specimens. *Journal of Clinical Microbiology* **35**, 446–449.

Ishihara, R. and Hayashi, Y. (1968). Some properties of ribosomes from the sporoplasm of *Nosema bombycis*. *Journal of Invertebrate Pathology* **11**, 377–385.

Issi, I.V. (1986). Microsporidia as a phylum of parasitic protozoa. *Doklady Akademi Nauk SSSR (Leningrad)* **10**, 6–136.

Kamaishi, T., Hashimoto, T., Nakamura, Y., Nakamura, F., Murata, S., Okada, N., Okamoto, D., Shimizu, M. and Hasegawa, M. (1996a). Protein phylogeny of translation elongation factor EF-1α suggests microsporidians are extremely ancient eukaryotes. *Journal of Molecular Evolution* **42**, 257–263.

Kamaishi, T., Hashimoto, T., Nakamura, Y., Masuda, Y., Nakamura, F., Okamoto, D., Shimizu, M. and Hasegawa, M. (1996b). Complete nucleotide sequences of the genes encoding translation elongation faction 1α and 2 from a microsporidian parasite, *Glugea plecoglosii*: implications for the deepest branching of eukaryotes. *Journal of Biochemistry* **120**, 1095–1103.

Katiyar, S.K., Gordon, V.R., McLaughlin, G.L. and Edlind, T.D. (1994). Antiprotozoal activities of benzimidazoles and correlations with β-tubulin sequence. *Antimicrobial Agents and Chemotherapy* **38**, 2086–2090.

Katiyar, S.K., Visvesvara, G.S. and Edlind, T.D. (1995). Comparisons of ribosomal RNA sequences from amitochondrial protozoa: implications for processing, mRNA binding and paromomycin susceptibility. *Gene* **152**, 27–33.

Katzwinkel-Wladarsch, S., Lieb, M., Helse, W., Loscher, T. and Rinder, H. (1996). Direct amplification and species determination of microsporidian DNA from stool specimens. *Tropical Medicine and International Health* **1**, 373–378.

Kawakami, Y., Inoue, T., Kikuchi, M., Takayanagi, M., Sunairi, M., Ando, T. and Ishihara, R. (1992). Primary and secondary structures of 5S ribosomal RNA of *Nosmea bombysis* (Nosematidae, Microsporidia) *Journal of Sericultural Science, Japan* **61**, 321–327.

Kawakami, Y., Inoue, T., Ito, K., Kitamizu, K., Hanawa, C., Ando, T., Iwano, H. and Ishihara, R. (1994). Identification of a chromosome harboring the small subunit ribosomal RNA gene of *Nosema bombycis*. *Journal of Invertebrate Pathology* **64**, 147–148.

Keeling, P.J. and Doolittle, W.F. (1996). α-Tubulin from early diverging eukaryotic lineages and the evolution of the tubulin family. *Molecular Biology and Evolution* **13**, 1297–1305.

Kent, M.L., Hervio, D.M., Docker, M.F. and Devlin, R.H. (1996). Taxonomy studies and diagnostic tests for myxosporean and microsporidian pathogens of salmonid fishes utilizing ribosomal DNA sequence. *Journal of Eukaryotic Microbiology* **43**, 98S–99S.

Kock, N.P., Petersen, H., Fenner, T., Sobottka, I., Schmetz, C., Deplazes, P., Pieniazek, N.J., Albrecht, H., Schottelius, J. (1997) Species-specific identification of microsporidia in stool and intestinal biopsy specimens by the polymerase chain reaction. *European Journal of Clinical Microbiology and Infectious Diseases* **16**, 369–376.

Kokoskin, E., Gyorkos, T.W., Camus, A., Cedilotte, L., Purtill, T. and Ward, B.J. (1994). Modified technique for efficient detection of microsporidia. *Clinical Microbiology* **32**, 1074–1075.

Kotler, D.P., Ciang, T.T., Garro, M.L., and Orenstein, J.M. (1994). Light microscopic diagnosis of microsporidiosis in patients with AIDS. *American Journal of Gastroenterology* **89**, 540–544.

Kudo, R. (1921). On the nature of structures characteristic of cnidosporian spores. *Transactions of the American Microscopy Society* **40**, 59–74.

Lacy, C.J., Clarke, A.M., Fraser, P., Metcalfe, T., Bonsor, G. and Curry, A. (1992). Chronic microsporidian infection of the nasal mucosae, sinuses and conjuctivae in HIV disease. *Genitourinary Medicine* **68**, 179–181.
Langley, R.C., Cali, A. and Somberg, E.W. (1987). Two dimensional electrophoretic analysis of spore proteins of the microsporida. *Journal of Parasitology* **73**, 910–918.
Larsson, J.I.R. (1986). Ultrastructure, function, and classification of microsporidia. *Progress in Protistology* **1**, 325–390.
Leipe, D.D., Gunderson, J.H., Nerad, T.A. and Sogin, M.L. (1993). Small subunit ribosomal RNA of *Hexamita inflata* and the quest for the first branch in the eukaryotic tree. *Molecular and Biochemical Parasitology* **59**, 41–48.
Levine, N.D., Corliss, J.O., Cox, F.E.G., Deroux, G., Grain, J., Honigberg, B.M., Leedale, G.F., Loeblich, A.R., Lom, J., Lynn, D., Merinfeld, E.G., Page, C., Poljansky, G., Sprauge, V., Vavra, J. and Wallace, F.G. (1980). A newly revised classification of the protozoa. *Journal of Protozoology* **27**, 37–58.
Li, J., Katiyar, S.K., Hamelin, A., Visvesvara, G.S. and Edlind, T.D. (1996). Tubulin genes from AIDS-associated microsporidia and implications for phylogeny and benzimidazole sensitivity. *Molecular and Biochemical Parasitology* **78**, 289–295.
Lom, J. (1972). On the structure of the extruded microsporidian polar filament. *Zeitschrift fur Parasitenkunde* **38**, 200–213.
Lowder, C.Y., Meisler, D.M., McMahon, J.T., Longworth, D.L. and Rutherford, I. (1990). Microsporidia infection of the cornea in a man seropositive for human immunodeficiency virus. *American Journal of Opthalmology* **109**, 242–244.
Lucas, S.B., Papadaki, L., Conlon, C., Sewankambo, N., Goodgame, R. and Serwadda, D. (1989). Diagnosis of intestinal microsporidiosis in patients with AIDS. *Journal of Clinical Pathology* **42**, 885–890.
Luna, V.A., Stewart, B.K., Bergeron, D.L., Clausen, C.R., Plorde, J.J. and Fritsche, T.R. (1995). Use of the fluorochrome calcofluor white in the screening of stool specimens for spores of microsporidia. *American Journal of Clinical Pathology* **103**, 656–659.
Malone, L.A. and McIvor, C.A. (1995). DNA probes for two microsporidia, *Nosema bombycis* and *Nosema costelytrae*. *Journal of Invertebrate Pathology* **65**, 269–273.
Mansfield, K.G., Carville, A., Shvetz, D., MacKey, J., Tzipori, S. and Lackner, A.A. (1997). Identification of *Enterocytozoon bieneusi*-like microsporidian parasite in simian immunodeficiency virus-inoculated macaques with hepatobillary disease. *American Journal of Pathology* **150**, 1395–1405.
McDougall, R.J., Tandy, M.W., Boreham, R.E., Stenzel, D.J. and O'Donoghue, P.J. (1993). Incidental finding of a microsporidian parasite from an AIDS patient. *Journal of Clinical Microbiology* **31**, 436–439.
Metcalfe, T.W., Doran, R.M.L., Rowlands, P.L., Curray, A. and Lacey, C.J. (1992). Microsporidial keratoconjuctivitis in a patient with AIDS. *British Journal of Ophthalmology* **76**, 177–178.
Moura, H., DaSilva, J.L., Sodre, F.C., Brasil, P., Walmo, D., Wahlquist, S., Wallace, S., Croppo, G.P. and Visvesvara, G.S. (1996). Gram-chromotrope: a new technique that enhances detection of microsporidial spores in clinical samples. *Journal of Eukaryotic Microbiology* **43**, 94S–95S.
Munderloh, U.G., Kurtti, T.J. and Ross, S. (1990). Electrophoretic characterization of chromosomal DNA from two microsporidia. *Journal of Invertebrate Pathology* **56**, 243–248.
Niederkorn, J.Y., Shadduck, J.H. and Weidner, E. (1980). Antigenic cross-reactivity among different microsporidian spores as determined by immunofluorescence. *Journal of Parasitology* **66**, 675–677.
Ombrouck, C., Romestand, B., da Costa, J.M., Desportes-Livage, I., Datry, A., Coste,

F., Bouix, G. and Gentilini, M. (1995). Use of cross-reactive antigens of the microsporidian *Glugea atherinae* for the possible detection of *Enterocytozoon bieneusi* by Western blot. *American Journal of Tropical Medicine and Hygiene* 52, 89–93.

Ombrouck, C., Ciceron, L. and Desportes-Livage, I. (1996). Specific and rapid detection of microsporidia in stool specimens from AIDS patients by PCR. *Parasite* 3, 85–86.

Ombrouck, C., Ciceron, L., Biligui, S., Brown, S., Marechal, P., van Gool, T., Datry, A., Danis, M. and Desportes-Livage, I. (1997). Specific PCR assay for direct detection of intestinal microsporidia *Enterocytozoon bieneusi* and *Encephalitozoon intestinalis* in fecal specimens from human immunodeficiency virus-infected patients. *Journal of Clinical Microbiology* 35, 652–655.

Orenstein, J.M., Seedor, J., Friedberg, D.N., Stenson, S.M., Tierno, P.M., Charles, N.C., Meisler, D.M., Lowder, C.Y., McMahon, J.T., Longworth, D.L. Rutherford, I., Yee, R.W., Martinez, A., Tio, F. and Held, K. (1990a). Microsporidian keratoconjunctivitis in patients with AIDS. *Morbidity and Mortality Weekly Report (CDC)* 39, 188–189.

Orenstein, J., Chiang, J., Steinberg, W., Smith, P.D., Rotterdam, H. and Kotler, D.P. (1990b). Intestinal microsporidiosis as a cause of diarrhea in human immunodefiency virus-infected patients: a report of 20 cases. *Human Pathology* 21, 475–481.

Pieniazek, N.J., daSilva, A.J., Slemenda, S.B., Visvesvara, G.S., Kurtti, T.J. and Yasunaga, C. (1996). *Nosema trichoplusiae* is a synonym of *Nosema bombycis* based on the sequence of the small subunit ribosomal RNA coding region. *J Inv Pathol* 67(3), 316–317.

Pinnolis, M., Egbert, P.R., Font, R.L. and Winter, F.C. (1981). Nosematosis of the cornea: case report including electron microscopic studies. *Archives of Ophthalmology* 99, 1044–1047.

Ragan, M.A. (1988). Ribosomal RNA and the major lines of evolution: a perspective. *BioSystems* 21, 177–187.

Rastrelli, P.D., Didier, E. and Yee, R.W. (1994). Microsporidial keratitis. *Ophthalmologic Clinics of North America* 7, 617–633.

Rijpstra, A.C., Canning, E.U., Van Ketel, R.J., Eeftinck-Schattenkerk, J.K. and Laarman, J.J. (1988). Use of light microscopy to diagnose small intestinal microsporidiosis in patients with AIDS. *Journal of Infectious Diseases* 157, 827–831.

Ryan, N.J., Sutherland, G., Coughlan, K., Globan, M., Doultree, J., Marshall, J., Baird, R.W., Pedersen, J. and Dwyer, B. (1993). A new trichrome-blue stain for detection of microsporidial species in urine, stool, and nasopharyngeal specimens. *Journal of Clinical Microbiology* 31, 3264–3269.

Schottelius, J., Lo, Y. and Schmetz, C. (1995). *Septata intestinalis* and *Encephalitozoon cuniculi*: cross-reactivity between two microsporidian species. *Folia Parasitolica (Praha)* 42, 169–172.

Schuitema, A.R.J., Hartskeerl, R.A., Van Gool, T., Laxminarayan, R. and Terpstra, W.J. (1993). Application of the polymerase chain reaction for the diagnosis of microsporidiosis. *AIDS* 7, supplement 3, S57–S61.

Schuitema, A.R.J., Sarfati, C., Liguory, O., Hartskeerl, R.A., Deroun, F. and Molina, J.M. (1996). Detection and species identification of intestinal microsporidia by polymerase chain reaction in duodenal biopsies from human immunodeficiency virus-infected patients. *Journal of Infectious Diseases* 174, 874–877.

Schwartz, D.A., Visvesvara, G.S., Diesenhouse, M.C., Weber, R., Font, R.L., Wilson, L.A., Corrent, G., Serdarevic, O.N., Rosberger, D.F., Keenen, P.C., Rossniklaus, H.E., Hewan-Lowe, K. and Bryan R.T. (1993). Pathologic features and immunofluorescent antibody demonstration of ocular microsporidiosis (*Encephalitozoon hellem*) in seven patients with acquired immunodeficiency syndrome. *American Journal of Ophthalmology* **115**, 285-292.

Schwartz, D.A., Bryan, R.T., Weber, R. and Visvesvara, G.S. (1994). Microsporidiosis in HIV positive patients: current methods for diagnosis using biopsy, cytologic, ultrastructural, immunological, and tissue culture techniques. *Folia Parasitolica (Praha)* **41**, 101-109.

Shadduck J.A. (1969). *Nosema cuniculi*: in vitro isolation. *Science* **166**, 516-517.

Shadduck J.A. and Orenstein, J.M. (1993). Comparative pathology of microsporidiosis. *Archives of Pathology and Laboratory Medicine* **117**, 1215-1219.

Shadduck, H.A., Meccoli, R.A., Davis, R. and Font, R.L. (1990). Isolation of a microsporidian from a human patient. *Journal of Infectious Diseases* **162**, 773-776.

Shah, G.K., Pfister, D., Probst, L.E., Ferrieri, P. and Holland, E. (1996). Diagnosis of microsporidial keratitis by confocal microscopy and the chromatrope stain. *American Journal of Ophthalmology* **121**, 89-91.

Silveira, H. and Canning, E.U. (1995). *Vittaforma corneae* N. Comb of the human microsporidium *Nosema corneum* Shadduck, Meccoli, Davis and Font, 1990, based on its ultrastructure in the liver of experimentally infected athymic mice. *Journal of Eukaryotic Microbiology* **42**, 158-165.

Singh, M., Kane, G.J., Mackinlay, L., Quaki, I., Yap, E.H., Ho, B.C., Ho, L.C. and Lim, K.C. (1982). Detection of antibodies to *Nosema cuniculi* (Protozoa: Microsporidia) in human and animal sera by the indirect fluorescent antibody technique. *Southeast Asia Journal of Tropical Medicine and Public Health* **13**, 110-113.

Sogin, M.L., Gunderson, J.H., Elwood, H.J., Alonso, R.A. and Peattie, D.A. (1989). Phylogenetic meaning of the kingdom concept: an unusual ribosomal RNA form *Giardia lamblia*. *Science* **4887**, 75-77.

Sprague, V. and Vavra, J. (1977). Systematics of the microsporidia. *Comparative Pathobiology* **2**, 31-335.

Sprague, V., Becnel, J.J. and Hazard, E.I. (1992). Taxonomy of the phylum Microspora. *Critical Reviews in Microbiology* **18**, 285-395.

Tiner, J.D. (1988). Birefringent spores differentiate *Encephalitozoon* and other microsporidia from coccidia. *Veterinary Pathology* **25**, 227-230.

van Gool, T., Hollister, W.S., Schattenkerk, J.E., Van den Bergh Weerman, M.A., Terpstra, W.J., Van Ketel, R.J., Reiss, P. and Canning, E.U. (1990). Diagnosis of *Enterocytozoon bieneusi* microsporidiosis in AIDS patients by recovery of spores from faeces. *Lancet* **336**, 697-698.

van Gool, T., Snijders, F., Reiss, P., Eeftinck Schattenkerk, J.K., van den Bergh Weerman, M.A., Bartelsman, J.F., Bruins, J.J., Canning, E.U. and Dankert, J. (1993). Diagnosis of intestinal and disseminated microsporidial infections in patients with HIV by a new rapid fluorescence technique. *Journal of Clinical Pathology* **46** 694-699.

van Gool, T., Canning, E.U., Gilis, H., vna den Bergh Weerman, M.A., Eeftinck Schattenkerk, J.K. and Dankert, J. (1994). *Septata intestinalis* frequently isolated from stool of AIDS patients with a new cultivation method. *Parasitology* **109**, 281-289.

Vavra, J. (1976). Structure of microsporidia. In: *Comparative Pathobiology*, Vol. 1 (L.A. Bulla and T.C. Cheng, eds), pp. 1-86. New York: Plenum Press.

Vavra, J., Dahbiova, R., Hollister, W.S. and Canning, E.U. (1993a). Staining of microsporidian spores by optical brighteners with remarks on the use of brighteners for the diagnosis of AIDS associated human microsporidiosis. *Folia Parasitologica (Praha)* **40**, 267-272.

Vavra, J., Nohynkova, E., Machala, L. and Spala, J. (1993b). An extremely rapid method for detection of microsporidia in biopsy materials from AIDS patients. *Folia Parasitologica (Praha)* **40**, 273–274.

Velasquez, J.N., Carnevale, S., Guarnera, E.A., Labbe, J.H., Chertcoff, A., Cabrera, M.G. and Rodriguez, M.I. (1996). Detection of the microsporidian parasite *Enterocytozoon bieneusi* in specimens from patients with AIDS by PCR. *Journal of Clinical Microbiology* **34**, 3230–3232.

Visvesvara, G.S., Da Silva, A.J., Croppo, C.P., Pieniazek, N.J., Slemednda, S., Leitch, C.J., Ferguson, D., Wallace, S., Tyrrel, L. and Medor, J. (1994a). Continuous cultivation, serologic and molecular characterization of *Septata intestinalis* from an AIDS patient with disseminated microsporidiosis. In: *Society of Protozoologists 47th Annual Meeting*, Cleveland, OH, abstract C4.

Visvesvara, G.S., Leitch, J.J., da Silva, A.J., Croppo, G.P., Moura, H., Wallace, S., Slemenda, S.B., Schwartz, D.A., Moss, D., Bryan, R.T. and Pieniazek, N.J. (1994b). Polyclonal and monoclonal antibody and PCR-amplified small-subunit rRNA identification of a microsporidian, *Encephalitozoon hellem*, isolated from an AIDS patient with disseminated infection. *Journal of Clinical Microbiology* **32**, 2760–2768.

Visvesvara, G.S., da Silva, A.J., Croppo, G.P., Pieniazek, N.J., Leitch, G.J., Ferguson, D., de Moura, H., Wallace, S., Slemenda, S.B., Tyrrell, I., Moore, D.F. and Meador, J. (1995a). *In vitro* culture and serologic and molecular identification of *Septata intestinalis* isolated from urine of a patient with AIDS. *Journal of Clinical Microbiology* **33**, 930–936.

Visvesvara, G.S., Leitch, G.J., Pieniazek, N.J., Da Silva, A.J., Wallace, S., Slemenda, S.B., Weber, R., Schwartz, D.A., Gorelkin, L., Wilcox, C.M. and Bryan, R.T. (1995b). Short-term *in vitro* culture and molecular analysis of the microsporidian, *Enterocytozoon bieneusi*. *Journal of Eukaryotic Microbiology* **42**, 506–510.

Visvesvara, G.S., Leitch, G.J., Wallace, S., Seaba, C., Erdman, D. and Ewing, E.P., jr (1996). Adenovirus masquerading as microsporidia. *Journal of Parasitology* **82**, 316–319.

Vivares, C., Biderre, C., Duffieux, F., Peyretaillade, E., Peyret, P., Metenier, G. and Pages, M. (1996). Chromosomal localization of five genes in *Encephalitozoon cuniculi* (Microsporidia). *Journal of Eukaryotic Microbiology* **43**, 97S.

Vossbrinck, C.R. and Woese, C.R. (1986). Eukaryotic ribosomes that lack a 5.8S RNA. *Nature* **320**, 287–288.

Vossbrinck, C.R., Maddox, J.V., Friedman, S., Debrunner-Vossbrinck, B.A. and Woese, C.R. (1987). Ribosomal RNA sequence suggests microsporidia are extremely ancient eukaryotes. *Nature*, **326**, 411–414.

Vossbrinck, C.R., Baker, M.D., Didier, E.S., Debrunner-Vossbrinck, B.A. and Shadduck, J.A. (1993). Ribosomal DNA sequences of *Encephalitozoon hellem* and *Encephalitozoon cuniculi*: species identification and phyogenetic construction. *Journal of Eukaryotic Microbiology* **40**, 354–362.

Walker, C.G. (1983). In: *Earth's Earliest Biosphere* (J.W. Schopf, ed.), pp. 280–289. Princeton, NJ: Princeton University Press.

Waller, T. and Bergquist, R.N. (1982). Rapid simultaneous diagnosis of toxoplasmosis and encephalitozoonosis in rabbits by carbon immunoassay. *Laboratory Animal Science* **32**, 515–517.

Weber, R., Bryan, R.T., Owen, R.L., Wilcox, C.M., Gorelkin, L. and Visvesvara, G.S. (1992). Improved light-microscopical detection of microsporidia spores in stool and duodenal aspirates. The Enteric Opportunistic Infections Working Group. *New England Journal of Medicine* **326**, 161–166.

Weber, R., Kuster, H., Visvesvara, G.S., Bryan, R.T., Schwartz, D.A. and Luthy, R. (1993). Disseminated microsporidiosis due to *Encephalitozoon hellem*: pulmonary

colonization, microhematuria, and mild conjunctivitis in a patient with AIDS. *Clinical Infectious Diseases* **17**, 415–419.
Weber, R., Bryan, R.T., Schwartz, D.A. and Owen, R.L. (1994a). Human microsporidial infections. *Clinical Microbiology Reviews* **7**, 426–461.
Weber, R., Sauer, B., Spycher, M.A., Deplazes, P., Keller, R., Ammann, R., Briner, J. and Luthy, R. (1994b). Detection of *Septata intestinalis* in stool specimens and coprodiagnostic monitoring of successful treatment with albendazole. *Clinical Infectious Diseases* **19**, 342–345.
Weidner, E. (1972). Ultrastructural study of microsporidian invasion into cells. *Zeitschrift fur Parasitenkunde* **40**, 227–242.
Weir, G.O. and Sullivan, J.T. (1989). A fluorescence screening technique for microsporia in histological sections. *Transactions of the American Microscopical Society* **108**, 208–210.
Weiser, J. (1977). Contribution to the classification of microsporidia. *Vestnil Ceskoslovenske Spolecnosti Zoologicke* **41**, 308–320.
Weiss, L.M., Cali, A., Levee, E., LaPlace, D., Tanowitz, H., Simon, D. and Wittner, M. (1992). Diagnosis of *Encephalitozoon cuniculi* infection by Western blot and the use of cross-reactive antigens for the possible detection of microsporidiosis in humans. *American Journal of Tropical Medicine and Hygiene* **47**, 456–462.
Weiss, L.M., Zhu, X., Cali, A., Tanowitz, H.B. and Wittner, M. (1994). Utility of microsporidian rRNA in diagnosis and phylogeny: a review. *Folia Parasitolica (Praha)* **41**, 81–90.
WHO Parasitic diseases surveillance (1983). Antibody to *Encephalitozoon cuniculi* in man. *World Health Organization Weekly Epidemiology Record* **58**, 30–32.
Wittner, M., Tanowitz, H.B. and Weiss, L.M. (1993). Parasitic infection in AIDS patients: cryptosporidiosis, isosporiasis, microsporidiosis, cyclosporiasis. *Infectious Disease Clinics of North America* **7**, 569–586.
Yee, R.W., Tio, F.O., Maritnes, J.A., Held, K.S., Shadduck, J.A. and Didier, E.S. (1991). Resolution of microsporidial eptithelial keratopathy in a patient with AIDS. *Ophthalmology* **98**, 196.
Zhu, X., Wittner, M., Tanowitz, H.B., Cali, A. and Weiss, L.M. (1993a). Small subunit rRNA sequence of *Septata intestinalis*. *Nucleic Acids Research* **21**, 4846.
Zhu, X., Wittner, M., Tanowitz, H.B., Cali, A. and Weiss, L.M. (1993b). Nucleotide sequence of the small ribosomal RNA of *Encephalitozoon cuniculi*. *Nucleic Acid Research* **21**, 1315.
Zhu, X., Wittner, M., Tanowitz, H.B., Kotler, D., Cali, A. and Weiss, L.M. (1993c). Small subunit rRNA sequence of *Enterocytozoon bieneusi* and its potential diagnostic role with use of the polymerase chain reaction. *Journal of Infectious Diseases* **168**, 1570–1575.
Zhu, X., Wittner, M., Tanowitz, H.B., Cali, A. and Weiss, L.M. (1994). Ribosomal RNA sequences of *Enterocytozoon bieneusi*, *Septata intestinalis* and *Ameson michaelis*: phylogenetic construction and structural correspondence. *Journal of Eukaryotic Microbiology* **41**, 204–209.
Zierdt, C.H., Gill, V.J. and Zierdt, W.S. (1993). Detection of microsporidian spores in clinical samples by indirect fluorescent-antibody assay using whole-cell antisera to *Encephalitozoon cuniculi* and *Encephalitozoon hellem*. *Journal of Clinical Microbiology* **31**, 3071–3074.

PART 3
Cyclospora cayetanensis **and related species**

Cyclospora cayetanensis

Ynes R. Ortega,[1] Charles R. Sterling[1] and Robert H. Gilman[2]

[1]*Department of Veterinary Science and Microbiology, University of Arizona, Tucson, AZ 85721, USA,* [2]*Department of International Health, School of Hygiene, Johns Hopkins University, Baltimore, MD 21205, USA, and Departamento de Patologia, Universidad Peruana Cayetano Heredia, Lima, Peru*

1. Introduction 400
2. Diagnosis and Purification 401
3. Molecular Biology 404
4. Life Cycle 405
5. Histopathology... 406
6. Immunology.. 408
7. Clinical Signs of Infection...................................... 408
8. Treatment.. 408
9. Cyclosporiasis in AIDS Patients 409
10. Epidemiology... 410
    10.1. Water-borne outbreaks 411
    10.2. Food-borne outbreaks 413
11. Pathogenesis 413
12. Cyclosporiasis in Peru 413
13. The Future 414
    References 414

Cyclospora cayetanensis *is a coccidian pathogen in humans. Cyclosporiasis is characterized by mild to severe nausea, anorexia, abdominal cramping, and watery diarrhea. Cyclospora has now been described from patients with protracted diarrheal illness in North, Central and South America, the Caribbean, Africa, Bangladesh, south-east Asia, Australia, England, and eastern Europe, and is characterized by marked seasonality. Routes of transmission are still unknown, although the fecal–oral route, either directly or via water, is probably the major one. A recent outbreak in the USA suggested transmission of* Cyclospora *by ingestion of contaminated berries.* Cyclospora *oocysts can be detected by phase contrast microscopy, modified acid-fast staining, autofluorescence, and amplification by the polymerase chain reaction. Oocysts are not sporulated when excreted in the feces, and sporulated oocysts are needed for infection. Each sporulated oocyst contains two sporocysts and each sporocyst contains two sporozoites. Humans seem to be the only host for this parasite. Histopathological examination of jejunal biopsies from infected individuals showed mild to moderate acute inflammation of the lamina propria and surface epithelial disarray. Parasitophorous vacuoles containing sexual and asexual forms of* Cycl. cayetanensis *were located in the cytoplasm of epithelial cells.* Cyclospora *infections can be treated succesfully with trimethoprim-sulfamethoxazole.*

1. INTRODUCTION

Cyclosporans were probably first noted by Eimer in 1870 in the intestine of moles. The genus was created in 1881 by Schneider for a parasite, *Cyclospora glomerica*, described from a myriapod. Schaudinn (1902) published the first life cycle study, describing *Cycl. caryolitica* which developed in the intestinal epithelium of moles and produced severe enteritis. Cyclosporan parasites have since been described from moles, rodents, insectivores, snakes and, recently, humans.

Cyclospora-like organisms were first observed in humans in 1979. Oocysts described from three individuals in Papua New Guinea remarkably resembled incompletely sporulated oocysts of *Cyclospora*, but were thought to represent a new species of *Isospora* because each of the two sporocysts observed were thought to contain four sporozoites (Ashford, 1979). From 1985 on, organisms 8–10 μm in size, staining red with modified acid-fast stains and autofluorescing under ultraviolet (UV) light, were reported with increasing frequency from humans worldwide. They were described frequently as CLBs (cyanobacter-like bodies, coccidian-like bodies) because they were thought to resemble cyanobacteria (blue green algae) when viewed under the microscope. Examination of expatriate residents and tourists visiting Nepal confirmed that CLBs were responsible for diarrheal illness. Initial attempts to sporulate these new organisms were unsuccessful, however, and identities other than coccidian were suggested in an attempt to provide a clue as to what was causing diarrheal illness in these patients (Long *et al.*, 1990, 1991; Hoge, 1993).

In 1989, studies were undertaken in Lima, Peru to define the role of *Cryptosporidium parvum* and other enteropathogens in diarrheal disease outbreaks in children. An organism with characteristics resembling CLBs, which had also been seen in older Peruvians in 1985 and 1987, was observed in feces of diarrheic children with no other detectable enteropathogen. The CLBs from these children were subsequently purified and

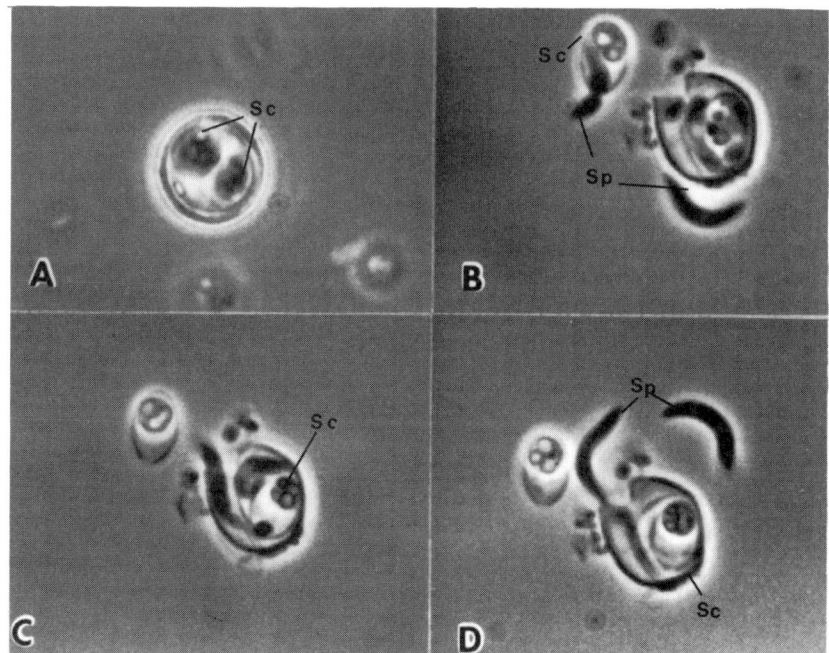

Figure 1 Excystation of a sporulated *Cyclospora cayetanensis* oocyst. Sc, sporocysts. Reproduced, with permission, from Ortega *et al.* (1993).

experiments were conducted to determine their identity. The general appearance on microscopic examination suggested the organism could be a coccidian. Attempts to cultivate it in standard bacteriological and algal media were unsuccessful. Sporulation eventually was obtained in potassium dichromate, demonstrating the presence of two sporocysts with two sporozoites each, thus placing the parasite in the coccidian genus *Cyclospora*. The morphological characteristics, sporulation, patient symptoms, and refractoriness to conventional antimicrobial therapy linked this newly described *Cyclospora* to previous reports of CLB infections in humans from different parts of the world. This assumption was verified in our laboratory by sporulating and excysting CLBs sent to us from different parts of the USA and elsewhere (Ortega *et al.*, 1993) (Figure 1).

2. DIAGNOSIS AND PURIFICATION

Cyclospora oocysts stain variably using a modified acid-fast technique. They stain best using the modified carbol–fuchsin technique. Oocysts also can be stained using the Kinyoun, Ziel–Neelsen, and safranin methods, but do not stain well with iron hematoxylin, Grocott–Gomori methenamine-silver nitrate, iodine, or periodic acid–Schiff stains (Long et al, 1991; Garcia and Bruckner, 1994).

Oocysts autofluoresce green under u.v. epi-illumination using a 450–490 nm dichroic mirror exciter filter, and blue when using a 365 nm dichroic mirror filter (Figure 2). Unsporulated oocysts are found in fresh stool samples. Oocysts can be concentrated by sequential differential centrifugation or by employing formalin ethyl acetate and sucrose flotation in Sheather's solution. Percoll® can also be used to obtain purified oocysts. Under ideal conditions, up to 40% of the oocysts will sporulate within 7–13 days when maintained in potassium dichromate at room temperature. Two sporozoites are released from each of the two sporocysts in sporulated oocysts which have been mechanically ruptured and exposed to trypsin and sodium taurocholate. At present, this is the only way of positively identifying the oocysts as belonging to the genus Cyclospora.

The species name for this parasite was derived from the university where it was initially studied — Universidad Peruana Cayetano Heredia. Oocysts are spheroidal, 8–10 μm in diameter and have a bilayered wall 113 nm thick. The outer wall is 63 nm and rough and the inner wall layer is 50 nm and smooth. Each oocyst contains two ovoidal sporocysts (4.0 × 6.3 μm). Stieda and substieda bodies are present. There are two sporozoites in each

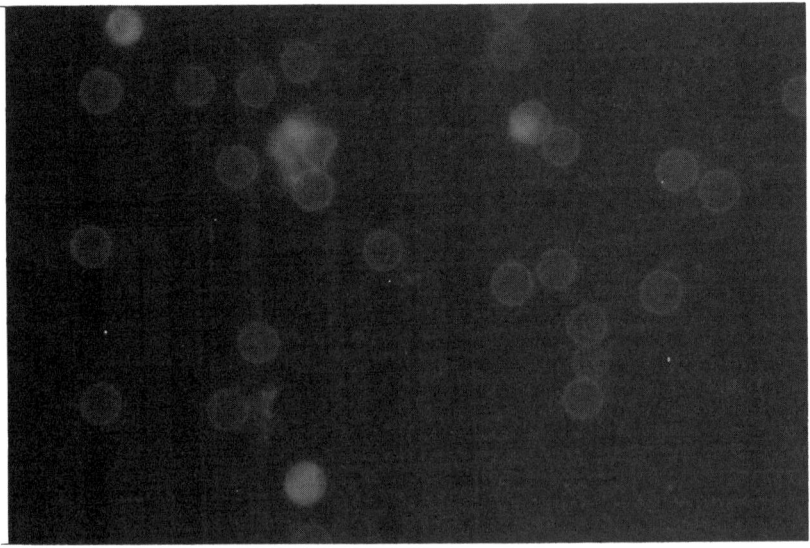

Figure 2 Autofluorescence of *Cyclospora cayetanensis* oocysts.

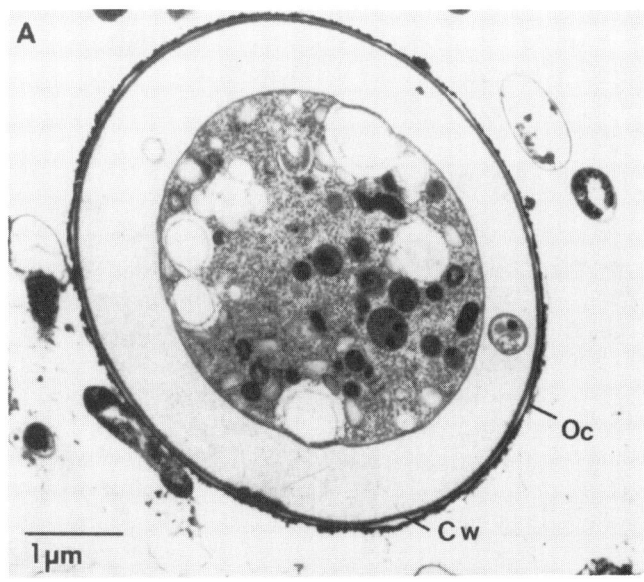

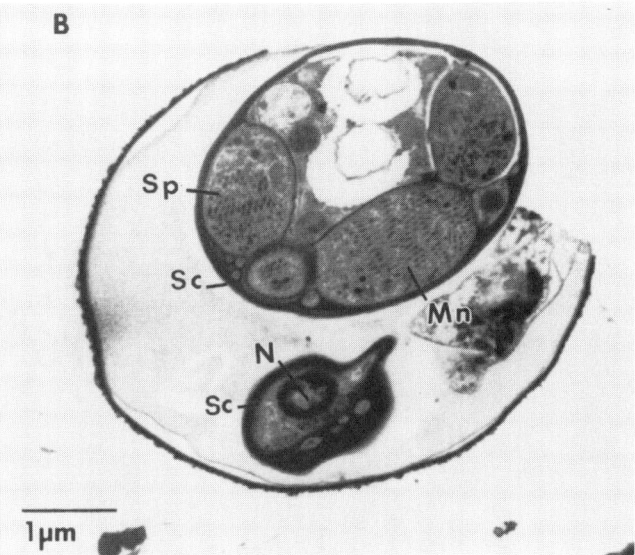

Figure 3 Transmission electron micrographs of *Cyclospora cayetanensis* oocysts. (A) Unsporulated; (B) sporulated. Cw, cyst wall; Mn, micronemes; N, nucleus; Oc, oocyst; Sc, sporocysts; Sp, sporozoites. Rhoptries can be seen. Reproduced, with permission, from Ortega *et al.* (1993).

sporocyst (1.2 3 9.0 µm). Sporozoites have the typical structure of coccidian sporozoites, including an apical complex, rhoptries, a nucleus, and micronemes (Ortega et al., 1994) (Figure 3).

3. MOLECULAR BIOLOGY

Taxonomically, *Cyclospora* has been placed in the subphylum Apicomplexa, subclass Coccidiasina, order Eucoccidiorida, family Eimeriidae. The small subunit rRNA coding region from *Cycl. cayetanensis* oocysts was purified, amplified and sequenced. Phylogenetic studies by Relman et al. (1996) have demonstrated that *Cyclospora* is closely related to parasites of the genus *Eimeria*.

A nested polymerase chain reaction assay has been developed by amplifying the *Cyclospora* 18S rDNA, and might be useful for detecting *Cyclospora* oocysts in human fecal samples (Yoder et al., 1996). Whether or not this assay will be sensitive enough to identify *Cyclospora* in environmental samples is not yet known.

Further work on the DNA of this parasite has demonstrated that the DNA 'signature' is different from that of *Cryptosporidium*. This DNA signature may be useful in the future as a tool for studying the epidemiology of *Cyclospora* as well as differentiating between strains (Figure 4).

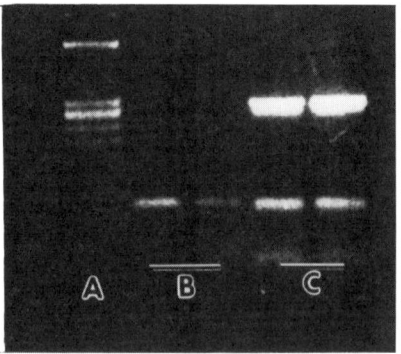

Figure 4 PCR amplification of *Cyclospora cayetanensis* DNA: (A) marker; (B) first amplification; (C) nested amplification.

4. LIFE CYCLE

The life cycle of cyclosporans varies according to the host from which they have been described. *Cycl. caryolytica* of the mole can apparently complete both asexual and sexual development within the nucleus of enterocytes. Oocyst sporulation of this parasite requires 4–5 days in the external environment. *Cyclospora talpae*, on the other hand, develops asexually within the nucleus of monocytic cells in the liver of moles, while sexual development occurs within the nucleus of epithelial cells lining the bile ducts. Complete sporulation of this species requires 12–14 days in the external environment. Exogenous sporulation of *Cycl. cayetanensis* closely resembles that of *Cycl. talpae* (see Ortega *et al.*, 1994).

Parasites with definite coccidian characters have been observed within jejunal enterocytes obtained from biopsy samples of patients excreting CLB-like oocysts (Bendall *et al.*, 1993). Gastrointestinal biopsies revealed the presence of intracellular parasites in both immunocompetent patients and those with the acquired immune deficiency syndrome (AIDS) (Bendall *et al.*, 1993; Sun *et al.*, 1996). In 16 of 17 immunocompetent Peruvian patients, parasitophorous vacuoles containing sexual and asexual forms of *Cycl. cayetanensis* were located at the luminal end of epithelial cells in jejunal biopsies. Merozoites of two different sizes were observed, suggesting two types of meront. These findings are in keeping with the coccidian identity of *Cycl. cayetanensis*. Since both the sexual and asexual stages are

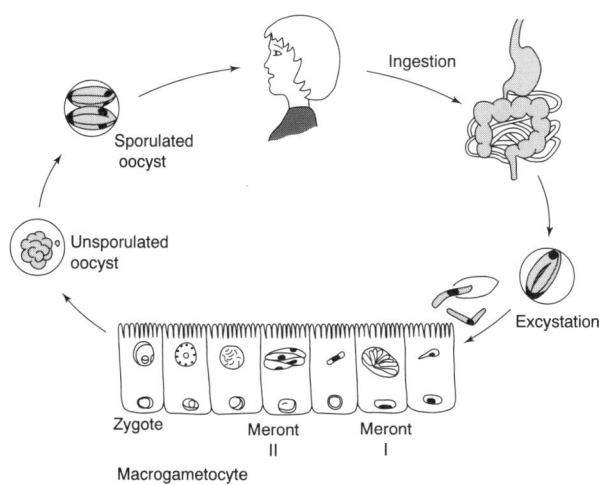

Figure 5 Proposed life cycle of *Cyclospora cayetanensis* in humans.

present in the same host, it is possible that humans are the sole host for this new coccidian (unpublished data) (Figure 5).

The complete life cycle of this parasite needs to be defined and, if possible, animal models developed. It also remains to be determined if humans are the only natural host. Organisms similar to CLBs have been observed in chimpanzees, suggesting that other animals may be susceptible to this infection (Ashford *et al.*, 1993).

5. HISTOPATHOLOGY

Impaired D-xylose absorption has been observed in individuals with cyclosporiasis, implying proximal small intestine involvement. Duodenal and jejunal biopsies have shown varying degrees of jejunal villous blunting, atrophy, and crypt hyperplasia. Villous/crypt ratios have ranged from 0.6 : 1 to 1.5 : 1 (normal 3 : 1 to 4 : 1). Endoscopies of infected patients demonstrated moderate to marked erythema of the distal duodenum. Abnormalities observed have included mild to moderate acute inflammation of the lamina propria and surface epithelial disarray. Five of nine infected patients had an increased number of plasma cells in the lamina propria (Connor *et al.*, 1993). In patients in Nepal, but not in Peru, the surface epithelium showed focal vacuolation, loss of brush border, and a change in the epithelial cells from columnar to cuboidal shape (unpublished data).

Jejunal aspirates have yielded oocysts trapped in mucus. In one jejunal biopsy sample, parasites were found within an intracytoplasmic vacuole located near the luminal end of the enterocyte. An increased number of intraepithelial leukocytes and mild villous blunting have also been observed. Gastric antral biopsy and colonic biopsy did not show any histopathologic change or organisms.

In a series of 17 Peruvian patients, no definitive abnormality was detected by endoscopy. Microscopical examination of the fourth portion of the duodenum or the first part of the jejunum demonstrated shortening and widening of the intestinal villi. The widening was due to diffuse edema and infiltration of the villus mucosa by a mixed inflammatory infiltrate. There were numerous plasma cells, lymphocytes, and frequent eosinophils. In addition, there was extensive infiltration of lymphocytes into the surface epithelium, which was particularly prominent at the tip of the shortened villi. Reactive hyperemia with dilatation and congestion of the villar capillaries was also seen. In some cases, there appeared to be increased eosinophilic extracellular matrix formation within the distorted villi. Focally, the underlying crypt epithelium showed reactive hyperplastic changes. Sexual and asexual parasites were observed in parasitophorous

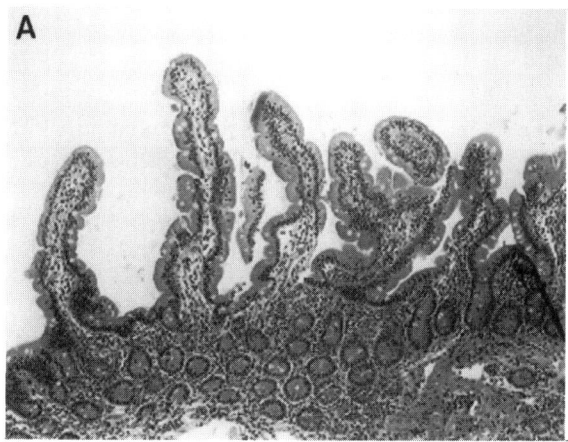

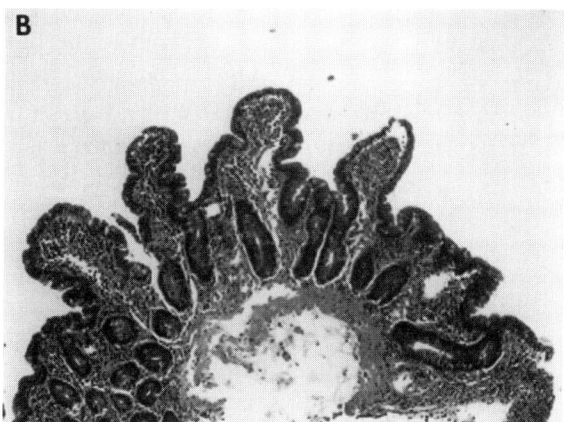

Figure 6 Transverse sections of jejunal mucosa, (A) uninfected and (B) infected with *Cyclospora cayetanensis*. Reproduced with permission from Ortega *et al.*, (1997).

vacuoles located at the luminal end of jejunal epithelial cells (unpublished data) (Figure 6).

6. IMMUNOLOGY

Cyclospora oocysts do not react with monoclonal antibodies specific to *Cryp. parvum* or *Giardia lamblia*. Yet Western blots reveal multiple antigens shared between *Cyclospora* and *Cryptosporidium*. Convalescent serum obtained from patients passing oocysts had a 10-fold increase in immunoglobulin M (IgM) antibody compared to acute-phase serum from the same patients (Wurtz, 1994). In Peruvian shanty towns, where sanitation is poor, children are infected early in life and have more than one infection episode. Teenagers and adults living in these areas are rarely infected, suggesting that they have developed immunity to the parasite.

7. CLINICAL SIGNS OF INFECTION

Onset of illness caused by *Cyclospora* has been reported as abrupt in 68% of adult patients and gradual in 32%. Symptoms persist for an average of 7 weeks. The symptoms mimic those of cryptosporidiosis, and include mild to severe nausea, anorexia, abdominal cramping, and watery diarrhea. Most adult patients reported weight loss of up to 5–10%. Diarrhea alternating with constipation has commonly been reported. Some patients have had flatulent dyspepsia and, less frequently, joint pain and night sweats. Bloody diarrhea accompanied by abdominal pain and tenesmus has been reported in one case, a child in Bangladesh (Albert *et al.*, 1994). The mechanisms of pathogenesis of *Cycl. cayetanensis* have yet to be defined, but tissue damage and jejunitis have been reported.

8. TREATMENT

To date, the only successful antimicrobial treatment for *Cyclospora* is trimethoprim–sulfamethoxazole (TMP–SMX) (Soave and Johnson, 1995). In the initial report from Peru, TMP–SMX brought about the cessation of symptoms and oocyst excretion in five patients (one adult and four children) after treatment for a mean of 4 days (Madico *et al.*, 1993). Fourteen controls had a mean oocyst excretion time of 14 days. The usefulness of TMP–SMX was confirmed in a 'double-blind' randomized placebo-controlled trial involving 40 expatriates living in Nepal who were infected with *Cyclospora*. Oocysts and symptoms cleared after 7 days of treatment in 94% of the patients receiving TMP–SMX. After 7 days of cross-over therapy with TMP–SMX, 91% of the patients in the placebo group also cleared the infection (Hoge *et al.*, 1995a). The efficacy of the drug was also confirmed in an open-label study in Haiti in which patients infected with the human immunodeficiency virus (HIV) were treated with TMP–SMX four times a day for 10 days, followed by prophylaxis with TMP–SMX three times a week (Pape et al., 1994). In Peru, in a 'double-blind' randomized trial in children infected with Cyclospora, 10 children treated with TMP–SMX had a significantly decreased duration of oocyst excretion compared to nine placebo-treated controls (mean ± SD = 4.8 ± 1 days vs. 12.1 ± 6 days, respectively). Antimicrobial drugs that have failed to treat Cyclospora infection include metronidazole, quinolone derivatives such as ciprofloxin and norfloxin, tetracycline, and pyrimethamine (Wurtz, 1994).

It is important to examine fecal samples using appropriate stains and by making accurate measurements, since oocysts of *Cycl. cayetanensis* (8–10 μm) can easily be confused with those of *C. parvum* (4–6 μm). Suspected positive samples can also be viewed by u.v. microscopy, since the oocysts of *Cycl. cayetanensis* autofluoresce (Figure 2). Differentiating *Cryptosporidium* and *Cyclospora* is important, since treatment is available for the latter but not for the former. A course of daily treatment for 7 days in adults and 3–5 days in children with 5 mg/kg TMP–SMX is currently recommended for *Cycl. cayetanensis* infection (Madico et al., 1993; Bartlett, 1996).

9. CYCLOSPORIASIS IN AIDS PATIENTS

Patients with AIDS may also be infected with *Cycl. cayetanensis* (see Hart et al., 1990; Long et al., 1990; Wittner et al., 1993; Pape et al., 1994; Scaglia et al.; 1994; Sifuentes-Osorio et al., 1995; Maggi et al., 1995) and, based on histological evidence, appear to harbor a larger number of parasites than do immunocompetent individuals infected with the same organism. In a series of AIDS patients studied in Lima, however, fewer than 1% of 126 individuals who presented with diarrhea were infected with *Cyclospora*. This mirrors the experience with AIDS patients in North America and Europe, and is probably due to the frequent use of TMP–SMX for *Pneumocystis carinii* prophylaxis. The high prevalence of *Cycl. cayetanensis* (10%) in adult AIDS patients in Haiti may be due to the infrequent use of TMP–SMX prophylaxis in that country (Pape et al., 1994).

10. EPIDEMIOLOGY

Organisms similar in appearance to *Cyclospora* have now been described from patients with protracted diarrheal illness in North, Central and South

America, the Caribbean, Africa, Bangladesh, south-east Asia, Australia, England, and eastern Europe (Ashford, 1979; Kaminski, 1991; Shlim et al., 1991; Pollok et al., 1992; Lebbad and Linder, 1993; Markus and Frean, 1993; McDougall and Tandy, 1993; Ortega et al., 1993; Rijpstra and Laarman, 1993; Smith, 1993; Villard et al., 1993; Wurtz et al., 1993; Bendall and Chiodini, 1994; Berlin et al., 1994; Chiodini, 1994; Junod et al., 1994; Pape et al., 1994; Petithory et al., 1994; Piales et al., 1994; Pratdesaba et al., 1994; Gascon et al., 1995; Hoge et al., 1995b; Ooi et al., 1995; Booy, 1996). Most infections have been reported in patients who were tourists or expatriate visitors to these countries (Figure 7).

Other commensal and parasitic organisms have been seen in association with *Cyclospora* infection. The most frequently reported have been *Entamoeba histolytica*, *Giardia duodenalis* (= *Giardia lamblia*), and *Cryp. parvum*. Helminths and commensal protozoa have also been observed.

Routes of transmission are still undocumented, although the fecal–oral route, either directly or via water, is probably the major one. Transmission indeed is a great mystery, since the recent epidemic in the USA seems to have been spread by the ingestion of raspberries which were imported from

Figure 7 Map of the world showing places where *Cyclospora* has been reported (the size of the dots is not proportional to the number of cases reported).

Guatemala. Oocysts are not sporulated when excreted, and sporulated oocysts are needed for infection. The speed at which sporulation occurs probably depends on a variety of environmental factors, including temperature and humidity.

Seasonality of infection is extremely strong. In Peru, in over 6 years of studying *Cyclospora* infection, nearly all infections were found to occur between December and July. It is extremely rare to find infection at any other time of the year. In the USA, the two major epidemics of 1995 and 1996 occurred from May to July. In Nepal, infection and illness occurred most frequently from May to August. The reasons for this frank seasonality need to be defined.

Increased awareness of the presence of *Cycl. cayetanensis* is likely to yield invaluable information about its relationship with disease and its route of transmission. Epidemiological evidence strongly suggests a role for water in *Cyclospora* transmission. The prolonged sporulation time, 1–2 weeks, further supports this contention. Reports of water-borne disease outbreaks due to parasites and other etiologic agents have increased in the USA for the past 20 years, despite regulations and measures to ensure the safety of drinking and recreational water (Ortega *et al.*, 1992).

In spite of these improvements, however, agents responsible for waterborne disease outbreaks are still identified in only 50% of the occurrences. Other sources of infection, including travelers or indigenous people infected with *Cycl. cayetanensis* and coming from different countries, contaminated food (Connor and Shlim, 1995; Ortega *et al.*, in press), and possibly insects as transport hosts, cannot yet be excluded. Epidemiological studies will be invaluable in helping to clarify these issues and to delineate strategies for prevention and treatment of infection with *Cycl. cayetanensis*.

10.1 Water-borne Outbreaks

Contaminated water has been suspected in several *Cyclospora* epidemics, including the latest in the USA, which was attributed to raspberries. In this outbreak, the raspberries had been sprayed with insecticide, possibly diluted with contaminated surface water. How many parasites are needed to infect an individual is still not known, nor is it known how long the organism can survive under different environmental conditions.

In the high season of 1992 in Nepal, patients with CLBs were more likely to have contracted *Cyclospora* when drinking untreated water 1 week before visiting the clinic (17 of 93 cases vs. 5 of 94 controls (odds ratio [OR] = 3.98, 95% confidence interval [95% CI] = 1.29–13.14; $P = 0.012$). Also, patients who consumed reconstituted milk were more likely to be infected with *Cyclospora* (11 of 93 vs 2 of 94; OR = 6.17, 95% CI = 1.22–42.17; P = 0.02) (Hoge *et al.*, 1993).

Water has been involved in several other outbreaks. The best documented was in Nepal, where 12 of 14 persons at a British military facility in Pokhara were infected with cyclosporiasis (Rabold *et al.*, 1994). *Cyclospora* oocysts were demonstrated in drinking water from river and municipal water sources. Other evidence incriminating water includes the following. In Utah, a man became infected after cleaning his flooded basement. His house was near a dairy farm and much of the basement flooding was attributed to water runoff from the farm following heavy rain (Hale *et al.*, 1994). In another incident, an 8 years old child became ill and passed CLBs in his feces 1 week after swimming in Lake Michigan, USA. Water samples taken from the inlet of the Chicago municipal water supply system showed the presence of organisms resembling CLBs, but their identity was not confirmed (Wurtz, 1994). In 1990, an outbreak involving 20 individuals, most of whom were resident hospital physicians, occurred in a Chicago hospital (Anon., 1991). Epidemiological evidence suggested that water from a roof-top reservoir was responsible for the infections: 12 of 59 people (20%) who drank tap water became infected (relative risk [RR] = 11.6; 95% CI = 2.7–103) (Huang *et al.*, 1995).

Most reports dealing with infection have come from cities or regions of countries which are predominately coastal, near both fresh and salt water (Table 1; Figure 7). Despite the implications of water in transmission, organisms confirmed as *Cyclospora* have rarely been identified from water samples. Ideal methods for the detection of *Cyclospora* in water still have not been defined and it is not known whether present methods are suitable for its isolation.

Table 1 Association of *Cyclospora* and water.

Location	Cases	Type of water	Reference
Chicago	20	Water reservoir	Anon. (1991)
Nepal	17	Untreated water (in reconstituted milk)	Hoge *et al.* (1993)
Chicago	1	Lake water	Wurtz *et al.* (1993)
Nepal	6–8	River/municipal water[a]	Rabold *et al.* (1994)
Utah	1	Farm	Hale *et al.* (1994)
Massachussetts	1	Well water	Ooi *et al.* (1995)

[a] '*Cyanobacter*-like bodies' were demonstrated in water samples.

10.2 Food-borne Outbreaks

Previous reports have incriminated contaminated lettuce, but the presence of the organism was not confirmed. In Peru, vegetables at markets in shanty towns have been shown to be contaminated with oocysts of *Cycl. cayetanensis*. Contamination was not common, occurring on about 2% of vegetables studied. In contrast, 15% of the vegetables were contaminated with *Cryp. parvum*. Washing vegetables experimentally contaminated with *Cycl. cayetanensis* oocysts removed less than 15% of the cysts (Ortega *et al.*, in press).

Two outbreaks occurred in the USA in 1996 affecting more than 1000 individuals. Some of the outbreaks were clustered, but isolated cases have also been observed. In Southern Carolina, 37 of 64 persons attending a party (58%) developed diarrhea, and seven of these patients had *Cyclospora* detected in their stools. Eating raspberries (RR = 5.6; 95% CI = 2.3–13.7), strawberries (RR = 2.2; 95% CI = 1.0–5.1), and potato salad (RR = 1.9; 95% CI = 1.3–2.7) were statistically associated with this outbreak (Centers for Disease Control and Prevention, 1996a,b).

11. PATHOGENESIS

How *Cycl. cayetanensis* causes diarrhea is unknown (Garavelli, 1994). In early reports, the brush border was stated to be disturbed, but in our electron micrographs of biopsy samples from infected patients the brush border appeared normal. The number of parasites found in the tissue was small, especially in relation to the marked inflammatory reaction present in the lamina propria.

12. CYCLOSPORIASIS IN PERU

In Peru, the epidemiology of *Cyclospora* infections is affected by socioeconomic class. In the shanty towns, *Cyclospora* prevalence reaches a peak in young children (aged 2–4 years); the infection is almost never detected after 11 years of age, suggesting that in this setting immunity becomes complete by adolescence. In this population, where fecal–oral contamination is intense and early infection common, the occurence of age-dependent immunity is also suggested by the fact that symptomatic infections are usually seen in young children and by the absence of patent *Cycl. cayetanensis* infection in adults. In contrast, in middle to upper class individuals who live in dwellings with good sanitation, children rarely appear to become infected and infections occur mainly in adults. The prevalence in symptomatic middle to upper class adults was 0.5%.

The relative mildness of symptoms in pediatric infections in an endemic zone contrasts markedly with the prolonged diarrhea experienced by expatriate adults visiting an endemic region and by the upper and middle class adults living in Lima. Immunologically naive adults appear to have a higher propensity for severe disease than do children living in highly endemic zones.

13. THE FUTURE

Unless positive identification of *Cyclospora* is made by sporulation and excystation, it may be preferable to refer to unidentified organisms resembling *Cyclospora* as CLBs. In this instance, however, the term CLB should be used to denote *Cyclospora*-like bodies. At present, sporulation and excystation procedures are the only reliable methods of positively identifying this organism. Specific detection tests based on the use of monoclonal antibodies or other specific probe technologies are being developed. The use of more specific identification techniques will enable us to define better the epidemiology of *Cyclospora* infection and to determine the extent to which water and other routes play a role in its transmission.

As with other coccidian parasites, there is a need to develop systems for the cultivation *in vitro* of *Cycl. cayetanensis* to help elucidate the life cycle and identify possible mechanisms of pathogenesis.

REFERENCES

Albert, J.M., Kabir, I., Azim, T., Hossain, A., Ansaruzzaman, M. and Unicomb, L. (1994). Diarrhea associated with *Cyclospora* sp. in Bangladesh. *Diagnostic Microbiology of Infectious Diseases* **19**, 47–49.

Anon. (1991). Diarrheal diseases—outbreaks associated with cyanobacteria (blue–green algae)-like bodies, United States of America and Nepal. *Weekly Epidemiological Record* **66**, 241–248.

Ashford, R.W. (1979). Occurrence of an undescribed coccidian in man in Papua New Guinea. *Annals of Tropical Medicine and Parasitology* **73**, 497–500.

Ashford, R.W., Warhurst, D.C. and Reid, G.D.F. (1993). Human infection with cyanobacterium-like bodies. *Lancet* **341**, 1034.

Bartlett, J.G. (1996). *Pocket Book of Infectious Disease Therapy*. Baltimore: Williams & Wilkins.

Bendall, R.P. and Chiodini, P.L. (1994). Epidemiology of human *Cyclospora*

infection in the UK. In: *Protozoan Parasites and Water* (W.B. Betts, D. Casemore, C. Fricker, H. Smith and J. Watkins, eds), pp. 26–29. Cambridge: Royal Society of Chemistry.
Bendall, R.P., Lucas, S., Moody, A., Tovey, G. and Chiodini, P.L. (1993). Diarrhoea associated with *Cyanobacterium*-like bodies: a new coccidian enteritis of man. *Lancet* **341**, 590–592.
Berlin, O.G., Novak, S.M., Porschen, R.K., Long, E.G., Stelma, G.N. and Schaefer, F., III (1994). Recovery of *Cyclospora* organisms from patients with prolonged diarrhea. *Clinical Infectious Diseases* **18**, 606–609.
Booy, R. (1996). Are *Cyclospora* an important cause of diarrhoea in Bangladesh? *Archives of Disease in Childhood* **74**, 90.
Centers for Disease Control and Prevention (1996a). Outbreaks of *Cyclospora cayetanensis* infection — United States, 1996. *Morbidity and Mortality Weekly Report* **45**, 549–551.
Centers for Disease Control and Prevention (1996b). Update: outbreaks of *Cyclospora cayetanensis* infection — United States and Canada. *Morbidity and Mortality Weekly Report* **45**, 611–612.
Chiodini, P.L. (1994). A 'new' parasite: human infection with *Cyclospora cayetanensis*. *Transactions of the Royal Society of Tropical Medicine and Hygiene* **88**, 369–371.
Connor, B.A. and Shlim, D.R. (1995). Foodborne transmission of *Cyclospora*. *Lancet* **346**, 1634.
Connor, B.A., Shlim, D.R., Scholes, J.V., Rayburn, J.L., Reidy, J. and Rajah, R. (1993). Pathologic changes in the small bowel in nine patients with diarrhea associated with a coccidia-like body. *Annals of Internal Medicine* **119**, 377–382.
Eimer, T. (1870). *Ueber die ei- und kugelförmigen sogenannten Psorospermien der Wirbelthiere*, pp. 1–58. Würzburg: A. Stuber's Verlangshandlung.
Garavelli, P.L (1994). Pathogenicity of *Cyclospora* sp. *Parasite* **1**, 94.
Garcia, L.S. and Bruckner, D.A. (1994). Intestinal protozoa: coccidia and microsporidia. In: *Diagnostic Medical Parasitology*, pp. 61–72. Washington: ASM Press.
Gascon, J., Corachan, M., Bombi, J.A., Valls, M.A. and Bordes, J.M. (1995). *Cyclospora* in patients with traveller's diarrhea. *Scandinavian Journal of Infectious Diseases* **27**, 511–514.
Hale, D., Aldeen,W. and Carroll, K. (1994). Diarrhea associated with cyanobacteria-like bodies in an immunocompetent host. An unusual epidemiological source. *Journal of the American Medical Association* **271**, 144–145.
Hart, R.H., Ridinger, M.T., Soundarajan, Peters, C.S., Swiatlo, A.L. and Kocka, F.E. (1990). Novel organism associated with chronic diarrhea in AIDS. *Lancet* **335**: 169–170.
Hoge, C.W., Shlim, D.R., Rajah, R., Triplett, J., Shear, M., Rabold, J.G. and Echeverria, P. (1993). Epidemiology of diarrhoeal illness associated with coccidian-like organism among travellers and foreign residents in Nepal. *Lancet* **341**, 1175–1179.
Hoge, C.W., Shlim, D.R., Ghimire, M., Rabold, J.G., Pandey, P., Walch, A., Rajah, R., Gaudio, P. and Echeverria, P. (1995a). Placebo-controlled trial of co-trimoxazole for *Cyclospora* infections among travellers and foreign residents in Nepal. *Lancet* **245**, 691–693.
Hoge, C.W., Echeverria, P., Rajah, R., Jacobs, J., Malthouse, S., Chapman, E., Jimenez, L.M. and Shlim, D.R. (1995b). Prevalence of *Cyclospora* species and

other enteric pathogens among children less than 5 years of age in Nepal. *Journal of Clinical Microbiology* **33**, 3058–3060.

Huang, P., Weber, J.T., Sosin, D.M., Griffin, P.M., Long, E.G., Murphy, J., Kocka, F., Peters, C. and Kallick, C. (1995). The first reported outbreak of diarrheal illness associated with *Cyclospora* in the United States. *Annals of Internal Medicine* **123**, 409–414.

Junod, C., Deluol, A.M., Cosnes, J. and Bauer, P. (1994). *Cyclospora*, nouvelle coccidie agent de diarrhées des voyageurs: 11 observations. *Press Médicale* **23**, 1312.

Kaminski, R. (1991). Cuerpos semejantes a cyanobacteria associados con diarrea en Honduras. *Revista Medica Hondurena* **59**, 179–182.

Lebbad, M. and Linder, E. (1993). Nyupptackt organism bakom diarresjukdom. *Lakartidningen* **90**, 951–952.

Long, E.G., Ebrahimzadeh, A., White, E.H., Swisher, B. and Callaway, C.S. (1990). Alga associated with diarrhea in patients with acquired immunodeficiency syndrome and in travelers. *Journal of Clinical Microbiology* **28**, 1101–1104.

Long, E.G., White, E.H., Carmichael, W.W., Quinlisk, Rajah, R., Swisher, B.L., Daugharty, H. and Cohen, M.T. (1991). Morphologic and staining characteristics of a cyanobacterium-like organism associated with diarrhea. *Journal of Infectious Diseases* **164**, 199–202.

Madico, G., Gilman, R.H., Miranda, E., Cabrera, L. and Sterling, C.R. (1993). Treatment of *Cyclospora* infections with co-trimoxazole. *Lancet* **342**, 122–123.

Maggi, P., Brandonisio, O., Larocca, A.M.V., Rollo, M., Panaro, M.A., Marani, A., Marzo, R., Angarano, G. and Pastore, G. (1995). *Cyclospora* in AIDS patients: not always an agent of diarrheic syndrome. *Microbologica* **18**, 73–76.

Markus, M.B. and Frean, J.A. (1993). Occurrence of human *Cyclospora* infection in sub-Saharan Africa. *South African Medical Journal* **83**, 862–863.

McDougall, R.J. and Tandy, M.W. (1993). Coccidian/*Cyanobacterium*–like bodies as a cause of diarrhea in Australia. *Pathology* **25**, 375–378.

Ooi, W.W., Zimmerman, S.K. and Needham, C.A. (1995). *Cyclospora* species as a gastrointestinal pathogen in immunocompetent hosts. *Journal of Clinical Microbiology* **33**, 1267–1269.

Ortega, Y.R., Sterling, C.R., Marshall, M.M. and Gilman, R.H. (1992). *Cyclospora*: the potential for water-borne transmission. In *AWWA Proceedings, Water Quality Technology Conference. Toronto, Ontario*, pp: 465–468. Denver: AWWA.

Ortega, Y.R., Sterling, C.R., Gilman, R.H., Cama, V.A. and Diaz, F. (1993). *Cyclospora* species — a new protozoan pathogen of humans. *New England Journal of Medicine* **328**, 1308–1312.

Ortega, Y.R., Sterling, C.R. and Gilman, R.H. (1994). A new coccidian parasite (Apicomplexa: Eimeriidae) from humans. *Journal of Parasitology* **80**, 625–629.

Ortega, Y.R., Roxas, C.R., Gilman, R.H., Miller, N.J., Cabrera, L., Taquiri, C. and Sterling, C.R. (in press). Isolation of *Cryptosporidium parvum* and *Cyclospora cayetanensis* oocysts from vegetables collected from markets of an endemic region in Peru. *American Journal of Tropical Medicine and Hygiene*.

Pape, J.W., Verdier, R.-I., Boncy, M., Boncy, J. and Johnson, W.D. (1994). *Cyclospora* infection in adults infected with HIV. Clinical manifestations, treatment, and prophylaxis. *Annals of Internal Medicine* **121**, 654–657.

Petithory, J.-C., Junod, C.H., Ardoin, F. and Jousserand, P. (1994). *Cyclospora* sp.: une nouvelle coccidie parasite de l'homme. *Revue Française des Laboratoires* **271**, 11–14.

Piales, I., Coursin, T., Deluol, A.-M., Poirot, J.L. and Tandeau de Marsac, N. (1994). Search for *Cyanobacterium*-like bodies in the stools of travellers with diarrheal illness. *Journal of Eukaryotic Microbiology* **41**, 58S.

Pollok, R.C., Bendall, R.P., Moody, A., Chiodini, P.L. and Churchill, D.R. (1992). Traveller's diarrhea associated with cyanobacterium-like bodies. *Lancet* **340**, 556–557.

Pratdesaba, R.A., Velasquez, T. and Torres, M.F. (1994). Occurrence of *Isospora belli* and cyanobacterium-like bodies in Guatemala. *Annals of Tropical Medicine and Parasitology* **88**, 449–450.

Rabold, J.G., Hoge, C.W., Shlim, D.R., Kefford, C., Rajah, R. and Echeverria, P. (1994). *Cyclospora* outbreak associated with chlorinated drinking water. *Lancet* **344**, 1360–1361.

Relman, D.A., Schmidt, T.S., Gajadhar, A., Sogin, M., Cross, J., Yoder, K., Sethabutr, O. and Echeverria P. (1996). Molecular phylogenetic analysis of *Cyclospora*, the human intestinal pathogen, suggests that it is closely related to *Eimeria* species. *Journal of Infectious Diseases* **173**, 440–445.

Rijpstra, A.C. and Laarman, J.J. (1993). Repeated findings of unidentified small *Isospora*-like coccidia in faecal specimens from travelers returning to The Netherlands. *Tropical and Geographical Medicine* **45**, 280–282.

Scaglia, M., Gatti, S., Bassi, P., Viale, P.I., Novati, S. and Ranieri, S. (1994). Intestinal coinfection by *Cyclospora* sp. and *Cryptosporidium parvum*: first report in an AIDS patient. *Parasite* **1**, 387–390.

Schaudinn, F. (1902). Studien über krankheitserregende Protozoen. I. *Cyclospora caryolytica* Schaud., der Erreger der perniciosen Enteritis des Maulwurfs. *Arbeiten aus dem Kaiserlichen Gesundheitsamte* **18**, 378–416.

Schneider, A. (1881). Sur les psorospermies oviformes ou coccidies, espèces nouvelles ou peu connues. *Archives de Zoologie Expérimentale et Générale* **9**, 387–404.

Shlim, D.R., Cohen, M.T., Eaton, M., Rajah, R., Long, E.G. and Ungar, B.L.P. (1991). An alga-like organism associated with an outbreak of prolonged diarrhea among foreigners in Nepal. *American Journal of Tropical Medicine and Hygiene* **45**, 383–389.

Sifuentes-Osorio, J., Porras-Cortes, G., Bendall, R.P., Morales-Villarreal, F., Reyes-Teran, G. and Ruiz-Palacios, G.M. (1995). *Cyclospora cayetanensis* infection in patients with and without AIDS: biliary disease as another clnical manifestation. *Clinical Infectious Diseases* **21**, 1092–7.

Smith, P.M. (1993). Traveller's diarrhoea associated with a cyanobacterium-like body. *Medical Journal of Australia* **158**, 724.

Soave, R. and Johnson, W.D. (1995). *Cyclospora*: conquest of an emerging pathogen. *Lancet* **345**, 667–668.

Sun, T., Illardi, C.F., Asnis, D., Bresciani, A.R., Goldenberg, S., Roberts, B. and Teichberg, S. (1996). Light and electron microscopic identification of *Cyclospora* species in the small intestine. Evidence of the presence of asexual life cycle in human host. *American Journal of Clinical Pathology* **105**, 216–220.

Villard, O., Himy, R., Brogard, C. and Kremer, M. (1993). Syndrome diarrhéique associé à la presence de cyanobacterium-like bodies. *Gastroenterologie Clinique et Biologique* **17**, 401–402.

Wittner, M., Tanowitz, H.B. and Weiss, L.M. (1993). Parasitic infections in AIDS patients. *Infectious Disease Clinics of North America* **7**, 569–586.

Wurtz, R.M. (1994). *Cyclospora*: a newly identified intestinal pathogen of humans. *Clinical Infectious Diseases* **18**, 620–623.

Wurtz, R.M., Kocka, F.E., Peters, C.S., Weldon-Linne, C.M., Kuritza, A. and Yungbluth, P. (1993). Clinical characteristics of seven cases of diarrhea associated with a novel acid-fast organism in the stool. *Clinical Infectious Diseases* **16**, 136–138.

Yoder, K.E., Sethabutr, O. and Relman, D. (1996). PCR-based detection of the intestinal pathogen *Cyclospora*. In: *PCR Protocols for Emerging Infectious Diseases*. (D.H. Persing, ed.), pp. 169–175. Washington, DC: ASM Press.

Index

acquired immunity, resolution of cryptosporidiosis 95–6
actin, *Cryptosporidium parvum* 172–3
acute cryptosporidiosis 56, 206
  protective immunity, animal studies 98, 99
adoptive transfer, immune cells, *Cryptosporidium parvum* infection 96
agricultural waste, water supply contamination 246, 270–1
AIDS
  AIDS-related cholangitis and acalculous cholecystitis 329–30, 331
  cryptosporidiosis
    antibody status 124–5
    chronic 57, 124–5, 246
    fulminant 57
    hepatobiliary involvement 61
    hyperimmune bovine colostrum immunoglobulin studies 140–1
    intestinal relapse 61
    nitazoxanide treatment 210
    oocyst shedding, chronic 208, 246
    paromomycin treatment 209
    respiratory 60
  cyclosporiasis 409
  microsporidiosis 284, 322
    hepatobiliary involvement 329–30, 331
    intestinal disease prevalence 323–4
    intestinal injury 325, 337–8
    species causing intestinal disease 323
air-borne transmission, cryptosporidiosis 54, 59–60
albendazole, microsporidiosis treatment 341, 357
alborixin, cryptosporidiosis 69
aminoglycosides, *Cryptosporidium* studies 27–8
  see also paromomycin
anchoring disk, microsporidia 288

animal models
  Cryptosporidium parvum 195–205
    agents for treatment and prevention 196–203, 211
    CD4 cells in protective immunity 98, 99
    genotype and disease severity 232
    immunity 94, 98, 99, 108
    oocyst viability, animal infectivity 262–3
    pathophysiology 203–5
  microsporidia
    cell-mediated immune responses 307–8
    pathogenicity 326
  see also calves, cryptosporidiosis: calf model; murine models: *Cryptosporidium parvum*; piglet diarrhea model, *Cryptosporidium parvum*; SIV-infected
annular ring, *Cryptosporidium* 24
anti-cryptosporidial agents, specific 66–70
  Cryptosporidium parvum, resistance to 66, 206
antibodies
  AIDS patients, antibody status 124–5
  antibody response and CD4 count 201
  antibody-drug combination therapy 143
  cryptosporidiosis antibody-based immunotherapy 121–43
    early studies 130–1
    future prospects 141–3
    human studies 138–41
    laboratory investigations 131–8
    mechanisms of action, theoretical 126–7
    monoclonal antibodies *see* monoclonal antibodies
    orally administered 128–9, 137
    polyclonal antibodies 132–6
    practical consideration 125–30
    prophylaxis 132, 143
    rationale 122–5
  immunodeficiency states 123
  microsporidiosis 305–6

antigens
  apicomplexans, stage-specific 127
  Cryptosporidium
    expressed in multiple stages 174
    oocyst/sporozoite proteins 155
antimotility agents, cryptosporidiosis 65
antioxidant enzymes, *Cryptosporidium* 176
apicomplexan parasites
  intracellular locations 20–6
    parasitophorous vacuole
      invasion and intracellular strategies 20
    intracellular stages 173–4
    invasive stages 159–60
    oocysts 154
  parasitophorous vacuole formation,
    Plasmodium, *Toxoplasma* and
    Cryptosporidium 21–3
  stage specific antigen expression 127
apoptosis, *Cryptosporidium parvum* infected cells 195
apple cider, cryptosporidiosis outbreak 52, 244
arginine aminopeptidase, sporozoite membrane protein 171–2
ASTM method, *Cryptosporidium* detection in water 248, 250–1
asymptomatic carriage, cryptosporidiosis 55–6
autofluorescence, *Cyclospora cayetanensis* oocysts 402
autoinfection, *Cryptosporidium parvum* 8, 154
azithromycin, cryptosporidiosis 60, 67–8

B lymphocytes, protective immunity, *Cryptosporidium parvum* 97
Babesia spp. 20–1
backpack mouse hybridoma model 202–3
β-tubulin
  amino acid sequence, microsporidial phylogeny 361–2
  genes, microsporidia 357
biliary disease *see* hepatobiliary disease
bovine colostrum 132–6
  early case reports 130–1
  immunoglobulins
    resistance to proteolysis 128
    resistance to stomach acid 129
  major studies 135
  serum and cellular constituents 133
  see also hyperimmune bovine colostrum
bowel changes, microsporidiosis 330–1
Brazil, cryptosporidiosis and HIV 42

breast feeding and cryptosporidiosis 54, 123–4, 131

calcium carbonate flocculation procedure 251–2, 252–3
calves, cryptosporidiosis 246
  calf model 94, 108
  pathophysiologic studies 204–5
  calf-to-human transmission 231
  oocyst
    production for research 189
    shedding 246
  passive immunity 132–3
cartridge filters, oocyst detection in water 250–1, 252
cats, *Cryptosporidium parvum* infection 18–19
CD4 cell count
  and antibody response 201
  chronic cryptosporidiosis 124
  and disease severity
    cryptosporidiosis 56, 62, 97
    microsporidiosis 308, 326, 330
  and susceptibility, microsporidiosis 304
CD4 cells, protective immunity 97–9, 101
CD8 cells, protective immunity 99–100
cell culture, *Cryptosporidium parvum* 190–5
  oocyst viability determination 263
cell-mediated immune responses
  cryptosporidiosis 95–110, 124
    and malnutrition 41, 49
  microsporidiosis 307–8
cellular components, innate immunity, *Cryptosporidium parvum* 92–5
chemofluorescent optical brightening agents, microsporidia 377–9
children, cryptosporidiosis
  asymptomatic carriage 55
  day-care centres 40, 52
  and malnutrition 41, 49, 56
chlorine treatment, *Cryptosporidium* oocysts 249, 269
chronic cryptosporidiosis 57, 206–7
  AIDS patients 57, 124–5, 246
  antibody immunodeficiency states 123
  CD4 cells in protective immunity, animal studies 98–9
  gastrointestinal architectural changes 206–7
  hepatobiliary involvement 61
clarithromycin, activity against *Cryptosporidium* 67–8

classification and taxonomy, *Cryptosporidium parvum* 10–17
  genetic markers, stability of 16–17
  phenotypic and genotypic variation, *Cryptosporidium parvum* isolates 14, 16
  preliminary molecular taxonomy 12–14
computer-based video-image technique, *in vitro* drug quantitation, *Cryptosporidium parvum* 193
COWP gene, *Cryptosporidium parvum* 178
cross-infectivity in vertebrates, *Cryptosporidium* spp. 11
cross-reactivity, microsporidia 370, 375
cryptopain 172
cryptosporidiosis 37–72
  acquired immunity 95–6
  acute 56, 98, 99, 206
  chronic *see* chronic cryptosporidiosis
  diagnosis 70–1
  epidemiology 39–50, 123–4
  fulminant 57
  gastrointestinal histopathology 136, 206–7
  hepatobiliary disease 55–6, 61–2, 97, 128
  interferon-γ in resolution 101–2, 110, 113
  immune responses *see* immune responses: *Cryptosporidium parvum*
  incubation period 56
  intestinal disease 55–9
    pathophysiology 57–9
    prevalence and incidence 40
  laboratory investigations and chemotherapy 187–212
  outbreaks 242–4
    see also Milwaukee, cryptosporidiosis outbreak, 1993; water-borne cryptosporidiosis: outbreaks
  respiratory disease 54, 59–60, 68
  transmission 50–5, 59–60, 231–2, 244
    see also water-borne cryptosporidiosis
  treatment 64–5
    antiviral (anti-HIV) therapy 63
    immunotherapy 66
    specific agents 66–70
      paromomycin *see* paromomycin, cryptosporidiosis
    specific antigens 66–70, 121–43
    supportive care 49, 63
Cryptosporidium baileyi 10, 11, 38, 60, 225
Cryptosporidium felis 18–19
Cryptosporidium meleagridis 10
Cryptosporidium muris 10, 17, 28, 38, 225

Cryptosporidium nasorum 10
Cryptosporidium parvum 38
  classification and taxonomy 10–17, 225
  developmental stages 152
  differentiation from *Cyclospora cayetanensis* 409
  genetic heterogeneity and PCR detection 223–35
  isolate propagation 225–6
  life cycle and control 8–10
  natural history 5–29
  oocysts *see* oocysts, *Cryptosporidium parvum*
  specific cell-mediated immunity 95–110
  transmission *see* transmission: cryptosporidiosis
  see also cryptosporidiosis
Cryptosporidium serpentis 10, 11
Cryptosporidium spp. 38
  antibody interactions with 127
  extracellular stages, targets for effector mechanisms 126–7
  heterogeneity 225–6
  host–parasite interactions, molecular basis 151–78
  infected cell 24
  infectious dose 247
  interspecies versus intraspecies relationships 15
  intracellular
    location 19–20, 21
    proteins 174–7
    stages, morphology and function 173
  invasion 22–3
    proteins of invasive stages 163–73
  oocysts *see* oocysts, *Cryptosporidium parvum*
  stages susceptible to antibodies 126
  taxonomy 5, 6, 10–17
  topology, implications 26–8
  water-borne *see* water-borne cryptosporidiosis
Cuba, cryptosporidiosis and HIV 42
Cyclospora cayetanensis 399–414
  diagnosis and purification 401–4
  differentiation from *Cryptosporidium* 409
  histopathology 406–7
  immunology 408
  life cycle 405–6
  molecular biology 404–5
  taxonomy 404

cyclosporiasis 399, 400
  AIDS patients 409
  clinical signs 408
  epidemiology 410–11
  outbreaks 411–13
  pathogenesis 413
  Peru 413–14
  transmission 113, 411–13
  treatment 408–9
cystein protease, *Cryptosporidium* 172
cytokines
  immunotherapy, *Cryptosporidium parvum* 111–12
  innate immunity, *Cryptosporidium parvum* 89–92, 113
  microsporidiosis, tumor necrosis factor α in 310
  specific cell-mediated immunity, *Cryptosporidium parvum* 101–10
cytoskeletal and structural proteins, *Cryptosporidium* 172–3

dairy farms 246
  see also water-borne cryptosporidiosis: watershed management
day-care centres, cryptosporidiosis 40, 52
dehydroepiandrosterone (DHEA), cryptosporidiosis 66, 112
Denmark, cryptosporidiosis and HIV 42
dense granules 20
  Cryptosporidium parasitophorous vacuole formation 23
  mAb recognition of antigen present in 170
developmental stages, *Cryptosporidium parvum*, host-parasite interactions 152
diabetes, cryptosporidiosis risk 49
diarrhea, cryptosporidiosis
  AIDS, survival rates 57
  calves, incidence 17, 246
  and malnutrition 41
  oral agents 67, 199, 210
  and paromomycin levels 68–9
  pathophysiology 57–9
  prevalence and peak incidence 39
  waterborne-related 7
  see also piglet diarrhea model, *Cryptosporidium parvum*
diethyldithiocarbamate, cryptosporidiosis 66, 112–13
dihydrofolate reductase–thymidylate synthase
  and folate metabolism 175
  gene sequence polymorphisms 230

disinfection, water treatment 269
disseminated disease, microsporidiosis 332
DNA, random amplification analysis, *C. parvum* isolates 227–8
domestic animals, cryptosporidiosis risk to humans 17–19
drinking water and cryptosporidiosis *see* water-borne cryptosporidiosis
droplet transmission, cryptosporidiosis 54
drug design and delivery
  Cryptosporidium 20
    aminoglycosides 28
    see also paromomycin
    implications of topology 27
    intracellular stages, potential targets 175, 176
    life cycle 9–10
    new approaches 205–8
    structural impediments to drug entry 66
  microsporidia 357
drug screening
  anticryptosporidial properties of current drugs 208–12
  in vitro assays 191–2
duodenal biopsies, *Cyclospora cayetanensis* 406

effector mechanisms, antibody action, cryptosporidiosis 126–7
egg yolk antibodies, cryptosporidiosis 136
Eimeria spp.
  intracellular stages 173–4
  oocyst wall 154
  protease expression 171
electron microscopy *see* transmission electron microscopy
ELISA, *Cryptosporidium parvum*
  in vitro parasite quantitation 192–3
  oocyst detection 190
elongation factor-1α
  Cryptosporidium parvum 176
  microsporidial phylogeny, maximum likelihood analysis 360–1
elongation factor-2, *Cryptosporidium parvum* 176–7
Encephalitozoon cuniculi 322
  disseminated disease 332
  GTTT repeats, variations of isolates 368
  host specificity 308, 309
  molecular biology 353
  nucleus 288
  spores 289

transmission 291–2, 304
Encephalitozoon hellum
  disseminated disease 332
  host specificity 308
  PCR primer sets 383
  transmission 299
Encephalitozoon intestinalis
  disseminated disease 332
  host specificity 308
  intestinal
    biopsy 373
    dysfunction 328–9
  morphologic features 335, 337
  PCR primer sets 383
  small intestine disease 330
  transmission 291, 299
  treatment 341
Encephalitozoon spp.
  germination 292
  merogony 293
  nucleus 288
  PCR primer sets 383
  sporogony 293, 295
  taxonomy 299–300
  transmission 291, 299
  ultrastructural features 372
endoplasmic reticulum, microsporidia 288–9
Enterocytozoon bieneusi
  associated cellular injury 337–8
  comparative rDNA analysis, phylogeny 366
  electron microscopy 333–5
  endoplasmic reticulum 289
  histological diagnosis 335, 337–9
  host specificity 308, 310
  intestinal dysfunction 328–9
  PCR primer sets 383–4
  small intestine disease 330
  taxonomy 298–9
  transmission 291, 298, 327
  treatment 341
Enterocytozoon spp. 293
  infection 295
  merogony 293
  nucleus 288
  sporogony 293, 295
  taxonomy 298–9
Environmental Protection Agency 264
enzymes in metabolic and synthetic pathways, *Cryptosporidium* 174–6
epidemiology
  cryptosporidiosis 39–50, 123–4
    global epidemiology 39–40

HIV infected people 40–1
malnutrition 41, 48
other groups at risk 49–50
cyclosporiasis 410–11
microsporidiosis 323–7
erythromycin, activity against
  Cryptosporidium 67
Ethiopia, cryptosporidiosis and HIV 42
eukaryotes, microsporidial phylogeny 358–63
Europe, studies, cryptosporidiosis and HIV 8, 43, 45
ex vivo analysis, T lymphocytes, immunity to
  . *Cryptosporidium parvum* 105–8
excystation
  Cryptosporidium parvum 154, 262
  Cyclospora cayetanensis 401
extracellular stages, *Cryptosporidium* 9
  as potential targets for effector mechanisms 126–7
extracytoplasmic location, *Cryptosporidium parvum* 19–20, 173, 205
extraintestinal phase, *Cryptosporidium parvum* 8

feeder organelle, *Cryptosporidium* 173
  formation 23–6
  membrane 19, 20, 28–9
    impediment to drug entry 66
    transport, features suggesting 26–7
filtration
  oocyst detection in water 247–8, 250–3
  water treatment 51–2, 71–2, 268
flat-bed membranes, oocyst detection in water 251
flotation method, oocyst separation from other debris 253–4
flow cytometry, *Cryptosporidium* oocyst detection 256–8
fluid, electrolyte and nutritional therapies
  cryptosporidiosis 49, 63
  microsporidiosis 341
fluorescent activated cell sorter (FACS) 256, 257
fluorescent *in situ* hybridization, oocyst detection in water 259–60
folate metabolism and dihydrofolate reductase–thymidylate synthase, *Cryptosporidium* 175, 230
food-borne outbreaks
  cryptosporidiosis 52–3, 244
  cyclosporiasis 53, 411, 413
France, cryptosporidiosis and HIV 43
fulminant cryptosporidiosis 57

Gal/GalNAc-specific lectin 170–1
γδT cells 94, 127
gastrointestinal
   flora and resistance to *Cryptosporidium parvum* 89
   rapid transit times, oral therapy efficacy 67, 199, 210
   genetic analysis, oocyst viability determination 263–4
geneticin, *Cryptosporidium* studies with 27–8
genomic C and G content, microsporidia, phylogeny 359–60
genotype
   and disease severity, animal studies 232
   variations, *Cryptosporidium parvum* 14, 16, 225–6, 227–33
      genetic markers, analysis of 16–17
      relevance of polymorphisms 224–5
germination, microsporidia in mammals 292
global epidemiology, human cryptosporidiosis 39–40
Glugea atherinae
   karyotype 353
   nucleus 288
glycosylated oocyst/sporozoite proteins, *Cryptosporidium* 155, 156
GP900, *Cryptosporidium parvum* 134, 165, 166
   antibodies to 168
ground water 267
   associated outbreaks, cryptosporidiosis 244, 245
gut epithelium infected with *Cryptosporidium parvum* 25
   histopathological changes 136, 206–7

hemA, *Cryptosporidium parvum* 176
hepatobiliary disease
   cryptosporidiosis 61–2
      antibody administration 128
      asymptomatic biliary carriage 55–6
      HIV and mortality 97
      symptoms and CD4 cell counts 62
   microsporidiosis 329–30, 331–2
heterogeneity in *Cryptosporidium* 225–6
   see also genotype
high molecular weight glycoproteins, *Cryptosporidium* 165–70
HIV
   cryptosporidiosis 40–1
      anti-viral therapy 63
      asymptomatic carriage 55

      studies published since 1990 42–8
      survival 57, 97
      transient 56
      vaccine responses 125
   cyclosporiasis, treatment with TMP-SMX 409
   see also CD4 cell count
hospitals, cryptosporidiosis transmission 52
host
   resistance
      cryptosporidiosis 123–5
      microsporidiosis 305
   specificity
      Cryptosporidium 10–11
      microsporidia 308–9
host–parasite interactions
   Cryptosporidium 151–78
   microsporidia 303–9, 371
household transmission, cryptosporidiosis 52
human immunoglobulins, oral administration 136
human studies
   anti-*Cryptosporidium* antibodies 138–41
   Cryptosporidium parvum reactive cells in peripheral blood, immunocompetent and HIV infected individuals 108, 110
   cryptosporidiosis and HIV 42–8
humoral immune responses
   cryptosporidiosis 123–5
   microsporidia 305–7
hyperimmune bovine colostrum 69–70
   antibodies to GP900 168
   degradation in GI tract 128
   hyperimmune bovine colostrum immunoglobulin studies, AIDS patients 140–1
   human studies 140–1
   and paromomycin, efficacy and disease severity 134–5
   production 133
   specific antigen-stimulated immunoglobulins 142
hyperimmune polyclonal antibodies, egg yolk 136
hypersensitivity responses, microsporidial infection 305, 309

ICR method, *Cryptosporidium* detection in water 248, 250–1
interferonγ, cryptosporidiosis
   immunotherapy 111
   protective immunity 101–4

resistance to 89–90, 113
resolution of 101–2, 110, 113
IgM antibody, diagnosis of microsporidiosis 370
immune responses
  Cryptosporidium parvum
    humoral 123–5
    innate 89–95
    specific cell-mediated 41, 49, 95–110, 124
  microsporidia 330
    cell-mediated 307–8
    humoral 305–7
immunodeficiency states, cryptosporidiosis risk 49–50, 123
  see also AIDS; HIV
immunodetection, microsporidia 375–6
immunofluorescence microscopy, *Cryptosporidium* oocyst detection 256
immunoglobulins G and M 138
immunomagnetic separation, oocysts from other debris 254–6
immunomodulators, cryptosporidiosis 66, 112–13
immunosuppression and cryptosporidiosis 49
  biliary or pancreatic 62
  fulminant 57
  respiratory 60
  see also AIDS; HIV
immunotherapy, cryptosporidiosis 66, 69–70, 121–43
immunotoxins 142
in vitro cultivation
  Cryptosporidium parvum 190–5
  microsporidia 295–6, 369–70
in vivo neutralization studies, immunity to *Cryptosporidium parvum* 105
incubation period, cryptosporidiosis 56
infant mortality, cryptosporidiosis 49
infectious dose, *Cryptosporidium parvum* 247
infectivity, assay of oocysts viability 193–4
information collection rule, 1996 265–6
innate immunity, *Cryptosporidium parvum* 89–95
  cellular components 92–5
  cytokines 89–92
insect-borne transmission, cryptosporidiosis 54–5
interleukins
  IL-2 90, 105
  IL-4 105
  IL-5 105, 110

IL-10 110
IL-12 91, 111–12
intestinal epithelium, *Cryptosporidium parvum* infected 206–7
  animal studies 25, 136
intestinal injury, enteric pathogens 206–7, 325, 327–30
intracellular
  locations
    Cryptosporidium 19–20, 173, 205
    other apicomplexans 20–6
  stages, *Cryptosporidium parvum* 9–10, 173, 174–7
intraepithelial lymphocytes, gut epithelium 127
invasion, apicomplexans 20, 21, 22, 159–60
  Cryptosporidium parvum
    invasive stages 158–9
    proteins of 160–73
ionophores, cryptosporidiosis 69
isoenzyme analysis, *Cryptosporidium* 226
isolate propagation, *Cryptosporidium parvum* 225–6
Italy, cryptosporidiosis and HIV 43
Ivory Coast, cryptosporidiosis and HIV 44

jejunal biopsies
  Cyclospora cayetanensis 406
  microsporidia 328

Las Vegas cryptosporidiosis epidemic, 1996: 51
lasalocid, cryptosporidiosis 69
lectins
  host-parasite interactions 170–1
  oocysts 156
letrazuril, cryptosporidiosis 69
light microscopy, microsporidiosis 335, 337–9
low molecular weight proteins, *Cryptosporidium* 160–4
  20 to 25 kDa 164–5

mAbs *see* monoclonal antibodies
macrolides, activity against *Cryptosporidium* 67
maduramycin, cryptosporidiosis 69
malabsorption, cryptosporidiosis 58
  oral agents 63, 67, 68
malarial parasitophorous vacuole membranes 21, 26
Malaysia, cryptosporidiosis and HIV 44

malnutrition, cryptosporidiosis 41, 49
  acute, and childhood death 56
  chronic 57
  see also malabsorption, cryptosporidiosis
mast cells, innate immunity, *Cryptosporidium parvum* 95
membrane filtration, *Cryptosporidium* detection in water 252
merogony, microsporidia 293
merozoites, *Cryptosporidium parvum* 158–9
metabolic enzymes, *Cryptosporidium parvum*, potential targets for drug design 207–8
Mexico, cryptosporidiosis and HIV 44
micronemes 20, 23
microsporidia, species infecting mammals 286–7
  biology 283–310
  development 294
  as enteric pathogens 324–7, 328
    establishing cause–effect relationship 324
    evidence for and against 325–7
  growth *in vitro* 295–6, 369–70
  histology 373–5
  host–parasite relationships 303–9
  life cycle 290–6
  morphology 285–90
  parasite survival 304–5
  phylogeny 358–69
  taxonomy 296–303
microsporidian
  genes in GenBank, 1996 354–6
  genomic size 353, 357
microsporidiosis
  biology 353–7
  clinical syndromes associated 321–42
    biliary, and biliary cryptosporidiosis 61
    diagnosis 333–40, 369–84
    hepatobiliary disease 329–30, 331–2
    intestinal injury 327–30
    small intestine disease 330–1
  epidemiology 323–7
  molecular biology 353–7
  treatment 341
Milwaukee, cryptosporidiosis outbreak, 1993: 8, 47, 51, 52, 61–2, 97, 242, 245
MLN cells, proliferative response to *Cryptosporidium parvum* 107, 109, 110
molecular diagnostic techniques, microsporidiosis 340
monoclonal antibodies
  antigen differentiation, *Cryptosporidium parvum* isolates 226–7

clinical diagnostic techniques 71
immunotherapy, cryptosporidiosis 137–8, 141–2
oocyst separation from other debris 255–6
recognition, *Cryptosporidium* proteins
  high molecular weight 168–70
  invasive stages 161
  low molecular weight 165
  oocyst wall 156, 158
  secretory IgA production 202
mRNA analysis, immune response to *Cryptosporidium parvum* 105
multisystem failure, microsporidiosis 332
municipal water treatment, current practices 267–8
murine models
  Cryptosporidium parvum 88
    immune responses 89–95, 96–7, 98–9, 107, 109, 110
    oocyst viability 262–3
    treatment and prevention, evaluating agents 196–8
  microsporidial infection
    host specificity 309

natural killer (NK) cells
  innate immunity, *Cryptosporidium parvum* 93–4
  microsporidial infection 308
neutralizing studies *in vivo* and *in vitro*, anti-cryptosporidial antibodies 132
New York, watershed management 270–1
nitazoxanide (NTZ), cryptosporidiosis 9, 70, 210–12
nitric oxide, *Cryptosporidium parvum* protective immunity 110
Nosema corneum see Vittaforma corneae
Nosema costelytrae 288
Nosema furnaclais 353
Nosema pyrausta 353
Nosema spp.
  infection 295
  merogony 293
  nucleus 288
  sporogony 293, 295
  taxonomy 301–2
  transmission 292
  ultrastructure 372

octreotide, cryptosporidiosis 59, 65
ocular microsporidial infections
  HIV 374–5

immunocompetent hosts 374
oocysts, *Cryptosporidium parvum* 8, 9
  chlorine resistance 249
  detection 234
    stool samples 190, 235
    in water 249–64
  infectious doses 247
  morphology and function 11, 153–4
  production for research 188–90
  proteins 154–8
  shedding 17–19, 246–7
    animal models of infection 91, 102, 103, 104
  surface labeling 156
  surface water, survival in 40
  thin-shell stage 154, 190
  viability determination 193–4, 260–4
  see also water-borne cryptosporidiosis
oocysts, *Cyclospora cayetanensis* 399–400, 402
  excystation 401
  stains 401–2
oocysts, other apicomplexans 154
opiates, cryptosporidiosis 65
oral agents, cryptosporidiosis
  antibodies 128–9, 136, 137, 138
  and malabsorption 63, 67, 68
  rapid intestinal transit time 67, 199, 210
outer envelope membrane, *Cryptosporidium* 24
ozone, water treatment 269–70

parasite
  quantitation *in vitro*, *Cryptosporidium parvum* 192–3
  survival, microsporidia 304–5
parasite–host relationships
  Cryptosporidium 151–78
  microsporidia 303–9
parasite-specific immune response to
  Cryptosporidium parvum, GKO mice 107, 110
parasitophorous vacuole membrane (PVM), *Cryptosporidium* 19–20, 24
parasitophorous vacuole (PV)
  Cryptosporidium 19, 21–3, 26, 159, 173
  Encephalitozoon 299–300
  Plasmodium 21, 26
  trafficking through 26
paromomycin, cryptosporidiosis 9, 68
  biliary infection 62
  drug screening *in vitro*, control 191
  respiratory infection, inhaled 60, 68

studies 27–8
passive antibody immunotherapy *see* antibodies: cryptosporidiosis antibody-based immunotherapy
PCR-RFLP, *Cryptosporidium parvum* polymorphisms 230–1
person-to-person transmission, cryptosporidiosis 50, 52, 244
Peru, cyclosporiasis 413–14
phenotypic variability, *Cryptosporidium parvum* 14, 16, 226–7
phylogeny, microsporidia 358–60
piglet diarrhea model, *Cryptosporidium parvum* 199–200
  nitazoxanide treatment efficacy 211
  pathophysiologic studies 203–4
Plasmodium knowlesi 9
Plasmodium spp.
  intracellular
    location 21
    stages 173–4
  invasive stages 159
  isolates, interspecies versus intraspecies relationships 15
  parasitophorous vacuole 21, 26
  protease expression 171
Pleistophora spp.
  infection 295
  merogony 293
  nucleus 288
  sporogony 293
  taxonomy 300–1
  transmission 292, 301
  ultrastructural characteristics 372
pneumococcal vaccines, antibody response 125
polar cap, microsporidia 288
polar filament/tubule, microsporidia 285, 288, 327, 371
polarized light, examination under, microsporidia 339
polaroplast, microsporidia 288
polyamine biosynthesis, *Cryptosporidium* 175
polyclonal antibodies 132–6
polyether ionophores, cryptosporidiosis 69
polymerase chain reaction
  Cryptosporidium parvum
    fingerprints, isolates 229
    oocyst detection 190, 258–9
  Cyclospora 404
  microsporidia
    diagnosis of microsporidiosis 379–84
    primers 365, 380–1

populations at risk, cryptosporidiosis 123
  see also AIDS; HIV
posterior vacuole, microsporidia 289
primer sets, PCR detection, microsporidia
  365, 382–4
proliferative responses to *C parvum* 107–8
  calf studies 108
  GKO mice 107, 109, 110
  human studies 108, 110
proteases, expression of surface,
  apicomplexans 171–2
protein disulfide isomerase, *Cryptosporidium*
  175

RAPD analysis, *Cryptosporidium parvum*
  isolates 227–8
rapid sand filtration, water treatment 268
raspberries, cyclosporiasis 53, 411, 413
rat model, *Cryptosporidium parvum*
  evaluation, agents for treatment and
    prevention 196, 197
  immune responses 98
  immunosuppressed 199
reinfection, cryptosporidiosis transmission 50
renal failure, microsporidiosis 305, 332
reservoir storage of water 267–8, 268
respiratory cryptosporidiosis 54, 59–60, 68
respiratory enzymes, *Cryptosporidium
  parvum* 208
rhoptries, apicomplexan parasites 20
ribosomal
  PCR fingerprints, *Cryptosporidium parvum*
    isolates 229
  RNA (rRNA)
    microsporidia, phylogeny 358–60, 364–8
    nucleotide sequences, *Cryptosporidium
      parvum*, *Cryptosporidium muris*,
      Cryptosporidium baileyi and
      Cryptosporidium wrairi 12–14, 15
    small subunit (SSU) gene
      Cryptosporidium parvum heterogeneity
        228–9
      microsporidia 354–6, 379
ribosomes, microsporidia 288–9

Safe Drinking Water Act (SWDA), 1974: 264
saquinavir, cryptosporidiosis 63
sclerosing cholangitis and papillary stenosis,
  cryptosporidiosis 61
SDS PAGE, Percoll-purified oocyst walls,
  Cryptosporidium 156
seasonal trends

  cryptosporidiosis 40
  cyclosporiasis 411
secondary
  cycles of infection, microsporidia 295
  transmission, cryptosporidiosis 244
secretory IgA 202
  backpack mouse hybridoma model 202–3
  oral administration 137, 138
sedimentation, removal of particles from
  water 268
*Septata intestinalis see Encephalitozoon
  intestinalis*
serine protease, *Cryptosporidium* 172
serological diagnosis, microsporidiosis 370
sexual transmission, cryptosporidiosis 53
similarity values, rRNA nucleotide sequences,
  Cryptosporidium 12–14, 15
SIV-infected
  macaques, cryptosporidiosis model 200–1
  rhesus monkey model
    Cryptosporidium parvum infection 200–2
    microsporidial pathogenicity 326
slow sand filtration, water treatment 268
small subunit (SSU) rRNA gene sequences
  Cryptosporidium parvum heterogeneity
    228–9
  microsporidia 354–6, 379
source water, occurrence of *Cryptosporidium*
  247–8
Spain, cryptosporidiosis and HIV 45
speciation, *Cryptosporidium parvum* 10–12
spiramycin, activity against *Cryptosporidium*
  67
spores, microsporidia 289–90, 371
  detection in stool 340
sporogony
  Enterocytozoon bieneusi 333–4, 335
  microsporidia in mammals 293–5
sporozoites
  Cryptosporidium parvum 8, 9, 158, 159
  incubated with hyperimmune bovine
    colostrum immunoglobulin 134
  Cyclospora 402–3
Spraguea lophii, karyotype 353
stage specific antigen expression,
  apicomplexans 127
stains
  Cryptosporidium parvum 71, 190, 261–2
  Cyclospora 401–2
  microsporidia 339–40, 373–4, 376–9
stool samples
  Cryptosporidium parvum 190, 235

microsporidia 340, 382–3
strain-to-strain antigenic variation, *Cryptosporidum* 127
sucrose flotation methods, *Cryptosporidium* 71
sulfonamides, cryptosporidiosis 70
supportive care, cryptosporidiosis 63
surface and apical complex proteins, *Cryptosporidium* 160–72
surface labeling of oocysts, *Cryptosporidium* 156
surface water
 Cryptosporidium parvum oocysts survival in 40
 cryptosporidiosis outbreaks associated 244–5
surface water treatment rule 1989 264–5
 enhanced 265, 266
swimming pools, fecal contamination 51
swine, *Cryptosporidium parvum* infection 18
 see also piglet diarrhea model, *Cryptosporidium parvum*
Syrian hamster model, *Cryptosporidium parvum* infection 94, 97

T cells, immune responses to *Cryptosporidium parvum*
 ex vivo analysis 105–8
 γδT cells 94, 127
 specific cell-mediated 96–100
 see also CD4 cell count
Taiwan, cryptosporidiosis and HIV 45
tap water, cryptosporidiosis transmission 50
taxonomy
 Cryptosporidium parvum 5, 6, 10–17
 Cyclospora cayetanensis 404
 microsporidia 296–303
Thailand, studies, cryptosporidiosis and HIV 45
Theileria parva 20
Thelohania spp.
 merogony 293
 nucleus 288
 sporogony 293
 taxonomy 302–3
thin-walled oocysts, *Cryptosporidium* 154, 190
tissue culture, microsporidia 369–70
Tumor necrosis factor α
 Cryptosporidium parvum, innate immunity 91–2
 microsporidiosis 310

topoisomerase II, *Cryptosporidium* 176
Toxoplasma gondii
 cell invasion 9
 cytoskeletal proteins 172
 intracellular
  location 21
  stages 173–4
 invasive stages 159
 parasitophorous vacuole, trafficking through 26
 membrane formation 21–22
Trachipleistophora spp.
 merogony 293
 nucleus 288
 sporogony 293
 taxonomy 301
 ultrastructural features 372
transient disease, cryptosporidiosis 56, 98, 99, 205
transmission
 cryptosporidiosis 50–5, 59–60, 231–2, 244
  see also water-borne cryptosporidiosis
 microsporidia 291–2, 298, 299, 301, 304, 327
transmission electron microscopy
 Cyclospora cayetanensis 403
 microsporidia 333–5, 336, 337, 338, 364
 diagnosis 328, 371–3
trimethoprin–sulfamethoxazole (TMP–SMX), cyclosporiasis 408–9
tubulin, *Cryptosporidium parvum* 173

UK
 regulatory status and environmental laws 266
 cryptosporidiosis and HIV 46
ultraviolet
 light and heat, water treatment 270
 microscopy, microsporidia 377–9
USA
 regulatory status and environmental laws, water-borne cryptosporidiosis 264–6
 cryptosporidiosis and HIV 46–7

vaccine
 development, *Cryptosporidium parvum* 201–2
 responses, HIV infected people 125
Vairimorpha spp. 288, 353
Venezuela, studies, cryptosporidiosis and HIV 48

villus enterocytes infected with *Enterocytozoon bieneusi* 334
vital dye methodology, oocyst viability determination 261
Vittaforma corneae 302
Vittaforma spp.
 merogony 293
 nucleus 288
 sporogony 293, 295
 taxonomy 302
vortex-flow filtration, *Cryptosporidium* detection in water 253

water-borne cryptosporidiosis
 drinking water contamination 8, 46, 50, 51–2, 224
 oocyst detection in water 249–64
  collection methods 250–3
  occurrence in source water 247–8
  separation from other debris 253–6
 outbreaks 242–4
  sources of infectious parasites 246–7
  type of source water 244–5
  see also Milwaukee, cryptosporidiosis outbreak 1993
 regulatory status and environmental laws 264–6
 swimming pools 51
 water treatment 267–72
  chlorine resistance 249, 269
  filtration 51–2, 71–2, 268
  theoretical considerations 271–2
 waterfowl, oocyst dissemination 246–7
 watershed management 270–1
water-borne cyclosporiasis 411–12
Western blotting, phenotypic differences, *Cryptosporidium parvum* 227

X-linked hyper-IgM syndrome 123

zidovudine 63
zinc supplementation, cryptosporidiosis 49
zoite, *Cryptosporidium parvum*
 high molecular weight surface and apical complex proteins 165–6
 low molecular weight surface proteins 160–4
 see also invasion, apicomplexans: *Cryptosporidium parvum* invasive stages
zoonotic transmission, cryptosporidiosis 17–19, 53–4, 246